T5-CUN-745

Methods in Enzymology

Volume 317

RNA–LIGAND INTERACTIONS

Part A

Structural Biology Methods

METHODS IN ENZYMOLOGY

EDITORS-IN-CHIEF

John N. Abelson   Melvin I. Simon

DIVISION OF BIOLOGY
CALIFORNIA INSTITUTE OF TECHNOLOGY
PASADENA, CALIFORNIA

FOUNDING EDITORS

Sidney P. Colowick and Nathan O. Kaplan

Methods in Enzymology

Volume 317

RNA–Ligand Interactions

Part A
Structural Biology Methods

EDITED BY

Daniel W. Celander

LOYOLA UNIVERSITY
CHICAGO, ILLINOIS

John N. Abelson

CALIFORNIA INSTITUTE OF TECHNOLOGY
PASADENA, CALIFORNIA

ACADEMIC PRESS
San Diego London Boston New York Sydney Tokyo Toronto

This book is printed on acid-free paper. 

Academic Press
A Harcourt Science and Technology Company
525 B Street, Suite 1900, San Diego, California 92101-4495, USA
http://www.academicpress.com

Academic Press Limited
24-28 Oval Road, London NW1 7DX, UK
http://www.hbuk.co.uk/ap/

International Standard Book Number: 0-12-182218-4

PRINTED IN THE UNITED STATES OF AMERICA
00 01 02 03 04 05 MM 9 8 7 6 5 4 3 2 1

Table of Contents

Section I. Semisynthetic Methodologies

A. RNA Synthetic Methods

B. Derivatization of RNA

Section II. RNA Structure Determination

A. X-Ray Crystallography

B. Nuclear Magnetic Resonance Spectroscopy

C. Electron Microscopy

Section III. Techniques for Monitoring RNA Conformation and Dynamics

A. Solution Methods

B. Electrophoretic and Spectroscopic Methods

Section IV. Modeling Tertiary Structure

Contributors to Volume 317

Article numbers are in parentheses following the names of contributors.
Affiliations listed are current.

RAJENDRA K. AGRAWAL (18, 19), *Howard Hughes Medical Institute, Health Research, Inc., at the Wadsworth Center, and Department of Biomedical Sciences, State University of New York, Albany, New York 12201-0509*

RUSS B. ALTMAN (28), *Departments of Medicine and Computer Science, Stanford University, Stanford, California 94305-5479*

MICHAEL A. BADA (28), *Stanford University, Stanford, California 94305-5479*

PETER BAYER (14), *MRC Laboratory of Molecular Biology, Cambridge CB2 2QH, England*

LEONID BEIGELMAN (3), *Ribozyme Pharmaceuticals, Inc., Boulder, Colorado 80301*

GREGOR BLAHA (19), *AG Ribosomen, Max-Planck-Institut für Molekulare Genetik, D-14195 Berlin, Germany*

MARC BOUDVILLAIN (10), *Howard Hughes Medical Institute, Columbia University, New York, New York 10032*

MICHAEL BRENOWITZ (22), *Department of Biochemistry, Center for Synchrotron Biosciences, Albert Einstein College of Medicine, Bronx, New York 10461*

JOHN M. BURKE (25), *University of Vermont, Burlington, Vermont 05405*

NILS BURKHARDT (17), *Gebäude 405, BAYER AG-Wuppertal, Abteilung MST, D-42096 Wuppertal, Germany*

JAMIE H. CATE (12), *Whitehead Institute, Cambridge, Massachusetts 02142-1479*

ROBERT CEDERGREN (27), *Département de Biochimie, Université de Montréal, Montréal, Québec H3C 3J7, Canada*

MARK R. CHANCE (22), *Department of Physiology and Biophysics, Center for Synchrotron Biosciences, Albert Einstein College of Medicine, Bronx, New York 10461*

VI T. CHU (10), *Department of Biochemistry and Molecular Biophysics, Columbia University, New York, New York 10032*

MARIA COSTA (29), *Centre de Génétique Moléculaire du CNRS, F-91190 Gif-sur-Yvette, France*

DONALD M. CROTHERS (9), *Department of Chemistry, Yale University, New Haven, Connecticut 06520-8107*

MICHAEL L. DERAS (22), *Department of Biophysics, Johns Hopkins University, Baltimore, Maryland 21218-2864*

JENNIFER A. DOUDNA (12), *Department of Molecular Biophysics and Biochemistry, Yale University, New Haven, Connecticut 06520-8114*

FRITZ ECKSTEIN (5), *Max-Planck-Institut für Experimentelle Medizin, D-37075 Göttingen, Germany*

XINGWANG FANG (24), *Department of Biochemistry and Molecular Biology, University of Chicago, Chicago, Illinois 60637*

JOACHIM FRANK (18, 19), *Howard Hughes Medical Institute, Health Research, Inc., at the Wadsworth Center, and Department of Biomedical Sciences, State University of New York, Albany, New York 12201-0509*

BARBARA L. GOLDEN (8), *Department of Biochemistry, Purdue University, West Lafayette, Indiana 47906-1153*

ROBERT A. GRASSUCCI (18), *Howard Hughes Medical Institute, Health Research, Inc., at the Wadsworth Center, Albany, New York 12201-0509*

PETER HAEBERLI (3), *Ribozyme Pharmaceuticals, Inc., Boulder, Colorado 80301*

PAUL J. HAGERMAN (26), *Department of Biochemistry and Molecular Genetics, University of Colorado Health Sciences Center, Denver, Colorado 80262*

MARK R. HANSEN (15), *Department of Chemistry and Biochemistry, University of Colorado, Boulder, Colorado 80309-0215*

PAUL HANSON (15), *Department of Chemistry and Biochemistry, University of Colorado, Boulder, Colorado 80309-0215*

AMY B. HEAGLE (18), *Howard Hughes Medical Institute, Health Research, Inc., at the Wadsworth Center, Albany, New York 12201-0509*

ECKHARD JANKOWSKY (10), *Department of Biochemistry and Molecular Biophysics, Columbia University, New York, New York 10032*

ALEXANDER KARPEISKY (3), *Ribozyme Pharmaceuticals, Inc., Boulder, Colorado 80301*

ANNE B. KOSEK (6), *Departments of Molecular Biophysics and Biochemistry, and Chemistry, Yale University, New Haven, Connecticut 06520-8114*

JON LAPHAM (9), *Department of Chemistry, Yale University, New Haven, Connecticut 06520-8107*

FABRICE LECLERC (27), *Department of Chemistry and Chemical Biology, Harvard University, Cambridge, Massachusetts 02138*

DAVID M. J. LILLEY (23), *Department of Biochemistry, University of Dundee, Dundee DD1 4HN, United Kingdom*

BELSIS LLORENTE (27), *Centro de Química Farmacéutica, Atabey, Habana, Cuba*

STEPHEN R. LYNCH (16), *Department of Structural Biology, Stanford University School of Medicine, Stanford, California 94305-5126*

CHRISTIAN MASSIRE (29), *Institut de Biologie Moléculaire et Cellulaire du CNRS, F-67084 Strasbourg, France*

JASENKA MATULIC-ADAMIC (3), *Ribozyme Pharmaceuticals, Inc., Boulder, Colorado 80301*

DAVID B. MCKAY (11), *Department of Structural Biology, Stanford University School of Medicine, Stanford, California 94305*

FRANÇOIS MICHEL (29), *Centre de Génétique Moléculaire du CNRS, F-91190 Gif-sur-Yvette, France*

MELISSA J. MOORE (7), *Department of Biochemistry, W. M. Keck Institute for Cellular Visualization, Brandeis University, Waltham, Massachusetts 02454*

JAMES B. MURRAY (13), *Department of Chemistry and Biochemistry, and Center for the Molecular Biology of RNA, University of California, Santa Cruz, California 95064*

KNUD H. NIERHAUS (17, 19), *AG Ribosomen, Max-Planck-Institut für Molekulare Genetik, D-14195 Berlin, Germany*

LORI ORTOLEVA-DONNELLY (6), *Departments of Molecular Biophysics and Biochemistry, and Chemistry, Yale University, New Haven, Connecticut 06520-8114*

TAO PAN (20), *Department of Biochemistry and Molecular Biology, University of Chicago, Chicago, Illinois 60637*

ARTHUR PARDI (15), *Department of Chemistry and Biochemistry, University of Colorado, Boulder, Colorado 80309-0215*

PAWEL PENCZEK (18), *Howard Hughes Medical Institute, Health Research, Inc., at the Wadsworth Center, and Department of Biomedical Sciences, State University of New York, Albany, New York 12201-0509*

JOSEPH D. PUGLISI (16), *Department of Structural Biology, Stanford University School of Medicine, Stanford, California 94305-5126*

ANNA MARIE PYLE (10), *Department of Biochemistry and Molecular Biophysics, and Howard Hughes Medical Institute, Columbia University, New York, New York 10032*

CHARLES C. QUERY (7), *Department of Cell Biology, Albert Einstein College of Medicine, Bronx, New York 10461*

CORIE Y. RALSTON (22), *Department of Physiology and Biophysics, Center for Synchrotron Biosciences, Albert Einstein College of Medicine, Bronx, New York 10461*

MICHAEL I. RECHT (16), *Department of Structural Biology, Stanford University School of Medicine, Stanford, California 94305-5126*

BEATRIX RÖHRDANZ (17), *AG Ribosomen, Max-Planck-Institut für Molekulare Genetik, D-14195 Berlin, Germany*

SEAN P. RYDER (6), *Department of Molecular Biophysics and Biochemistry, Yale University, New Haven, Connecticut 06520-8114*

STEPHEN A. SCARINGE (1), *Dharmacon Research, Inc., Boulder, Colorado 80301*

BIANCA SCLAVI (22), *Department of Physiology and Biophysics, Center for Synchrotron Biosciences, Albert Einstein College of Medicine, Bronx, New York 10461*

LINCOLN G. SCOTT (2), *The Scripps Research Institute, La Jolla, California 92037*

WILLIAM G. SCOTT (13), *Department of Chemistry and Biochemistry, and Center for the Molecular Biology of RNA, University of California, Santa Cruz, California 95064*

VALERIE M. SHELTON (24), *Department of Chemistry, University of Chicago, Chicago, Illinois 60637*

TOBIN R. SOSNICK (24), *Department of Biochemistry and Molecular Biology, University of Chicago, Chicago, Illinois 60637*

RUI SOUSA (4), *Department of Biochemistry, University of Texas Health Science Center, San Antonio, Texas 78284-7760*

CHRISTIAN M. T. SPAHN (17, 19), *Wadsworth Center, New York State Department of Health, New York, New York 12201-0509*

ULRICH STELZL (19), *AG Ribosomen, Max-Planck-Institut für Molekulare Genetik, D-14195 Berlin, Germany*

SCOTT A. STROBEL (6), *Departments of Molecular Biophysics and Biochemistry, and Chemistry, Yale University, New Haven, Connecticut 06520-8114*

MICHAEL SULLIVAN (22), *Department of Physiology and Biophysics, Center for Synchrotron Biosciences, Albert Einstein College of Medicine, Bronx, New York 10461*

DAVID SWEEDLER (3), *Ribozyme Pharmaceuticals, Inc., Boulder, Colorado 80301*

THOMAS J. TOLBERT (2), *The Scripps Research Institute, La Jolla, California 92037*

DANIEL K. TREIBER (21), *The Scripps Research Institute, La Jolla, California 92037*

FRANCISCO J. TRIANA-ALONSO (17), *Centro de Investigaciones Biomédicas, Universidad de Carabobo, LaMorita, Maracay, Venezuela*

GABRIELE VARANI (14), *MRC Laboratory of Molecular Biology, Cambridge CB2 2QH, England*

LUCA VARANI (14), *MRC Laboratory of Molecular Biology, Cambridge CB2 2QH, England*

L. CLAUS S. VÖRTLER (5), *Max-Planck-Institut für Experimentelle Medizin, D-37075 Göttingen, Germany*

NILS G. WALTER (25), *Department of Chemistry, University of Michigan, Ann Arbor, Michigan 48109-1055*

JOSEPH E. WEDEKIND (11), *Department of Structural Biology, Stanford University School of Medicine, Stanford, California 94305*

ERIC WESTHOF (29), *Institut de Biologie Moléculaire et Cellulaire du CNRS, F-67084 Strasbourg, France*

JAMES R. WILLIAMSON (2, 21), *The Scripps Research Institute, La Jolla, California 92037*

SARAH A. WOODSON (22), *Department of Biophysics, Johns Hopkins University, Baltimore, Maryland 21218-2864*

Preface

A decade has passed since *Methods in Enzymology* addressed methods and techniques used in RNA processing. As has been evident since its inception, research in RNA processing progresses at a rapid pace. Its expansion into new areas of investigation has been phenomenal with novel discoveries being made in a variety of subspecialty areas. The subfield of RNA–ligand interactions concerns research problems in RNA structure, in the molecular recognition of structured RNA by diverse ligands, and in the mechanistic details of RNA's functional role following ligand binding. At the beginning of this new millennium, we celebrate the explosive development of exciting new tools and procedures whereby investigators explore RNA structure and function from the perspective of understanding RNA–ligand interactions.

New insights into RNA processing are accompanied with improvements in older techniques as well as the development of entirely new methods. Previous *Methods in Enzymology* volumes in RNA processing have focused on basic methods generally employed in all RNA processing systems (Volume 180) or on techniques whose applications might be considerably more specific to a particular system (Volume 181). RNA–Ligand Interactions, Volumes 317 and 318, showcase many new methods that have led to significant advances in this subfield. The types of ligands described in these volumes certainly include proteins; however, ligands composed of RNA, antibiotics, other small molecules, and even chemical elements are also found in nature and have been the focus of much research work. Given the great diversity of RNA–ligand interactions described in these volumes, we have assembled the contributions according to whether they pertain to structural biology methods (Volume 317) or to biochemistry and molecular biology techniques (Volume 318). Aside from the particular systems for which these techniques have been developed, we consider it likely that the methods described will enjoy uses that extend beyond RNA–ligand interactions to include other areas of RNA processing.

This endeavor has been fraught with many difficult decisions regarding the selection of topics for these volumes. We were delighted with the number of chapters received. The authors have taken great care and dedication to present their contributions in clear language. Their willingness to share with others the techniques used in their laboratories is apparent from the quality of their comprehensive contributions. We thank them for their effort and appreciate their patience as the volumes were assembled.

Daniel W. Celander
John N. Abelson

METHODS IN ENZYMOLOGY

VOLUME I. Preparation and Assay of Enzymes
Edited by SIDNEY P. COLOWICK AND NATHAN O. KAPLAN

VOLUME II. Preparation and Assay of Enzymes
Edited by SIDNEY P. COLOWICK AND NATHAN O. KAPLAN

VOLUME III. Preparation and Assay of Substrates
Edited by SIDNEY P. COLOWICK AND NATHAN O. KAPLAN

VOLUME IV. Special Techniques for the Enzymologist
Edited by SIDNEY P. COLOWICK AND NATHAN O. KAPLAN

VOLUME V. Preparation and Assay of Enzymes
Edited by SIDNEY P. COLOWICK AND NATHAN O. KAPLAN

VOLUME VI. Preparation and Assay of Enzymes (*Continued*)
Preparation and Assay of Substrates
Special Techniques
Edited by SIDNEY P. COLOWICK AND NATHAN O. KAPLAN

VOLUME VII. Cumulative Subject Index
Edited by SIDNEY P. COLOWICK AND NATHAN O. KAPLAN

VOLUME VIII. Complex Carbohydrates
Edited by ELIZABETH F. NEUFELD AND VICTOR GINSBURG

VOLUME IX. Carbohydrate Metabolism
Edited by WILLIS A. WOOD

VOLUME X. Oxidation and Phosphorylation
Edited by RONALD W. ESTABROOK AND MAYNARD E. PULLMAN

VOLUME XI. Enzyme Structure
Edited by C. H. W. HIRS

VOLUME XII. Nucleic Acids (Parts A and B)
Edited by LAWRENCE GROSSMAN AND KIVIE MOLDAVE

VOLUME XIII. Citric Acid Cycle
Edited by J. M. LOWENSTEIN

VOLUME XIV. Lipids
Edited by J. M. LOWENSTEIN

VOLUME XV. Steroids and Terpenoids
Edited by RAYMOND B. CLAYTON

VOLUME XVI. Fast Reactions
Edited by KENNETH KUSTIN

Volume XVII. Metabolism of Amino Acids and Amines (Parts A and B)
Edited by Herbert Tabor and Celia White Tabor

Volume XVIII. Vitamins and Coenzymes (Parts A, B, and C)
Edited by Donald B. McCormick and Lemuel D. Wright

Volume XIX. Proteolytic Enzymes
Edited by Gertrude E. Perlmann and Laszlo Lorand

Volume XX. Nucleic Acids and Protein Synthesis (Part C)
Edited by Kivie Moldave and Lawrence Grossman

Volume XXI. Nucleic Acids (Part D)
Edited by Lawrence Grossman and Kivie Moldave

Volume XXII. Enzyme Purification and Related Techniques
Edited by William B. Jakoby

Volume XXIII. Photosynthesis (Part A)
Edited by Anthony San Pietro

Volume XXIV. Photosynthesis and Nitrogen Fixation (Part B)
Edited by Anthony San Pietro

Volume XXV. Enzyme Structure (Part B)
Edited by C. H. W. Hirs and Serge N. Timasheff

Volume XXVI. Enzyme Structure (Part C)
Edited by C. H. W. Hirs and Serge N. Timasheff

Volume XXVII. Enzyme Structure (Part D)
Edited by C. H. W. Hirs and Serge N. Timasheff

Volume XXVIII. Complex Carbohydrates (Part B)
Edited by Victor Ginsburg

Volume XXIX. Nucleic Acids and Protein Synthesis (Part E)
Edited by Lawrence Grossman and Kivie Moldave

Volume XXX. Nucleic Acids and Protein Synthesis (Part F)
Edited by Kivie Moldave and Lawrence Grossman

Volume XXXI. Biomembranes (Part A)
Edited by Sidney Fleischer and Lester Packer

Volume XXXII. Biomembranes (Part B)
Edited by Sidney Fleischer and Lester Packer

Volume XXXIII. Cumulative Subject Index Volumes I–XXX
Edited by Martha G. Dennis and Edward A. Dennis

Volume XXXIV. Affinity Techniques (Enzyme Purification: Part B)
Edited by William B. Jakoby and Meir Wilchek

Volume XXXV. Lipids (Part B)
Edited by John M. Lowenstein

VOLUME 91. Enzyme Structure (Part I)
Edited by C. H. W. HIRS AND SERGE N. TIMASHEFF

VOLUME 92. Immunochemical Techniques (Part E: Monoclonal Antibodies and General Immunoassay Methods)
Edited by JOHN J. LANGONE AND HELEN VAN VUNAKIS

VOLUME 93. Immunochemical Techniques (Part F: Conventional Antibodies, Fc Receptors, and Cytotoxicity)
Edited by JOHN J. LANGONE AND HELEN VAN VUNAKIS

VOLUME 94. Polyamines
Edited by HERBERT TABOR AND CELIA WHITE TABOR

VOLUME 95. Cumulative Subject Index Volumes 61–74, 76–80
Edited by EDWARD A. DENNIS AND MARTHA G. DENNIS

VOLUME 96. Biomembranes [Part J: Membrane Biogenesis: Assembly and Targeting (General Methods; Eukaryotes)]
Edited by SIDNEY FLEISCHER AND BECCA FLEISCHER

VOLUME 97. Biomembranes [Part K: Membrane Biogenesis: Assembly and Targeting (Prokaryotes, Mitochondria, and Chloroplasts)]
Edited by SIDNEY FLEISCHER AND BECCA FLEISCHER

VOLUME 98. Biomembranes (Part L: Membrane Biogenesis: Processing and Recycling)
Edited by SIDNEY FLEISCHER AND BECCA FLEISCHER

VOLUME 99. Hormone Action (Part F: Protein Kinases)
Edited by JACKIE D. CORBIN AND JOEL G. HARDMAN

VOLUME 100. Recombinant DNA (Part B)
Edited by RAY WU, LAWRENCE GROSSMAN, AND KIVIE MOLDAVE

VOLUME 101. Recombinant DNA (Part C)
Edited by RAY WU, LAWRENCE GROSSMAN, AND KIVIE MOLDAVE

VOLUME 102. Hormone Action (Part G: Calmodulin and Calcium-Binding Proteins)
Edited by ANTHONY R. MEANS AND BERT W. O'MALLEY

VOLUME 103. Hormone Action (Part H: Neuroendocrine Peptides)
Edited by P. MICHAEL CONN

VOLUME 104. Enzyme Purification and Related Techniques (Part C)
Edited by WILLIAM B. JAKOBY

VOLUME 105. Oxygen Radicals in Biological Systems
Edited by LESTER PACKER

VOLUME 106. Posttranslational Modifications (Part A)
Edited by FINN WOLD AND KIVIE MOLDAVE

VOLUME 107. Posttranslational Modifications (Part B)
Edited by FINN WOLD AND KIVIE MOLDAVE

VOLUME 108. Immunochemical Techniques (Part G: Separation and Characterization of Lymphoid Cells)
Edited by GIOVANNI DI SABATO, JOHN J. LANGONE, AND HELEN VAN VUNAKIS

VOLUME 109. Hormone Action (Part I: Peptide Hormones)
Edited by LUTZ BIRNBAUMER AND BERT W. O'MALLEY

VOLUME 110. Steroids and Isoprenoids (Part A)
Edited by JOHN H. LAW AND HANS C. RILLING

VOLUME 111. Steroids and Isoprenoids (Part B)
Edited by JOHN H. LAW AND HANS C. RILLING

VOLUME 112. Drug and Enzyme Targeting (Part A)
Edited by KENNETH J. WIDDER AND RALPH GREEN

VOLUME 113. Glutamate, Glutamine, Glutathione, and Related Compounds
Edited by ALTON MEISTER

VOLUME 114. Diffraction Methods for Biological Macromolecules (Part A)
Edited by HAROLD W. WYCKOFF, C. H. W. HIRS, AND SERGE N. TIMASHEFF

VOLUME 115. Diffraction Methods for Biological Macromolecules (Part B)
Edited by HAROLD W. WYCKOFF, C. H. W. HIRS, AND SERGE N. TIMASHEFF

VOLUME 116. Immunochemical Techniques (Part H: Effectors and Mediators of Lymphoid Cell Functions)
Edited by GIOVANNI DI SABATO, JOHN J. LANGONE, AND HELEN VAN VUNAKIS

VOLUME 117. Enzyme Structure (Part J)
Edited by C. H. W. HIRS AND SERGE N. TIMASHEFF

VOLUME 118. Plant Molecular Biology
Edited by ARTHUR WEISSBACH AND HERBERT WEISSBACH

VOLUME 119. Interferons (Part C)
Edited by SIDNEY PESTKA

VOLUME 120. Cumulative Subject Index Volumes 81–94, 96–101

VOLUME 121. Immunochemical Techniques (Part I: Hybridoma Technology and Monoclonal Antibodies)
Edited by JOHN J. LANGONE AND HELEN VAN VUNAKIS

VOLUME 122. Vitamins and Coenzymes (Part G)
Edited by FRANK CHYTIL AND DONALD B. MCCORMICK

VOLUME 123. Vitamins and Coenzymes (Part H)
Edited by FRANK CHYTIL AND DONALD B. MCCORMICK

VOLUME 124. Hormone Action (Part J: Neuroendocrine Peptides)
Edited by P. MICHAEL CONN

VOLUME 125. Biomembranes (Part M: Transport in Bacteria, Mitochondria, and Chloroplasts: General Approaches and Transport Systems)
Edited by SIDNEY FLEISCHER AND BECCA FLEISCHER

VOLUME 126. Biomembranes (Part N: Transport in Bacteria, Mitochondria, and Chloroplasts: Protonmotive Force)
Edited by SIDNEY FLEISCHER AND BECCA FLEISCHER

VOLUME 127. Biomembranes (Part O: Protons and Water: Structure and Translocation)
Edited by LESTER PACKER

VOLUME 128. Plasma Lipoproteins (Part A: Preparation, Structure, and Molecular Biology)
Edited by JERE P. SEGREST AND JOHN J. ALBERS

VOLUME 129. Plasma Lipoproteins (Part B: Characterization, Cell Biology, and Metabolism)
Edited by JOHN J. ALBERS AND JERE P. SEGREST

VOLUME 130. Enzyme Structure (Part K)
Edited by C. H. W. HIRS AND SERGE N. TIMASHEFF

VOLUME 131. Enzyme Structure (Part L)
Edited by C. H. W. HIRS AND SERGE N. TIMASHEFF

VOLUME 132. Immunochemical Techniques (Part J: Phagocytosis and Cell-Mediated Cytotoxicity)
Edited by GIOVANNI DI SABATO AND JOHANNES EVERSE

VOLUME 133. Bioluminescence and Chemiluminescence (Part B)
Edited by MARLENE DELUCA AND WILLIAM D. MCELROY

VOLUME 134. Structural and Contractile Proteins (Part C: The Contractile Apparatus and the Cytoskeleton)
Edited by RICHARD B. VALLEE

VOLUME 135. Immobilized Enzymes and Cells (Part B)
Edited by KLAUS MOSBACH

VOLUME 136. Immobilized Enzymes and Cells (Part C)
Edited by KLAUS MOSBACH

VOLUME 137. Immobilized Enzymes and Cells (Part D)
Edited by KLAUS MOSBACH

VOLUME 138. Complex Carbohydrates (Part E)
Edited by VICTOR GINSBURG

VOLUME 139. Cellular Regulators (Part A: Calcium- and Calmodulin-Binding Proteins)
Edited by ANTHONY R. MEANS AND P. MICHAEL CONN

VOLUME 140. Cumulative Subject Index Volumes 102–119, 121–134

VOLUME 141. Cellular Regulators (Part B: Calcium and Lipids)
Edited by P. MICHAEL CONN AND ANTHONY R. MEANS

VOLUME 142. Metabolism of Aromatic Amino Acids and Amines
Edited by SEYMOUR KAUFMAN

VOLUME 143. Sulfur and Sulfur Amino Acids
Edited by WILLIAM B. JAKOBY AND OWEN GRIFFITH

VOLUME 144. Structural and Contractile Proteins (Part D: Extracellular Matrix)
Edited by LEON W. CUNNINGHAM

VOLUME 145. Structural and Contractile Proteins (Part E: Extracellular Matrix)
Edited by LEON W. CUNNINGHAM

VOLUME 146. Peptide Growth Factors (Part A)
Edited by DAVID BARNES AND DAVID A. SIRBASKU

VOLUME 147. Peptide Growth Factors (Part B)
Edited by DAVID BARNES AND DAVID A. SIRBASKU

VOLUME 148. Plant Cell Membranes
Edited by LESTER PACKER AND ROLAND DOUCE

VOLUME 149. Drug and Enzyme Targeting (Part B)
Edited by RALPH GREEN AND KENNETH J. WIDDER

VOLUME 150. Immunochemical Techniques (Part K: *In Vitro* Models of B and T Cell Functions and Lymphoid Cell Receptors)
Edited by GIOVANNI DI SABATO

VOLUME 151. Molecular Genetics of Mammalian Cells
Edited by MICHAEL M. GOTTESMAN

VOLUME 152. Guide to Molecular Cloning Techniques
Edited by SHELBY L. BERGER AND ALAN R. KIMMEL

VOLUME 153. Recombinant DNA (Part D)
Edited by RAY WU AND LAWRENCE GROSSMAN

VOLUME 154. Recombinant DNA (Part E)
Edited by RAY WU AND LAWRENCE GROSSMAN

VOLUME 155. Recombinant DNA (Part F)
Edited by RAY WU

VOLUME 156. Biomembranes (Part P: ATP-Driven Pumps and Related Transport: The Na,K-Pump)
Edited by SIDNEY FLEISCHER AND BECCA FLEISCHER

VOLUME 157. Biomembranes (Part Q: ATP-Driven Pumps and Related Transport: Calcium, Proton, and Potassium Pumps)
Edited by SIDNEY FLEISCHER AND BECCA FLEISCHER

VOLUME 158. Metalloproteins (Part A)
Edited by JAMES F. RIORDAN AND BERT L. VALLEE

VOLUME 227. Metallobiochemistry (Part D: Physical and Spectroscopic Methods for Probing Metal Ion Environments in Metalloproteins)
Edited by JAMES F. RIORDAN AND BERT L. VALLEE

VOLUME 228. Aqueous Two-Phase Systems
Edited by HARRY WALTER AND GÖTE JOHANSSON

VOLUME 229. Cumulative Subject Index Volumes 195–198, 200–227

VOLUME 230. Guide to Techniques in Glycobiology
Edited by WILLIAM J. LENNARZ AND GERALD W. HART

VOLUME 231. Hemoglobins (Part B: Biochemical and Analytical Methods)
Edited by JOHANNES EVERSE, KIM D. VANDEGRIFF, AND ROBERT M. WINSLOW

VOLUME 232. Hemoglobins (Part C: Biophysical Methods)
Edited by JOHANNES EVERSE, KIM D. VANDEGRIFF, AND ROBERT M. WINSLOW

VOLUME 233. Oxygen Radicals in Biological Systems (Part C)
Edited by LESTER PACKER

VOLUME 234. Oxygen Radicals in Biological Systems (Part D)
Edited by LESTER PACKER

VOLUME 235. Bacterial Pathogenesis (Part A: Identification and Regulation of Virulence Factors)
Edited by VIRGINIA L. CLARK AND PATRIK M. BAVOIL

VOLUME 236. Bacterial Pathogenesis (Part B: Integration of Pathogenic Bacteria with Host Cells)
Edited by VIRGINIA L. CLARK AND PATRIK M. BAVOIL

VOLUME 237. Heterotrimeric G Proteins
Edited by RAVI IYENGAR

VOLUME 238. Heterotrimeric G-Protein Effectors
Edited by RAVI IYENGAR

VOLUME 239. Nuclear Magnetic Resonance (Part C)
Edited by THOMAS L. JAMES AND NORMAN J. OPPENHEIMER

VOLUME 240. Numerical Computer Methods (Part B)
Edited by MICHAEL L. JOHNSON AND LUDWIG BRAND

VOLUME 241. Retroviral Proteases
Edited by LAWRENCE C. KUO AND JULES A. SHAFER

VOLUME 242. Neoglycoconjugates (Part A)
Edited by Y. C. LEE AND REIKO T. LEE

VOLUME 243. Inorganic Microbial Sulfur Metabolism
Edited by HARRY D. PECK, JR., AND JEAN LEGALL

VOLUME 244. Proteolytic Enzymes: Serine and Cysteine Peptidases
Edited by ALAN J. BARRETT

Section I

Semisynthetic Methodologies

A. RNA Synthetic Methods
Articles 1 through 3

B. Derivatization of RNA
Articles 4 through 10

[1] Advanced 5′-Silyl-2′-Orthoester Approach to RNA Oligonucleotide Synthesis

By STEPHEN A. SCARINGE

Introduction

The need for routine syntheses of RNA oligonucleotides has grown rapidly during the 1990s as research reveals the increasing breadth of RNA's biological functions.[1] RNA synthesis can be accomplished for most applications via either biochemical methods, e.g., T7 transcription, or chemical methods. Transcription and chemical methodologies complement each other well and enable a wide range of RNAs to be synthesized. Current chemical synthetic methods[2] enable the synthesis of RNA in acceptable yields and quality, but none are as routine and dependable as DNA. The need for an improved oligoribonucleotide synthesis technology has continued to persist. This article describes a recently developed technological advance in the chemical synthesis of RNA oligonucleotides utilizing a novel 5′-*O*-silyl ether protecting group in conjunction with an acid-labile 2′-*O*-orthoester.[3] Using this technology, numerous RNA oligonucleotides have been routinely synthesized in high yields and of unprecedented quality.

The advantageous properties of 5′-silyl-2′-orthoester RNA chemistry make it possible to synthesize RNA oligonucleotides with a quality only previously observed in DNA. The ribonucleoside phosphoramidites couple in >99% stepwise yields in less than 90 sec. Consequently, yields are routinely 1.5–3 times that observed with older RNA synthesis chemistries. At the same time, the overall purity is significantly increased. For some applications, the RNA is of sufficient purity to use without further processing. After synthesis of an RNA oligonucleotide, the 2′-orthoester pro-

[1] S. Altman, *Proc. Natl. Acad. Sci. U.S.A.* **90,** 10898 (1993); B. A. Sullenger and T. R. Cech, *Science* **262,** 1566 (1993); T. Cech, *Curr. Opin. Struct. Biol.* **2,** 605 (1992); N. Usman and R. Cedergren, *Trends Biochem. Sci.* **17,** 334 (1992).

[2] F. Wincott, A. DiRenzo, C. Shaffer, S. Grimm, D. Tracz, C. Workman, D. Sweedler, C. Gonzalez, S. Scaringe, and N. Usman, *Nucleic Acids Res.* **23,** 2677 (1995); N. Usman, K. K. Ogilvie, M.-Y. Jiang, and R. J. Cedergren, *J. Am. Chem. Soc.* **109,** 7845 (1987); T. Wu, K. K. Ogilvie, and R. T. Pon, *Nucleic Acids Res.* **17,** 3501 (1989); T. Tanaka, S. Tamatsukuri, and M. Ikehara, *Nucleic Acids Res.* **14,** 6265 (1986); J. A. Hayes, M. J. Brunden, P. T. Gilham, and G. R. Gough, *Tetrahedron Lett.* **26,** 2407 (1985); M. V. Rao, C. B. Reese, V. Schehlman, and P. S. Yu, *J. Chem. Soc. Perkin Trans. I,* 43 (1993).

[3] S. A. Scaringe, F. E. Wincott, and M. H. Caruthers, *J. Am. Chem. Soc.* **129,** 11820 (1998).

0076-6879/00 $30.00

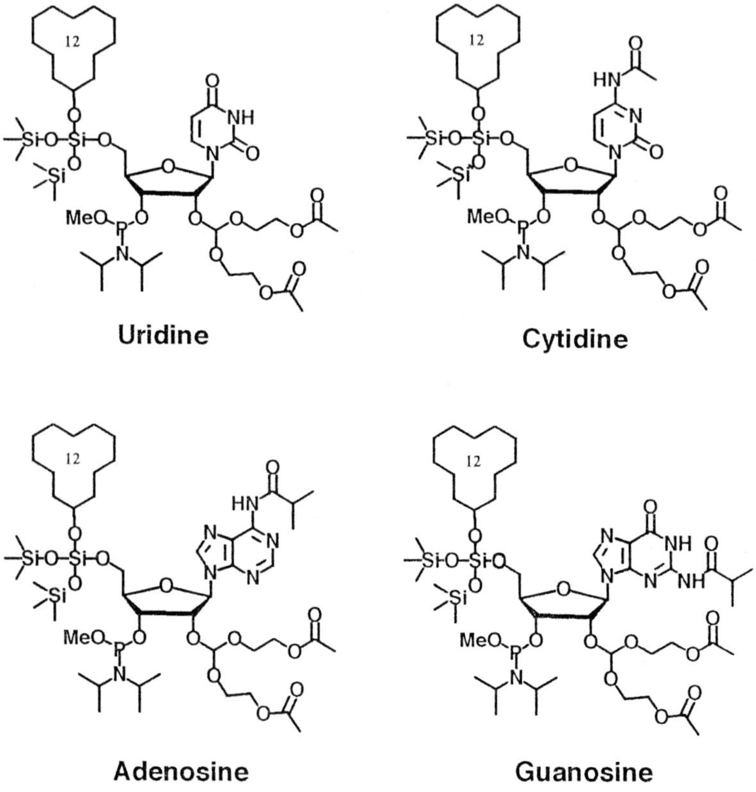

Uridine

Cytidine

Adenosine

Guanosine

FIG. 1. Protected ribonucleoside phosphoramidites for 5′-silyl-2′-orthoester RNA synthesis chemistry.

tected RNA is water soluble and significantly more stable to degradation than the final fully deprotected RNA product. These features of the 2′-orthoester group enable the RNA to be easily handled in aqueous solutions. Furthermore, the 2′-orthoester groups interrupt secondary structure. This property has made it possible to analyze and purify RNA oligonucleotides of every sequence to date regardless of secondary structure. This includes 10- to 15-base-long homopolymers of guanosine. Finally, when the RNA is ready for use, the 2′-orthoester groups are completely removed in less than 30 min under extremely mild conditions in common aqueous buffers. These unique properties of the 5′-silyl ether and 2′-orthoester protecting groups have made it possible to routinely synthesize high-quality RNA oligonucleotides.

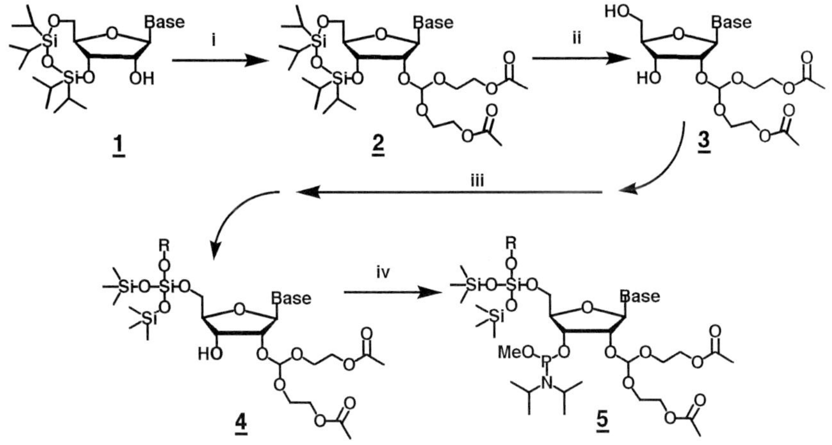

FIG. 2. General synthesis scheme from TIPS-protected nucleoside to fully protected nucleoside phosphoramidite, where R is cyclododecyl and Base is either *N*-isobutyryladenine, *N*-acetylcytosine, *N*-isobutyrylguanine, or uracil [reaction (i); tris(2-acetoxyethoxy)orthoformate, pyridinium toluene sulfonate; reaction (ii); TEMED-HF, acetonitrile; reaction (iii); DOD-Cl, imidazole, THF; reaction (iv); bis(*N*,*N*-diisopropylamine)methoxyphosphine, tetrazole, DCM].

Synthesis of Nucleoside Phosphoramidites

The 5′-hydroxyl, 2′-hydroxyl, and amine protecting groups used in 5′-silyl-2′-orthoester chemistry continue to be refined and optimized. At this time the ribonucleoside phosphoramidites use the bis(trimethylsiloxy)-cyclododecyloxysilyl ether (DOD) protecting group on the 5′-hydroxyl and the bis(2-acetoxyethoxy)methyl (ACE) orthoester protecting group on the 2′-hydroxyl (Fig. 1). The exocyclic amines are protected with the following acyl groups: acetyl for cytidine and isobutyryl for adenosine and guanosine. Synthesis of these compounds proceeds according to the general outline in Fig. 2 [reactions (i)–(iv)].

The *N*-acyl-5′-*O*-3′-*O*-tetraisopropyldisiloxanyl-protected ribonucleoside starting materials (**1**) (*N*-acyl-TIPS nucleosides) can be synthesized according to the literature[4] or obtained commercially (Aldrich, Milwaukee, WI, or Monomer Sciences, Huntsville, AL). The remaining reactions can be effected utilizing the following generalized protocols.

[4] G. S. Ti, B. L. Gaffney, and R. A. Jones, *J. Am. Chem. Soc.* **104,** 1316 (1982); W. T. Markiewicz and M. Wiewiorowski, *Nucl. Acids Res. Spec. Pub.* **4,** 185, (1978); W. T. Markiewicz, E. Biala, R. W. Adamiak, K. Grzeskowiak, R. Kierzek, A. Kraszewski, J. Stawinski, and M. Wieworowski, *Nucl. Acids Res. Symp.* **7,** 115 (1980).

*Synthesis of 2′-O-ACE Protected Nucleoside (**2**): Reaction (i)*

The ACE orthoester is introduced onto the 2′-hydroxyl by reacting the *N*-acyl-TIPS nucleoside with the trisorthoformate reagent under acid catalysis. The 2′-hydroxyl displaces one of the alcohols on the orthoformate reagent (Fig. 3) to produce the desired product (**2**). As described later, the reaction proceeds under high vacuum to remove the 2-acetoxyethanol by-product and drive the reaction forward. An improved method for introducing the 2′-*O*-ACE orthoester is currently being developed and will be reported shortly.

Procedure. *N*-acyl-TIPS-nucleoside (**1**) (1 equivalent, 10 mmol) is reacted neat with tris(2-acetoxyethoxy) orthoformate (322.31 g/mol, 5.6 equivalent, 18.04 g) and pyridinium *p*-toluene sulfonate (251.31 g/mol, 0.2 equivalent, 0.50 g) at 55° for 3 hr under high vacuum (<0.015 mm Hg). The reaction is cooled to room temperature, neutralized with *N,N,N′,N′*-tetramethylethylenediamine (TEMED) (150 ml/mol, 0.5 equivalent, 0.75 ml), diluted with 50 ml dichloromethane (DCM) and 150 ml hexanes, and purified on 300 g silica gel (Merck–VWR Scientific) with a hexane/ethyl acetate gradient. Column chromatography removes the neutralized catalyst but does not yield pure product because the excess orthoformate reagent generally eluted with the nucleoside product. However, the excess reagent does not interfere with the following reaction and it is easily removed during purification of the next nucleoside intermediate (**3**). Therefore, the semipurified product is carried through to the next reaction.

Removal of 5′-3′-TIPS Protecting Group: Reaction (ii)

The 5′-3′-TIPS group is removed with fluoride ions, e.g., tetrabutylammonium fluoride or amine hydrofluoride salts. These salts can chromatograph with the product and complicate purification. (Tetrabutylammonium fluoride and triethylammonium hydrofluoride are known to cause this problem.) Therefore, a very polar amine salt of hydrofluoric acid is used to ensure that during chromatography these salts do not elute with the product.

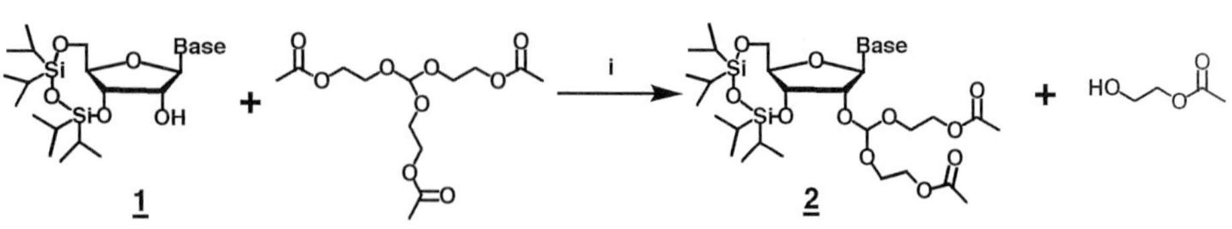

FIG. 3. Reaction of protected nucleoside with tris(2-acetoxyethoxy) orthoformate. Base is either *N*-isobutyryladenine, *N*-acetylcytosine, *N*-isobutyrylguanine, or uracil.

Procedure. To TEMED (150 ml/mol, 5 equivalent, 7.50 ml) in acetonitrile (CH_3CN) (100 ml) is slowly added 48% hydrofluoric acid (36 ml/mol, 3.5 equivalent, 1.26 ml) at 0°. This solution is then added to compound **2**. The reaction proceeds at room temperature with mixing. After 6 hr, the CH_3CN is removed under vacuum, but not to dryness. The residue is resuspended in 100 ml of DCM and purified on 300 g of silica gel with an ethyl acetate/methanol gradient. The overall yield from **1** [reactions (i) and (ii)] was 40–70%. The 2′-ACE uridine nucleoside is a clear oil and the remaining three nucleosides are white foams.

5′-O-Silylation: Reaction (iii)

The steric hindrance of the bis(trimethylsiloxy)cyclododecyloxysilyl chloride (DOD-Cl) silylating reagent permits the silyl chloride to react preferentially with the primary 5′-hydroxyl group over the secondary 3′-hydroxyl. The silyl chloride will react with the 3′-hydroxyl but factors such as a slow rate of addition and low temperature enhance the selectivity and increase yields.

Procedure. To a solution of 2′-*O*-ACE-nucleoside (**3**) (1 equivalent, 10 mmol) and imidazole (68.08 g/mol, 4 equivalent, 2.72 g) in tetrahydrofuran (50 ml) at 0° is added bis(trimethylsiloxy)cyclododecyloxysilyl chloride (DOD-Cl) (424 g/mol, 1.5 equivalent, 6.36 g in 20 ml tetrahydrofuran) over 30 min with stirring. The reaction is worked up by adding 70 ml of ethyl acetate, washing with saturated sodium chloride and drying the organic phase over sodium sulfate. The solvent is removed from the organic phase and the residue resuspended in 50 ml DCM and 150 ml hexanes. The 5′-silyl-2′-ACE nucleoside product (**4**) is purified on 300 g silica gel with a hexane/ethyl acetate gradient in the presence of 20% acetone. The products are isolated as oils or oily foams in 75–85% yields.

3′-O-Phosphitylation: Reaction (iv)

The final nucleoside phosphoramidite products are synthesized using the bis(*N*,*N*,-diisopropylamine)methoxyphosphine method.[5]

Procedure. To a solution of a 5′-*O*-silyl-2′-*O*-ACE-nucleoside (1 equivalent, 10 mmol) in 25 ml of dry dichloromethane is added bis(*N*,*N*-diisopropylamine)methoxyphosphine (262 g/mol, 1.5 equivalent, 3.93 g) and then tetrazole (70 g/mol, 0.8 equivalent, 0.56 g) with stirring. After 4 hr the reaction is washed with saturated sodium chloride and the organic phase dried over sodium sulfate. The nucleoside phosphoramidite product (**5**) is purified on 300 g silica gel with a hexane/dichloromethane gradient

[5] A. D. Barone, J.-Y. Tang, and M. H. Caruthers, *Nucleic Acids Res.* **12,** 4051 (1984).

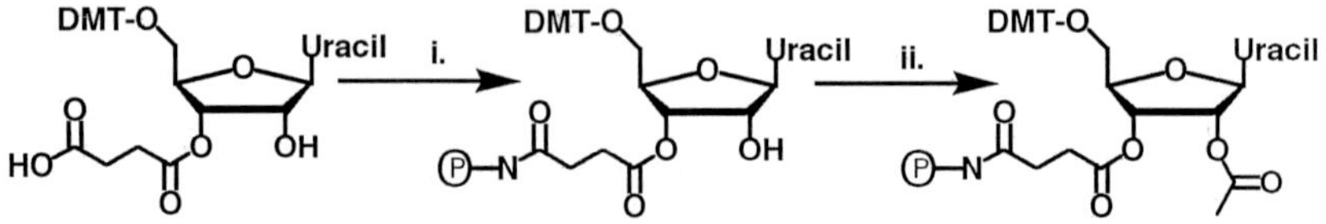

FIG. 4. Loading polystyrene support (P) with ribonucleoside succinate [step (i): DEC, pyridine; step (ii): acetic anhydride, *N*-methylimidazole, CH_3CN].

in the presence of 10% triethylamine. The purified products are isolated as clear oils in 80–90% yields.

Derivatization of Solid Supports

The use of fluoride ions to remove the 5′-silyl ether protecting group during each base addition cycle precludes the use of silica-based supports, e.g., control pore glass (CPG). We have tested many polystyrene-based supports of which several worked well. However, we have found the aminomethylpolystyrene from Pharmacia (Piscataway, NJ) to yield the best results to date. The polystyrene is loaded with ribonucleoside succinates (Fig. 4) using conventional procedures.[6] 5′-*O*-DMT protected ribonucleoside succinates are employed so that the extent of loading can be easily quantified by assaying for the DMT cation.[7]

The extent of loading affects the synthesis quality because the length of the oligoribonucleotide to be synthesized is limited by increased nucleoside loadings on the support. We have found that oligoribonucleotides up to 36–38 bases in length can be routinely synthesized with an optimal loading of 5–6 μmol per gram of support.

Procedure. Aminomethylpolystyrene (5 g) and nucleoside succinate (1 mmol) are dried by coevaporation with pyridine almost to dryness. Pyridine (50 ml) is added followed by 1-(3-dimethylaminopropyl)-3-ethylcarbodiimide hydrochloride (DEC) (210 g/mol, 10 mmol, 2.10 g) (Acros) and the slurry gently shaken. Aliquots are removed periodically to assay for nucleoside loading.[7] When the desired loading of 5–6 μmol per gram of polystyrene is reached, the solution is filtered and the resin washed with pyridine and CH_3CN (2 times 60 ml each solvent). The unreacted amines on the resin are acetylated for 15 min with a solution of 10 ml acetic anhydride, 10 ml *N*-methylimidazole in CH_3CN (80 ml). The resin is next washed with CH_3CN, water, acetone, DCM, methanol, and finally ethyl

[6] R. T. Pon, *in* "Methods in Molecular Biology: Protocols for Oligonucleotides and Analogs" (S. Agrawal, ed.), Vol. 20, pp. 465–495. Humana Press, Totowa, New Jersey, 1993.

[7] M. J. Gait, "Oligonucleotide Synthesis—a Practical Approach." IRL Press, Oxford, 1984.

ether (2 times 60 ml each solvent). Prior to use, the 5′-*O*-DMT group is removed with 3% (v/v) dichloroacetic acid in dichloromethane.

Oligonucleotide Synthesis

In 5′-silyl-2′-ACE RNA oligonucleotide synthesis, each nucleotide is added sequentially (3′ to 5′ direction) via solid phase synthesis using the cycle of reactions illustrated in Fig. 5. First, the ribonucleoside phosphoramidite and activator are added [step (i) in Fig. 5], coupling the second

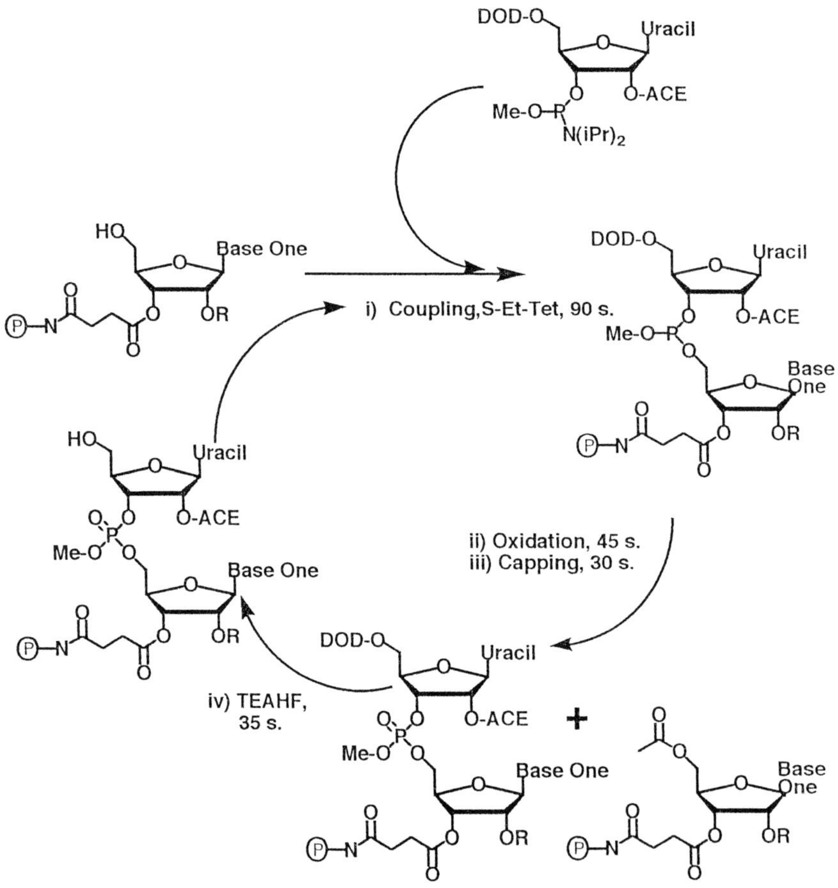

FIG. 5. Outline of 5′-silyl-2′-ACE RNA synthesis cycle [step (i): couple next nucleoside with S-ethyl-tetrazole catalyst, 90 sec; step (ii): oxidize phosphorus linkage, 45 sec; step (iii): cap unreacted 5′-hydroxyl groups, 30 sec; step (iv): 5′-deprotection with fluoride ions, 35 sec.]

base onto the 5′ end of the growing oligoribonucleotide. The support is washed and the P(III) linkage is oxidized to the more stable and ultimately desired P(V) linkage [step (ii)]. Unreacted 5′-hydroxyl groups are capped with acetic anhydride to yield 5′-acetyl esters [step (iii)]. At the end of the nucleotide addition cycle, the 5′-silyl group is cleaved with fluoride ions [step (iv)], and the cycle is repeated for each subsequent ribonucleotide addition.

Preparation of Oligonucleotide Synthesis Reagents

Amidites. The ribonucleoside phosphoramidites are diluted to a standard concentration of 0.1 *M* in dry acetonitrile. Because these compounds are oils, care must be taken to ensure that the compound is completely dissolved. The solutions are filtered through a 0.2-μm filter before use. These compounds are water sensitive; anhydrous conditions are recommended when handling these reagents.

Coupling Activator. A 0.17 *M* solution of *S*-ethyltetrazole (Glen Research, Sterling, VA, or American International Chemical, Natick, MA) in dry acetonitrile is used to catalyze the coupling reaction. Anhydrous conditions are recommended while making and handling this reagent.

5′-Deprotection. A 1.1 *M* hydrofluoride/2.9 *M* triethylamine (TEA) solution in dimethylformamide (DMF) is employed for 5′-deprotection and prepared as follows. To a solution of DMF (60 ml) and TEA (40 ml), slowly add 48% hydrofluoric acid (Mallinckrodt) (3.9 ml). This solution is transferred to a Nalgene polypropylene bottle for use on the synthesizer. (*A glass bottle cannot be used with this reagent.*) This reagent is prepared fresh and has a shelf-life of up to 1 week.

Oxidation. A ~1 *M* solution of *tert*-butyl hydroperoxide in toluene[8] is used for oxidation of the phosphorous internucleotide linkage after coupling. This reagent is prepared as follows. Toluene (750 ml) and 70% *tert*-butyl hydroperoxide (120 ml) (Lancaster) are shaken vigorously in a 1-liter separatory funnel. The layers are allowed to separate (~2 hr) and the lower aqueous phase discarded. The toluene phase can be stored up to 6 months at −20° or up to 4 days at room temperature.

Capping. Standard acetic anhydride and catalyst capping reagents are used. The two solutions are prepared in separate bottles. (During synthesis, these reagents are mixed in the reaction column.) The first solution is 10% (v/v) acetic anhydride (10 ml) in CH_3CN (90 ml). The second capping solution is 10% (v/v) *N*-methylimidazole (10 ml) in CH_3CN (90 ml). These reagents have a shelf-life of up to 6 months at room temperature.

[8] J. G. Hill, B. E. Rossiter, and K. B. Sharpless, *J. Org. Chem.* **48,** 3607 (1983).

Solvents. The following solvents are used in oligonucleotide synthesis: acetonitrile (anhydrous grade), DMF (HPLC grade), and water (HPLC grade).

Synthesis Columns. Milligen/Biosearch style 1.0-μmol size columns (Prime Synthesis, Aston, PA) are packed with 28–30 mg of polystyrene support for each 0.2-μmol synthesis. Prior to synthesis, the 5′-DMT group on the nucleoside succinate derivatized polystyrene is removed with 3% dichloroacetic acid in dichloromethane.

Synthesizer Instrumentation

5′-Silyl-2′-ACE RNA chemistry has been tested on two commercially available synthesizers, the Gene Assembler Plus (Pharmacia, Piscataway, NJ) and the 380B (PE-ABI, Foster City, CA). The Gene Assembler requires no modifications to use this chemistry. The 380B requires some modifications as described later. We currently use the 380B for the majority of our syntheses.

Adaptation of 380B Synthesizer. To enable a 380B instrument to be used for 2′-ACE chemistry, two general modifications are required: (1) replacement of the glass flow restrictors and (2) replumbing of the gas pressure lines to the solvent reservoirs. As mentioned earlier in the reagent section, the bottle for the 5′-deprotection reagent is replaced with an all-plastic one.

The 380B employs a set of glass flow restrictors to ensure equal flow rates to three columns in parallel. However, the glass material is incompatible with the 5′-desilylation reagent. These flow restrictors are replaced as follows. Remove the lines from the bottom Luer fitting of the columns to the valve block. For each column replace these lines with a 25-cm length of 0.012-inch i.d. Teflon tubing (Varian, Walnut Creek, CA). Next, between the upper Luer fitting and the top valve block is a short length of Teflon tubing. Disconnect this tubing from the valve block and insert an 8.5-cm length of 0.007-in. i.d. Tefzel tubing (Alltech, Deerfield, IL). The 0.007-in. i.d. tubing serves as the new flow restrictor. Flow rates are measured as acetonitrile is delivered to all three columns simultaneously and the 0.007-in. i.d. tubing is trimmed to provide equal flow rates to each column. (We have found that syntheses are best with the flow restrictor located after the synthesis column.) With CH_3CN in bottle 18, regulator A is adjusted to deliver CH_3CN to all three columns, through the flow restrictors, and out to waste at a rate of 1.8–1.9 ml/min per column.

The addition of several new reagents necessitates changes to the gas delivery system. The 5′-deprotection reagent, oxidation solution, DMF, and water are relatively viscous and require higher pressures for adequate flow rates. On the 380B, the 5′-deprotection is on bottle position 14, oxidation on position 15, DMF on position 17, and water on position 13. The gas delivery lines for these four positions are rerouted to the high-pressure regulator, i.e., regulator D. With DMF in bottle 17, the regulator D pressure is then adjusted to deliver DMF to a single column, through the flow restrictor, and to the waste at a rate of 1.8–1.9 ml/min.

Other commercial synthesizers, e.g., the Expedite series and model 394, are currently being investigated for their compatibility with 5′-silyl-2′-ACE chemistry. It is anticipated that modifications to the 394 will resemble those done on the model 380B. Because the Expedite model is pump driven, it is anticipated that fewer modifications will be required. However, the Expedite instrument employs a ceramic reagent channel block and the compatibility of the 5′-deprotection reagent with ceramic materials is unknown. Use of the Expedite synthesizer may require a PEEK or other polymer-based reagent channel block.

Oligonucleotide Synthesis Methods

Table I summarizes the base addition cycle for three parallel 0.2-μmol syntheses on a 380B instrument. Prior to synthesis, the 5′-DMT group on the nucleoside succinate derivatized polystyrene is removed with 3% dichloroacetic acid in dichloromethane.

TABLE I
SYNTHESIS CYCLE ON 380B INSTRUMENT

Reaction/wash	Reagents/solvents (0.2 μmol scale)	Time (sec)
Coupling	1:1 mixture of 0.1 *M* nucleoside phosphoramidite in CH_3CN (15 equivalent) and 0.17 *M* *S*-ethyltetrazole in CH_3CN (25 equivalent)	90
Wash	CH_3CN	45
Wash	Water	20
Wash	CH_3CN	45
Oxidation	1 *M* *tert*-Butyl hydroperoxide in toluene	45
Wash	CH_3CN	45
Capping	1:1 mixture of 10% acetic anhydride in CH_3CN and 10% *N*-methylimidazole in CH_3CN	30
Wash	CH_3CN	45
Wash	DMF	10
5′-Deprotection	1.1 *M* HF/3 *M* TEA in DMF	35
Wash	DMF	10
Wash	CH_3CN	60

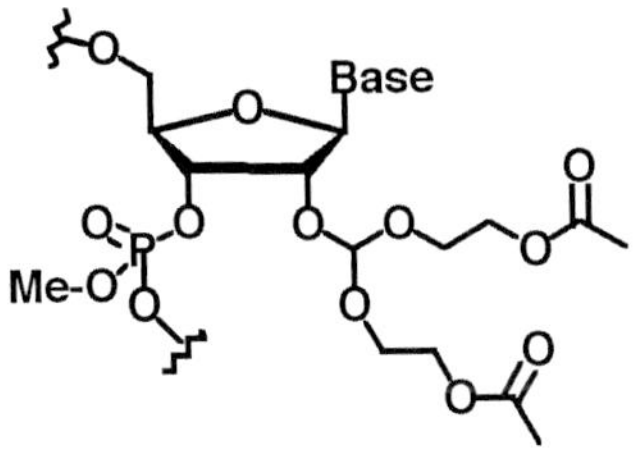

FIG. 6. General structure of support-bound RNA oligonucleotide after the final coupling.

Postsynthesis Processing and Analysis

Following synthesis, the methyl protecting groups on the phosphates are removed while the oligonucleotide is still on the support. The RNA is then cleaved from the support under basic conditions and released into solution with the concomitant deprotection of the exocyclic amines and deacetylation of the 2′-ACE groups. The RNA can be analyzed by high-performance liquid chromatography (HPLC) without further processing at this point. Prior to 2′-deprotection the RNA is thoroughly dried *in vacuo*. The residue is easily 2′-deprotected in common aqueous buffers. The fully deprotected RNA, while still in the final aqueous buffer, can be loaded directly onto an HPLC or polyacrylamide gel (PAGE). It is also possible to leave the RNA in the buffer and carry it forward to the next application, e.g., 5′-kinasing for labeling.

Phosphate Deprotection

At the end of the oligoribonucleotide synthesis the RNA consists of the general structure shown in Fig. 6. The first postsynthesis process is to remove the methyl protecting groups on the phosphates while the RNA is still attached to the supports. Historically, thiophenol has been the reagent employed. However, we routinely use an improved odorless alternative, disodium 2-carbamoyl-2-cyanoethylene 1,1-dithiolate (S_2Na_2).[9]

Procedure. The demethylation reagent is prepared fresh prior to use by dissolving S_2Na_2 (200 mg) in DMF (2 ml). This reagent is syringed back and forth through the column for 30 min. The column is subsequently washed with water (5 ml) and acetone (5 ml). The column is dried and the polystyrene support transferred to a 4-ml glass vial (VWR) with a Teflon cap (VWR) for the next reaction.

Synthesis of S_2Na_2. Sodium (23 g/mol, 1 equivalent, 0.2 mol, 4.6 g) is carefully dissolved in 125 ml ethanol at 0°. Cyanoacetamide (84.08 g/mol,

[9] B. J. Dahl, K. Bjergarde, L. Henriksen, and O. Dahl, *Acta Chem. Scand.* **44,** 639 (1990).

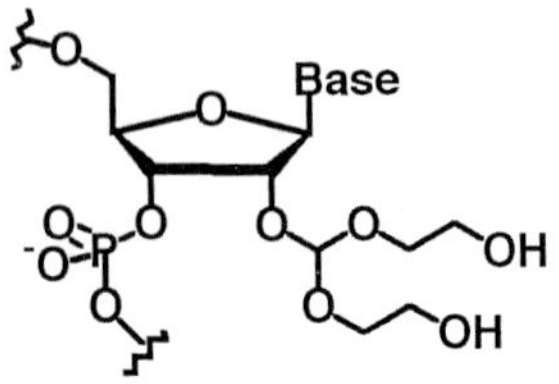

FIG. 7. 2′-Protected RNA following methylamine deprotection.

1 equivalent, 0.2 mol, 16.8 g) is suspended in a separate 50 ml of ethanol (200 proof). After the sodium has dissolved, it is added dropwise to the cyanoacetamide/ethanol suspension, which remained a suspension even after addition. Carbon disulfide (1 equivalent, 0.2 mol, 12.2 ml) is then added all at once and the solution is stirred for 1 hr. The solution turns yellow.

Sodium (1 equivalent, 4.6 g) is dissolved in 125 ml ethanol (at 0° and then added dropwise to the proceeding reaction 1 hr after the addition of carbon disulfide. The solution becomes cloudy. This is stored at 4° overnight to allow product to precipitate. The precipitate is filtered and washed with ethanol. The solid is easily redissolved in 100 ml of 80% (v/v) methanol, 20% water; then 300 ml ethanol is added. Crystallization begins within 1 hr and the solution is stored at 4° to complete the crystallization of the product. Yield of the trihydrate product (molecular weight 258.14) is 36.4 g, 70%.

Cleavage and Basic Deprotection

The partially protected RNA is next treated with a strong basic solution to simultaneously remove the exocyclic amine protecting groups, remove the acetyl groups on the 2′-ACE group, and cleave the RNA from the support and into solution. Either 40% methylamine[10,11] or ammonium hydroxide can be used, although we have observed slightly better results with the methylamine reagent. The structure of the protected oligoribonucleotide following this reaction is illustrated in Fig. 7.

Procedure. To the support in a 4-ml vial is added 2 ml of 40% methylamine in water (ACROS). The vial is sealed tightly with a Teflon-lined cap and heated to 55° for 10 min. The vial is cooled to room temperature. The oligoribonucleotide is now deprotected except at the 2′ position (Fig.

[10] M. P. Reddy, N. B. Hanna, and F. Farooqui, *Tetrahedron Lett.* **25,** 4311 (1994).

[11] F. Wincott, A. DiRenzo, C. Shaffer, S. Grimm, D. Tracz, C. Workman, D. Sweedler, C. Gonzalez, S. Scaringe, and N. Usman, *Nucleic Acids Res.* **23,** 2677 (1995).

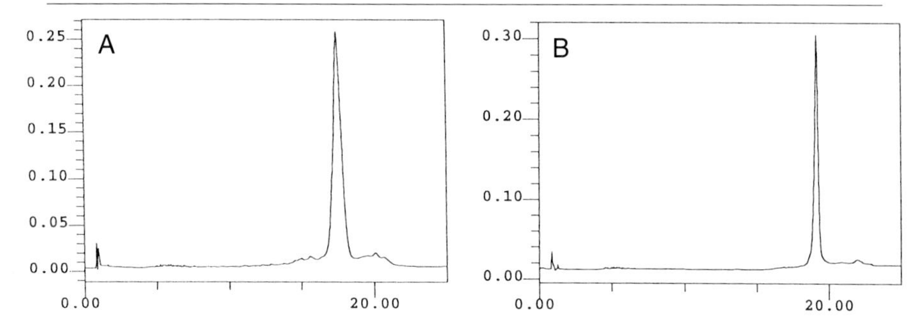

FIG. 8. Anion-exchange HPLC analysis of (A) crude 2′-protected 35-mer, (A = 15, C = 3, G = 10, U = 7), and (B) crude fully deprotected 35-mer. *X* axis is time in minutes. *Y* axis is AU260. (Same gradient in both analyses.)

7) and is ready for analysis. Alternatively, the RNA can be dried *in vacuo,* e.g., Speed-Vac, and 2′-deprotected prior to analysis.

HPLC Analysis

At this point the 2′-ACE protected RNA is water soluble and can be analyzed via anion-exchange HPLC or PAGE. HPLC analyses of a 35-mer both 2′-protected and fully deprotected are illustrated in Fig. 8. The purity of this synthesis is representative of the unprecedented quality possible with 5′-silyl-2′-ACE RNA chemistry.

Of further significance is that the 2′-protecting groups appear to interrupt secondary structure. This is clearly demonstrated in the HPLC analysis (Fig. 9) of the synthesis of a 10-mer homopolymer of guanosine.

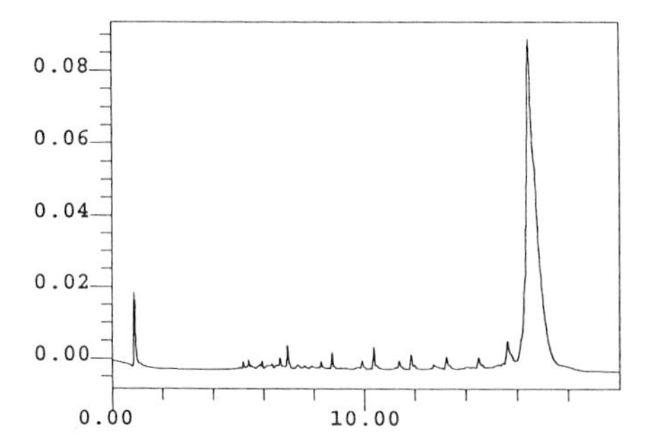

FIG. 9. Anion-exchange HPLC analysis of crude 2′-protected 10-mer homopolymer of guanosine (GGG GGG GGG G). *X* axis is time in minutes. *Y* axis is AU260.

TABLE II
REPRESENTATIVE HPLC GRADIENT

Time (min)	Percent buffer A	Percent buffer B
0	100	0
1	100	0
4	65	35
24	35	65
25	0	100

HPLC Procedure. An aliquot (5–20 μl depending on length and scale) of the oligoribonucleotide in methylamine is injected on an HPLC. A gradient from 0% buffer B to 100% buffer B is then run using the following conditions:

Buffer A: 98% 10 m*M* sodium perchlorate, 2% acetonitrile
Buffer B: 98% 300 m*M* sodium perchlorate, 2% acetonitrile
Column: Dionex DNAPac column (25 cm $\times$ 4 mm) (Dion Ex Corp., Sunnyvale, CA)
Flow: 1.5 ml/min
Temperature: 55°
Detection wavelength: 260 nm

The actual rate of change in the gradient depends on the length and base composition of each oligonucleotide. We suggest beginning with a 3% per minute gradient, then adjusting the gradient to ramp up more quickly to a shallower gradient to elute the main product. For example, the gradient used in Fig. 8 is outlined in Table II. The results in Fig. 8 were obtained using an earlier version of the Dionex NucleoPac column. Profiles and elution times are different on the newer Dionex DNAPac column.

2′-*O*-Deprotection

Complete cleavage of the 2′-*O*-protecting groups is effected using extremely mild acidic conditions (pH 3, 10 min, 60°) followed by adjusting the pH to ~8.0 for an additional 15 min. Orthoesters are hydrolyzed under acid catalysis by the mechanism in Fig. 10. A mixture of two products is formed: the 2′-hydroxyl and the formyl derivative.

The formyl groups are hydrolyzed at pH greater than 7.0 to yield the desired 2′-hydroxyl.[12] Therefore, the 2′-orthoester deprotection protocol

[12] B. E. Griffin, M. Jarman, C. B. Reese, and J. E. Sulston, *Tetrahedron, Lett.* **23,** 2301 (1967); H. P. M. Fromageot, B. E. Griffin, C. B. Reese, and J. E. Sulston, *Tetrahedron Lett.* **23,** 2315 (1967).

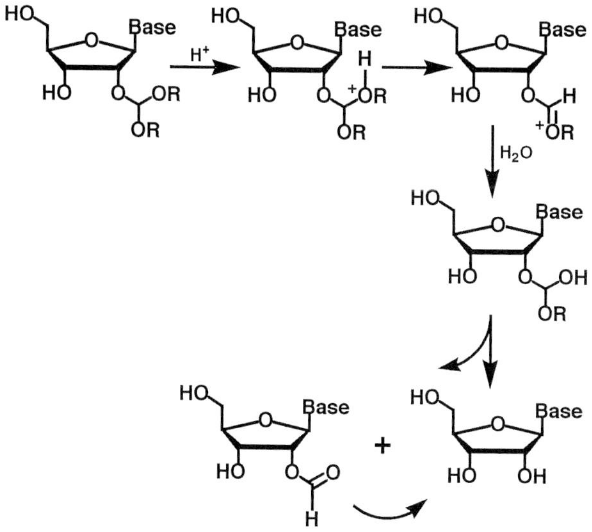

FIG. 10. Generalized mechanism for acid-catalyzed hydrolysis of 2′-orthoester protected ribonucleosides.

is comprised of two steps. After 2′-deprotection the RNA is easily degraded by RNases. Therefore, it is important to observe sterile conditions.

Procedure. The 2′-protected oligonucleotide is first dissolved in a pH 3.0 buffer (200 m*M* acetic acid adjusted to pH 3.0 with sodium hydroxide) and incubated for 10 min at 55–60° to hydrolyze the orthoesters. A second buffer at pH 8.7 (300 m*M* Tris base adjusted to pH 8.7 with hydrochloric acid) is then added and incubated at 60° (15 min) to effect hydrolysis of any 2′-formyl groups and yield the fully deprotected RNA product. (All tubes, pipette tips, and solutions are sterilized prior to use.)

Conclusion

5′-*O*-silyl-2′-*O*-ACE technology is a powerful tool for the reliable synthesis of high-quality RNA oligonucleotides. Coupling yields of >99% in less than 90 sec result in fast synthesis cycles that produce RNA of exceptional quality in high yields. After synthesis, the water-soluble 2′-protected RNA is stable to degradation and easily handled. When ready for use, the RNA is then easily 2′-deprotected in mild aqueous buffers, which are compatible with subsequent biological applications. The HPLC results in Fig. 8 are representative of the quality of RNA that can be synthesized with 5′-silyl-2′-ACE chemistry. Of further significance is that the 2′-protecting groups appear to disrupt secondary structure (Fig. 9), thereby permitting every synthesis to be conclusively analyzed on a routine basis. This is of tremendous value because it now appears possible to analyze the synthesis of sequences that possess strong secondary structure and are difficult to analyze when fully deprotected. These advantageous properties of 5′-silyl-2′-ACE chemistry are now enabling RNA syntheses to be routine and dependable.

[2] Preparation of Specifically ^{2}H- and ^{13}C-Labeled Ribonucleotides

By LINCOLN G. SCOTT, THOMAS J. TOLBERT, and JAMES R. WILLIAMSON

Introduction

The routine production of uniformly isotopically labeled proteins and RNA remains critical to current nuclear magnetic resonance(NMR) methodology. Through the use of ^{13}C and ^{15}N labeling and multidimensional heteronuclear NMR experiments, studies of 25-kDa proteins and 30 nucleotide containing RNAs are routine.[1–4] But as the size of the macromolecule increases, the number of observable nuclei and the line widths increase, while sensitivity decreases, making assignments difficult and ambiguous.[1,5–7] Deuteration of isotopically labeled proteins has generally been used to edit and enhance the quality of heteronuclear NMR spectra for large proteins, thus partially overcoming the aforementioned problems. Random fractional deuteration reduces line widths, resulting in more efficient magnetization transfer, better spectral resolution, and increased sensitivity.[8–10] But the benefits of fractional deuteration generally break down as the proteins under study continue to increase in size. Specific deuteration reduces the number of observable nuclei, resulting in narrower line widths, increased

[1] A. Pardi, *Methods Enzymol.* **261,** 350 (1995).

[2] J. Nowakowski and I. J. Tinoco, *Biochemistry* **35,** 2577 (1996).

[3] F. H. Allain, C. C. Gubser, P. W. Howe, K. Nagai, D. Neuhaus, and G. Varani, *Nature (London)* **380,** 646 (1996).

[4] A. S. Brodsky and J. R. Williamson, *J. Mol. Biol.* **267,** 624 (1997).

[5] G. Wagner, *J. Biomol. NMR* **3,** 375 (1993).

[6] T. Dieckermann and J. Feigon, *Curr. Opin. Struct. Biol.* **4,** 745 (1994).

[7] G. Varani, F. Aboul-ela, and F. H. Allain, *Prog. NMR Spectrosc.* **29,** 51 (1996).

[8] Y. Oda, H. Nakamura, T. Yamazaki, K. Nagayama, M. Yoshida, S. Kanaya, and M. Ikehara, *J. Biomol. NMR* **2,** 137 (1992).

[9] V. L. Hsu and I. M. Armitage, *Biochemistry* **31,** 12778 (1992).

[10] T. J. Tolbert and J. R. Williamson, *J. Am. Chem. Soc.* **118,** 7929 (1994).

0076-6879/00 $30.00

spectral resolution and sensitivity, but with a loss of information. Successful NMR studies of proteins as large as 60 kDa have been carried out using a combination of specific deuteration and heteronuclear labeling.[11–13] Here we present a detailed procedure for the production of specifically labeled ribonucleotide triphosphates (NTPs) as precursors for the preparation of RNA for study by NMR spectroscopy. In principle, the same advantages should be enjoyed for specific deuteration of RNAs.

Isotopically labeled RNAs are prepared in transcription reactions using labeled NTPs and T7 RNA polymerase.[14,15] Uniformly labeled NTPs are readily produced by phosphorylation of nucleotides isolated from bacterial cultures.[15,16] While some specific isotopic labeling patterns have been produced by bacterial growth on specifically labeled substrates,[17] a general biochemical synthesis of labeled NTPs offers the advantage of a diversity of isotopic labeling patterns that can be created without metabolic scrambling. Coupling the enzymes from the glycolysis and pentose phosphate pathways with those of nucleotide biosynthesis and salvage, isotopically labeled glucose can be converted into NTPs (Table I).

The strategy for enzymatic synthesis (Fig. 1) involves conversion of glucose into 5-phospho-D-ribosyl-α-1-pyrophosphate (PRPP),[18–20] which is then converted to three of the nucleoside triphosphates (ATP, GTP, and UTP).[8] CTP is prepared from UTP in a second enzymatic reaction catalyzed by CTP synthase (*PYRG*). This second reaction was found necessary because CTP could not be produced *in situ* because *PRYG* is inhibited by the conditions required for the efficient formation of ATP, GTP, and UTP.[8,21,22] In only two enzymatic reactions, higher yields of all four NTPs can be achieved per gram of glucose, making this strategy both more efficient and economical than previous methods.

The conversion of glucose into NTPs begins with the phosphorylation of glucose catalyzed by hexokinase (*HXK*). Glucose 6-phosphate (G6P) is then taken in a tandem oxidation catalyzed by glucose-6-phosphate dehy-

[11] S. Grzesiek, J. Anglister, H. Ren, and A. Bax, *J. Am. Chem. Soc.* **115,** 4369 (1993).
[12] M. A. Markus, K. T. Dayie, P. Matsudaira, and G. Wagner, *J. Magn. Reson. B* **105,** 192 (1994).
[13] X. Shan, K. H. Gardner, D. R. Muhandiram, N. S. Rao, C. H. Arrowsmith, and L. E. Kay, *J. Am. Chem. Soc.* **118,** 6570 (1996).
[14] J. F. Milligan and O. C. Uhlenbeck, *Methods Enzymol.* **180,** 51 (1989).
[15] J. R. Wyatt, M. Chastain, and J. D. Puglisi, *BioTechniques* **11,** 764 (1991).
[16] R. T. Batey, J. L. Battiste, and J. R. Williamson, *Methods Enzymol.* **261,** 300 (1995).
[17] D. W. Hoffman and J. A. Holland, *Nucleic Acids Res.* **23,** 3361 (1995).
[18] D. W. Parkin and V. L. Schramm, *Biochemistry* **26,** 913 (1987).
[19] K. A. Rising and V. L. Schramm, *J. Am. Chem. Soc.* **116,** 6531 (1994).
[20] G. E. Lienhard and I. A. Rose, *Biochemistry* **3,** 190 (1964).
[21] C. W. Long and A. B. Pardi, *J. Biol. Chem.* **242,** 4715 (1967).
[22] P. M. Anderson, *Biochemistry* **22,** 3285 (1983).

TABLE I

GENERAL ENZYMATIC REACTION SCHEME AND ENZYMES USED FOR SYNTHESIS OF NTPs[a]

Glucose → → → → PRPP → ATP, UTP, GTP; UTP → CTP

Enzyme[b]	Abbreviation	EC	Source	Vendor
Hexokinase	*HXK*	2.7.1.1	Baker's yeast	Sigma
Phosphoglucose isomerase (Glucose-6-phosphate isomerase)	*PGI1*	5.3.1.9	Baker's yeast	Sigma
Glucose-6-phosphate dehydrogenase	*ZWF*	1.1.1.49	*L. mesenteroides*	Sigma
Phosphogluconate dehydrogenase	*GDN*	1.1.1.44	*Torula* yeast	Sigma
Phosphoriboisomerase (Ribose-5-phosphate isomerase)	*RPI1*	5.3.1.6	Spinach	Sigma
Phosphoribosylpyrophosphate synthetase (Ribose-phosphatic pyrophosphokinase)	*PRSA*	2.7.6.1	*E. coli*	Sigma
Adenine phosphoribosyltransferase	*APT*	2.4.2.7	JM109/pTTA6	—
Uracil phosphoribosyltransferase	*UPP*	2.4.2.9	JM109/pTTU2	—
Xanthine-guanine phosphoribosyltransferase	*GPT*	2.4.2.22	JM109/pTTG2	—
Nucleoside-monophosphate kinase	*NMPK*	2.7.4.4	Bovine liver	Sigma
Myokinase (Adenylate kinase)	*ADK*	2.7.4.3	Rabbit muscle	Sigma
Guanylate kinase	*GMK*	2.7.4.8	Porcine brain	Sigma
3-Phosphoglycerate mutase	*YIBO*	5.4.2.1	Rabbit muscle	Sigma
Enolase (Phosphopyruvate hydratase)	*ENO*	4.2.1.11	Baker's yeast	Sigma
Pyruvate kinase	*PYKF*	2.7.1.40	Rabbit muscle	Sigma
Glutamate dehydrogenase	*GLUD*	1.4.1.3	Bovine liver	Sigma
CTP synthase	*PYRG*	6.3.4.2	JM109/pMW5	—
L-Lactate dehydrogenase	*LDH*	1.1.1.27	Rabbit muscle	Sigma

[a] Included is the acronym, identifying EC number, source, and vendor for each enzyme.
[b] IUBMB recommended name given in parentheses if different from enzyme name listed.

drogenase (*ZWF*) and 6-phosphogluconic dehydrogenase (*GND*) forming first 6-phosphogluconic-γ-lactone (6PG), and then ribulose 5-phosphate (Ru5P). Isomerization of Ru5P to ribose 5-phosphate (R5P) with ribose-5-phosphate isomerase (*RPI1*) followed by phosphorylation of the C-1 position of R5P by 5-phospho-D-ribosyl-α-1-pyrophosphate synthetase (*PRSA*) generates PRPP, which is then used to make NTPs. Coupling the free base to PRPP is catalyzed by the specific phosphoribosyltransferase for adenine, guanine, or uracil forming AMP, GMP, and UMP, respectively. Nucleotide monophosphates are converted to the corresponding triphosphate using myokinase (*ADK*), nucleoside monophosphate kinase (*NMPK*), guanylate kinase (*GMK*), and pyruvate kinase (*PYKF*).

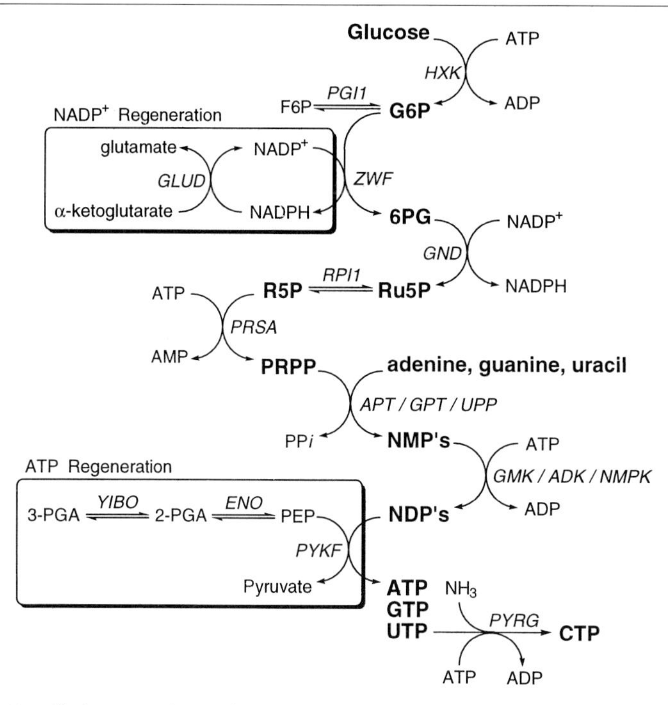

FIG. 1. Detailed enzymatic reaction scheme for the conversion of glucose to the four NTPs used to make RNA. Glucose, as well as all intermediates preceding the four NTPs are shown in boldface type, while enzymes used are denoted in italics.

This synthesis requires five equivalents of ATP, and two equivalents of $NADP^+$. The ATP required was generated from phosphoenolpyruvate (PEP), by the action of the *PYKF*/*ADK* coupled enzyme system that converts AMP into ATP. The PEP was in turn generated *in situ* from excess sodium 3-phosphoglycerate (3-PGA) with the 3-phosphoglycerate mutase (*YIBO*)/enolase (*ENO*) coupled enzyme system.[23,24] The $NADP^+$ required to oxidize glucose 6-phosphate (G6P) and 6-phosphogluconate (6PG) was regenerated by reductive amination of excess α-ketoglutarate and ammonia with NADPH catalyzed by glutamic dehydrogenase (*GLUD*).[19]

A number of advantages are associated with the synthesis of NTPs using the glycolysis and pentose phosphate pathways. First, glucose can be obtained in a variety of isotopic labeled forms available commercially. Second, all but four of the enzymes required to convert glucose into NTPs

[23] B. L. Hirschbein, F. P. Mazenod, and G. M. Whitesides, *J. Org. Chem.* **47,** 3765 (1982).
[24] E. S. Simon, S. Grabowski, and G. M. Whitesides, *J. Org. Chem.* **55,** 1834 (1989).

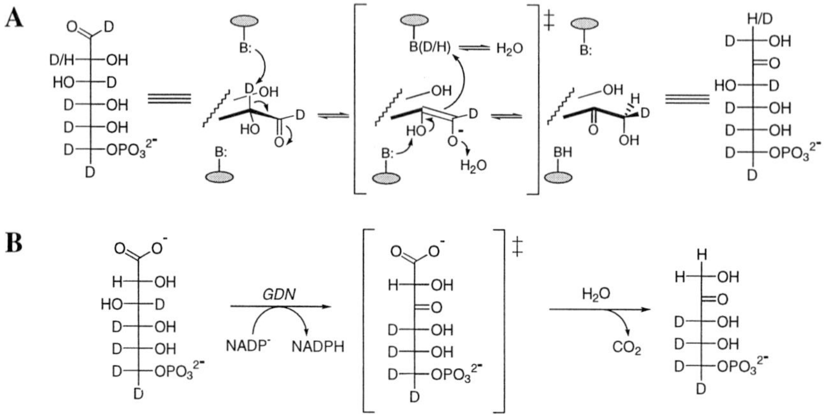

FIG. 2. (A) General mechanism for glucose-6-phosphate isomerase (*PGI1*) and ribose-5-phosphate isomerase (*PRI1*) showing the sugar deuterium atoms which exchange with solvent while residing on a protein base and (B) the general mechanism for the conversion of 6-phosphogluconate (6PG) to ribulose 5-phosphate (Ru5P) catalyzed by 6-phosphogluconate dehydrogenase (*GND*).

are commercially available. The remaining enzymes not commercially available—adenine phosphoribosyltransferase (*APT*),[8] xanthine-guanine phosphoribosyltransferase (*GPT*),[8] uracil phosphoribosyltransferase (*UPP*),[25,26] and *PYRG*[8] must be purified from overproducing *Escherichia coli* strains (see later discussion).

The enzymatic scheme we employ also provides the opportunity to exchange the C-1′ position of NTPs with hydrogen or deuterium atoms from solvent, catalyzed by glucose-6-phosphate isomerase (*PGI1*). In addition, the C-2′ position of NTPs can be exchanged with solvent atoms using both 6-phosphogluconic dehydrogenase (*GND*) and ribose-5-phosphate isomerase (*RPI1*).[18,19] Glucose-6-phosphate isomerase catalyzes the C-1 aldose–C-2 ketose (G6P–F6P) isomerization through an enedioate intermediate (Fig. 2A). In an identical mechanism, ribose-5-phosphate isomerase (*RPI1*) catalyzes the C-1 aldose (R5P) to the C-2 ketose (Ru5P) isomerization. The accepted mechanism involves a basic residue on the isomerase removing the C-2 hydrogen (or deuterium) atom from the sugar and placing it on the C-1 position. While the hydrogen (or deuterium) atom abstracted is resident on the protein prior to ketose formation, exchange with solvent can occur, resulting in equilibration of the C-2 position of the sugar with solvent.

[25] P. S. Andersen, J. M. Smith, and B. Mygind, *Eur. J. Biochem.* **204,** 51 (1992).
[26] U. B. Rasmussen, B. Mygind, and P. Nygaard, *Biochem. Biophys. Acta* **881,** 268 (1986).

Exchanging the C-2′ position of NTPs with solvent occurs when *GND* removes the hydrogen (or deuterium) atom from the C-3 position of 6-phosphogluconate (6PG) forming an γ-ketocarboxylic acid intermediate (Fig. 2B). During the decarboxylation, a solvent atom is placed at the C-1 position forming ribulose 5-phosphate (Ru5P).[20] In the reverse of the general mechanism in Fig. 2A, the former solvent atom is then removed from the C-1 position by *RPI1* and placed at the C-2 position, producing ribose 5-phosphate (R5P).

The use of glucose as a starting material and the biosynthetic pathways allow many different patterns of isotopic labeling to be created with only slight variations in the general reaction scheme for the synthesis of any RNA. BY changing the type of isotopically labeled glucose, using or not using *PGI1* in the reaction, or conducting the reaction in D_2O or H_2O, a wide variety of different labeling patterns can be created. Here we demonstrate the methodology utilizing $^{13}C,^{2}H$-uniformly labeled glucose and *PGI1* in H_2O, to create [1′,2′,3′,4′,5′-$^{13}C_5$, 3′,4′,5′-$^{2}H_4$]NTPs. This particular labeling pattern conserves important sequential NOE assignments between base hydrogen atoms and the H1′/H2′ of ribose while affording all the benefits of deuteration previously described.

Note: This procedure had been optimized to a great extent; departures from the desired protocol should be avoided without due consideration. It is also highly recommended that these techniques be reproduced with unlabeled materials prior to embarking on a labeled preparation.

Method

General Purification of Phosphoribosyltransferases

Reagents

EZMix LB broth (Sigma, St. Louis, MO)
Ampicillin (Sigma)
1 *M* Tris-HCl buffer, pH 7.8 (Mallinckrodt, Paris, KY)
2-Mercaptoethanol (Aldrich, Milwaukee, WI)
Streptomycin sulfate (Sigma)
Ammonium sulfate (Mallinckrodt)
DEAE-650M Toyopearl resin (Supelco, Bellefonte, PA)
Potassium chloride (Mallinckrodt)
Glycerol (Fisher Scientific, Pittsburgh, PA)
Isopropyl-β-thiogalactoside (Sigma)
Adenine phosphoribosyltransferase strain JM109/pTTA6

Xanthine-guanine phosphoribosyltransferase strain JM109/pTTG2
Uracil phosphoribosyltransferase strain JM109/pTTU2

Equipment

SLA-3000 rotor for Sorvall RC5C plus low-speed centrifuge
Sonic disrupter 550 (Fisher Scientific, Pittsburgh PA)
10-kDa MWCO dialysis membrane (Spectrum Medical Industries, Houston, TX)
14- × 2.5-cm Econo-column (Bio-Rad, Hercules, CA)
Innova 4330 shaker incubator (New Brunswick Scientific, Edison, NJ)

Procedure

1. LB media is prepared by autoclaving 1 liter of a solution containing 20 g of EZMix LB broth.[27] After autoclaving, sterile filtered ampicillin 50 μg/ml is added to the solution.

2. Isopropyl-β-thiogalactoside (IPTG) inducible overexpressing strains of adenine phosphoribosyltransferase (*APT*) JM109/pTTA6, xanthine-guanine phosphoribosyltransferase (*GPT*) JM109/pTTG2, or uracil phosphoribosyltransferase (*UPP*) JM109/pTTU2, are grown for 8 hr in a 5-ml culture containing 50 μg/ml of ampicillin.

3. The 5-ml culture is used to inoculate the 1-liter culture of LB medium containing 50 μg/ml of ampicillin. The culture is grown at 37° with shaking for 12 hr, then induced with IPTG (0.234 g/liter) and grown for another 6 hr.

4. Cells are then harvested by centrifugation at 6000*g* (5960 rpm for a SLA-3000 rotor) in 500-ml centrifuge bottles for 15 min at 4°. The cell pellet is suspended in 20 ml of 65 m*M* Tris-HCl buffer (pH 7.8) containing 5 m*M* 2-mercaptoethanol, and the cells are lysed at 4° with thirty 30-sec sonication bursts with a 2.5-min interval between bursts at a power setting of 7. The solution is transferred to a 50-ml centrifuge tube and the cellular debris removed by centrifugation at 31,000*g* (13,550 rpm for a SLA-3000 rotor) at 4° for 30 min.

5. The nucleic acids are precipitated by adding 20% streptomycin sulfate (w/v) to the protein supernatant. After stirring for 15 min, the resulting precipitate is removed by centrifugation at 31,000*g* (13,550 rpm for a SLA-3000 rotor) at 4° for 30 min.

6. The resulting supernatant is fractionated by adding 70% ammonium sulfate (w/v) and stirring at 4° for 30 min. The resulting precipitate is removed by centrifugation at 31,000*g* (13,550 rpm for a SLA-3000 rotor) at 4° for 30 min.

[27] D. J. Clark and O. Maaløe, *J. Mol. Biol.* **23,** 99 (1967).

7. The ammonium sulfate pellet is dissolved in 10 ml of 65 m*M* Tris-HCl buffer (pH 7.8) containing 5 m*M* 2-mercaptoethanol and transferred to a 10-kDa MWCO dialysis membrane and dialyzed 24 hr against 4 liter of the same buffer.

8. To prepare the DEAE column, 5 g of DEAE-650M Toyopearl resin is washed thoroughly with 3 *M* KCl, then equilibrated with water and packed in a 14- × 2.5-cm Econo-column at 4°.

9. The protein solution is applied to the column and subjected to a 500-ml linear gradient (3 ml/min) from 0 to 300 m*M* KCl in 65 m*M* Tris-HCl buffer (pH 7.8) containing 5 m*M* 2-mercaptoethanol, collecting 5 ml fractions. Column fractions containing phosphoribosyltransferase *APT* (eluting at approximately 150 m*M* KCl), *GPT* (eluting at approximately 140 m*M* KCl), or *UPP* (eluting at approximately 140 m*M* KCl) are detected by the spectrophotometric assays described later, or in the case of *GPT*, by SDS–PAGE gel electrophoresis.

10. Fractions containing the desired phosphoribosyltransferase are combined, concentrated by ammonium sulfate precipitation, and dialyzed as previously described in steps 6 and 7. The dialyzed phosphoribosyltransferases are stored at −20° as a 50% (v/v) glycerol stock, made by adding an equal volume of glycerol.

Adenine Phosphoribosyltransferase Activity Assay

Reagents

NADH (Sigma)
Phosphoenolpyruvate (Sigma)
Magnesium chloride (Mallinckrodt)
PRPP (Sigma)
Adenine hydrochloride (Sigma)
ATP (Sigma)
1 *M* Tris-HCl buffer, pH 7.8
L-Lactate dehydrogenase (EC 1.1.1.27)
Pyruvate kinase (EC 2.7.1.40)
Myokinase (EC 2.7.4.3)
Adenine phosphoribosyltransferase (EC 2.4.2.7)

Equipment

U-2000 spectrophotometer (Hitachi, San Jose, CA)

Procedure

1. Adenine phosphoribosyltransferase (*APT*) activity is measured by coupling the APT reaction to NADH oxidation with the enzymes myoki-

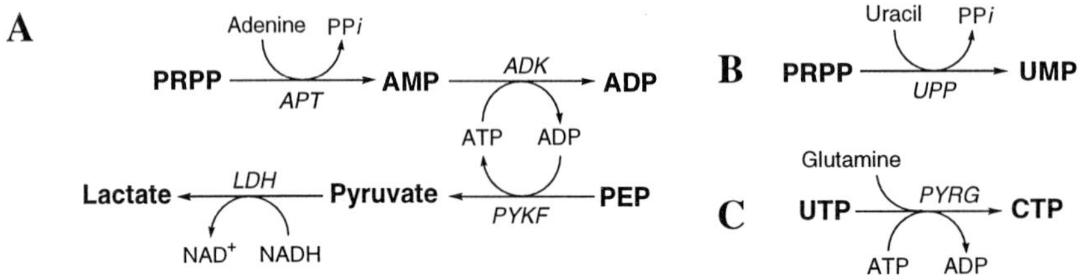

FIG. 3. (A) The adenine phosphoribosyltransferase (*APT*) assay utilizing myokinase (*ADK*), pyruvate kinase (*PYKF*), and L-lactate dehydrogenase (*LDH*) to couple APT activity with NADH oxidation. NAD^+ production is measured by the change in absorbance at 340 nm ($\Delta\varepsilon_{340} = 6220\ cm^{-1}\ M^{-1}$). (B) The uracil phosphoribosyltransferase (*UPP*) assay based on the change in extinction coefficient at 271 nm between uracil and UMP ($\Delta\varepsilon_{271} = 2763\ cm^{-1}\ M^{-1}$). (C) The CTP synthase (*PYRG*) assay based on the change in extinction coefficient at 291 nm between UTP and CTP ($\Delta\varepsilon_{291} = 1338\ cm^{-1}\ M^{-1}$). Substrates are denoted in boldface type and enzymes are denoted in italics.

nase (*ADK*), pyruvate kinase (*PYKF*), and L-lactate dehydrogenase (*LDH*) as shown in Fig. 3A.[28] The conversion of NADH to NAD^+ is monitored at 340 nm.

2. The assay solution (1 ml) contains 0.2 m*M* NADH, 1 m*M* PEP, 10 m*M* $MgCl_2$, 1.5 m*M* PRPP, 1.5 m*M* adenine hydrochloride, 3m*M* ATP, 50 m*M* Tris-HCl buffer, pH 7.8, 2 units of *LDH,* 2 units of *PYKF,* and 2 units of *ADK.*

3. A 20-μl aliquot of *APT* solution is added to start the assay, and the absorbance change at 340 nm is monitored as a function of time. The activity is obtained using Eq. (1), where $\Delta\varepsilon$ is the change in extinction coefficient of 6220 $cm^{-1}\ M^{-1}$ at 340 nm for the conversion of NADH to NAD^+:

$$\text{Units of activity} = \frac{\mu\text{mol}}{\text{min}} = \frac{l \cdot \Delta A \cdot 10^6}{t \cdot \Delta\varepsilon \cdot c} \quad (1)$$

where l is the volume of the reaction assay in liters, c is the path length in centimeters, t is the time in minutes, and ΔA is the change in the absorbance observed at 340 nm. From 1 liter of culture approximately 350 units of APT are obtained.

Note: When measuring the ΔA, we use the regions of the kinetic trace showing a linear change in absorbance over 1 min and take the average of three to five measurements.

Comments: It is common to observe an initial slight decrease in ab-

[28] A. Gross, O. Abril, J. M. Lewis, S. Geresh, and G. M. Whitesides, *J. Am. Chem. Soc.* **105,** 7428 (1983).

sorbance when *LDH, PYKF,* and *ADK* are added to the assay mixture. This is probably due to a trace amount of ADP present in the ATP added or nonspecific ATPase activity in the enzyme solution.

Uracil Phosphoribosyltransferase Activity Assay

Reagents

Magnesium chloride (Mallinckrodt)
PRPP (Sigma)
Uracil (Sigma)
1 *M* Tris-HCl buffer, pH 7.8 (Mallinckrodt)
Uracil phosphoribosyltransferase (EC 2.4.2.9)

Equipment

U-2000 spectrophotometer (Hitachi)

Procedure

1. A spectrophotometric assay for uracil phosphoribosyltransferase (*UPP*) activity has been developed to detect UPP activity by monitoring the change in absorbance at 271 nm that occurs when uracil is converted to UMP (Fig. 3B).
2. The assay solution (1 ml) contains 5 m*M* $MgCl_2$, 1.5 m*M* PRPP, 0.1 m*M* uracil, and 50 m*M* Tris-HCl, pH 7.8, buffer.
3. A 20-μl aliquot of *UPP* solution is added to start the assay, and the absorbance change at 271 mm is monitored as a function of time. The activity is obtained using Eq. (1) as shown for APT, except that $\Delta\varepsilon$ is the change in extinction coefficient of 2763 cm^{-1} M^{-1} at 271 nm for the conversion of uracil to UMP. From 1 liter of culture approximately 40 units of *UPP* are obtained.

Comments: It is important to have a relatively low concentration of uracil in the assay solution. If high levels of uracil are present, then the change in absorbance is difficult to observe.

Purification of CTP Synthase

Reagents

EZMix LB broth (Sigma)
Ampicillin (Sigma)
1 *M* Tris-HCl, pH 7.8 (Mallinckrodt)
Glutamine (Sigma)
2-Mercaptoethanol (Aldrich)

Streptomycin sulfate (Sigma)
Ammonium sulfate (Mallinckrodt)
DEAE-650M Toyopearl resin (Supelco)
1 *M* potassium phosphate (Mallinckrodt)
CTP synthase strain JM109/pMW5

Equipment

SLA-3000 rotor for Sorvall RC5C plus low-speed centrifuge
Sonic disrupter 550 (Fisher Scientific)
10-kDa MWCO dialysis membrane (Spectrum Medical Industries)
14- × 2.5-cm Econo-column (Bio-Rad)
Innova 4330 shaker incubator (New Brunswick Scientific)

Procedure

1. Overexpressing strain CTP synthase (*PYRG*) JM109/pMW5 is grown for 8 hr in a 5-ml LB culture containing 50 μg/ml of ampicillin.
2. The 5-ml culture is used to inoculate a 1-liter culture of LB medium containing 50 μg/ml of ampicillin and incubated with shaking at 37° for 16 hr.
3. The cells are then harvested, lysed, and prepared for purification in a manner similar to that described previously for the phosphoribosyltransferase proteins, but with a 20-ml buffer solution of 50 m*M* Tris-HCl (pH 7.8) containing 20 m*M* glutamine.[22]
4. The nucleic acids are precipitated, and the supernatant fractionated with ammonium sulfate in a manner similar to that described previously for the phosphoribosyltransferase proteins, but with 45% ammonium sulfate (w/v) solution.
5. The ammonium sulfate pellet is dissolved in 10 ml of 25 m*M* Tris buffer (pH 7.8) containing 5 m*M* glutamine, transferred to a 10-kDa MWCO dialysis membrane, and dialyzed 24 hr against 4 liter of the same buffer.
6. A DEAE column is prepared by washing thoroughly 5 g of DEAE-650M Toyopearl resin with 1 *M* potassium phosphate buffer pH 7.5, then equilibrating with a 50 m*M* potassium phosphate buffer, pH 7.5, containing 70 m*M* 2-mercaptoethanol and 5 m*M* glutamine, then packed in a 14- × 2.5-cm column at 4°.
7. The protein solution is applied to the column and subjected to a 500-ml linear gradient (3 ml/min) of 50–300 m*M* potassium phosphate buffer, pH 7.5, containing 70 m*M* 2-mercaptoethanol and 5 m*M* glutamine collecting 5-ml fractions. Column fractions containing *PYRG* (eluting at approximately 170 m*M* potassium phosphate) are detected by the spectrophotometric assay described later.
8. Fractions containing *PYRG* are combined, ammonium sulfate precip-

itated, dialyzed, and stored at −20° as a 50% glycerol stock as previously described.

CTP Synthase Activity Assay

Reagents

Magnesium chloride (Mallinckrodt)
Glutamine (Sigma)
ATP (Sigma)
GTP (Sigma)
UTP (Sigma)
1 *M* Tris-HCl buffer pH 7.8
CTP synthase (EC 6.3.4.2)

Equipment

U-2000 spectrophotometer (Hitachi)

Procedure

1. A spectrophotometric assay for CTP synthase (*PYRG*) activity monitors the change in absorbance at 291 nm that occurs when UTP is converted to CTP (Fig. 3C).
2. The assay solution (1 ml) contains 10 m*M* $MgCl_2$, 10 m*M* glutamine, 1 m*M* ATP, 1 m*M* UTP, 0.2 m*M* GTP in 50 m*M* Tris-HCl, pH 7.8, buffer.
3. A 20-μl aliquot of *PYRG* solution is added to start the assay, and the absorbance change at 291 nm is monitored as a function of time. The activity is obtained using Eq. (1) as shown for APT, except that $\Delta\varepsilon$ is the change in extinction coefficient of 1338 cm^{-1} M^{-1} at 291 nm for the conversion of UTP to CTP. From 1 liter of culture approximately 38 units of *PYRG* are obtained.

Comments: CTP synthase is subject to product inhibition, therefore, one will observe a leveling off of absorption over time as CTP is being formed. Glutamine, ammonium sulfate, or ammonium chloride can all be used as a source of ammonia. It is recommended that the activity assay be performed under conditions similar to those to be used for large-scale synthesis of CTP.

Preparation of ATP, GTP, and UTP from Glucose

Reagents

Amberlite IR120Plus Acidic Resin (Sigma)
Ampicillin (Sigma)

Dithiothreitol (Mallinckrodt)
ATP (Sigma)
1 *M* potassium phosphate buffer (pH 8.0)
1 *M* magnesium chloride (Mallinckrodt)
1 *M* ammonium chloride (Mallinckrodt)
Adenine hydrochloride (Sigma)
Guanine (Sigma)
Uracil (Sigma)
D-[1,2,3,4,5,6$-{}^{13}C_6,{}^{2}H_7$]glucose (Cambridge Isotopes Laboratory, Andover, MA)
α-Ketoglutarate (Sigma)
β-Nicotinamide adenine dinucleotide phosphate (Sigma)
Barium 3-phosphoglycerate (Sigma)
Hexokinase (EC 2.7.1.1)
Phosphoglucose isomerase (EC 5.3.1.9)
Glucose-6-phosphate dehydrogese (EC 1.1.1.49)
6-Phosphogluconic dehydrogenase (EC 1.1.1.44)
Phosphoriboisomerase (EC 5.3.1.6)
Phosphoribosylpyrophosphate synthetase (EC 2.7.6.1)
Adenine phosphoribosyltransferase (EC 2.4.2.7)
Uracil phosphoribosyltransferase (EC 2.4.2.9)
Xanthine-guanine phosphoribosyltransferase (EC 2.4.2.22)
Nucleoside-monophosphate kinase (EC 2.7.4.4)
Myokinase (EC 2.7.4.3)
Guanylate kinase (EC 2.7.4.8)
3-Phosphoglycerate mutase (EC 5.4.2.1)
Enolase (EC 4.2.1.11)
Pyruvate kinase (EC 2.7.1.40)
Glutamate dehydrogenase (EC 1.4.1.3)

Equipment

Stir plate, model 320 (VWR, San Diego, CA)
250-ml three-neck flask (VWR, San Diego, CA)

Procedure

1. The sodium form of 3-phosphoglycerate is prepared by stirring 5 g of barium 3-phosphoglycerate, 40 g of Amberlite IR120Plus resin (prewashed with three 25-ml portions of H_2O), in 50 ml H_2O for 2 hr.

2. The resin is removed by filtration and washed three times with 5 ml H_2O. The pH of the combined filtrates is adjusted to 7.5 with 1 *M* NaOH. The final concentration of sodium 3-phosphoglycerate is approximately 0.17 *M*.

3. To a three-neck flask, sodium 3-phosphoglycerate (1.3 mmol), α-ketoglutarate (0.29 g, 2.0 mmol), and NH_4Cl (2.8 mmol) is dissolved in 40-ml of solution containing 10 m*M* $MgCl_2$, 20 m*M* dithiothreitol (DTT), and 50 m*M* potassium phosphate buffer (pH 7.5), and the pH adjusted (if needed) to 7.5 with the addition of 1 *M* NaOH.

4. After the pH is adjusted, ATP (5 μmol) and D-[1,2,3,4,5,6-$^{13}C_6$,$^2H_7$]glucose (0.078 g, 0.4 mmol) are added to the mixture.

5. The phosphorylation of glucose and isomerization (exchanging the C-2 position) is started by adding 150 units of *YIBO,* 50 units of *ENO,* 75 units of *PYKF,* 35 units of *ADK,* 50 units of *HXK,* and 75 units of *PGII* (see Table I for enzyme acronyms).

6. After 36 hr, the phosphorylation of glucose appears to be complete (by HPLC analysis), and 25 units of *GLUD,* 10 units of *ZWF,* 5 units of *GND,* 100 units of *RPI1,* 1 unit of *PRSA,* 2 units of *APT,* 50 units of *YIBO,* 25 units of *ENO,* 25 units of *PYKF,* 25 units of *ADK,* $NADP^+$ (0.009 g, 12 μmol), 3-PGA (2.5 mmol), adenine hydrochloride (0.017 g, 0.1 mmol), uracil (0.022 g, 0.2 mmol), and guanine (0.015 g, 0.1 mmol) are added to begin formation of [1′,2′,3′,4′,5′-$^{13}C_5$,3′,4′,5′-$^2H_4$]ATP.

7. After approximately 40% [1′,2′,3′,4′,5′-$^{13}C_5$,3′,4′,5′-$^2H_4$]ATP formation has occurred (as determined by HPLC analysis), 2 units of *UPP,* 2 units of *GPT,* 1 unit of *NMPK,* 2 units of *GMK,* 50 units of *YIBO,* 25 units of *ENO,* 25 units of *PYKF,* 25 units of *ADK,* and 3-PGA (2.5 mmol) are added to begin formation of [1′,2′,3′,4′,5′-$^{13}C_5$,3′,4′,5′-$^2H_4$]GTP and [1′,2′,3′,4′,5′-$^{13}C_5$,3′,4′,5′-$^2H_4$]UTP.

8. When generation of NTPs has concluded, the reaction is frozen for storage and later purification by boronate chromatography.

Comments: When exchange reactions are to be performed with *PGI1,* to avoid incomplete exchange, it is important that all the glucose is phosphorylated, and that ample time is allowed for the exchange to occur prior to PRPP synthesis. Approximately 60% H1′ exchange with solvent is observed in reactions carried out at room temperature for 2 days. In contrast, 100% exchange of the H1′ is typically observed when the reactions are heated at 34° for the same period of time. If heating is to be used, it is important that all the glucose be phosphorylated prior to elevating the temperature, and the reaction cooled to room temperature before continuing with the synthesis of NTPs. Many of the enzymes lose activity at elevated temperature. In reactions that are to be conducted for an extended period of time (i.e., days), it is advisable to add ampicillin (50 μg/ml) to prevent bacterial growth in the reactions. It is also necessary to monitor and maintain the solution pH between 7.5 and 8.0 by adding dropwise 1 *M* NaOH or 1 *M* HCl as needed.

Troubleshooting: An initial lag phase of 1–5 days (depending on the amount of ATP used) is commonly observed as ATP is being formed. This is due to the high concentrations of ADP being formed in the generation of PRPP, which in turn inhibit the function of PRPP synthetase. This lag phase often hides problems in the synthesis, such as low enzyme activity or a missing enzyme or reagent. It is therefore important for one to be sure that all needed enzymes and reagents are added. It is often helpful to use a checklist and record that each reagent has been added. If no reaction is observed after several days, we have often found that adding all the enzymes again will restart the reaction.

Monitoring of Reaction Mixtures

Reagents

0.045 *M* ammonium formate, pH 4.6 (buffer A)
0.5 *M* NaH_2PO_4, pH 2.7 (buffer B)
85% Phosphoric acid (Fisher Scientific)
Concentrated formic acid (Fisher Scientific)

Equipment

25- × 4.6-mm Nucleotide column 303NT405 (Vydac, Hesperia, CA)

Procedure

1. Buffer A, 0.045 *M* ammonium formate, pH 4.6, is prepared by adding 2.84 g of ammonium formate to 700 ml of H_2O and adjusting the pH to 4.6 by dropwise addition of phosphoric acid. Then H_2O is added to a final volume of 1 liter.
2. Buffer B, 0.5 *M* NaH_2PO_4, pH 2.7, is prepared by adding 68.99 g of NaH_2PO_4 to 700 ml of H_2O and adjusting the pH to 2.7 by dropwise addition of formic acid. Then H_2O is added to a final volume of 1 liter.
3. The nucleotide forming reactions are monitored by HPLC on a 25- × 4.6-mm Vydac 303NT405 nucleotide column equilibrated with five column volumes of buffer A. Nucleotides are eluted with a linear gradient from 100% buffer A to 100% buffer B over 10 min at a flow rate of 1 ml/min, with detection at 260 or 254 nm (Fig. 4A).

Comments: It has been observed that age, as well as pH and salt concentration play a dramatic role in column resolution. It is therefore recommended that care be taken in the preparation of both the mobile phases and equilibration of the column. Vydac 303NT405 nucleotide columns are no longer commercially available. However, the solid phase used in the Vydac column is a 300-Å quaternary ammonium cation material that is available from many vendors in a variety of forms.

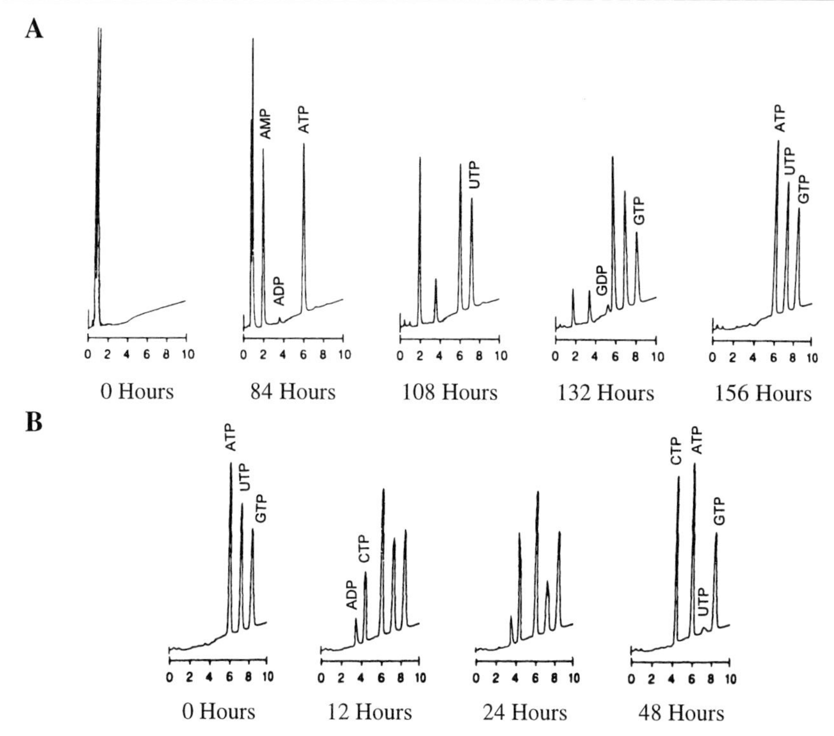

FIG. 4. (A) HPLC chromatograms of the reaction time course of an enzymatic nucleotide synthesis. The free bases are observed in the first chromatogram. After 84 hr, significant adenine ribonucleotides are observed and, eventually, complete conversion of the bases to NTPs is observed. (B) HPLC chromatograms of the reaction time course of an enzymatic conversion synthesis of UTP to CTP. UTP is observed in the first chromatogram, a steady increase in CTP is observed, and ultimately most of the UTP is converted to CTP.

Purification of Ribonucleotides from Enzymatic Reaction Pools

Reagents

Affi-Gel 601 boronate-derivatized polyacrylamide gel (Bio-Rad)
Triethylamine (Fisher Scientific)
1 *M* triethylammonium bicarbonate (TEABC), pH 9.5
Distilled water, pH 5.0
Ethanol (Fisher Scientific)

Equipment

20- × 5-cm Econo-column (Bio-Rad, Westbury, NY)
Rotavapor R110 (Brinkmann)
Filter paper 5 (Whatman, Clifton, NJ)

Procedure

1. Carbon dioxide is conveniently generated by placing dry ice in a properly stoppered filtration flask. The CO_2 rapidly sublimes and is diverted into the solution from the side arm of the flask via a Tygon tube attached to a fritted tube.

2. 1 *M* triethylammonium bicarbonate (TEABC), pH 9.5, is prepared by bubbling CO_2 through 141 ml of triethylamine in 700 ml of H_2O at 4° until the pH dropped to pH 9.5. Then H_2O is added to a final volume of 1 liter. To adjust the pH of 1 *M* triethylamine to 9.5 requires 1–2 hr typically when CO_2 is bubbled through a fritted tube.

3. Water, pH 5.0, is prepared by bubbling CO_2 through 900 ml of H_2O at 4° until the pH drops to 5.0. Then H_2O is added to a final volume of 1 liter. To adjust the pH of water to 5.0 takes less than 1 hr when CO_2 is bubbled through a fritted tube.

4. To prepare the affinity chromatography column, 5 g Affi-Gel 601 is hydrated with five column volumes of CO_2 acidified water and packed in a 20- × 5-cm Econo-column at 4°. The column is then equilibrated with five column volumes of 1 *M* TEABC at 4°.

5. The nucleotide forming reactions are concentrated *in vacuo* and then dissolved in a minimum amount of 1 *M* TEABC. Once the residue dissolves, the solution is allowed to stand at room temperature for 15–30 min while a white precipitate usually forms (precipitated proteins), which is then removed by filtration.

6. The nucleotide filtrate is applied to the column and then washed with 1 *M* TEABC while collecting 5 ml fractions until the A_{260} of the eluant drops below 0.1.[29] The proteins, salts, and other impurities wash through the boronate column, while the NTPs remains covalently bound to the boronate resin.

7. To elute the bound material, the column is washed with CO_2 acidified water until the A_{260} of the eluant drops below 0.1, which usually occurs within 100 ml after the start of elution.

8. The purified nucleotide triphosphates are concentrated *in vacuo* and finally dried to remove excess TEABC (which will interfere with subsequent enzymatic synthesis and transcription reactions) *in vacuo* with three 25-ml volumes of ethanol.

Comments: The boronate chromatography procedure allows quantitative and reproducible separation of ribonucleotides from the proteins and other materials of the reaction mixture. There are three caveats, however,

[29] H. Schott, E. Rudloff, P. Schmidt, R. Roychoudhury, and H. Kossel, *Biochemistry* **12,** 932 (1973).

to achieving these results. First, boronate chromatography is best performed at 4° because NTPs have a reduced affinity for the boronate column at room temperature. If the chromatography is performed at room temperature, it is important to save the flow-through, which can be subjected again to boronate chromatography to recover additional NTPs. Second, a wide column bed was found to be important for good flow rates because the resin volume changes with pH and ionic strength. This property provides a visual gauge for the progress of elution. Immediately after elution is initiated with CO_2 acidified water, the column trix begins to swell slightly. During this time, the NTPs do not appreciably elute off the column. After further washing, the column matrix begins to shrink until it reaches about half of its original volume. During this time the solid phase turns visibly darker and NTPs elute off the column. The completion of this process correlates to the completion of nucleotide elution from the column. Third, it is very important to load significantly less than the advertised binding capacity to minimize ribonucleotides eluting in the wash. Typically, we load approximately 250 mg of NTPs per 5 g of boronate at a time onto the column.

In all cases where a Rotavapor is used to remove solvent, lyophilization can be substituted. We recommend not heating the NTP solution above 37° when removing the solvent. Prolonged heating causes dephosphorylation to occur. In addition, it is possible to automate the boronate column chromatography through the use of a programmable FPLC system with fraction collector ability.

Preparation of CTP from UTP

Reagents

Sodium 3-phosphoglycerate (Sigma)
Half of a purified nucleotide mixture from the synthesis of ATP, GTP, and UTP
1 *M* magnesium chloride (Mallinckrodt)
1 *M* ammonium chloride (Mallinckrodt)
Dithiothreitol (Mallinckrodt)
CTP synthase (EC 6.3.4.2)
3-Phosphoglycerate mutase (EC 5.4.2.1)
Enolase (EC 4.2.1.11)
Myokinase (EC 2.7.4.3)
Pyruvate kinase (EC 2.7.1.40)

Equipment

Stir plate, model 320 (VWR)
500-ml Three-neck flask (VWR)

Procedure

1. Into a 500-ml three-neck flask is placed half of the purified nucleotide mixture from the previous reaction containing [1′,2′,3′,4′,5′-$^{13}C_5$,3′,4′,5′-$^2H_4$]ATP (0.05 mmol), [1′,2′,3′,4′,5′-$^{13}C_5$-3′,4′,5′-$^2H_4$]GTP (0.05 mmol), and [1′,2′,3′,4′,5′-$^{13}C_5$, 3′,4′,5′-$^2H_4$]UTP (0.1 mmol).

2. To the flask, NH_4Cl (10 mmol), sodium 3-phosphoglycerate (0.5 mmol), and 200 ml of a solution containing 5 m*M* $MgCl_2$, 1 m*M* dithiothreitol, pH 7.5, are added.

3. The reaction is started with the addition of 100 units of *YIBO,* 50 units of *ENO,* 50 units of *PYKF,* 50 units of *ADK,* and 3 units of *PYRG* (see Table I for enzyme acronyms).

4. The reaction is allowed to run for 48 hr, while being monitored by HPLC as described previously (Fig. 4B). When all UTP has been consumed, the reaction is stopped and purified by boronate chromatography as previously described.

Comments: All comments and troubleshooting points that applied to the formation of ATP, GTP, and UTP also hold true here. In addition, it is important that concentrations of UTP for this reaction be kept below 1 m*M* to reduce CTP product inhibition of CTP synthase.

In Vitro Transcription of TAR HIV-2 RNA

Procedure. RNA is synthesized in an optimized 40-ml *in vitro* T7 RNA polymerase transcription[30] with 2 m*M* [1′,2′,3′,4′,5′-$^{13}C_5$-3′,4′,5′-$^2H_4$]NTPs, 4 m*M* K^+ HEPES, pH 8.1, 0.1 m*M* spermidine, 10 m*M* DTT, 4.5 m*M* $MgCl_2$, 0.001% Triton X-100, 80 mg/ml polyethylene glycol (8000 molecular weight), 450 n*M* each DNA strand, and 0.07 mg/ml T7 RNA polymerase, incubated for 3 hr at 37°.

Comments: The transcription reactions need to be reoptimized for each new preparation of NTPs, because varying amounts of salts such as Mg^{2+} may copurify with the NTPs during the preparation. Low transcription yields may also be due to excess TEABC present in the NTPs after purification.

Applications

A complete overview of the methodology for using isotopically labeled RNA in NMR structure determination is presented elsewhere. Here we present examples of heteronuclear and homonuclear NMR experiments.

[30] J. F. Milligan, D. R. Groebe, G. W. Witherell, and O. C. Uhlenbeck, *Nucleic Acids Res.* **15,** 8783 (1987).

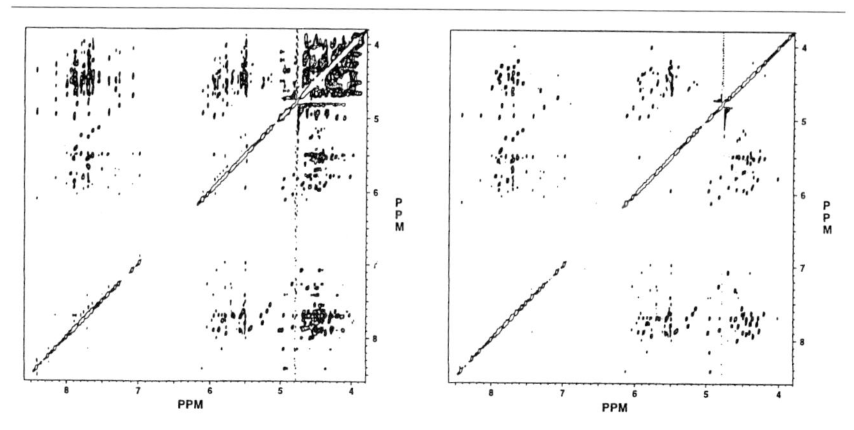

FIG. 5. Spectral simplification by deuteration demonstrated in ^{1}H, ^{1}H-NOESY spectra of TAR HIV-2 RNA (30-mer). (Left) Unlabeled RNA and (right) [1′,2′,3′,4′,5′-$^{13}C_5$,3′,4′,5′-$^{2}H_4$]RNA.

Nuclear Overhauser Enhancement Spectroscopy

Most of the strong NOEs in a NOESY spectrum of RNA (Fig. 5, left) arise from ribose–ribose and ribose–base NOEs.[7,31] Simplification of the NOESY spectrum (Fig. 5, right) is achieved through deuteration of the 3′−5′ positions of ribose, while maintaining a great deal of valuable information, such as ring conformation, and sequential assignment information between NOE base hydrogen atoms and H1′ of ribose. The selective spectral editing shown here will be useful for simplification of complicated spectra of large RNAs and RNA–protein complexes.

Heteronuclear Single Quantum Coherence

Sequential assignments between NOE base hydrogen atoms and H1′ of ribose are an important first step for assignments of RNA molecules.[7,31] The constant-time heteronuclear single quantum coherence (CT-HSQC) experiment[32,33] is useful not only in assigning $^{1}J_{CH}$ hydrogen atoms, but also removes the $^{1}J_{CC}$ couplings.

The chemical shifts of ribose are fairly well dispersed with the exception of the 2′ and 3′ positions, which overlap both in ^{1}H and ^{13}C chemical shifts (Fig. 6, left). The crosspeaks of H3′-C3′ overlap with those of the H2′-C2′ region, making the unambiguous assignment difficult. Deuteration of the

[31] A. Pardi, *Methods Enzymol.* **261,** 350 (1995).
[32] J. Santoro and G. C. King, *J. Magn. Reson.* **97,** 202 (1992).
[33] G. W. Vuister and A. Bax, *J. Magn. Reson.* **98,** 428 (1992).

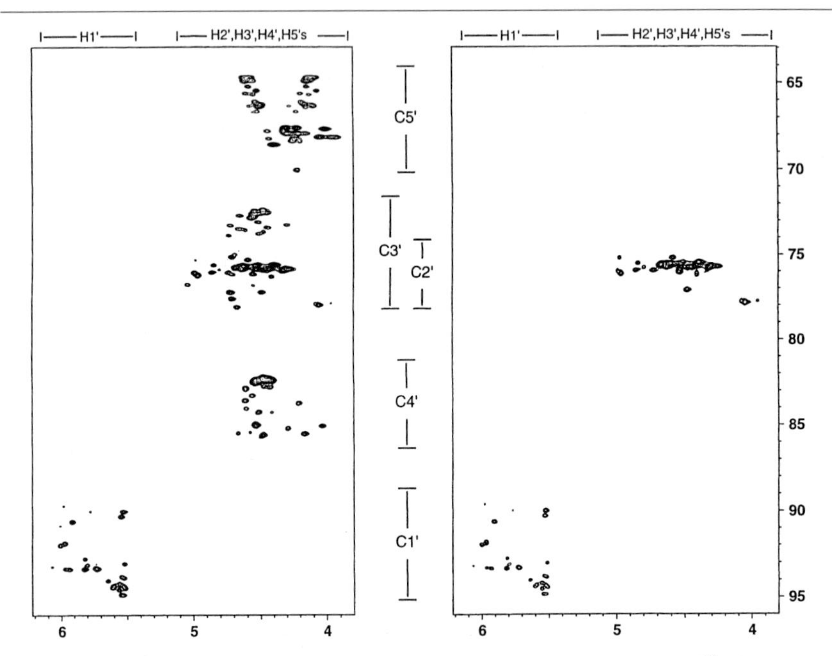

FIG. 6. ^{1}H, {^{13}C}-CT-HSQC spectra of TAR HIV-2 RNA (30-mer). (Left) [$^{13}C_5$]RNA and (right) [1′,2′,3′,4′,5′-$^{13}C_5$-3′,4′,5′-$^2H_4$]RNA.

C3′ position of ribose in the [1′,2′,3′,4′,5′-$^{13}C_5$,3′,4′,5′-$^2H_4$]ribose-RNA (Fig. 6, right) makes the identification of the H2′-C2′ crosspeaks much easier. Utilization of ^{13}C labeling in RNA from glucose offers the possibility of combining the advantages of multidimensional heteronuclear NMR with the spectral editing advantages of specific deuteration.

Conclusion

The enzymatic synthesis of specifically labeled NTPs for the *in vitro* transcription of RNA strategy described here is very efficient, flexible, and economical. The amount of labeled glucose required to make an NMR sample is low when compared to other methods of isotopically labeling macromolecules from glucose.

Acknowledgments

This work was supported by grants provided by the Alfred P. Sloan Foundation and the National Institutes of Health (GM-53757). We would also like to thank Dr. JoAnne Stubbe, Dr. Kaj Frank Jensen, Dr. Charles L. Turnbough, Jr., and Dr. Howard Zalkin for providing plasmids and overexpressing strains.

[3] Base-Modified Phosphoramidite Analogs of Pyrimidine Ribonucleosides for RNA Structure–Activity Studies

By LEONID BEIGELMAN, JASENKA MATULIC-ADAMIC, ALEXANDER KARPEISKY, PETER HAEBERLI, and DAVID SWEEDLER

Introduction

This chapter describes the design, synthesis, and utilization of some base-modified analogs of natural ribonucleosides originally developed in our laboratory for structure–activity studies of the hammerhead ribozyme. Subsequently, through collaborations with other research groups, these compounds were applied to study RNA–ligand interactions in the hairpin ribozyme, group II intron, VP vaccinia virus and the development of a "subtraction mutagenesis" approach with abasic modifications in the hammerhead system.

Organic synthesis is sometimes a challenging task for trained biologists and even for chemists working in a biochemical environment. We, therefore, decided to provide detailed procedures, highlighting potential problems. In multistep synthesis of analogs of nucleic acids one always strives to achieve the highest yield on each step in order to get a sufficient amount of target compound for biological studies at the end of the synthesis. Often, success in this endeavor depends on personal style and level of technical skills, but synthetic procedures described here were repeated by different members of our research group and summer students several times on different scales with very close results.

It is essential to master two basic techniques before starting the synthesis: (1) monitoring the reactions by thin-layer chromatography (TLC) and (2) purification of mixtures by flash chromatography. For TLC we usually use silica gel 60 F_{254} coated precut glass plates, 2 × 5 cm, with a layer thickness of 0.25 mm (VWR) or silica gel 60 F_{254} aluminum-backed sheets, 20 × 20 cm, with a layer thickness of 0.25 mm. Both plates contain a fluorescent indicator, therefore nucleosides and their derivatives can be visualized by a short-wavelength UV lamp (for example, hand-held UV lamp VWR). In our hands both TLC plates perform comparably. Occasionally glass plates provide better separation, but they are more expensive. For flash chromatography, we use Ace Glass columns with related adapters and O rings. We also use silica gel 60 (particle size, 0.040–0.063 mm; mesh, 230–400) from Merck. It is important to discard sorbent after chromatography in a

0076-6879/00 $30.00

reasonable time (2–5 hr) because O rings have the tendency to expand, particularly in ethyl acetate–hexane systems, which makes the opening of columns quite challenging. For solvent delivery, we recommend gear pumps (Cole-Parmer, Chicago, IL) because they provide flexibility in changing flow rates even for large columns.

For many preparations (especially phosphitylation and organometallic reactions) anhydrous conditions are critical. This can be achieved by rigorously drying starting materials (we use overnight drying on oil vacuum pump Welch 1402 with efficient condenser trap at −78°) and utilizing dry solvents, either commercial (Sure/seal dry solvent line from Aldrich, Milwaukee, WI) or distilled in the laboratory.[1]

Finally, it is well known that base analogs of nucleosides can be toxic and/or carcinogenic. We did not have a chance to test this for the compounds described here, but for safety reasons it is advisable to assume that these nucleosides may have such properties. Please handle all intermediates and target compounds in a fume hood wearing a laboratory coat, safety glasses, and disposable gloves.

Design and Synthesis of Modified Phosphoramidites

In this section we describe the preparation of modified nucleosides summarized in Fig. 1. The most frequently used method for incorporation of modified nucleosides in synthetic RNAs is via solid-phase oligonucleotide synthesis based on phosphoramidite chemistry.[2] This method utilizes repetitive coupling of appropriately protected monomeric nucleotides to the growing oligonucleotide chain in the 3′ to 5′ direction. After assembly of the chain is completed, removal of the phosphate, base, and 2′-OH protecting groups results in target RNA with or without incorporated modified monomeric units. To incorporate a specifically modified nucleotide at a desired position, it is necessary to prepare a monomeric building block protected at the phosphate, base (if necessary), and/or 2′-OH with protecting groups which are compatible with the general methodology of incorporation of regular unmodified monomers. Thus, synthesis of the target monomer includes the standard procedure for introduction of 5′-*O*-dimethoxytrityl group[3] (general procedure A), 2′-*O*-*tert*-butyldimethylsilyl group[4] (general

[1] D. D. Perrin and W. L. Armarego, "Purification of Laboratory Chemicals," 3rd Ed. Pergamon, New York, 1988.

[2] N. Usman, K. K. Ogilvie, M.-Y. Jiang, and R. J. Cedergren, *J. Am. Chem. Soc.* **109,** 7845 (1987).

[3] M. Smith, D. H. Rammler, I. H. Goldberg, and H. G. Khorana, *J. Am. Chem. Soc.* **84,** 430 (1962).

[4] K. K. Ogilve, K. L. Sadana, E. A. Thompson, M. A. Quilliam, and B. Westmore, *Tetrahedron Lett.* **15,** 2861 (1974).

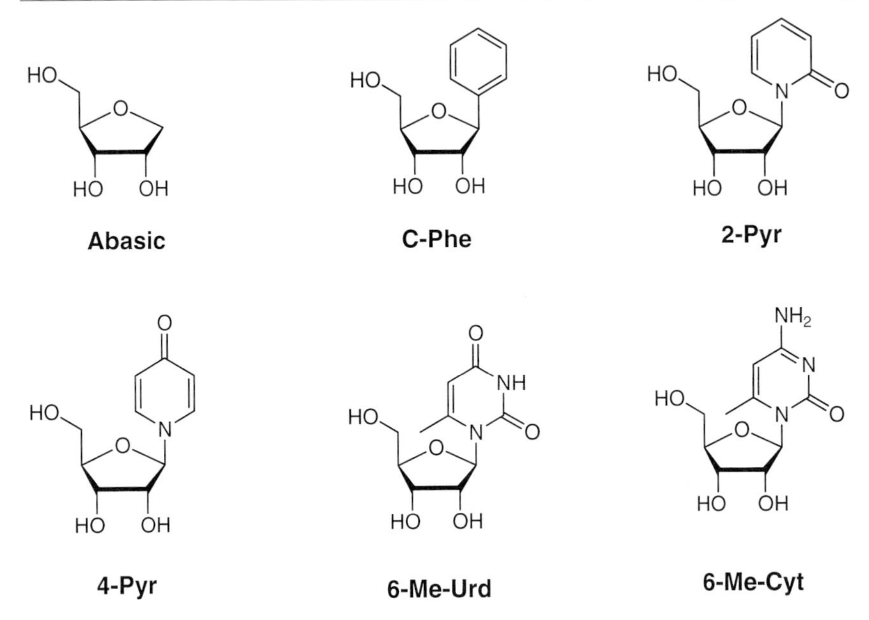

FIG. 1. Modified compounds discussed in this chapter. Abasic, riboabasic nucleoside; C-Phe, *C*-phenylriboside; 2-Pyr, 2-pyridinone riboside; 4-Pyr. 4-pyridinone riboside; 6-Me-Urd, 6-methyluridine; 6-Me-Cyt, 6-methylcytidine.

procedure B), and introduction of 3′-diisopropylaminocyanoethylphosphoramidite[5] (general procedure C), and protection of the base[6] (where applicable).

Riboabasic Phosphoramidite (**6**)

Procedure 1 (see below) and Fig. 2 (structures **1**–**6**) outline the synthesis of riboabasic phosphoramidite. This compound was originally designed and synthesized in our laboratory to elucidate the importance of natural bases in stem–loop II of the hammerhead ribozyme.[7] Substitution of pyrimidine or purine bases by a hydrogen atom in this analog provides a baseline probe to establish the involvement of removed bases in functionally important interactions.

Synthesis of riboabasic phosphoramidite **6** is straightforward; preparation of phenylthioglycoside **(2)** from inexpensive 1-*O*-acetyl-2,3,5-tri-*O*-

[5] N. Sinha, J. Biernat, and H. Koster, *Tetrahedran Lett.* **24,** 5843 (1983).
[6] S. L. Beaucage and R. P. Iyer, *Tetrahedron Lett.* **48,** 2251 (1992).
[7] L. Beigelman, A. Karpeisky, and N. Usman, *Bioorg. Med. Chem. Lett.* **4,** 1715 (1994).

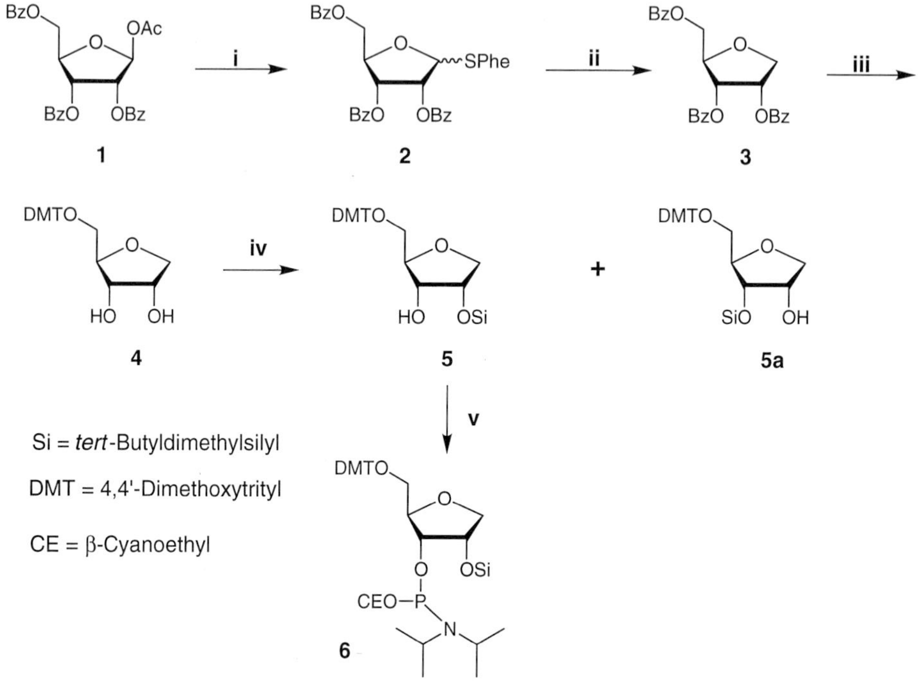

Si = *tert*-Butyldimethylsilyl

DMT = 4,4'-Dimethoxytrityl

CE = β-Cyanoethyl

FIG. 2. Preparation of riboabasic derivative suitable for oligonucleotide synthesis. (Figure should be used in conjunction with procedure 1.) Reagents: (i) Thiophenol, boron trifluoride etherate; (ii) tributyltin hydride, benzoyl peroxide; (iii) sodium hydroxide, Dowex 50 (Pyr^+), 4,4′-dimethoxytrityl chloride, pyridine; (iv) silver nitrate, *tert*-butyldimethylsilyl chloride, tetrahydrofuran/pyridine; (v) 2-cyanoethyl-*N,N*-diisopropylchlorophosphoramidite, *N*-methylimidazole, *N,N*-diisopropylethylamine.

benzoyl-D-ribofuranose (**1**) in the presence of $BF_3 \cdot (C_2H_5)_2O$ proceeds smoothly by the described procedure.[8] The next step is a radical reduction of phenylthioglycoside (**2**) and its success is critical for the entire procedure. We found that a 1:6:2 ratio of substrate: tributyltin hydride (Bu_3SnH): benzoyl peroxide (Bz_2O_2) reproducibly provides the target 1-deoxy intermediate **3** in 60–65% yield. Slow addition (over 1 hr) of a radical initiator Bz_2O_2 is also important as well as its addition to a boiling reaction mixture of the substrate with Bu_3SnH. Subsequent removal of benzoyl protecting groups and standard dimethoxytritylation is combined in a "one-pot" procedure that allows for the reliable isolation of hydrophobic intermediate **4** by flash chromatography. Standard silylation[9] in the presence of $AgNO_3$

[8] R. I. Ferrier and R. H. Furnaux, *Carbohydr. Res.* **52,** 63 (1976).

[9] G. H. Hakimelahi, Z. A. Proba, and K. K. Ogilvie, *Can. J. Chem.* **60,** 1106 (1982).

results in the formation of a 4:1 ratio of 2′ and 3′ isomers along with a small amount of the 2′,3′-bis-silylated isomer. This mixture is easily separated by flash chromatography in hexane–ethyl acetate 2:1 eluent. Standard phosphitylation[10] of the 2′-*O*-Si 2-O-tert-butyldimethylsilyl (2′-O-TBDMSi) isomer (**5**) is complete in 2–3 hr in the presence of *N*-methylimidazole. We recommend this procedure instead of phosphitilation in the presence of collidine[11] because of easy workup and purification of the final phosphoramidite.

C-Phenyl phosphoramidite (**17**)

To determine the importance of stacking and hydrophobic interactions in the hammerhead system, we have designed and synthesized[12] *C*-phenyl phosphoramidite (**17**) (Fig. 3, structures **1–17**). In this analog the pyrimidine base is substituted by a phenyl ring, therefore this nucleotide is incapable of forming hydrogen bonding contacts but can support stacking and hydrophobic interactions.

Our synthetic approach to 1-deoxy-1-phenyl-β-D-ribofuranose (**11**) is based on the work[13] by Czernecki and Ville for highly stereoselective preparation of 1-deoxy-1-phenyl-β-D-glucopyranose from protected D-glucopyranolactone and phenyllithium. The choice of protecting groups for the starting ribonolactone is critical for this approach: they must be compatible with highly reactive organometallic reagent and also withstand the strongly acidic conditions during reduction with $(C_2H_5)_3SiH/BF_3 \cdot (C_2H_5)_2O$. We first developed a synthesis starting from 5-*O*-*tert*-butyldiphenylsilyl-2,3-*O*-isopropylidene-D-ribono-1,4-lactone (**13**) (sequence **12–13–14–15–16–17**) and later refined this approach using a shorter route through 2,3,5-tri-*O*-benzyl-D-ribono-1,4-lactone (**8**) (sequence **8–11–16–17**). Both approaches provide the target phosphoramidite in comparable yields and are reproducible. Nevertheless, handling of phenyllithium and boron tribromide requires precautions because these reagents can react violently in contact with water.

Synthesis from 2,3,5-Tri-O-Benzyl-D-Ribono-1,4-Lactone

The development of acid-catalyzed benzylation of aldonolactones by Jensen *et al.*[14] allowed for one-step preparation of 2,3,5-tri-*O*-benzyl-D-

[10] T. Tuschl, M. M. P. Ng, W. Pieken, F. Benseler, and F. Eckstein, *Biochemistry* **32,** 658 (1993).
[11] S. A. Scaringe, C. Franclyn, and N. Usman, *Nucleic Acids Res.* **18,** 5433 (1990).
[12] J. Matulic-Adamic, L. Beigelman, S. Portmann, M. Egli, and N. Usman, *J. Org. Chem.* **61,** (11), 3909 (1996).
[13] S. Czernecki and G. Ville, *J. Org. Chem.* **54,** 610 (1989).
[14] H. S. Jensen, G. Limberg, and C. Pedersen, *Carbohydr. Res.* **302,** 109 (1997).

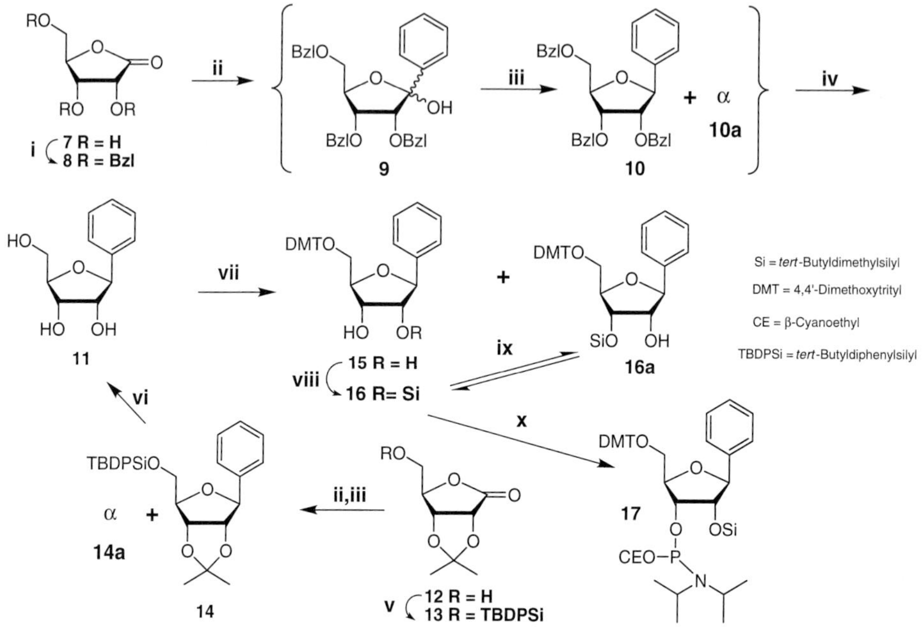

FIG. 3. Preparation of 1-deoxy-1-phenyl-β-D-ribofuranose derivative suitable for oligonucleotide synthesis. (Figure should be used in conjunction with procedure 2.) Reagents: (i) Triflic acid, benzyl 2,2,2-trichloroacetimidate; (ii) phenyllithium −78° (iii) triethylsilane, boron trifluoride etherate, −40°; (iv) boron tribromide, −78°; (v) butyl diphenylsilyl chloride, imidazole, pyridine; (vi.a) tetrabutylammonium fluoride; (vi.b) 70% AcOH, 100°; (vii) 4.4′-dimethoxytrityl chloride, pyridine; (viii) silver nitrate, *tert*-butyl dimethylsilyl chloride, tetrahydrofuran/pyridine; (ix) 5% triethylamine in methanol; (x) 2-cyanoethyl-*N,N*-diisopropylchlorophosphoramidite, *N*-methylimidazole, *N,N*-diisopropylethylamine.

ribono-1,4,-lactone (**8**) in 90% yield from commercially available D-ribono-1,4-lactone. Condensation of protected lactone **8** with phenyllithium leads to a mixture of lactols **9** which are reduced without isolation to an $\alpha:\beta$ mixture of protected *C*-nucleosides (**10** and **10a**). According to ^{1}H NMR, these two reactions provide predominantly the desired β anomer **10** with ratio $\alpha:\beta$ of 1:4 for the mixture. Monitoring of the first step by TLC requires several developments of the TLC plate in the provided system to achieve adequate resolution. We also recommend charring TLC plates at >100° for visualizing carbohydrate precursors after development and dipping in a 1% (v/v) solution of H_2SO_4 in ether. This technique proves useful for visualizations of closely moving compounds with weak chromophores (i.e., benzylated sugars).

The mixture of *C*-phenylnucleosides **10** and **10a** can be separated and then **10** can be debenzylated to provide the target free nucleoside **11**. In our hands the reverse sequence—debenzylation of the mixture of **10** and **10a** followed by separation of free nucleosides—proved to be more reproducible since separation of **10** and **10a** is quite tedious. Combining these three reactions in one procedure allows for only one chromatographic purification at the end and reproducibly provides gram quantities of *C*-phenylriboside **11**. Subsequent standard tritylation and silylation lead to the mixture of 2′- and 3′-silyl isomers **16** and **16a** in a ~1 : 1 ratio with a small amount of bis-silylated product. Careful separation of these products by flash chromatography affords the faster running 3′-*O*-Si isomer **16a** followed by the 2′-*O*-si isomer **16**. It is worth noting that the 3′-*O*-Si isomer in this case is faster moving than the 2′-*O*-Si isomer, unlike the majority of *tert*-butyl dimethylsilylated 5′-*O*-dimethoxytrityl (DMT) derivatives of ribonucleosides.

Since the ratio of silyl isomers during silylation of the 5′-*O*-DMT intermediate is close to 1 : 1, it is worthwhile to reisomerize the isolated undesired 3′-*O*-Si isomer to the mixture of 2′-and 3′-*O*-Si compounds. The routinely used[15] mixture of 5–10% pyridine in methanol overnight does not produce any isomerization, however, 5% $(C_2H_5)_3N$ in methanol, in 1 hr, results in a 1 : 1 mixture of 2′ : 3′ isomers, which are separated again to provide an additional amount of target 2′-*O*-Si isomer **16**. These isomerization conditions are general and we recommend this procedure for equalibration of 3′-*O*-Si isomers of different analogs during large-scale (5–10 g) preparations or for occasions when unfavorable 3′-*O*-regioselectivity is observed. This simple approch allows for maximizing the yields of the desired 2′-*O*-Si intermediates.

Standard phosphitylation of **16** completes synthesis of phosphoramidite **17**.

Synthesis from 2,3-O-Isopropylidene-D-Ribono-1,4-Lactone

Commercially available 2,3-*O*-isopropylidene-D-ribono-1,4-lactone is also an attractive starting material for the preparation of phenylriboside (**11**) because the 2,3-*O*-isopropylidene group is compatible with organometallic reactions and can withstand acidic conditions during reduction with $(C_2H_5)_3SiH/BF_3 \cdot (C_2H_5)_2O$.[16] Extensive experimentation identified *tert*-butyl diphenylsilyl group as the protection of choice for the 5-OH. The less acid stable *tert*-butyl dimethylsilyl protecting group was completely cleaved dur-

[15] K. K. Ogilvie *in* "Nucleosides and Nucleotides and Their Biological Applications," p. 209. Academic Press, San Diego, 1983.

[16] J. A. Piccirilli, T. Krauch, L. J. MacPherson, and S. A. Benner, *Helv. Chem. Acta.* **74,** 397 (1991).

ing acidic reduction conditions and led to C-5-OH–C-1 cyclization.[12] We provide later a large-scale procedure for the introduction of the *tert*-butyl diphenylsilyl group at the 5-OH of (**12**) although the final product **13** recently became commercially available (Lancaster, England).

The conversion of **13** to the target *C*-nucleoside **11** follows the same approach as described for benzylated lactone **8**: condensation with phenyllithium provides a mixture of lactols that is reduced without isolation to an α,β mixture of protected nucleosides **14** and **14a** (ratio 3:1). At this stage, the β anomer is isolated by chromatography and two-step deprotection provides the target nucleoside (**11**). The number of steps in this approach is slightly higher and the observed β,α-regioselectivity in the reduction of intermediate lactols is lower (3:1 versus 4:1) than in the case of the benzylated lactone (**8**). At the same time the overall yield of target nucleoside (**11**) is essentially the same (33% versus 31%) and separation of **14** from **14a** is easier than isolation of **11** from α,β mixture. It is also worth mentioning that the structure of nucleoside **11** was unambiguously determined by X-ray crystallography.[12] The structure elucidation based only on NMR data is not very reliable in *C*-nucleoside series and can result in wrong assignments.[17]

Phosphoramidites of 2-Pyridinone Riboside (22) and 4-Pyridinone Riboside (27)

We have designed and synthesized the phosphoramidites of 2-pyridinone riboside (**22**) and 4-pyridinone riboside (**27**) to probe the importance of 2- and 4-keto functions at position U7 of the hammerhead domain (Fig. 4, structures **18–27**).

The critical step in the preparation of these two analogs is the condensation of silylated bases with protected ribofuranose (**1**) in the presence of a Lewis acid (Vorbrüggen reaction[18]). This versatile reaction allows one to synthesize many different nucleoside analogs.[19] Sugar-protected analogs obtained by Vorbrüggen reaction can be converted into building blocks suitable for oligonucleotide synthesis after standard deprotection and subsequent introduction of 5′-*O*-dimethoxytrityl group, 2′-*O*-*tert*-butyl dimethylsilyl group and 3′-phosphitylation.

Pyridin-2-one and pyridin-4-one ribonucleosides were first prepared by Pischel and Wagner[20] by condensation of silver salts of 2- and 4-hydroxypyr-

[17] N. C. Chaudhuri and E. T. Kool, *Tetrahedron Lett.* **36,** 1795 (1995).

[18] H. Vorbrüggen and B. Bennua, *Tetrahedron Lett.* **19,** 1339 (1978).

[19] E. Lukevics and A. Zablocka, "Nucleoside Synthesis, Organosilicon Methods." Ellis Horwood, 1991.

[20] H. Pischel and G. Wagner, *Arch. Pharm.* (*Weinheim*) **300,** 602 (1967).

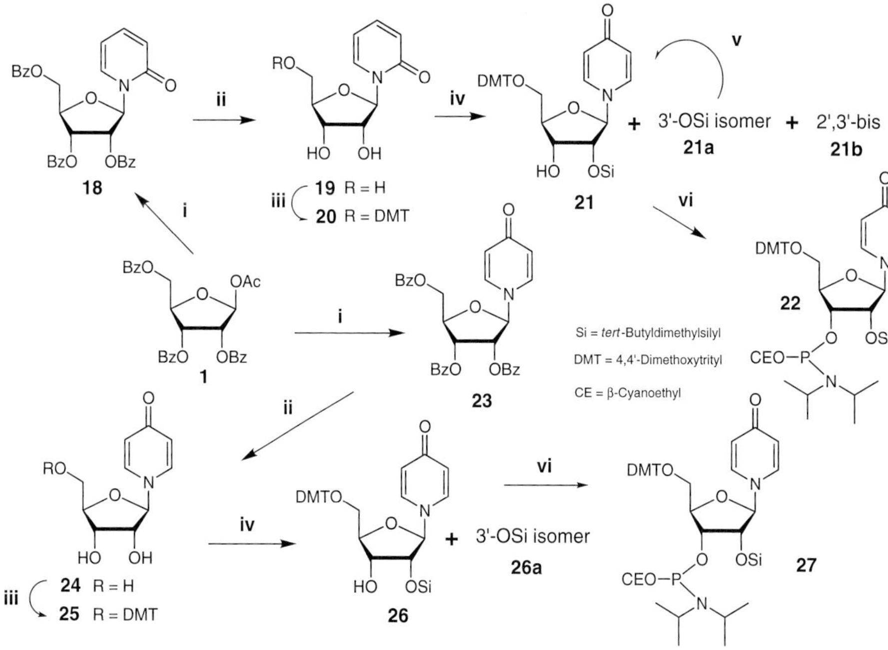

FIG. 4. Preparation of 1-(β-D-ribofuranosyl)-pyridine-2(4)-one derivatives suitable for oligonucleotide synthesis. (Figure should be used in conjunction with procedure 3.) Reagents: (i) 2-(or 4-) Hydroxypyridine, *N,O*-bis (trimethylsilyl)acetamide, trimethylsilyl trifluoromethanesulfonate; (ii) sodium methoxide, methanol, Dowex 50 (Py^+); (iii) 4,4′-dimethoxytrityl chloride, pyridine, 4-(dimethyamino)pyridine; (iv) silver nitrate, *tert*-butyl dimethylsilyl chloride, tetrahydrofuran/pyridine; (v) 3% triethylamine in methanol; (vi) 2-cyanoethyl-*N,N*-diisopropylchlorophosphoramidite, *N*-methylimidazole, *N,N*-diisopropylethylamine.

idine with 1-chloro-2,3,5-tri-*O*-benzoyl-D-ribofuranose to afford *O*-glycosides, followed by O,N rearrangement in boiling toluene in the presence of $HgBr_2$. The same compounds were also synthesized by the Hilbert–Johnson reaction of 2- and 4-ethoxypyridines with a 1-chloro sugar.[20] Later Niedballa and Vorbrüggen[21] applied the silyl Hilbert–Johnson reaction catalyzed by $SnCl_4$ to the synthesis of pyridinone nucleosides. While silylated 2-hydroxypyridine reacted smoothly with 1-*O*-acetyl-2,3,5-tri-*O*-benzoyl-β-D-ribofuranose (**1**) to give a high yield of *N*-1-riboside **18**, the analogous reaction of 4-hydroxypyridine took place only under forced conditions and in moderate yield. An improvement in the synthesis of ribonucleosides was reported[22]

[21] U. Niedballa and H. Vorbrüggen, *J. Org. Chem.* **39,** 3668 (1974).

[22] H. Vorbrüggen, K. Krolikiewicz, and B. Bennua, *Chem. Ber.* **114,** 1234 (1981).

by switching the Friedel–Crafts catalyst from $SnCl_4$ to trimethylsilyl triflate (TMSTfl), which has lower Lewis acidity compared to $SnCl_4$. Consequently, higher yields of the desired *N*-1-nucleosides are obtained in the case of more basic silylated heterocycles like cytosine and 4-hydroxypyridine by using this catalyst.

We recommend the one-pot procedure[23] for the synthesis of pyridinone nucleosides **18** and **23** from silylated bases and the 1-*O*-acetyl sugar (**1**) in the presence of TMSTfl (Fig. 4).

This procedure proved to be particularly suitable for the synthesis of pyridin-2-one nucleoside **18** since silylated 2-hydroxypyridine is a volatile compound (bp 63°/12 mm Hg)[24] that is not easily dried by evaporation and coevaporation with toluene, a requirement when hexamethyldisilazane or $(CH_3)_3SiCl$ is used for the preparation of the silylated base. Protected nucleoside **18**, obtained in 98% yield after flash chromatography, was saponified using $NaOCH_3/CH_3OH$ to give **19** in 91% yield. Dimethoxytrityl protection of the 5′-OH under standard conditions [DMT-Cl, DMAP, $(C_2H_5)_3N$, Pyr] afforded, after chromatography, the 5′-*O*-DMT derivative **20** in 76% yield. Selective protection of the 2′-OH using *tert*-butyl dimethylsilyl chloride proceeded in the presence of $AgNO_3$ and pyridine in tetrahydrofuran (THF) to afford a mixture of 2′-*O*-TBDMSi, 3′-*O*-TBDMSi isomers and some 2′,3′-bis-*O*-TBDMSi compound. Separation of these products using flash chromatography afforded a faster running 2′-*O*-TBDMSi isomer **21** in 69% yield and slower running 3′-*O*-TBDMSi isomer **21a** in 17% yield. Phosphitylation of **21** using 2-cyanoethyl-*N,N*-diisopropylchlorophosphoramidite in the presence of *N,N*-diisopropylethylamine and *N*-methylimidazole yielded the desired 3′-*O*-phosphoramidite **22**.

Protected pyridin-4-one nucleoside **23** was prepared in 93% yield in a similar way by Vorbrüggen reaction. It is worth noting that the original procedure of Niedballa and Vorbrüggen,[25] in our hands, resulted in a poor yield of the desired *N*-1-nucleoside caused by the competing formation of the *O*-4-riboside and decomposition products. Debenzoylation of **23** using $NaOCH_3$ yielded **24** in 84% yield. 5′-*O*-Dimethoxytritylation under the conditions used for the synthesis of **19** yielded the 5′-*O*-dimethoxytrityl derivative in 67% yield. Selective protection of the 2′-OH with TBDMSi group proceeded as for pyridin-2-one nucleoside **20**, yielding the mixture of 2′-, 3′- and 2′,3′-bis substituted nucleosides. Careful separation of this mixture by column chromatography using 0.5–5% methanol in ethyl acetate

[23] L. W. Dudycz and G. E. Wright, *Nucleosides and Nucleotides* **3,** 34 (1984).

[24] L. Birkofer, A. Ritter, and H.-P. Kühlthau, *Chem. Ber.* **97,** 934 (1964).

[25] U. Niedballa and H. Vorbrüggen, "Nucleic Acid Chemistry" (L. B. Townsend and R. S. Tipson, eds.), Part 1, pp. 481–484. J. Wiley & Sons, New York, 1978.

for elution yielded the desired faster moving 2′-*O*-TBDMSi isomer **26** in 41% yield and slower 3′-*O* isomer in 36% yield. Phosphitylation proceeded smoothly to give 3′-*O*-phosphoramidite **27** in 94% yield.

Phosphoramidites of 6-Methyluridine (**31**) and 6-Methylcytidine (**36**)

Based on NMR data it was demonstrated[26] that incorporation of a bulky methyl group at C-6 of a natural pyrimidine ribonucleosides restricts the rotation of the base around the glycosidic bond due to steric clash of 6-methyl group with C-2′-H of ribofuranose moiety. Thus, these analogs, which are locked in *syn* conformation, can serve as probes for determination of importance of *syn–anti* equilibrium in RNA–ligand interactions.

The synthetic approach for preparation of phosphoramidites of 6-methyluridine (**31**) and 6-methylcytidine (**36**) is similar to the synthesis of pyridinone phosphoramidites (**22**) and (**27**) (Fig. 5, structures **28–31**). Two key steps in this preparation are Vorbrüggen condensation of protected furanose **1** with silylated 6-methyluracil and the conversion of 6-methyluridine nucleoside **28** to 6-methylcytidine nucleoside **32** through a triazolide intermediate. The later transformation is a commonly used method applicable to a wide variety of modified uridine nucleosides.

The Vorbrüggen glycosylation of 6-methyluracil is often used as a testing case for improvements in the glycosylation procedure: synthesis using $SnCl_4$ as a Lewis acid catalyst was reported[27] to provide the target N^1 nucleoside in 41% yield with considerable amounts of N^3 and N^1,N^3-bis by-products. The application of TMSiTfl as a new catalyst was reported to improve the yield of N^1 isomer formation to 71%, but still, N^3 and N^1,N^3-bis by-products were obtained.[28]

We found that a minor modification of the reported procedure—decreasing condensation temperature to 0° and dropwise addition of TMSiTfl–increases the yield of target N^1 nucleoside to 83% with traces of N^3 isomer and only 5–7% of N^1,N^3-bis by-product. Saponification of the protected nucleoside with NaOH can be combined with the subsequent dimethoxytritylation resulting in 5′-*O*-DMT intermediate **29** in 80% yield. Silylation ($AgNO_3$, TBDMSiCl, Py) of **29** results in a mixture of 2′, 3′-bis-Si (5%), 2′-*O*-Si (**30**) (40%), and 3′-*O*-Si (**30a**) (35%) isomers. These compounds are easily separable by flash chromatography. Standard phosphitylation completes the synthesis of target phosphoramidite **31**.

[26] M. P. Schweizer, E. B. Benta, J. T. Witkowski, and R. K. Robins, *J. Am. Chem. Soc.* **95,** 3370 (1987).

[27] U. Niedballa and H. Vorbrüggen, *J. Org. Chem.* **39,** 3660 (1974).

[28] H. Vorbrüggen and K. Krolikiewitcz, *Angew. Chem. Int. Ed. Ingl.* **14,** 421 (1975).

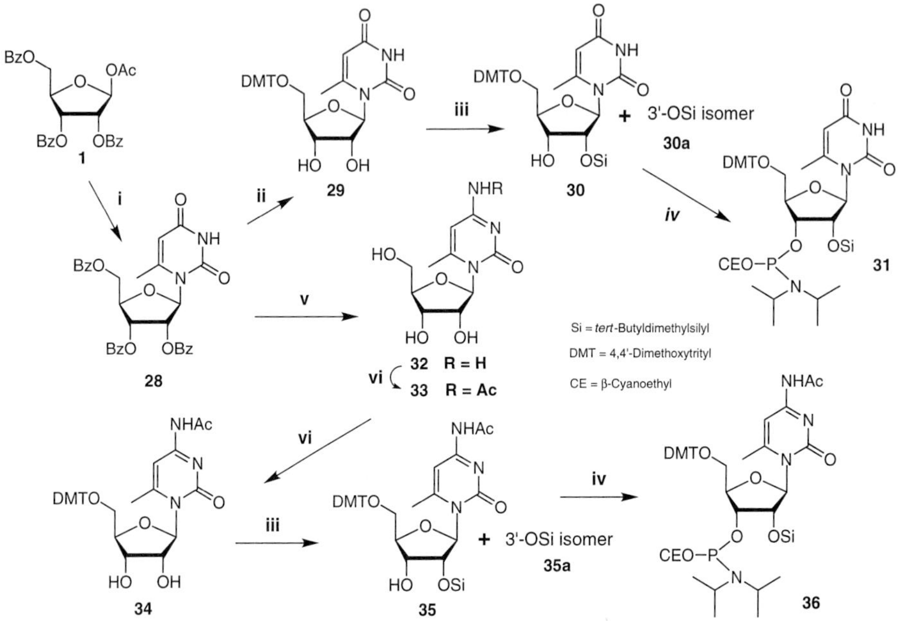

FIG. 5. Preparation of 6-methyuridine and 6-methyl-cytidine derivative suitable for oligonucleotide synthesis. (Figure should be used in conjunction with procedure 4.) Reagents: (i) Bis(trimethylsilyl)-6-methyluracil, trimethylsilyl trifluoromethanesulfonate, 0°; (ii.a) 2 *M* sodium hydroxide, Dowex 50 (Py^+), (ii.b) 4,4′-dimethoxytrityl chloride, pyridine; (iii) silver nitrate, *tert*-butyl dimethylsilyl chloride, tetrahydrofuran/pyridine; (iv) 2-cyanoethyl-*N,N*-diisopropylchlorophosphoramidite, *N*-methylimidazole, *N,N*-diisopropylethylamine; (v.a) 1,2,4-triazole, phosphorus oxychloride; (v.b) 29% aqueous ammonia; (vi.a) trimethylchlorosilane, acetic anhydride, (vi.b) methanol, water, (vi.c) 4,4′-dimethoxytrityl chloride, pyridine.

Conversion of protected 6-methyluridine into the corresponding cytidine derivative (**32**) is achieved through the C-4 triazolide intermediate being displaced by NH_4OH by the procedure originally developed by Sung[29] and Divakar and Reese.[30] These procedures are routinely used for uridine-to-cytidine conversion. Although it is tempting to reduce the seemingly high amount of triethylamine and triazole (10 and 9 equivalents to the substrate) used in this reaction, we are advising not to do so since such "economy" usually leads to the incomplete conversion to the C-4 triazolide intermediate and thus reduces the yield and complicates the isolation of cytidine nucleosides.

Selective acetylation and introduction of the dimethyoxytrityl group for

[29] W. L. Sung, *J. Org. Chem.* **47,** 3623 (1982).

[30] K. J. Divakar and C. B. Reese, *J. Chem. Soc. Perkin Trans.* **1,** 1171 (1982).

6-methylcytidine can be combined in a "one-pot" procedure using the transient protection methodology[31] without isolation of intermediate N^4-acetyl nucleosides. This results in the formation of 5′-*O*-DMT-N^4-acetylcytidine (**34**) in 74% yield. Subsequent silylation (TBDMSill, Py) of (**34**) results in a mixture of 2′-*O*-Si (**35**) (26%) and 3′-*O*-Si (**35a**) (31%) compounds, which are separable by flash chromatography. Standard phosphitylation finishes the synthesis of target phosphoramidite (**36**).

Incorporation of Modified Phosphoramidites into Oligoribonucleotides, Purification, and Characterization

All modified phosphoramidites described in this article are easily introduced at specific positions of the desired oligoribonucleotide using standard conditions of solid-phase oligoribonucleotide synthesis. We did not find it necessary to increase the coupling time or create special customized cycles for the introduction of these nucleotides into medium size (40-mer) oligoribonucleotides. All of the modified phosphoramidites couple in >97% yield as monitored by trityl cation release. The current state of the art allows for the reliable incorporation of modified nucleotides in such a fashion into 36- to 40- mer synthetic RNAs.[32,33] Synthesis of modified 70- to 80-mers is challenging but not impossible with increased coupling time and modified cycles.[34,35]

The modified phosphoramidites should be dissolved in anhydrous DNA/RNA synthesis-grade acetonitrile, usually at a concentration of 0.1 *M* and filtered through 25 × 0.5-mm PFTE Millex-SR filters (Millipore, Bedford, MA) to remove any particulate material that can clog the synthesizer's lines. We routinely filter acetonitrile solutions of modified phosphoramidites immediately after their preparation and then dry them *in vacuo* for cold storage under Ar.

We recommend use of a recently developed protocol for oligoribonucleotide synthesis and deprotection that utilizes *S*-ethyltetrazole[32,33] as an activator, CH_3NH_2[36] and $(C_2H_5)_3NH$–3HF–4-pyrrolidinone–dimethylsulfoxide (DMSO) for base and TBDMSi deprotection.[37] With

[31] R. Kierzek, *Nucleosides & Nucleotides* **14,** 255 (1995).

[32] F. Wincott, A. DiRenzo, C. Shaffer, S. Grimm, D. Tracz, C. Workman, D. Sweedler, C. Gonzalez, S. Scaringe, and N. Usman, *Nucleic Acids Res.* **23,** 2677 (1995).

[33] B. Sproat, F. Colonna, B. Mullah, D. Tsou, A. Andrus, and A. Hampel, *Nucleosides & Nucleotides* **14,** 255 (1995).

[34] M. Lyttle, P. Wright, N. Sinha, I. Bain, and A. Chamberlin, *J. Org Chem.* **56,** 4608 (1991).

[35] J. T. Goodwin, W. A. Stanick, and G. D. Glick, *J. Org Chem.* **59,** 7941 (1994).

[36] M. P. Reddy, N. B. Hanna, and F. Farooqui, *Tetrahedron Lett.* **35,** 4311 (1994).

[37] F. Wincott and N. Usman, *in* "Methods in Molecular Biology" (P. C. Turner, ed.), Vol. 74, p. 59. Humana Press, 1997.

S-ethyltetrazole(0.15 *M*) as an activator and a 600-min coupling time for all monomers, a 35–40% full-length material is achieved reliably on the 2.5 μM scale for a 36-mer. That corresponds to the isolation of 130–150 ODU of the crude material per synthesis.

Usually, it is advisable to perform synthesis in "trityl-on" mode to utilize reversed-phase "trityl-specific" purification followed by final removal of 5′-terminal dimethoxytrityl group by acidic treatment.[38] This reversed-phase "trityl-specific" purification often provides material sufficiently pure for most uses. For NMR or crystallographic studies, additional purification may be needed, in this case anion-exchange HPLC on Dionex media is recommended.[39]

Two complimentary methods can be used to characterize oligonucleotides containing modified nucleotides. In the first method,[40,41] the oligonucleotide is digested to individual nucleoside components by snake venom phosphodiesterase (or nuclease P1) and alkaline phosphatase. The mononucleosides thus formed are analyzed by reversed-phase HPLC and are compared with authentic standards that are characterized independently (usually by NMR). This procedure allows one to confirm the presence of a modified base and the absence of any side products that may form during synthesis and deprotection. It also can be performed in a quantitative manner by integrating relative peaks and calculating the mole fraction of modified and natural nucleosides, which should be close to the value determined from the sequence of the synthesized oligonucleotide.

When the introduced modification does not have a chromophore (as in case of the abasic residue) mass spectroscopic methods are very useful.[42,43] For oligonucleotides, the MALDITOF technique[44] allows one to determine the precise mass of the molecule, and when coupled with enzymatic digestion by 5′- and/or 3′-processive exonucleases, may be used for direct sequencing. This technique is clearly gaining popularity, although the instrumentation required to perform this analysis is quite expensive.

It is important to stress that in all studies which utilize modified oligonu-

[38] H. J. Fritz, R. Belagaje, E. L. Brown, R. H. Fritz, R. A. Jones, R. F. Lees, and H. G. Khorana, *Biochemistry* **17,** 1225 (1978).

[39] R. Vinayak, *in* "Methods in Molecular Biology" (P. C. Turner, ed.), Vol. 74, p. 51. Humana Press, 1997.

[40] J. S. Eadie, J. McBride, J. Lincoln, W. Efscavitch, L. B. Hoff, and R Cathcart, *Anal. Biochem.* **165,** 442 (1987).

[41] S. A. Strobel, T. R. Cech, N. Usman, and L. Beigelman, *Biochemistry* **33,** 13824 (1994).

[42] P. A. Limbach, *Mass Spectrom. Rev.* **15,** 297 (1996).

[43] R. W. Ball and L. C. Packman, *Anal. Biochem.* **246,** 185 (1997).

[44] T. A. Millican, G. A. Mock, M. A. Chauncey, T. P. Patel, M. A. Eaton, J. Cunning, S. D. Cutbush, S. Neidle, and J. Mann, *Nucleic Acids Res.* **12,** 7435 (1984).

cleotides, the oligonucleotide must be characterized by one, or better, by a combination of the above methods, since relying only on the chromatographic behavior of the modified oligonucleotide as proof of identity may be highly misleading.

Application of Oligoribonucleotides Containing Modified Pyrimidine Nucleotides (**6, 17, 22, 27, 31, 36**)

The majority of modified analogs described in this article were designed to study structure–activity relationships in the hammerhead ribozyme.[45] We began by investigating the importance of natural bases in the stem–loop II region of the hammerhead motif. To our surprise a straightforward tool that would allow us to test the importance of any base in RNA–RNA or RNA–protein interactions—the riboabasic nucleotide[7] was not yet synthesized at that time. A related 2′-deoxy abasic analog was described[44] in 1984.) Incorporation of abasic nucleotide **6** in stem–loop II of hammerhead motif and determination of the cleavage activity of generated mutants supported the hypothesis that the majority of the stem–loop II region serves a general structural role in maintaining or allowing a certain conformation in the catalytic core and that there are no specific base–base or base–metal interactions. Sequential replacement of nucleotides in the catalytic core of the hammerhead ribozyme with abasic analog **6** demonstrated a dramatic loss of activity for 9 out of 11 mutants.[46] Subsequent detailed analysis of this data in collaboration with D. Herschlag's group led to a "core folding" model[47] to explain peculiarities of hammerhead catalysis.

Later abasic modifications were used for the analysis of base and sugar requirements in internal loop B of the hairpin ribozyme[48]; in determination of factors involved in recognition of branch-site adenosine in the group II intron[49]; and in the analysis of recognition of the functional groups of pyrimidine nucleotide in the cleavage site of the hammerhead ribozyme.[50]

A promising alternative for the classical atomic mutagenesis approach,

[45] N. Usman, L. Beigelman, and J. A. McSwiggen, *Curr. Opin. Struct. Biol.* **1,** 527 (1996).

[46] L. Beigelman A. Karpeisky, J. Matulic-Adamic, C. Gonzalez, and N. Usman, *Nucleosides & Nucleotides* **14,** 907 (1995).

[47] A. Peracchi, A. Karpeisky, L. Maloney, L Beigelman, and D. Herschlag, *Biochemistry* **37,** 14765 (1998).

[48] S. Schmidt, L. Beigelman, A. Karpeisky, N. Usman, U. S. Sorensen, and M. J. Gait, *Nucleic Acids Res.* **24,** 573 (1996).

[49] Q. Lin, J. B. Green, A. Khodadadi, P. Haeberli, L. Beigelman, and A. M. Pyle, *J. Mol. Biol.* **267,** 163 (1997).

[50] N. Baidya, G. E. Ammons, J. Matulic-Adamic, A. Karpeisky, L. Beigelman, and O. C. Uhlenbeck, *RNA* **3,** 1135 (1997).

based on the rescue of abasic mutants with exogenous bases added *in trans,* was developed in collaboration with A. Peracchi and D. Herschlag. It was demonstrated that deleterious effects from introducing abasic nucleotide in the hammerhead core can, in some cases, be "rescued" by ablated base added *in trans.*[51] Utilization of commercially available base analogs in this approach allows one to rapidly determine the functional groups involved in important interactions at "rescuable" positions.[52]

The general applicability of *C*-phenyl-**17** and 2- or 4-pyridinone nucleotides (**22**, **27**) in mutagenesis analysis of RNA–RNA interactions was demonstrated for position N-7 of the hammerhead system[53] and for the cleavage site pyrimidine nucleotide.[50] 6-Methyluridine (**31**) nucleotide was used for nuclease stabilization[54] of position U4 in the hammerhead domain and together with other pyrimidine analogs to study interactions of VP 55, the vaccinia virus poly(A) polymerase with rU2-N15-RU recognition sites on oligonucleotide primers.[55]

Experimental Procedures

General Procedure A: Introduction of DMT Group

Nucleoside (5 mmol) is evaporated with anhydrous pyridine (2 $\times$ 50 ml) and dissolved in anhydrous pyridine (50 ml). To this stirred solution, 4-(dimethylamino)pyridine (0.05 equivalent), triethylamine (1.4 equivalent), and 4,4-dimethoxytrityl chloride (1.2 equivalent) are added under Ar. The reaction mixture is stirred at room temperature until TLC demonstrates the disappearance of starting material (2–12 hr). The reaction is then quenched by addition of 5 ml of anhydrous methanol and evaporated to dryness *in vacuo.* The residue is partitioned between saturated $NaHCO_3$ (100 ml) and CH_2Cl_2 (100 ml), the organic layer extracted with brine, a saturated solution of NaCl, (100 ml), dried (Na_2SO_4), and evaporated to dryness. Resulting residue is then purified by flash chromatography on silica gel with the appropriate solvent mixture as an eluent.

[51] A. Peracchi, L. Beigelman, N. Usman, and D. Herschlag, *PNAS* **93,** 11522 (1996).

[52] A. Peracchi, J. Matulic-Adamic, S. Wang, L. Beigelman, and D. Herschlag, *RNA* **4,** 1332 (1998).

[53] A. B. Burgin, Jr., C. Gonzalez, J. Matulic-Adamic, A. M. Karpeisky, N. Usman, J. A. McSwiggen, and L. Beigelman, *Biochemistry* **35,** 14090 (1996).

[54] L. Beigelman, A. Karpeisky, and N. Usman, *Nucleosides & Nucleotides* **14,** 895 (1995).

[55] Li Deng, L. Beigelman, J. Matulic-Adamic, A. Karpeisky, and P. D. Gershon, *J. Biol. Chem.* **272,** 31542 (1997).

General Procedure B: Introduction of TBDMS Group

To the stirred solution of the protected nucleoside (1 mmol) in 50 ml of dry THF is added pyridine (4 equivalent) followed by silver nitrate ($AgNO_3$) (2.4 equivalent). After 10 min *tert*-butyldimethylsilyl chloride (TBDMSiCl) (1.5 equivalent) is added and the reaction mixture stirred at room temperature for 12 hr. The resulting suspension is filtered into 100 ml of 5% aqueous $NaHCO_3$. The solution is extracted with dichloromethane (CH_2Cl_2) (2 × 100 ml). The combined organic layer is washed with brine, dried over Na_2SO_4, and evaporated *in vacuo*. The residue is purified by flash chromatography on silica gel with an appropriate solvent mixture as an eluent.

General Procedure C: Phosphitylation

To a solution of protected nucleoside (1 mmol) stirring at 0° under argon in anhydrous dichloromethane (20 ml) is added *N,N*-diisopropylethylamine (2.5 equivalent) followed by *N*-methylimidazole (1.0 equivalent) via syringe. After stirring for 15 min at 0°, 2-cyanoethyl-*N,N*-diisopropylchlorophosphoramidite (1.2 equivalent) is added dropwise via syringe to the stirred reaction mixture. The reaction mixture is warmed to room temperature and monitored by TLC (70% ethyl acetate/hexanes). After 2 hr at room temperature the reaction mixture is again cooled to 0° and subsequently quenched with 0.5 ml anhydrous ethanol. Hexane (20 ml) is then added to the reaction mixture, which is then pump loaded directly onto a flash silica column that has been previously equilibrated with 2% TEA in hexanes. A gradient of 10–70% ethyl acetate in hexane provides pure phosphoramidite as a white foam after appropriate fractions are pooled and evaporated *in vacuo*.

*Procedure 1: Preparation of 2'-O-tert-Butyldimethylsilyl-5'-O-dimethoxytrityl-3'-O-(2-cyanoethyl-N,N-diisopropylphosphoramidite)-1-deoxy-D-ribofuranose (**6**)*

THIOPHENYL-2,3,5-TRI-O-BENZOYL-D-RIBOFURANOSE (**2**). 1-*O*-Acetyl-2,3,5-tri-*O*-benzoyl-D-ribofuranose (**1**) (Aldrich) (10.0g 19.84 mmol) is dissolved in dry CH_2Cl_2 (200ml) and cooled to 0° (ice bath). Thiophenol (2.47 ml, 24 mmol) followed by $BF_3 \cdot (C_2H_5)_2O$ (3.43 ml, 28 mmol) is added dropwise over 15 min while maintaining temperature at 0°. Reaction mixture is maintained at 0° for 3 hr, when TLC (ethyl acetate/hexane, 1 : 4) demonstrated complete disappearance of starting material (R_f0.36) and formation of the product (R_f 0.52). Reaction mixture is diluted with CH_2Cl_2 (100 ml), washed by saturated $NaHCO_3$ (50 ml), H_2O (100 ml), and brine (50 ml). The organic layer is dried over Na_2SO_4 and the solvent removed *in vacuo*.

The residue is purified by flash chromatography on silica gel using CH_2Cl_2 as eluent to give 9.5 g (86%) of **2** as slightly yellow syrup.

2,3,5-TRI-O-BENZOYL-1-DEOXY-D-RIBOFURANOSE (**3**). Thiophenyl-2,3,5-tri-*O*-benzoyl-D-ribofuranose **2** (5.95 g, 10.73 mmol) is dissolved, under argon, in dry, degassed toluene (70 ml) and tributyltin hydride (17.16 ml, 63.79 mmol) is added. A solution of benzoyl peroxide (5.15 g, 21.46 mmol) in dry, degassed toluene (50 ml) is added to the refluxing reaction solution over 1 hr. The reaction mixture is then allowed to reflux under argon for an additional 6 hr. TLC (20% ethyl acetate/hexane) demonstrated complete disappearance of starting material (R_f 0.52) and formation of the product (R_f 0.49). The solvent is removed *in vacuo,* the residue dissolved in CH_2Cl_2 (100 ml), and washed sequentially with saturated $Na_2S_2O_3$ (50 ml), saturated $NaHCO_3$ (50 ml), and brine (50 ml). The organic layer is dried over Na_2SO_4 and evaporated to dryness. The crude mixture is purified by flash chromatography on silica gel using CH_2Cl_2 followed by 2% methanol CH_2Cl_2/as eluent. The appropriate fractions are collected, evaporated to dryness, and dissolved in a minimal volume of CH_2Cl_2. The CH_2Cl_2 solution is then added dropwise to cooled light petroleum ether (1 liter). The resulting precipitate is filtered to give 3.0 g (62.5%) of **3** as a white powder.

5-*O*-DIMETHOXYTRITYL-1-DEOXY-D-RIBOFURANOSE (**4**). To a cooled (−15°) solution of **3** (2.93 g, 6.5 mmol) in a mixture of pyridine (60 ml) and methanol (10 ml) is added an ice-cooled 2 *M* aqueous solution of sodium hydroxide (9.75 ml) with stirring. The reaction mixture is stirred at −10 to −15° for an additional 30 min and then neutralized to pH 7 with Dowex 50 (Pyr^+) by addition of small portions of resin. The resin is filtered and washed with H_2O : pyridine 4 : 1 (200 ml). The combined mother liquor and washings are evaporated to dryness and dried by multiple (at least 4) coevaporations with dry pyridine. The residue is redissolved in dry pyridine (75 ml), and dimethoxytrityl chloride (2.42 g, 7.15 mmol) is added and the reaction mixture left overnight at room temperature. The reaction is quenched with methanol (25 ml) and evaporated to dryness. The residue is dissolved in CH_2Cl_2 and washed with saturated $NaHCO_3$ and brine. The organic layer is dried over Na_2SO_4 and the solvent removed *in vacuo*. The residue is purified by flash chromatography on silica gel using 2% methanol in CH_2Cl_2 as eluent to give 2 g (70%) of **4**.

2′-*O*-*tert*-BUTYLDIMETHYLSILYL-5′-*O*-DIMETHOXYTRITYL-1-DEOXY-D-RIBOFURANOSE (**5**). To a stirred solution of **4** (1.85 g, 4.19 mmol) in dry THF (50 ml) is added pyridine (1.35 ml, 16.76 mmol) and $AgNO_3$ (1.71 g, 10.06 mmol). After 10 min, TBDMSiCl (0.76 g, 5.03 mmol) is added and the reaction mixture stirred at room temperature for 6 hr. The resulting suspension is filtered into 100 ml of 5% aqueous $NaHCO_3$. The solution is extracted with CH_2Cl_2 (2 × 100 ml). The combined organic layer is washed with brine, dried over Na_2SO_4, and evaporated to dryness. The residue is purified by flash chromatography on silica gel with hexane–ethyl acetate (2:1) as eluent to yield 1.4 g of the desired compound **5**, 0.37 g of 3-silyl isomer **5a**, and 0.26 g of the bissilyl derivative.

2′-*O*-*tert*-BUTYLDIMETHYLSILYL-5′-*O*-DIMETHOXYTRITYL-3′-*O*-(2-CYANOETHYL-*N*,*N*-DIISOPROPYLPHOPHORAMIDITE)-1-DEOXY-D-RIBOFURANOSE (**6**). Phosphitylation of **5** (1.16 g, 2.11 mmol) using general procedure C resulted in crude **6** purified by flash chromatography on silica gel using hexane–ethyl acetate (6:1) containing 1% $(C_2H_5)_3N$ as an eluent, yielding 1.3 g (82%) of pure **6** as a white foam.

*Procedure 2: Preparation of 2′-O-tert-Butyldimethylsilyl-5′-O-dimethoxytrityl-3′-O-(2-cyanoethyl-N,N-diisopropylphosphoramidite)-1′-deoxy-1′-phenyl-β-D-ribofuranose (**17**)*

2,3,5-TRI-*O*-BENZYL-D-RIBONO-1,4-LACTONE (**8**). To a solution of **7** (Aldrich) (5.0 g, 33.8 mmol) stirring at 0° under argon in anhydrous dioxane is added benzyl-2,2,2-trichloroacetimidate (37.0 ml, 253.2 mmol) via syringe. After 10 min, triflic acid (0.25 ml, 5.7 mmol) is added via syringe and the reaction is maintained at 0° for 1 hr and then for 16 hr at room temperature. TLC (25% ethyl acetate/hexane) indicated complete consumption of **7** R_f 0.05 and formation of (**8**) R_f 0.5. Dioxane is removed *in vacuo* and the resulting residue suspended in dichloromethane and filtered to remove excess trichloroacetimidate. The filtrate is washed with saturated aqueous sodium bicarbonate and the organic layer dried over sodium sulfate, filtered, and then evaporated *in vacuo*. Additional trichloroacetimidate is removed by filtration from 1:1 dichloromethane/hexane. The filtrate is flash chromatographed using a gradient of 10–25% ethyl acetate/hexane to give (**8**) still contaminated with small amount of trichloroacetimidate, which is removed by crystallization from CH_2Cl_2 at −15°. Evaporation of the filtrate *in vacuo* provided pure **8** (12.38 g, 90.8%).

1′-DEOXY-1′-PHENYL-β-D-RIBOFURANOSE (**11**). **Caution!** This procedure utilizes organometallic reagents that can react violently with water. First, run reaction on 1/10 of the scale below. If you are not sure about your skills, ask trained organic chemists to supervise you.

One-pot procedure. To a solution of **8** (18.43 g, 44.0 mmol) stirring at −78° in anhydrous THF (200 ml) under argon is added phenyllithium (27 ml, 48.44 mmol) dropwise via syringe. The reaction mixture is stirred at −78° for 2 hr followed by 4 hr at room temperature before being cooled to 0° and quenched with ice–water (200 ml). TLC (25% ethyl acetate/hexane) indicated the formation of a new product **9** R_f 0.52, which charred differently than (**2**) R_f 0.50. Extraction with diethyl ether (2 × 200 ml) followed by drying over sodium sulfate and evaporation *in vacuo* provides crude hemiacetal (**9**). This material is dissolved in anhydrous acetonitrile (300 ml) and cooled to −40° while stirring under argon. Triethylsilane (14.1 ml, 88 mmol) is added followed by dropwise addition of boron trifluoride etherate (6.14 ml, 48.44 mmol) over 30 min. After stirring at −40° for 1 hr, the reaction mixture is allowed to warm to 0° at which time saturated aqueous potassium carbonate solution (85 ml) is added to quench the reaction. TLC (10% ethyl acetate/hexane) indicated the disappearance of **9** R_f 0.22, and formation of intermediates **10** and **10a** R_f 0.50. Extraction with diethyl ether (2 × 200 ml) and drying over sodium sulfate followed by filtration and drying *in vacuo* provides intermediates **10** and **10a** (9.75 g, 20.3 mmol, 46% over two steps). To a solution of **10** and **10a** stirring at −78° under argon in anhydrous CH_2Cl_2 (200 ml), is added boron tribromide (1 *M* in CH_2Cl_2) (51.0 ml, 50.75 mmol) dropwise via syringe. The reaction is stirred for 2.5 hr at −78° then quenched with a 1 : 1 solution of methanol/ CH_2Cl_2 (50 ml). TLC (10% ethanol/CH_2Cl_2) indicated the disappearance of **10** R_f 0.95 and formation of product **11** R_f 0.49. Treatment with pyridine (50 ml) followed by evaporation *in vacuo* and flash chromatography (5–15% ethanol/CH_2Cl_2) provided **11** (2.91 g, 68%).

5-*O*-*tert*-Butyldiphenylsilyl-2,3-*O*-isopropylidene-D-ribono-1,4-lactone (**13**). To the mixture of 2,3-*O*-isopropylidene-D-ribono-1,4-lactone **12** (Aldrich) (6.0 g, 31.88 mmol) and imidazole (4.77 g, 70.06 mmol) in dry DMF (50 ml) *tert*-butyldiphenylsilyl chloride (10.5 ml, 40.38 mmol) is added and the mixture stirred at room temperature overnight. TLC (25% ethyl acetate/hexane) showed the disappearance of the starting material and formation of one major product. The reaction mixture is poured into iced water (100 ml) and then extracted with ether (3 × 100 ml). The combined extracts are washed with water, dried (Na_2SO_4), and evaporated to a syrup. Flash column chromatography using a 3–10% gradient of ethyl acetate in hexane yielded **13** as syrup that crystallized on standing at room temperature (11.7 g, 86%).

5′-*O*-*tert*-Butyldiphenylsilyl-2′,3′-*O*-isopropylidene-1′-deoxy-1′-phenyl-β-D-ribofuranose (**14**). **Caution!** This procedure utilizes organometallic reagents, which can react violently with water. First, run reaction 1/10 of the scale below. If you are not sure about your skills ask trained organic chemists to supervise you.

One-Pot Procedure. To a stirred solution of **13** (24 g, 56.26 mmol) in dry THF (150 ml) cooled to −78° is added dropwise phenyllithium (2 *M* solution in cyclohexane/ether, 70/30, 30 ml, 60 mmol). The reaction mixture is stirred at −78° for 2 hr, at room temperature for 4 hr, and then left at 5° overnight. It is then quenched with cold water (300 ml) and extracted with ether (3 × 200 ml). The combined organic layers are dried (Na_2SO_4), evaporated to a syrup, and coevaporated 2x times with dry toluene. The material is dissolved in dry acetonitrile (400 ml) and cooled to −40°. Triethylsilane (18 ml, 112.7 mmol) is added followed by dropwise addition of $BF_3 \cdot (C_2H_5)_2O$ (7.83 ml, 61.79 mmol). The mixture is stirred at −40° for 2 hr, warmed to room temperature, then quenched with saturated $NaHCO_3$ (200 ml). The aqueous mixture is extracted with ether (2 × 300 ml) and the organic layer washed with brine (200 ml), dried (Na_2SO_4), and evaporated to dryness. Flash chromatography purification using 0.5–10% gradient of ethyl acetate in hexanes afforded faster running **14** (9 g, 33%). The α-anomer **14a** is isolated too (3 g, 11%).

1′-DEOXY-1′-PHENYL-β-D-RIBOFURANOSE (**11**). Compound **14** (1 g, 2.1 mmol) is dissolved in THF (20 ml) and 1 *M* TBAF in THF (3 ml, 3 mmol) is added. The reaction mixture is stirred at room temperature for 30 min and then evaporated to syrup. The residue is applied on the silica gel column and eluted with toluene followed by 5–10% gradient of ethyl acetate in hexane. The 5-*O*-desilylated product was obtained as colorless foam (0.68 g, 96% yield). This material is dissolved in 70% aqueous acetic acid and heated at 100° (oil bath) for 30 min. Evaporation to dryness under reduced pressure and crystallization of the residual syrup from toluene afforded **11** (0.49 g, 94% yield).

3′-*O*-*tert*-BUTYLDIMETHYLSILYL-5′-*O*-DIMETHOXYTRITYL-1′-DEOXY-1′-PHENYL-β-D-RIBOFURANOSE (**16a**) AND 2′-*O*-*tert*-BUTYLDIMETHYLSILYL-5′-*O*-DIMETHOXYTRITYL-1′-DEOXY-1′-PHENYL-β-D-RIBOFURANOSE (**16**). Compound **11** (770 mg, 3.7 mmol) was 5′-*O*-dimethoxytritylated according to general procedure A to yield, after silica gel column chromatography (0.5–2% gradient of ethyl acetate in hexanes), 1.4 g (75% yield) of 5′-*O*-dimethoxytrityl derivative **15** as a yellowish foam. This material (1.4 g, 2.73 mmol) is treated with *tert*-butyldimethylsilyl chloride using general procedure B and products are purified by the silica gel column chromatography (1–2% gradient of ethyl acetate in hexanes) to afford a faster moving 3′-*O*-TBDMSi isomer **16a** as a foam (0.55 g, 32%). The slower migrating 2′-*O*-TBDMSi isomer **16** is then eluted to give, on evaporation, white foam (0.60 g, 35%).

The 3′-*O*-TBDMSi isomer **16a** is isomerized in solution of 5% $(C_2H_5)_3N$ in methanol (50 ml) for 1 hr at room temperature providing a 1 : 1 of mixture of 2′ and 3′ isomers **16** and **16a**. Column chromatography of this mixture under conditions described earlier allows isolation of an additional 0.25 g of **16**.

2′-*O*-*tert*-BUTYLDIMETHYLSILYL-5′-*O*-DIMETHOXYTRITYL-3′-*O*-(2-CYANOETHYL-*N*,*N*-DIISOPROPYLPHOSPHORAMIDITE)-1′-DEOXY-1′-PHENYL-β-D-RIBOFURANOSE (**17**). Compound **16** (0.87 g, 1.4 mmol) is phosphitylated using general procedure C and the product isolated by silica gel column chromatography using 0.5% ethyl acetate in toluene [(1% $(C_2H_5)_3N$] for elution (0.85 g, 74% yield).

Procedure 3: Preparation of 1-(5-O-Dimethoxytrityl-2-O-tert-Butyldimethylsilyl-β-D-ribofuranosyl)pyridin-2-one 3′-O-(2-Cyanoethyl-N,N-Diisopropylphosphoramidite) **(22)** *and 1-(5-O-Dimethoxytrityl-2-O-tert-butyldimethylsilyl-β-D-ribofuranosyl)pyridin-4-one-3′-O(2-Cyanoethyl-N,N-diisopropylphosphoramidite)* **(27)**

1-(2,3,5-Tri-*O*-benzoyl-*β*-d-ribofuranosyl)pyridin-2-one (**18**). 2-Hydroxypyridine (Aldrich) (2.09 g, 22 mmol), 1-*O*-acetyl-2,3,5-tri-*O*-benzoyl-*β*-D-ribofuranose (**1**) (10.08 g, 20 mmol), and *N,O*-bis(trimethylsilyl) acetamide (5.5 ml, 22 mmol) are dissolved in dry CH_3CN (100 ml) under argon at 70° (oil bath) and the mixture stirred for 10 min. Trimethylsilyl-trifluoromethane sulfonate (TMSTfl) (5.5 ml, 28.5 mmol) is added and the mixture stirred for 2 hr at which time the starting sugar has been consumed as judged by TLC (75% ethyl acetate/hexane). The mixture is then cooled to room temperature, diluted with $CHCl_3$ (200 ml), and extracted with saturated $NaHCO_3$ (50 ml). Organic layer is washed with brine (50 ml), dried (Na_2SO_4), evaporated to dryness, and residue purified by flash chromatography, eluting with the 25–75% gradient of ethyl acetate in hexane. Fractions containing product are combined and evaporated to dryness to yield 10.6 g (98%) of **18** as colorless oil.

1-(*β*-d-Ribofuranosyl)pyridin-2-one (**19**). Compound **18** (10.2 g, 18.90 mmol) is dissolved in 145 ml of 0.3 *M* sodium methoxide in methanol and the solution stirred at room temperature for 1 hr. Ion-exchange resin AG 50X8 in Pyr^+ form [obtained by passing the 50% aqueous pyridine through the column of AG 50X8 (H^+ form)] is added to neutralize the reaction mixture to pH 7. The resin is filtered off, then washed well with water and warm methanol. The filtrate is evaporated to dryness and the residue dissolved in water (200 ml) and extracted with ether (2 × 100 ml). The aqueous layer is evaporated to dryness and **19** is crystallized from ethyl acetate (3.9 g, 91%).

1-(5-*O*-Dimethoxytrityl-*β*-d-ribofuranosyl)pyridin-2-one (**20**). 1-(*β*-D-Ribofuranosyl)pyridin-2-one **19** (3.80 g, 16.72 mmol) is dried by co-evaporation with anhydrous pyridine (2 × 50 ml), then dissolved in dry pyridine (120 ml). 4-(Dimethylamino)pyridine (117 mg) and triethylamine (3.5 ml) are added, followed by 4,4′-dimethoxytrityl chloride (6.84 g, 20.2 mmol). The reaction mixture is stirred under argon at room temperature for 2 hr when TLC (CH_2Cl_2 90, CH_3OH 10) showed the complete conversion to **20**. Methanol (20 ml) is added and solvents are removed by evaporation. The residue is partitioned between saturated $NaHCO_3$ (250 ml) and $CHCl_3$ (250 ml). The organic layer is extracted with brine (100 ml), dried (Na_2SO_4), filtered, and evaporated to dryness. The residue is purified by flash chromatography using 0.5–10% gradient of methanol in CH_2Cl_2 to afford 6.7 g (76%) of **20** as a yellowish foam.

1-(5-*O*-Dimethoxytrityl-2-*O*-*tert*-butyldimethylsilyl-β-D-ribofuranosyl)pyridin-2-one (**21**). To the solution of **20** (5.4 g, 10.20 mmol) in dry THF (80 ml) dry pyridine (3.07 ml, 38 mmol) and $AgNO_3$ (2.03 g, 12 mmol) are added under argon. The reaction mixture is stirred at room temperature for 15 min, then TBDMSiCl (2.3 g, 15.3 mmol) is added and stirring continued overnight. TLC (ethyl acetate/hexane 1 : 1) showed disappearance of the starting material and formation of the mixture of the slower running 3′*O*-TBDMS derivative **21a** and faster running 2′-*O*-TBDMS derivative **21**. Some 2′,3′-bis-*O*-TBDMS compound **21b** is visible, too. (Another TLC system in which separation of products is good after several developments: 5% methanol in ethyl acetate. In this system **21**: R_f 0.35; **21a**: R_f 0.20; **21b**: R_f 0.50.) The mixture is filtered through the Celite pad into the saturated $NaHCO_3$ (200 ml) and the filtrate extracted with CH_2Cl_2 (2 × 200 ml). Combined organic extracts are dried (Na_2SO_4), filtered, evaporated to dryness, and purified by flash chromatography eluting with 20–50% gradient of ethyl acetate in hexane. We obtained 4.5 g (68.5%) of **21**, 1.1 g (16.7%) of 3′-*O*-TBDMS derivative **21a** and 0.45 g (5.8%) of 2′,3′-bis-*O*-TBDMS substituted derivative **21b**.

1-(5-*O*-Dimethoxytrityl-2-*O*-*tert*-butyldimethylsilyl-β-D-ribofuranosyl)pyridin-2-One 3′-*O*-(2-Cyanoethyl-*N*,*N*-diisopropylphosphoramidite) (**22**). Compound **21** (1.5 g, 2.33 mmol) is phosphitylated using general procedure C. Phosphoramidite **22** was obtained after purification by flash column chromatography. Elution is carried out with 15–50% gradient of ethyl acetate in hexane (1% triethylamine is added to the eluant) to afford **22** (a mixture of diastereoisomers) as a colorless foam (1.75 g, 89%).

1-(2,3,5-Tri-*O*-benzoyl-β-D-ribofuranosyl)pyridin-4-one (**23**). The same procedure is used as that for the preparation of protected nucleoside **18**, only 4-hydroxypyridine (Aldrich) is used instead of 2-hydroxypyridine and the reaction time is extended to 4 hr. TLC (10% methanol/CH_2Cl_2) showed the disappearance of starting sugar and formation of the nucleoside product. After standard aqueous workup the reaction mixture is purified by flash chromatography eluting with 2–10% gradient of methanol in CH_2Cl_2 to afford 10 g (93%) of **23**.

1-(β-D-Ribofuranosyl)pyridin-4-one (**24**). Base-catalyzed hydrolysis of **23** is conducted as described for the preparation of **19** from **18**. The final, chromatographically pure product (84% yield) resisted crystallization and was used in the next step without further purification.

1-(5-*O*-Dimethoxytrityl-β-D-ribofuranosyl)pyridin-4-one (**25**). Using the same dimethoxytritylation procedure as for the preparation of **20** from **19**, **25** is obtained from **24** in 67% yield.

1-(5-*O*-Dimethoxytrityl-2-*O*-*tert*-butyldimethylsilyl-β-d-ribofuranosyl)pyridin-4-one (**26**). Silylation of **24** (0.6 g) as described for preparation of **21** showed by TLC (20% methanol/ethyl acetate), in addition to the minor amount of starting material, the formation of three products (faster running 2′ isomer, slower running 3′ isomer and 2′,3′-bis substituted derivative). After the standard workup, the reaction mixture is purified by flash chromatography using ethyl acetate to elute the 2′,3′-bis silylated derivative, followed by 0.1–5% gradient of methanol in ethyl acetate to elute 2′-*O*-TBDMSi derivative **26** (0.30 g, 41% yield) and finally 3′-*O*-TBDMSi derivative **26a** (0.26 g, 36% yield).

To increase the amount of the desired isomer **26**, 3′-isomer **26a** (0.26 g) is dissolved in 5% triethylamine in methanol (50 ml/1 g of **26a**) and the solution is kept at room temperature overnight. The base-catalyzed migration of the silyl group yields a 1:1 mixture of **26** and **26a** as judged by TLC. Solvents are removed by evaporation and the residue subjected to flash chromatography as described earlier to yield an additional 90 mg of **26**, increasing its yield to 53%.

1-(5-*O*-Dimethoxytrityl-2-*O*-*tert*-butyldimethylsilyl-β-d-ribofuranosyl)pyridin-4-one 3′-*O*-(2-Cyanoethyl-*N,N*-diisopropylphosphoramidite) (**27**). Using the same procedure as for the preparation of **22**, except that neat ethyl acetate containing 1% triethylamine is used for chromatographic purification, **27** was obtained as a colorless foam in 94% yield.

*Procedure 4: Preparation of 5′-O-Dimethoxytrityl-2′-O-tert-butyldimethylsilyl-6-Methyluridine-3′-O-(2-cyanoethyl-N,N-diisopropylphosphoroamidite) (**31**) and 5′-O-Dimethoxytrityl-2′-O-tert-butyldimethylsilyl-N^4-acetyl-6-methylcytidine-3′-O-(2-cyanoethyl-N,N-diisopropylphosphoramidite) (**36**)*

2′,3′,5′-Tri-*O*-benzoyl-6-methyluridine (**28**). The suspension of 6-methyluracil (Aldrich) (2.77g, 21.96 mmol) in the mixture of hexamethyldisilazane (50 ml) and dry pyridine (50 ml) is refluxed for 3 hr. The resulting clear solution of bistrimethylsilyl derivative of 6-methyluracil was evaporated to dryness and coevaporated with dry toluene (2 × 50 ml) to remove traces of pyridine. To the solution of the resulting clear oil in dry acetonitrile 1-*O*-acetyl-2,3,5-tri-*O*-benzoyl-β-d-ribose (**1**) (10.1 g, 20 mmol) is added and the reaction mixture cooled to 0°. To the above stirred solution, trimethylsilyl trifluoromethanesulfonate (4.35 ml, 24 mmol) is added dropwise and the reaction mixture stirred for 1.5 hr at 0° and then 1 hr at room temperature. After that the reaction mixture is diluted with CH_2Cl_2 (100 ml) washed with saturated $NaHCO_3$ (50 ml) and brine (50 ml). The organic layer is dried (Na_2SO_4), filtered, and evaporated and the residue is purified by flash chromatography on silica gel with ethyl acetate–hexane (2:1) mixture as an eluent to give 9.5 g (83%) of the compound **28** and 0.8 g of the corresponding N^1,N^3-bis derivative.

5′-*O*-Dimethoxytrityl-6-methyluridine (**29**). To a solution of 2′,3′,5′-tri-*O*-benzoyl-6-methyluridine (**28**) (19.0 g, 33.3 mmol) stirred at −10° in pyridine (250 ml) and methanol (50 ml) is added a precooled solution of 2 *N* aqueous NaOH (75 ml). The reaction mixture is stirred at −10° for 20 min then neutralized by addition of Dowex 50 Py^+ resin. The reaction mixture is then filtered to remove the resin and the resin bed washed with 25% aqueous pyridine (1 liter) and the filtrate evaporated *in vacuo.* Crude 6-methyluridine (**2**) is coevaporated with anhydrous pyridine (5 × 50 ml) and then treated with DMTCl (12.4 g, 36.63 mmol) in anhydrous pyridine (700 ml) at room temperature while stirring under argon for 16 hr. TLC (10% ethanol/dichloromethane) indicated some remaining **28**; additional DMTCl (2.25 g, 0.2 equivalent) is added and the reaction quenched with ethanol (5 ml) after 2 hr at room temperature. Pyridine is removed *in vacuo* and the resulting crude residue partitioned between CH_2Cl_2 (200 ml) and saturated $NaHCO_3$ (100 ml). The organic layer is dried over Na_2SO_4, filtered, and flash chromatographed using a gradient of 0–10% ethanol in CH_2Cl_2 to give **29** (15.1 g, 80%).

5′-*O*-Dimethoxytrityl-2′-*O*-*tert*-butyldimethylsilyl-6-methyluridine (**30**). To a solution of **29** (15.1 g, 26.9 mmol) stirred at room temperature under argon in anhydrous THF (200 ml) is added $AgNO_3$ (11.0 g, 64.5 mmol) and pyridine (8.7 ml, 107.6 mmol). After 10 min, TBDMSiCl (5.0 g, 33.62 mmol) is added and the reaction mixture stirred for 18 hr at room temperature and shielded from light. TLC (70% ethyl acetate/hexane) indicated remaining **29**; additional $AgNO_3$ (2.75 g, 16.1 mmol), pyridine (4.35 ml, 53.8 mmol), and TBDMSiCl (2.5 g, 20.2 mmol) are added. The reaction is quenched with ethanol (10 ml) after stirring for an additional 6 hr at room temperature. The reaction mixture is filtered over Celite, diluted with CH_2Cl_2 (200 ml), and washed with saturated $NaHCO_3$ (100 ml). The aqueous layer is then back extracted with CH_2Cl_2 and the combined organic extracts dried over Na_2SO_4 filtered and evaporated *in vacuo.* Residue is purified by flash silica gel chromatography utilizing a gradient of 5–70% ethyl acetate/hexane to yield faster moving 2′-*O*-Siisomer **30** (5.8 g, 32%) and an equivalent amount of 3′-*O*-Si isomer **30a**, which can be isomerized as described for **16**.

5′-*O*-Dimethoxytrityl-2′-*O*-*tert*-Butyldimethylsilyl-6-methyluridine 3′-*O*-(2-cyanoethyl-*N,N*-diisopropylphosphoroamidite) (**31**). Compound **30** (5.6 g, 8.3 mmol) (1.5 g, 2.33 mmol) is phosphitylated using general procedure C. Phosphoramidite **31** is obtained after purification by flash column chromatography. Elution is carried out with 5–40% ethyl acetate/hexane (1% triethylamine is added to the eluant) to afford pure **31** (6.13 g, 84%).

6-Methylcytidine (**32**). Triethylamine (13.4 ml, 100 mmol) is added dropwise to a stirred ice-cooled mixture of 1,2,4-triazole (6.22 g, 90 mmol) and phosphorus oxychloride (1.89 ml, 20 mmol) in 50 ml of anhydrous acetonitrile. To the resulting suspension the solution of 2′,3′,5′-tri-*O*-benzoyl-6-methyluridine **28** (5.7g, 10 mmol) in 30 ml of acetonitrile is added dropwise and the reaction mixture stirred for 4 hr at room temperature. Then it is concentrated *in vacuo* to minimal volume (not to dryness). The residue is dissolved in chloroform (300 ml) and washed with water (100 ml), saturated $NaHCO_3$ (100 ml) and brine (100 ml). The organic layer is dried over Na_2SO_4, filtered, and the solvent removed *in vacuo*. The residue is dissolved in 100 ml of 1,4-dioxane and treated with 50 ml of 29% aqueous NH_4OH overnight. The solvents are removed *in vacuo*. The residue is dissolved in the mixture of pyridine (60 ml) and methanol (10 ml), cooled to −15°, and ice-cooled 2 *M* aqueous solution of NaOH is added under stirring. The reaction mixture is stirred at −10 to −15° for an additional 30 min and then neutralized to pH 7 with Dowex 50 (Py^+). The resin is filtered and washed with 200 ml of the H_2O–pyridine (4:1) mixture. The combined mother liquor and washings are evaporated to dryness. The residue is crystallized from aqueous methanol to give 1.6 g (62%) of 6-methylcytidine (**32**).

5′-*O*-Dimethoxytrityl-N^4-Acetyl-6-methylcytidine (**34**). To the solution of 6-methylcytidine **32** (1.4 g, 5.44 mmol) in dry pyridine 3.11 ml of trimethylchlorosilane is added and the reaction mixture stirred for 2 hr at room temperature. Then acetic anhydride (0.51 ml, 5.44 mmol) is added and the reaction mixture stirred for an additional 3 hr at room temperature. TLC showed disappearance of the starting material and the reaction is quenched with methanol (20 ml), ice-cooled, and treated with water (20 ml, 1 hr). The solvents are removed *in vacuo* and the residue dried by four coevaporations with dry pyridine. Finally, it is redissolved in dry pyridine (75 ml) and 4,4′-dimethoxytrityl chloride (2.2 g, 6.52 mmol) is added. The reaction mixture is stirred overnight at room temperature and quenched with methanol (20 ml). The solvents are removed *in vacuo*. The remaining oil is dissolved in CH_2Cl_2 (150 ml), washed with saturated $NaHCO_3$ (50 ml) and brine (50 ml). The organic layer is separated and evaporated to dryness. The residue is purified by flash chromatography on silica gel with the (3–5%) gradient of methanol in CH_2Cl_2 to give 2.4 g (74%) of the compound **34**.

5′-*O*-Dimethoxytrityl-2′-*O*-*tert*-butyldimethylsilyl-6-methylcytidine (**35**). Using general procedure B, **34** (1.8 g, 3.0 mmol) is silylated generating a mixture of 2′,3′-bis-Si derivative (0.18 g, 7%), 2′-*O*-Si isomer

35 (0,9 g, 41%), and 3′-*O*-Si isomer **35a** (0.9 g, 41%) separated by flash chromatography in ethyl acetate–hexane, 3:2. The 3′-*O*-Si isomer **35a** can be isomerized as described for **16**.

5′-*O*-DIMETHOXYTRITYL-2′-*O*-*tert*-BUTYLDIMETHYLSILYL-6-METHYLURIDINE 3′-(2-CYANOETHYL-*N,N*-DIISOPROPYLPHOSPHORAMIDITE) (**36**). Using general procedure C, **35** (0.5 g, 0.7 mmol) is phosphitylated providing phosphoramidite **36**. It is purified by flash chromatography using 1:1 mixture of ethyl acetate and hexane to afford pure **36** (0.5 g, 78%).

[4] Use of T7 RNA Polymerase and Its Mutants for Incorporation of Nucleoside Analogs into RNA

By RUI SOUSA

Introduction

The single-subunit RNA polymerases encoded by the T7 and Sp6 bacteriophage are widely used for preparing RNAs by *in vitro* transcription for a variety of applications.[1] A mutant of T7 RNA polymerase (RNAP) was recently identified that readily incorporates nucleotide monophosphates (NMPs) bearing noncanonical groups at the ribose 2′-position into RNA.[2] This mutant makes it facile to prepare 2′-modified RNAs for structural and structure–function studies,[3–6] as well as for other applications.[7] Modifications in the standard reaction conditions used for transcription with wild-type T7 RNAP further increase the utility of the mutant enzyme for preparing nucleic acids in which most or all of the 2′-hydroxyls are replaced by noncanonical substitutents.

Characteristics and Handling of Mutant Enzyme

The mutation changes active site tyrosine-639 to phenylalanine.[2] Tyrosine-639 is near the C terminus of an α-helix which corresponds to helix

[1] J. F. Milligan, D. R. Groebe, G. W. Witherell, and O. C. Uhlenbeck, *Nucleic Acids Res.* **17,** 8783 (1987).
[2] R. Sousa and R. Padilla, *EMBO J.* **14,** 4609 (1995).
[3] S. A. Strobel, L. Ortoleva-Donnelly, S. P. Ryder, J. H. Cate, and E. Moncoeur, *Nature Struct. Biol.* **5,** 60 (1998).
[4] L. Ortoleva-Donnelly, A. A. Szewczak, R. R. Guttell, and S. A. Strobel, *RNA* **4,** 498 (1998).
[5] S. A. Strobel and J. A. Doudna, *Trends Biochem. Sci.* **22,** 262 (1998).
[6] K. Raines and P. A. Gottlieb, *RNA* **4,** 340 (1998).
[7] M. Sioud and D. R. Sorensen, *Nature Biotech.* **16,** 556 (1998).

0076-6879/00 $30.00

O in the homologous DNAP I enzyme.[8,9] Both structural and structure–function studies have identified the C-terminal region of this conserved structural element as important in binding the ribose moiety of the NTP and in ribose 2′ and 3′ group discrimination in this superfamily of homologous DNAPs and RNAPs.[10–15] The phenotype of the Y639F mutation was first defined in a screen of a large number of T7 RNAP active site mutants,[16] but evidence for its unusual properties was first obtained by Makarova *et al.*,[17] who noted that transcripts generated *in vivo* by this enzyme yielded up to ~70% less protein per transcript than was obtained from transcripts made by the wild-type enzyme. The initial conclusion that this reflected a lack of fidelity in the mutant proved incorrect and, in fact, the fidelity, promoter specificity, and overall activity of the mutant appear indistinguishable from wild-type. Instead, characterization revealed that the mutation resulted in the specific loss of discrimination of the chemical character of the ribose 2′-substituent.[14] Presumably, the *in vivo* phenotype reflects incorporation of dNMPs into transcripts and a resultant deficit in their translatability.

Expression of Enzyme. The mutant enzyme has been sent to a large number of requesting laboratories in a vector (pDPT7) cloned into the protease-deficient BL21 strain (important for preparing T7 RNAP in intact form). This vector was constructed by subcloning the *Bam*HI fragment from pAR 1219, which was originally constructed in F. W. Studier's laboratory[18] into the pUC119 phagemid. In our experience T7 RNAP is expressed to higher levels by pDPT7 than by pAR 1219. This may reflect the higher copy number of the pUC119 derivative, the fact that the construction resulted in T7 RNAP being expressed from two tandem *lac* promoters, or both. In any case, while the increased expression levels facilitate preparation of the mutant enzyme, they may contribute to the instability of this plasmid (unpublished observations and Ref. 17). Because of this instability, we

[8] R. Sousa, Y. J. Chung, J. P. Rose, and B. C. Wang, *Nature* **364,** 593 (1993).

[9] R. Sousa, *Trends Biochem. Sci.* **21,** 186 (1996).

[10] S. Tabor and C. Richardson, *Proc. Natl. Acad. Sci. U.S.A.* **92,** 6339.

[11] M. Astatke, N. D. F. Grindley, and C. M. Joyce, *J. Mol. Biol.* **278,** 147 (1998).

[12] S. Doublie, S. Tabor, A. M. Long, C. C. Richardson, and T. Ellenberger, *Nature* **391,** 251 (1998).

[13] R. Guajardo and R. Sousa, *J. Mol. Biol.* **265,** 8 (1997).

[14] Y. Huang, F. Eckstein, R. Padilla, and R. Sousa, *Biochemistry* **36,** 8231 (1997).

[15] M. Astatke, N. D. F. Grindley, and C. M. Joyce, *J. Biol. Chem.* **270,** 1945 (1995).

[16] G. Bonner, D. Patra, E. M. Lafer, and R. Sousa, *EMBO J.* **11,** 3767 (1992).

[17] O. V. Makarova, E. M. Makarov, R. Sousa, and M. Dreyfus, *Proc. Natl. Acad. Sci. U.S.A.* **92,** 12250 (1995).

[18] P. Davanloo, A. H. Rosenberg, J. J. Dunn, and F. W. Studier, *Proc. Natl. Acad. Sci. U.S.A.* **81,** 2035 (1984).

always carry out the following procedure for expression of the polymerase: 12–18 hr before inoculating a culture, cells are streaked out on LB agar with 100 μg/ml ampicillin to obtain a plate with ~100–10,000 colonies. The plates should not be stored after growth. Instead, immediately before inoculation, the cells from a freshly grown plate are resuspended in 5–10 ml of media and this suspension is used to inoculate 1–4 liters of LB containing 100 μg/ml ampicillin. In 2–6 hr (depending on inoculum and culture size) the culture will have reached an 0 D_{600} of 0.3–0.6, and expression is induced by addition of 125 μg/ml isopropylthiogalactoside (IPTG).

This procedure ensures that the majority of the cells in the inoculum will be healthy and will have the plasmid. When liquid cultures or old plates are used for inoculation it often happens that a fraction of the cells will have lost the plasmid. These cells may remain in stasis, and when the cells containing the plasmid have secreted sufficient β-lactamase into the media to destroy the ampicillin, the cells lacking the plasmid will overgrow the culture. The described procedure routinely ensures high yields of the polymerase. If the procedure described is inconvenient the *Bam*HI fragment can be subcloned into the *Bam*HI site of pBR322 to regenerate a derivative of pAR 1219 carrying the mutant T7 RNAP gene, though yields of enzyme will be lower. (Such a construct carrying mutant Y639F is also available from the laboratory of Dr. Marc Dreyfus at the Ecole Normale Superieure, CNRS, Paris; and a construct expressing a His-tag version of the mutant allowing purification by Ni-agarose chromatography is available from Dr. William T. McAllister, SUNY Health Science Center, Brooklyn, NY.)

Purification of Mutant Enzyme. Four to six hours after ITPG addition the cells are harvested by centrifugation. The cell pellet is resuspended in chilled 25 m*M* ethylenediaminetetraacetic acid (EDTA), 20 m*M* Tris-Cl, pH 8.0, 10% (w/v) sucrose, 10 mg/ml lysozyme (use 10 ml of this buffer for every 1000 ml of culture) and incubated at 4° for 30 min. Freeze the suspension and thaw to crack cells. Test whether the suspension has become viscous, indicating lysis. If not, go through another freeze–thaw cycle. Treat the material gently at this point so as not to shear the genomic DNA. Spin at 4° (15 min at top speed in minifuge, or 15,000–20,000 for 30 min in a Sorvall SS-34 rotor for larger scales). Load supernatant onto a Whatman (Clifton, NJ) P-11 phosphocellulose column (~2–3 ml of packed resin for every 10 ml of supernatant) equilibrated with 25 m*M* EDTA, 10% (w/v) sucrose, 20 m*M* Tris-Cl, pH 8.0; wash with several column volumes of same buffer + 0.15 *M* NaCl; elute with either a step of buffer + 0.4 *M* NaCl or a gradient of buffer + 0.15 *M* NaCl to buffer + 0.5 *M* NaCl. T7 RNAP will elute between 0.3 and 0.4 *M* NaCl. Up to 30 mg protein per 1000 ml of culture may be obtained. The high expression level means that the enzyme is suitable for most purposes after this single chromatographic step. If greater purity is desired the protein can be further purified by anion-exchange chromatography on any strong anion-exchange resin (i.e., DEAE, Pharmacia Mono Q), and/or by sizing chromatography on gel exclusion media of appropriate molecular weight cutoff (Pharmacia, Piscataway, NJ Sephacryl S-200) as described.[19]

STORAGE. Dialyze eluate from phosphocellulose column into 0.5 *M* NaCl, 20 m*M* sodium phosphate, pH 8.0, 5 m*M* dithiothreitol (DTT), 1 m*M* EDTA, 50% glycerol for storage at −20°. In our experience the mutant polymerase is not as stable over long-term storage as the wild-type enzyme. This is not a serious problem as in neither case is the enzyme extremely labile, but it should be considered if a preparation that has been stored for many months begins to lose activity. If a preparation has lost activity after a few months try adding a fresh aliquot of DTT as it may be the case that the problem is oxidation of the reducing agent.

Use of Mutant Enzyme in Modified Transcription Buffers

In most cases the mutant can be used just as one would use the wild-type enzyme in standard transcription buffers. As long as rGTP is not replaced, activity in standard transcription buffer with a supercoiled template containing a consensus promoter is reduced only about twofold when a single rNTP is replaced by a dNTP, a 2′-NH_2-NTP, or a 2′-F-NTP.[2,14] However, more drastic activity reductions are obtained when two, three, or four of the rNTPs in the transcription reaction are replaced by NTPs with noncanonical 2′ groups.[2,14,20] Characterization has shown that this is probably due, at least in part, to a conformational effect on the transcript : template hybrid. Efficient transcript extension appears to require that the transcript : template hybrid in the active site assume an A-type conformation.[20] Since the mutation eliminates discrimination of the chemical character of the 2′ substituent, but does not eliminate the requirement for a particular conformation of the transcript : template hybrid, transcripts that are heavily modified with noncanonical 2′ groups become poor substrates for extension if the modifications favor assumption of a distinct (B-type) conformation in the transcript : template hybrid.

It may also be that conformational effects may interfere with the incorporation step. While the mutation may eliminate direct discrimination of the chemical character of the 2′ substituent, it does not eliminate discrimination of the effects of this group on the ribose conformation of the NTP. Ribose conformation is a function of the electronegativity of the 2′ substitu-

[19] R. Sousa, J. P. Rose, Y. J. Chung, E. M. Lafer, and B. C. Wang, *Proteins Struct. Func. Genet.* **5,** 266 (1989).

[20] Y. Huang, A. Beaudry, J. McSwiggen, and R. Sousa, *Biochemistry* **36,** 13718 (1997).

ent, with more electronegative substituents preferring the $C_{3'}$-endo conformer seen in A-form nucleic acid structures, while less electronegative substituents prefer the $C_{2'}$-endo conformer seen in B-type helices.[21,22] In experiments with a set of NTPs differing in their preferred ribose pucker and the H-bonding character of their 2′ substituents, the wild-type enzyme strongly preferred NTPs whose substituents could act as H-bond donors or acceptors, while the mutant enzyme displayed a weak, residual preference that followed the $C_{3'}$-endo content of the NTP ribose.[14]

In an effort to overcome these conformational obstacles to synthesis of nucleic acids heavily modified with noncanonical 2′ substituents, we have screened reagents known to stabilize A-form helical structures. These compounds include methanol and ethanol, trifluoroacetic acid, cobalt hexamine, spermine, and spermidine. Mn^{2+}, which may favor utilization of substrates or transcripts of noncanonical structure by a distinct mechanism,[20] has also been tested. We have also examined the use of reagents (nonionic detergents, use of acetate as the predominant counteranion, pyrophosphatase) which are not expected to specifically enhance use of noncanonical substrates, but which have been reported to generally enhance T7 RNAP activity.[23] This survey led to definition of an optimized buffer that we recommend for use with the Y639F T7 RNAP: 40 m*M* Tris–acetate pH 8.0, 10 m*M* magnesium acetate, 0.5 m*M* $MnCl_2$, 5 mM DTT, 0.1% Tween 20, 1 U/μl pyrophosphatase, 8 m*M* spermidine. Using this buffer and the Y639F T7 RNAP we have obtained high yields of transcripts in reactions in which three of four rNTPs are replaced with dNTPs, 2′-F-NTPs, or 2′-NH_2-NTPs, and have even obtained significant synthesis in reactions in which all 4 rNTPs were replaced with dNTPs.[24] This reaction buffer is useful with a range of different NTP substrates and templates, and can be conveniently used with the mutant enzyme just as one would use the wild-type enzyme in standard transcription buffer. However, in many cases it may be useful to further modify these conditions for use with particular sets of substrates. For those cases we offer the following considerations on how these compounds may affect the transcription reaction.

Spermidine and Spermine. Addition of these polyamines had by far the most stimulatory effect on the activity of the mutant enzyme in reactions in which three rNTPs were replaced by dNTPs. The concentration optima for the polyamines appeared to be a function of two opposed effects. Spermine or spermidine concentrations in excess of ~1 or ~10 m*M*, respec-

[21] W. Saenger, "Principles of Nucleic Acid Structure." Springer-Verlag, New York, (1984).
[22] W. Guschlbauer and K. Jankowski, *Nucleic Acids Res.* **8,** 1421 (1980).
[23] M. Maslak and C. T. Martin, *Biochemistry* **33,** 6918 (1994).
[24] R. Padilla and R. Sousa, *Nucleic Acids Res.* **27,** 1561 (1999).

tively, were generally inhibitory of activity as assessed in a reaction with four rNTPs and plasmid templates. However, these reagents decreased the degree to which activity was reduced in reactions with noncanonical substrates. The net effect led to up to about fivefold stimulation of activity in reactions with three rNTPs and plasmid templates, with concentration optima around 8 m*M* for spermidine and 1–2 m*M* for spermine. The stimulatory effects of the polyamines may be largely due to stabilization of an A-conformation in transcript : template hybrids carrying transcripts heavily substituted with noncanonical NMPs whose riboses favor a $C_{2'}$-endo conformation. It is also possible that the polyamines may enhance formation of the catalytically correct geometry in the template : transcript • NTP complex, especially if the NTP must stack on the transcript and base pair with the template so as to extend the helix conformation of the transcript : template hybrid. The stimulatory effect of the polyamines may therefore differ, depending on the transcript and NTP structure. Reactions with NTPs and transcripts incorporating NMPs that strongly favor the $C_{2'}$-endo conformer may be stimulated to a greater degree by high polyamine concentrations. For example, we have found that high polyamine concentrations are more stimulatory of reactions with 2′-NH_2-NTPs than in reactions with 2′-F-NTPs, while 2′-H-NTPs are intermediate in this respect. We interpret this to be due to the fact that 2′-NH_2-NTPs strongly favor $C_{2'}$-endo pucker, while 2′-F-NTPs favor $C_{3'}$-endo, and 2′-H-NTPs favor $C_{2'}$-endo, but to a lesser extent than the 2′-NH_2 substrates.

The polyamines are also more stimulatory in reactions in which three of four rNTPs are replaced with noncanonical substrates than in reactions in which only one or two are replaced. The recommended 8 m*M* spermidine concentration is convenient because it is stimulatory in reactions with three dNTPs but is not inhibitory in reactions with four rNTPs. However, it may be useful to test higher spermidine concentrations if substrates are encountered that result in poor synthesis, particularly if a very high degree of substitution is sought or if the preferred ribose pucker of the substrate strongly favors $C_{2'}$-endo (i.e., if the 2′ substituent is less electronegative than an OH group).

We have also found that the inhibitory effects of the polyamines are greatly reduced when short, synthetic templates are used than when plasmid templates are used. For example, with a 42-bp synthetic template we found that the reduced sensitivity to polyamine inhibition allowed use of spermine concentrations up to ~13 m*M* (optimal at ~3 m*M*) in reactions with three dNTPs. With this template the stimulatory effects of spermine were greater than those of spermidine. In contrast, with plasmid templates, spermine concentrations in excess of ~2 m*M* were strongly inhibitory. This may mean that the inhibitory effect of the polyamines is partly due to their tendency to cause nucleic acid aggregation, a tendency that will increase as the length and concentration of the template increases. It may therefore be useful to screen spermine concentrations in the 1–10 m*M* range (or spermidine concentrations in excess of 10 m*M*) when short, synthetic DNAs or purified restriction fragments as used as templates. The helix stabilizing effects of the polyamines may also contribute to stimulating activity by stabilizing the interaction between the template and short transcript during the initial transcription reaction, even when specific conformational effects are not relevant. Because of the effects of polyamines on aggregation, care should be taken not to mix templates directly with concentrated stock buffers containing polyamines and to warm such solutions to room temperature or 37° before mixing with nucleic acids. The other A-conformation stabilizing reagents tested had effects qualitatively similar, but quantitatively inferior, to the polyamines and are not as useful. Combinations of these compounds with or without polyamines were also tested, but no combination was superior to use of polyamines alone.

Nonionic Detergents, Use of Acetate as Counteranion, Addition of Pyrophosphatase. These reagents are not specifically directed against improving incorporation of noncanonical substrates, but have been reported to generally enhance T7 RNAP activity and are recommended for use with the mutant enzyme as specified earlier. Unless high concentrations of RNA are being prepared, the effect of pyrophosphatase is negligible, and because it is an expensive reagent, it is dispensable. However, when high concentrations of RNA (whether of canonical structure or not) are being made, the accumulation of pyrophosphate can prove inhibitory and the use of pyrophosphatase is recommended.[25]

Use of Mn^{2+} as Catalytic Cofactor. Enzymes that normally catalyze phosphoryl transfer reactions with Mg^{2+} will usually also work with Mn^{2+}, albeit with reduced efficiency. A large number of studies reveal that the Mn^{2+}-catalyzed reaction is more tolerant of noncanonical substrates or disrupted active site structures, and may be thought of as compatible with a wider range of catalytic group geometries than the Mg^{2+}-catalyzed reaction. This reduced stringency for catalytic group geometry may be useful, and partial or complete replacement of Mg^{2+} with Mn^{2+} should be tested if substrates are encountered that are poorly or nonuniformly incorporated in the reaction buffer specified earlier. The use of Mn^{2+} to improve the uniformity and rate of ddNMP incorporation (relative to dNMP incorporation) by

[25] N. Sasaki, M. Izawa, M. Watahiki, K. Ozawa, T. Tanaka, Y. Yoneda, S. Matsuura, P. Carninci, M. Muramatsu, Y. Okazaki, and Y. Hayashizaki, *Proc. Natl. Acad. Sci. U.S.A.* **95,** 3455 (1998).

T7 DNA polymerase is an example of this.[26] It should be appreciated that Mn^{2+} competes for the active site more effectively than Mg^{2+},[27] so that in mixtures of both ions Mn^{2+} will contribute disproportionately to catalysis.

Effects of Template Structure

Variation from Consensus in Initially Transcribed Sequence. T7 RNAP is sensitive to variations from consensus in the initially transcribed sequence (ITS) (+1 to +6 consensus: GGGAGA), with variations becoming increasingly detrimental the closer they are to +1. The effects of nonconsensus ITSs are exacerbated if combined with use of noncanonical substrates.[2] It has been found that such nonconsensus ITSs increase the NTP concentrations required to achieve maximal rates of transcription initiation.[28,29] These effects can be specific for a particular NTP. If it is necessary to use nonconsensus ITSs due to constraints on the transcript sequence and the resultant transcript yields are poor, then increases in the concentration of specific NTPs (up to 2 m*M* or more) should be tested. We typically use rNTPs at 0.5 m*M* and noncanonical substrates at 1–2 m*M*, but higher concentrations can be used, especially of the pyrimidines (purines show a stronger tendency to be inhibitory at high concentrations). When using such high NTP concentrations care should be taken to ensure that the Mg^{2+} concentration remains in excess of the total NTP concentration.

Use of Partially Single-Stranded, Nicked, or Gapped Templates. We have found that the use of template structures that facilitate promoter melting and/or stabilize the initial transcription complex can increase transcript synthesis rates in reactions in which noncanonical substrates are incorporated during the earliest stages of the initiation reaction. For example, on a promoter that initiates GGGAGACCGGAAU, the use of such templates enhances productive initiation rates when rGTP is replaced with dGTP (and especially when both rGTP and rATP are replaced with dGTP and dATP), but not in reactions with four rNTPs nor when rA, rC, and rU (but not rG) are replaced with the corresponding rNTPs.[24] This probably reflects enhancement of the poorly processive initial transcription reaction by lowering or eliminating the energetic barriers associated with promoter melting.[30] On the other hand, we have found that transcript elongation on partially single-stranded templates is less processive than on double–stranded templates, and that the latter stages of the initial transcription

[26] S. Tabor and C. Richardson, *Proc. Natl. Acad. Sci. U.S.A.* **86,** 4076 (1989).

[27] A. Y. M. Woody, S. S. Eaton, P. A. Osumi-Davis, and R. W. Woody, *Biochemistry* **35,** 144 (1996).

[28] R. Guajardo, P. Lopez, M. Dreyfus, and R. Sousa, *J. Mol. Biol.* **281,** 777 (1998).

[29] J. Villemain and R. Sousa, *J. Mol. Biol.* **281,** 793 (1998).

[30] J. Villemain, R. Guajardo, and R. Sousa, *J. Mol. Biol.* **273,** 956 (1997).

reaction (transcription from +6 to ∼+9 and promoter release) can be less efficient on such templates.[31] We interpret these observations in terms of the conflicting requirements of different steps in the transcription reaction: during the earliest stages of transcription initiation, reducing or lowering the barrier to promoter melting is helpful, however reannealing of the promoter contributes to releasing the polymerase from the promoter during the latter stages of the initiation reaction and template : nontemplate reannealing contributes to RNA displacement and formation of the correct elongation complex structure during transcript elongation.[31] In practical terms this means that template structures that facilitate promoter melting should be used when the use of noncanonical substrates may severely compromise the early steps (transcription from +1 to ∼+6) of the initiation reaction. If the transcript sequence is such that noncanonical substrates are not incorporated until the latter steps in initiation or only during elongation, or if long transcripts (>∼100 bases) are being made, then double-stranded templates should be used. By partially single-stranded templates we mean templates lacking a nontemplate strand downstream of −6 to −1; nicked templates are constructed by annealing a nontemplate strand composed of two noncovalently linked oligonucleotides (so that a nick is introduced in the nontemplate strand between −6 and −1); gapped templates are similarly constructed from two nontemplate oligonucleotides such that a gap of 1–5 bases in size is introduced somewhere in the −6 to −1 region.

Use of Noncanonical Substrates Modified at Other than Ribose 2′ Position or Use of Bulky 2′ Substituents

The Y639F mutation essentially eliminates discrimination of the chemical character of 2′ substituents similar to or smaller in volume than an OH or NH_2 group, but is less useful for incorporation of substrates modified at other positions or carrying bulky 2′ substituents. 2′-O-Me-NMPs, for example, are poorly incorporated and replacement of even one rNTP by a 2′-O-Me-NTP reduces activity by 10- to 100-fold in reactions with a plasmid template with a consensus promoter/ITS, the Y639F polymerase, and the optimized buffer specified earlier.[24] However, the reduced size of the side- chain in the Y639F mutant may provide some modest relaxation of steric clashes with bulkier substituents. In general, the mutation of larger active site side chains to smaller ones presents itself as an obvious strategy for improving incorporation of noncanonical substrates with bulkier than normal substituents. A difficulty is that such mutations may generally lower

[31] V. Gopal, L. G. Brieba, R. Guajardo, W. T. McAllister, and R. Sousa, *J. Mol. Biol.* **290,** 411 (1999).

catalytic activity to a degree that is not compensated for by a decreased selectivity against any particular substrate. This may be true for Y639M and Y639L, for example, which may provide more room for a bulky 2′ substituent, but which we have found to be less active than Y639F with 2′-O-Me-NTPs, probably because retention of the aromatic ring at this position is important for overall catalytic activity.[14] However, only direct testing will reveal whether particular mutant forms of the enzyme will be useful with specific substrates. Because the T7 RNAP active site has been extensively mutagenized,[14,16,24] such screening is feasible without necessarily requiring construction of new mutants. In carrying out such experiments the considerations outlined here on the use of polyamines to stabilize transcript : template : substrate interaction and geometry, on the use of Mn^{2+} to relax the stringency of the requirement for precise catalytic group alignment, and on the use of high substrate concentrations and templates with structures designed to facilitate initiation should be generally applicable.

Acknowledgment

Work in R.S.'s laboratory is supported by NIH grant GM-52522 and funds from the state of Texas ARP/ATP program.

[5] Phosphorothioate Modification of RNA for Stereochemical and Interference Analyses

By L. Claus S. Vörtler and Fritz Eckstein

Introduction

Nucleoside phosphorothioates are derived from the natural nucleoside phosphates by replacement of one of the two nonbridging phosphate oxygens by sulfur (Fig. 1).[1] They resemble the nucleoside phosphates very closely in overall size, negative charge, and geometry, but they differ in terms of metal ion coordination and in the somewhat larger van der Waals radius of sulfur over oxygen. In addition, because of the chirality of the phosphorus the internucleotidic phosphorothioate linkage exists as a pair of diastereomers. These small differences between phosphates and phosphorothioates have a number of consequences that render the phosphorothioate analogs useful to obtain information on questions concerning RNA

[1] F. Eckstein, *Ann. Rev. Biochem.* **54,** 367 (1985).

0076-6879/00 $30.00

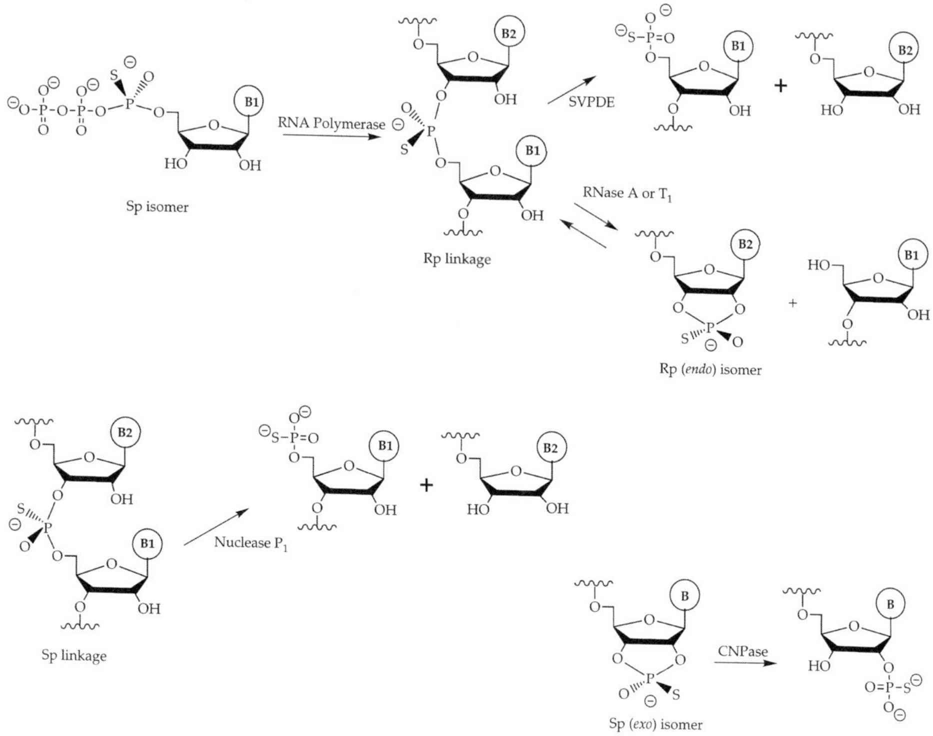

FIG. 1. Structure and configuration of nucleoside phosphorothioates and stereospecificity of some enzymatic interconversions. SVPDE, Snake venom phosphodiesterase; CNPase, cyclic nucleotide phosphodiesterase. [Adapted from S. Verma, N. K. Vaish, and F. Eckstein, *in* "Comprehensive Natural Products Chemistry" (Barton and Nakanishi, eds.). Elsevier Science, New York, 1998 with permission from Elsevier Science.]

structure and function. The main applications in the RNA field are the analysis of the stereochemical course of ribozyme-catalyzed reactions, the identification of phosphate groups important for RNA and ribozyme function, the coordination of metal ions to such phosphate groups, and, in combination with sugar- or base-modified nucleotides, as an aid for functional group analysis in transcripts. Examples of such applications have recently been reviewed in the context of modified oligonucleotides[2] and of ribonucleotide analogs.[3]

[2] S. Verma and F. Eckstein, *Ann. Rev. Biochem.* **67,** 99 (1998).

[3] S. Verma, N. K. Vaish, and F. Eckstein, *in* "Comprehensive Natural Products Chemistry" (D. Soell and S. Nishimura, eds.), Vol. 6, pg. 217–233. Elsevier Science Ltd., UK, 1998.

An important factor for the interpretation of results obtained with phosphorothioates is the knowledge of the configuration of the diastereomers. This rests on the configurational assignment of the endo (or R_p) isomer of uridine 2′,3′-cyclic phosphorothioate (Fig. 1) by X-ray structural analysis.[4] The configuration of all other phosphorothioates such as the internucleotidic phosphorothioate linkages or the nucleoside 5′-*O*-(1-thiotriphosphates) (NTPαS) is linked to this structure through enzymatic interconversions (Fig. 1).[1,3]

Nucleoside phosphorothioates can be incorporated into RNAs enzymatically by transcription or into oligoribonucleotides by chemical synthesis. The most commonly used enzyme for enzymatic incorporation is the T7 RNA polymerase[5] or its Y630F[6] mutant, both of which utilize the S_p NTPαS as substrate. Incorporation proceeds with inversion of configuration at phosphorus and thus the internucleotidic linkage is of the R_p configuration (Fig. 1).[1] Unfortunately, so far no mutant has been described that incorporates the R_p NTPαS, yielding the internucleotidic phosphorothioate of the S_p configuration. Thus, the usual procedure to obtain the S_p-phosphorothioate is to synthesize chemically an oligoribonucleotide with a mixture of diastereomers at one position, separate them by high-performance liquid chromatography (HPLC)[7,8] and ligate the isolated single diastereomer-containing oligomer into a longer RNA either with T4 RNA ligase or T4 DNA ligase using an oligodeoxynucleotide as a splint.[9]

Examples for the analysis of the stereochemical course of nucleotidyl transfer or hydrolysis reactions on RNA are the group I intron, the hammerhead and the hairpin ribozyme reactions. A single phosphorothioate of known configuration is incorporated at the cleavage site of the substrate and the phosphorothioate configuration of the product analyzed. This can be done in different ways. For the hammerhead and the hairpin reactions the cleavage product was degraded to the terminal nucleoside 2′,3′-cyclic phosphorothioate whose configuration was determined by comparison with an authentic sample.[7,10] In case of the spliceosome reaction, the stereopreference for one of the phosphorothioate diastereomers for degradation by RNase T1, snake venom phosphodiesterase (both specific for cleavage of the R_p diastereomer) and nuclease P (specific for the S_p diastereomer)

[4] W. Saenger and F. Eckstein, *J. Am. Chem. Soc.* **92,** 4712 (1970).
[5] A. Griffith, B. Potter, and I. Eperon, *Nucleic Acids Res.* **15,** 4145 (1987).
[6] R. Sousa and R. Padilla, *EMBO J.* **14,** 4609 (1995).
[7] G. Slim and M. Gait, *Nucleic Acids Res.* **119,** 1183 (1991).
[8] M. Koizumi and E. Ohtsuka, *Biochemistry* **30,** 5145 (1991).
[9] M. J. Moore and P. A. Sharp, *Nature* (*Lond.*) **365,** 364 (1993).
[10] H. van Tool, J. M. Buzayan, P. A. Feldstein, F. Eckstein, and G. Bruenig, *Nucleic Acids Res.* **118,** 1971 (1990).

was used for analysis (Fig. 2).[9,11] Similarly the stereochemical courses of the group I intron[12,13] and the group II intron reactions were determined.[14,15] All of the ribozyme-catalyzed reactions proceed with inversion of configuration of phosphorus, consistent with a single nucleophilic substitution reaction without a covalent intermediate.[1]

Despite the close similarity between phosphorothioates and phosphates, in a number of instances reaction rates for the phosphorothioates are reduced, indicating that the subtle change from oxygen to sulfur interferes with the normal progress of the reaction. This observation has led to the development of the phosphorothioate interference analysis to probe for the most sensitive phosphate positions in RNA transcripts. The basis is the random incorporation of any one of the four NTPαS to a small degree into a transcript, performing the reaction in question such as cleavage, binding or folding, separation of the reactive from the unreactive part, and analysis of the position of phosphorothioates by iodine cleavage of both parts[16,17] followed by gel electrophoretic analysis. Overrepresentation of a phosphorothioate in the unreactive molecules indicates interference with the reaction in question. This analysis is obviously restricted to the R_p diastereomer because the S_p is not available from transcription. However, the use of a mixture of diastereomers, prepared by chemical synthesis, and subsequent comparison with the results obtained with the R_p isomer allows for the identification of interference by the S_p isomer as well.[18]

Sulfur and oxygen also have different affinities for metal ions and this property can be used to identify whether metal ion coordination is responsible for the interference effect. This analysis is based on the soft–hard acid–base concept.[19] Sulfur as the softer ion coordinates preferentially with soft metal ions, whereas oxygen does so with hard metal ions. Mg^{2+} belongs to this latter class and thus coordinates strongly with one of the two nonbridging oxygens in a phosphate. If the *pro*-R_p oxygen is replaced by sulfur in the R_p-phosphorothioate, such coordination will be considerably weakened resulting in reduction or loss of the reaction in question if this interaction is crucial. However, use of a soft metal ion that will also coordinate to sulfur, such as Mn^{2+}, Co^{2+}, Zn^{2+}, or the even softer Cd^{2+}, should restore

[11] K. L. Maschhoff and R. A. Padgett, *Nucleic Acids Res.* **21,** 5456.
[12] J. A. McSwiggen and T. R. Cech, *Science* **244,** 679 (1989).
[13] J. Rajagopal, J. A. Doudna, and J. W. Szostak, *Science* **244,** 692 (1989).
[14] R. A. Padgett, M. Podar, S. C. Boulanger, and P. S. Perlman, *Science* **266,** 1685.
[15] G. Chanfreau and A. Jacquier, *Science* **266,** 1383 (1994).
[16] G. Gish and F. Eckstein, *Science* **240,** 1520 (1988).
[17] D. Schatz, R. Leberman, and F. Eckstein, *Proc. Natl. Acad. Sci. U.S.A.* **88,** 6132 (1991).
[18] R. Knöll, R. Bald, and J. P. Fürste, *RNA* **3,** 132 (1997).
[19] R. G. Pearson, *Science* **151,** 172 (1966).

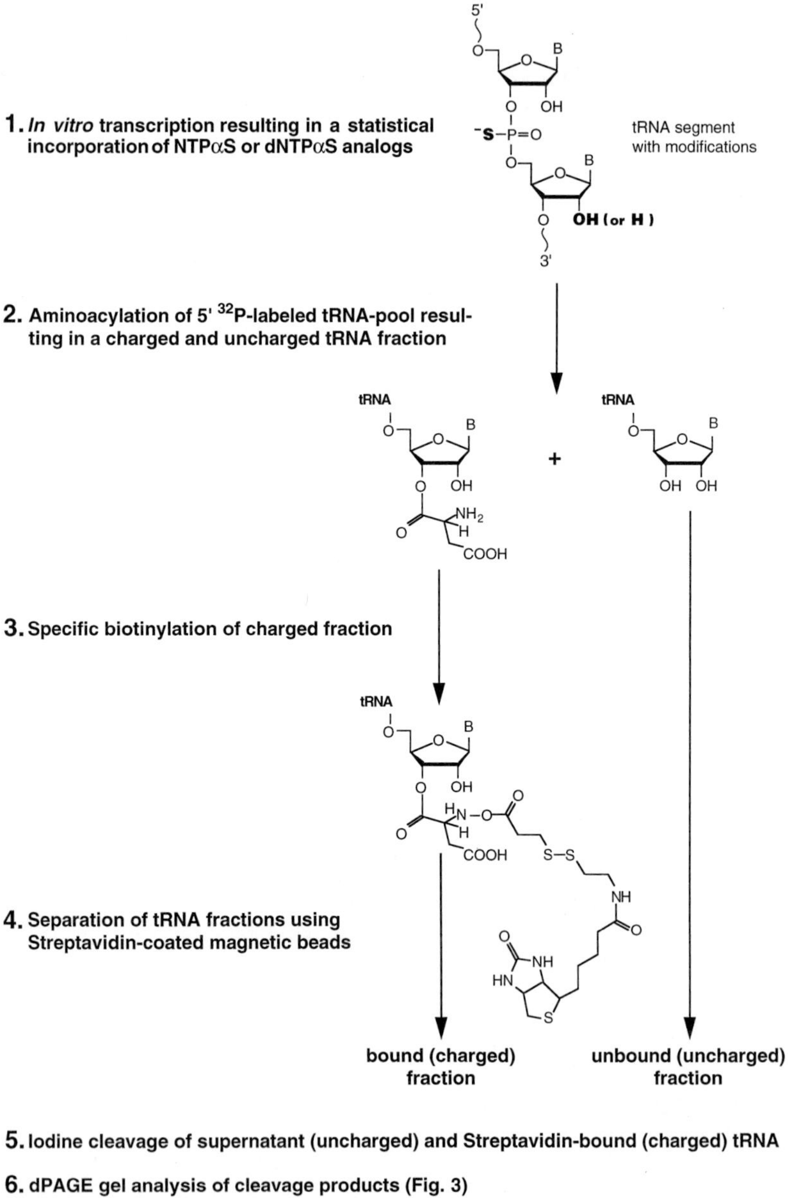

1. ***In vitro*** **transcription resulting in a statistical incorporation of NTPαS or dNTPαS analogs**

tRNA segment with modifications

2. Aminoacylation of 5' ^{32}P-labeled tRNA-pool resulting in a charged and uncharged tRNA fraction

3. Specific biotinylation of charged fraction

4. Separation of tRNA fractions using Streptavidin-coated magnetic beads

bound (charged) fraction

unbound (uncharged) fraction

5. Iodine cleavage of supernatant (uncharged) and Streptavidin-bound (charged) tRNA

6. dPAGE gel analysis of cleavage products (Fig. 3)

FIG. 2. Outline of the modification interference protocol. [Reprinted from C. S. Vörtler, O. Fedorova, T. Persson, and F. Eckstein, *RNA* **4,** 1444 (1998) with the permission of Cambridge University Press.]

at least some of the activity.[20,21] The use of Hg^{2+} has the additional benefit of producing a specific UV absorption signal when coordinated to sulfur and thus indicating such coordination directly.[22] This restoration of activity with soft metal ions is termed phosphorothioate rescue. A more detailed analysis, where possible, would use the S_p-phosphorothioate, where the R_p oxygen is available for coordination with Mg^{2+} and should thus form an active complex. This phenomenon is described as stereoisomer switch. This set of experiments would identify the *pro*-R_p oxygen as a position of metal ion coordination. An example of this situation is described for the group II intron where it has been combined with X-ray structure analysis.[23]

However, interference by the phosphorothioate in a given reaction could also be solely due to the 30% larger van der Waals radius of sulfur compared to oxygen. In this case no rescue would be expected. Such a situation has been demonstrated for an oligodeoxynucleotide–DNA polymerase I complex in the ground state.[24] The X-ray structure of the complex with the oligodeoxynucleotide with a R_p-phosphorothioate between the last two nucleotides is essentially superimposable with that of the native oligomer, including the precise position of the essential two metal ions. However, in the structure with the S_p isomer, the sulfur occupies a site where one of the metal ions was situated originally, thus displacing it. As a consequence, the second metal ion is also displaced. The larger size of sulfur has also been suggested to be responsible for the slowdown of the GTPγS hydrolysis rate of transducin as interfering with the formation of the transition state.[25]

The phosphorothioate interference analysis has recently been extended to the nucleotide analog interference mapping (NAIM) stragegy.[26–30] It is aimed at identifying nucleoside base functional groups or ribose 2′-hydroxyls essential for a reaction. It follows the same experimental setup as the phosphorothioate interference by employing the 5′-(1-*O*-thiotriphosphates) of the nucleoside analog for random incorporation into transcripts, separation of active from inactive species, iodine cleavage and localization

[20] M. Cohn, *Acc. Chem. Res.* **15,** 326 (1982).
[21] V. L. Pecoraro, J. D. Hermes, and W. W. Cleland, *Biochemistry* **23,** 5262 (1984).
[22] L. A. Cunningham, L. Li, and Y. Lu, *J. Am. Chem. Soc.* **120,** 4518 (1998).
[23] J. H. Cate, R. L. Hanna, and J. A. Doudna, *Nature Struct. Biol.* **4,** 553 (1997).
[24] C. A. Brautigam and T. A. Steitz, *J. Mol. Biol.* **277,** 363 (1998).
[25] J. P. Noel, H. E. Hamm, and P. B. Sigler, *Nature* **366,** 654 (1993).
[26] S. A. Strobel and K. Shetty, *Proc. Natl. Acad. Sci. U.S.A.* **94,** 2903 (1997).
[27] S. A. Strobel, L. Ortoleva-Donnelly, S. P. Ryder, J. H. Cate, and E. Moncoeur, *Nature Struct. Biol.* **5,** 60 (1998).
[28] L. Ortoleva-Donnelly, A. A. Szewczak, R. R. Gutell, and S. A. Strobel, *RNA* **4,** 498 (1998).
[29] A. V. Kazantsev and N. R. Pace, *RNA* **4,** 937 (1998).
[30] C. S. Vörtler, O. Fedorova, T. Persson, and F. Eckstein, *RNA* **4,** 1444 (1998).

TABLE I

Modified Nucleoside Triphosphates used for Interference Studies

Analog	RNA system studied	Reference
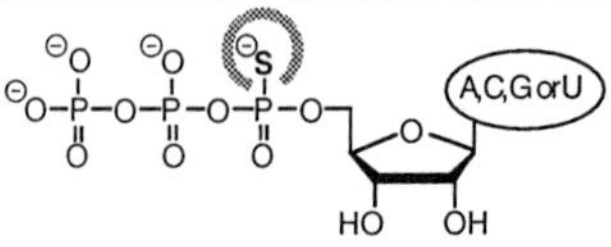 Nucleoside 5'-*O*-(1-thiotriphosphate)	Aminoacyl tRNA synthetase footprinting	(17, 47, 48)
	Group I intron splicing	(49, 50, 51)
	Small ribozyme activity	(52, 53, 54, 55)
	RNase P tRNA processing	(56, 57, 58)
	Group II introl splicing	(59)
	Ribosomal RNA, protein or ligand interaction	(60, 61, 62, 63, 64, 65)
	mRNA splicing	(66, 67)
	tRNA aminoacylation	(30)
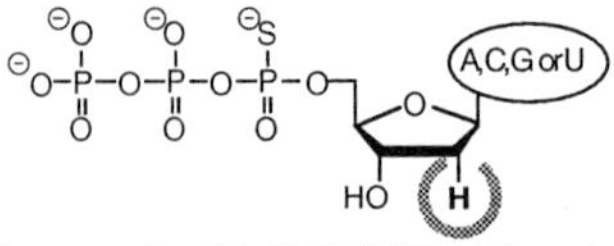 2'-Deoxynucleoside 5'-*O*-(1-thiotriphosphate)	RNase P tRNA processing	(58, 68, 69, 29)
	Ribosomal tRNA binding	(70)
	tRNA Aminoacylation	(30)
	Group I intron splicing	(26, 27, 28)
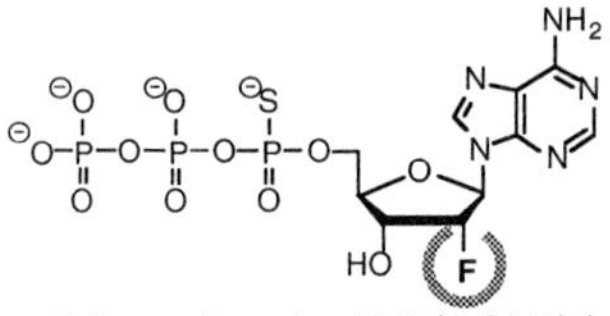 2'-Deoxy-2'-fluoroadenosine 5'-*O*-(1-thiotriphosphate)	Group I intron splicing	(26, 27, 28)
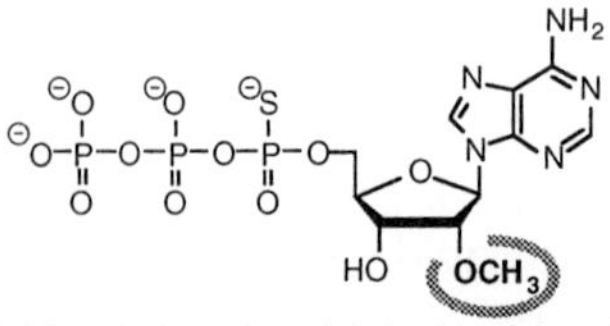 2'-*O*-Methyladenosine 5'-*O*-(1-thiotriphosphate)	Group I intron splicing	(26, 27, 28)
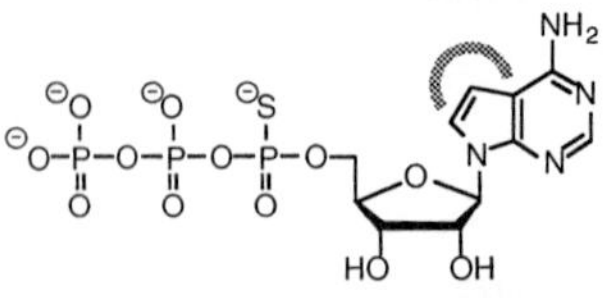 7-Deaza-adenosine 5'-*O*-(1-thiotriphosphate)	Group I intron splicing	(26, 27, 28)
	RNase P tRNA processing	(29)
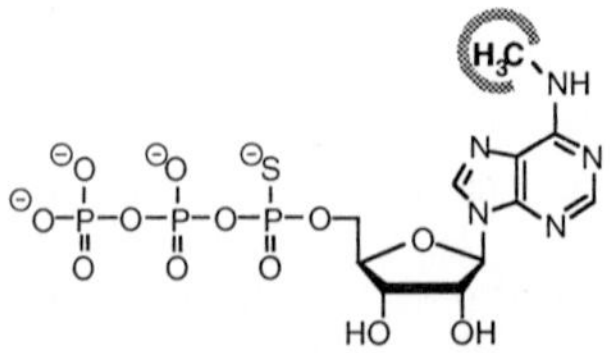 *N*-Methyladenosine 5'-*O*-(1-thiotriphosphate)	Group I intron splicing	(26, 27, 28)

TABLE I (*continued*)

Analog	RNA system studied	Reference
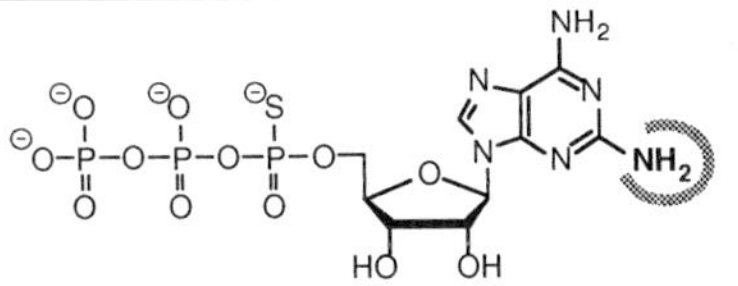 Diaminopurine nucleoside 5'-*O*-(1-thiotriphosphate)	Group I intron splicing	(26, 27, 28)
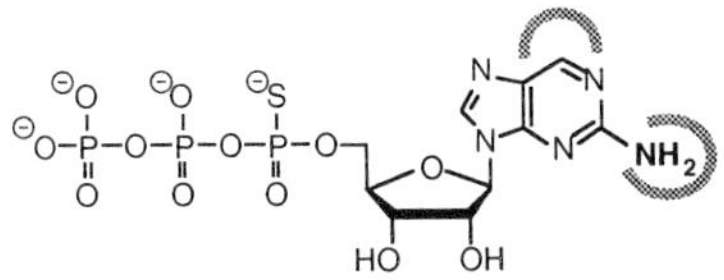 2-Aminopurine nucleoside 5'-*O*-(1-thiotriphosphate)	Group I intron splicing	(26, 27, 28)
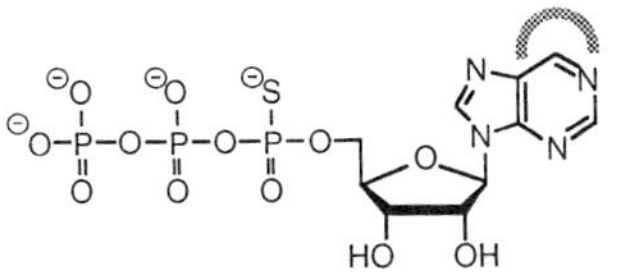 Purine nucleoside 5'-*O*-(1-thiotriphosphate)	Group I intron splicing	(26, 27, 28)

as well as quantification of the interference effect based on the cleavage pattern. However, since in the NAIM strategy the phosphorothioate group is linked to the modified nucleoside, interference can originate from one or both chemical modifications. It follows that interference due to the sugar or base modification alone has to be established by subtraction of the phosphorothioate interference by performing a separate experiment with the appropriate NTPαS. With different nucleoside analogs the importance of the guanosine 2-NH_2 group, the N-7 of adenosine, the cytosine 4-NH_2 group, and the 2'-hydroxyl group has been identified in several examples (Table I). The tRNA work is described in experimental detail later.

Chemical Synthesis of Phosphorothioate Oligoribonucleotides

Chemical synthesis of phosphorothioate oligoribonucleotides is carried out on any commercial DNA/RNA synthesizer using conventional ribonucleoside phosphoramidites but replacing the iodine oxidation step by a sulfurization step. The most common reagent for this reaction is 3*H*-1,2-benzodithiol-3-one 1,1-dioxide (the Beaucage reagent) but others can be used as well such as tetraethylthiuram disulfide (TETD), 1,2,4-dithiazolidine-3,5-dione (DtsNH), or 3-ethoxy-1,2,4-dithiazolin-5-one (EDITH).[31] The sulfurization reaction can be combined with the iodine oxidation such that oligonucleotides with a mixed phosphate/phosphorothioate backbone can be prepared. The oligonucleotide synthesis can also be performed by the H-phosphonate chemistry. This method turns all internucleotidic linkages into phosphorothioates.[32] In both cases the result is a mixture of diastereomers. This mixture can be resolved by reversed-phase HPLC (RP-HPLC) as long as there is only one phosphorothioate present in the oligonucleotide.[7] Efforts are being undertaken to develop stereospecific syntheses but so far the reagents are not commercially available.[33,34]

Synthesis of Nucleoside 5′-*O*-(1-Thiotriphosphates)

The synthesis of NTPαS can be performed by reaction of a 2′,3′-protected nucleoside, and of dNTPαS by using a 2′-protected nucleoside with salicylphosphorochloridite.[35] Alternatively, an unprotected nucleoside can be reacted with $PSCl_3$ and subsequent addition of pyrophosphate.[36] These procedures are also applicable to modified nucleosides. The common NTPαS and dNTPαS are commercially available from Amersham (Amersham, UK) or NEN/DuPont (Boston, MA).

Example for Nucleotide Analog Interference Study: Phosphorothioate and Ribose 2′-Hydroxyl Group Important for tRNA Aminoacylation[30]

A tRNA composed solely of 2′-deoxynucleotides is not chargeable under physiologic conditions,[37,38] indicating that at least some 2′-hydroxyl groups are important for this reaction. The aim of the study was to identify such tRNA 2′-hydroxyl groups by an interference assay, using the four dNTPαS. The phosphorothioate analogs were chosen because the presence of 2′-deoxynucleotide could easily be detected by the iodine cleavage

[31] Q. Xu, K. Musier-Forsyth, R. P. Hammer, and G. Barany, *Nucleic Acids Res.* **24,** 1602 (1996).
[32] G. Zon and W. J. Stec, *in* "Oligonucleotides and Analogues: A Practical Approach" (F. Eckstein, ed.), p. 87. IRL Press, Oxford University Press, 1991.
[33] W. J. Stec, B. Karwowski, M. Boczkowska, P. Guga, M. Koziolkiewicz, M. Sochacki, M. W. Wieczorek, and J. Blaszczyk, *J. Am. Chem. Soc.* **120,** 7156 (1998).
[34] R. P. Iyer, M.-J. Guo, D. Yu, and S. Agrawal, *Tetrahedron Lett.* **39,** 2491 (1998).
[35] J. Ludwig and F. Eckstein, *J. Org. Chem.* **54,** 631 (1989).
[36] A. Arabshahi and P. A. Frey, *Biochem. Biophys. Res. Commun.* **204,** 150 (1994).
[37] A. Khan and B. Roe, *Science* **241,** 74 (1988).
[38] J. Paquette, K. Nicghosian, G. Qi, N. Beauchemin, and R. Cedergren, *Eur. J. Biochem.* **189,** 259 (1990).

method[16,17] providing a positive signal rather than loss of a signal over a high background by limited alkaline hydrolysis. The *Escherichia coli* $tRNA^{Asp}$ is used for this study because it facilitates comparison of the interference data with the X-ray structure of its complex with the cognate synthetase (L. Moulinier and D. Moras, personal communication).

The sequence of reactions is summarized in Fig. 2. tRNA transcripts are prepared by runoff transcription from the corresponding plasmid using a certain percentage of one dNTPαS in the presence of the corresponding NTP. To scan the effect of all four nucleotides, four transcripts are prepared in which each of the four nucleotides is replaced by the analog. After 5′-^{32}P-end labeling and aminoacylation of the tRNA transcripts, the amino group of the aspartate attached to the tRNA is biotinylated. Charged biotinylated tRNAs are separated from uncharged nonbiotinylated on streptavidin magnetic beads. Both fractions are subjected to iodine cleavage and subsequently separated by denaturing polyacrylamide gel electrophoresis (dPAGE). After normalizing the radioactive intensity of the cleavage bands to account for loading errors in different lanes, a stronger band in the fraction of the uncharged tRNA indicates interference of the 2′-deoxynucleoside phosphorothioate at this position with aminoacylation. To separate the 2′-deoxy effect from the phosphorothioate effect, transcriptions had to be performed where one NTP is replaced in part by the corresponding NPTαS. A comparison of the effects with the two transcripts identifies positions with a 2′-deoxy and those with purely a phosphorothioate effect. A representative example of a gel analysis for the investigation with dCTPαS and CTPαS is given in Fig. 3. Such an analysis for all four dNTPαS scans the whole tRNA and gives a complete interference picture. A summary of results obtained with (d)CTPαS is shown in Fig. 4.

Sources of Enzymes, Chemicals, and Materials

The T7 RNA polymerase was overproduced and purified following the standard protocol.[6,39] A 90% pure preparation of *E. coli* aspartyl-tRNA synthetase as ammonium precipitate was a gift from Dr. Dino Moras (Strasbourg). Calf intestinal alkaline phosphatase including the 10× incubation buffer was from Boehringer Mannheim (Mannheim, Germany), T4 polynucleotide kinase and its buffer [700 m*M* Tris-HCl, pH 7.6, at 25°, 100 m*M* $MgCl_2$, 50 m*M* dithiothreitol (DTT)] were from NEB (Schwalbach, Germany), and RNase inhibitor was from MBI Fermentas (St. Leon-Rot, Germany). The plasmid AspUC containing the *E. coli* $tRNA^{Asp}$ sequence[40]

[39] J. Grodberg and J. J. Dunn, *J. Bacteriol.* **170,** 1245 (1988).

[40] M. Sprinzl and D. Gauss, *Nucleic Acids Res.* **12,** supplement, r59 (1984).

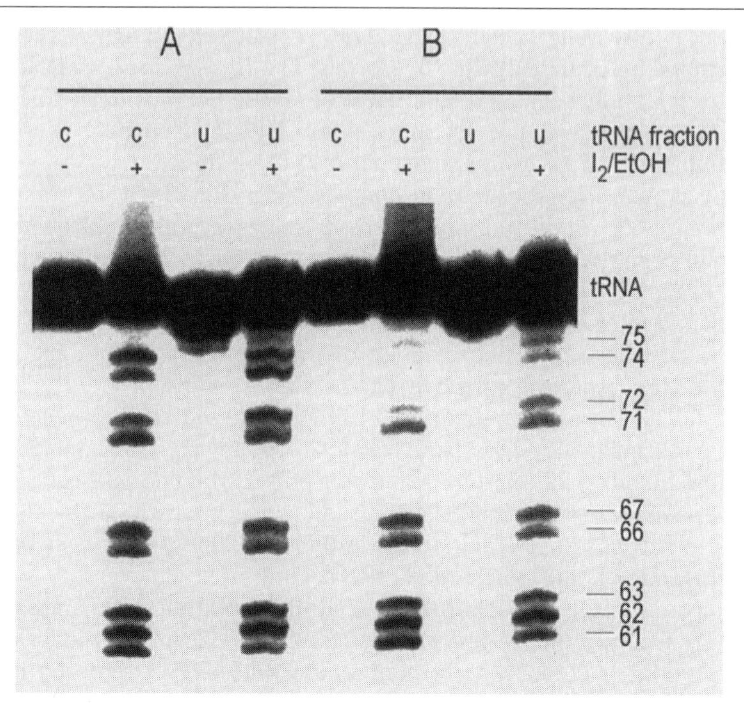

FIG. 3. PAGE analysis of charged and uncharged tRNA transcripts. (A) Transcripts prepared with 2.4 m*M* CTP and 600 μM CTPαS. (B) Transcripts prepared with 400 m*M* CTP and 600 μM dCTPαS. c, charged fraction; u, uncharged fraction; −, no iodine added; +, iodine added. [Reprinted from C. S. Vörtler, O. Fedorova, T. Persson, and F. Eckstein, *RNA* **4,** 1444 (1998), with the permission of Cambridge University Press.]

cloned between the *Bam*HI and *Eco*RI restriction sites of a pUC19 plasmid was constructed by Dr. G. Kozorek and A. Fahrenholz (Göttingen). It is linearized with *Bst*NI for transcription. Unmodified NTP were purchased from Boehringer Mannheim, [γ-^{32}P]ATP (>5000 Ci/mmol) from Amersham Buchler (Braunschweig, Germany), NTPαS, and dNTPαS from either Amersham Buchler (Braunschweig, Germany) or NEN/DuPont (Dreieich, Germany) or were synthesized according to Ludwig and Eckstein.[35] Molecular Biology Quality Glycogen was obtained from Merck (Darmstadt, Germany), acidic unbuffered water-saturated phenol from Roth (Karlsruhe, Germany), L-aspartate from Sigma (Deisenhofen, Germany), Sulfo-NHS-SS-biotin from Pierce (Rockford, IL) and streptavidin-coated magnetic beads from Dynal (Hamburg, Germany). The used water was glass distilled and autoclaved. Micropure ultrafiltration devices were from Amicon (Witten, Germany), Qiaquick spin nucleotide removal columns and buffers from Qiagen (Hilden, Germany).

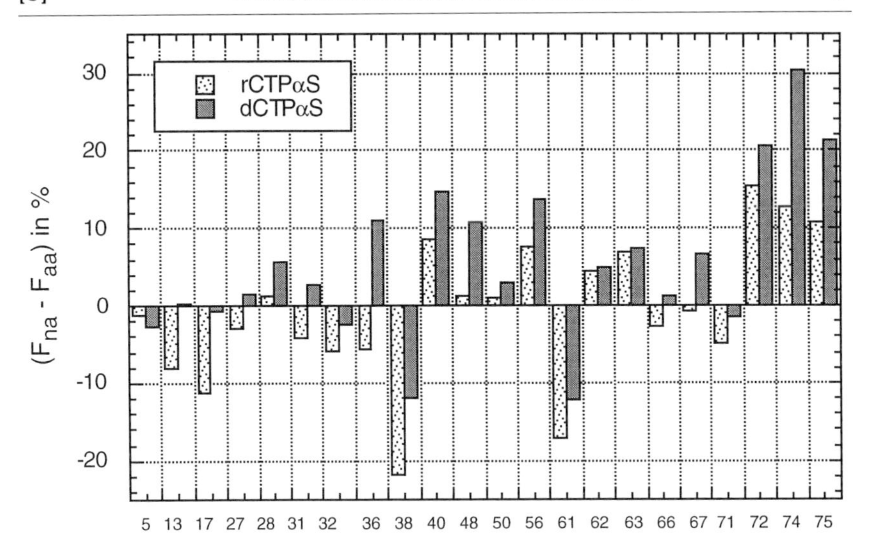

FIG. 4. Histograms showing the difference in intensities of phosphorothioate cleavage ladders from the charged versus the uncharged tRNA fraction for cytidine. [Reprinted from C. S. Vörtler, O. Fedorova, T. Persson, and F. Eckstein, *RNA* **4,** 1444 (1998), with the permission of Cambridge University Press.]

Incorporation of NTPαS and dNTPαS into RNA Transcripts with T7 RNA Polymerase

The random incorporation of nucleoside 5′-*O*-(1-thiotriphosphate) (NTPαS) into tRNA transcripts follows the protocol for incorporation of nucleoside 5′-triphosphates (NTPs).[41]

Linearized tRNA plasmid (1.5 μ*M*; 3 μg/μl)	20 μl
(NTPαS) (10 m*M*)	12 μl
Corresponding unmodified NTP (50 m*M*)	9.6 μl
Remaining three NTP (each 100 m*M*)	6 μl each
5× Transcription buffer	40 μl
H_2O	76.4 μl
T7 RNA polymerase (1080 U/μl)	24 μl
	200 total volume

The 5× transcription buffer contains 600 m*M* HEPES–KOH, pH 7.67, 5 m*M* spermidine, 50 m*M* $MgCl_2$, and 50 m*M* DTT. The water is glass distilled

[41] J. Milligan and O. C. Uhlenbeck, *Methods Enzymol.* **180,** 51 (1989).

and autoclaved. The 1:5 ratio of NTPαS:NTP (600 μM analog to 2.4 mM unmodified nucleoside triphosphate) gave a phosphorothioate substitution level of approximately 20%.

The 2′-deoxynucleoside 5′-*O*-(1-thiotriphosphate) (dNTPαS) are incorporated by replacement of the wild-type polymerase by the Y639F mutant[6] in the preceding protocol and with the exception that a 3:2 ratio of dNTPαS:NTP is required to obtain a 2′-deoxyphosphorothioate substituted tRNA of the same substitution level. Thus 36 μl of 10 mM dNTPαS and 4.8 μl of 50 mM unmodified NTP is used. However, it is recommended to determine the optimal ratio for each template. tRNA transcripts are gel purified on 20 × 40 × 0.1 cm 12% dPAGE gels containing 8 M urea and running the sample for 8 hr at 35 W to the bottom of the gel in order to have good resolution of the n, $n + 1$, $n + 2$ transcription products. The tRNA was UV-shadowed, the band of interest excised with a sterile razor blade, the cutout gel piece crushed, soaked in 2× its volume of 1 M sodium acetate, pH 4.6, frozen in liquid N_2, and placed on ice for 15 min. The supernatant is recovered by careful pipetting and replaced with a fresh sodium acetate solution added to the slurry. Repetition of this freeze–thaw cycle 3–4× recovers approximately 80% of the expected tRNA. The combined supernatants are freed of small gel pieces by ultrafiltration using a Micropure unit. The RNA is recovered as a pellet after addition of 2.5× the volume of ethanol, overnight storage at −20° and centrifugation. For further use it is dissolved in water and the concentration adjusted to 1 μg/μl.

Dephosphorylation of tRNA Transcripts

Dephosphorylation of the 5′-end of the transcript was required for subsequent 5′ labeling[42] and an additional heat-denaturation step aided this reaction. The solution of the tRNA transcript is first pipetted into a 1.5-ml Eppendorf tube containing the amount of water and dimethylformamide (DMF) indicated below, placed 1 min at 90° in a heating block and then immediately on ice. After another minute, the enzyme and buffer are added and the reaction mixture vortexed and incubated at 37° for 30 min.

Purified tRNA-transcript (40 μM, 1 μg/μl)	3 μl
DMF	1.3 μl
H_2O	15.7 μl

[42] J. Sambrock, E. F. Fritsch, and T. Maniatis, *in* "Molecular Cloning: A Laboratory Manual," 2nd Ed., p. 1.60. Cold Spring Harbor Laboratory Press, Cold Spring Harbor, New York, 1989.

—Heat denaturation as described—	
10× Commercial CIAP buffer	3.5 μl
Calf intestinal alkaline phosphatase (1 U/μl)	12 μl
RNase inhibitor (40 U/μl)	0.5 μl
	36 μl total volume

The 10× buffer contains 500 m*M* Tris-HCl, pH 7.5, and 1 m*M* EDTA. The enzyme is supplied in 25 m*M* Tris-HCl, pH 7.5, 1 m*M* $MgCl_2$, 0.1 m*M* $ZnCl_2$, and 50% glycerol. For workup the reaction mixture is run over a QiaQuick nucleotide removal column according to the manufacturer's instruction and eluted with 2× 20 μl H_2O, followed by a Speed-Vac concentration to approximately 10 μl.

^{32}P-Labeling of tRNA Transcripts

The standard polynucleotide kinase protocol is followed[43]:

5′-Dephosphorylation transcript	10 μl
10× Commercial T4 PNK buffer	2.5 μl
RNAse inhibitor (40 U/μl)	0.5 μl
T4 Polynucleotide kinase (10 U/μl)	2 μl
[γ-^{32}P]ATP (10 μCi/μl)	3 μl
H_2O	7 μl
	25 μl total volume

The reaction mixture is incubated at 37° for 30 min, 1 μl of a 10 m*M* ATP solution added, and then incubated for a further 10 min. The reaction mixture is loaded onto a prewarmed 20 × 40 × 0.04 cm 12% dPAGE gel containing 8 *M* urea and run for 2.5 hr at 30 W. After a 1-min exposure to an imaging plate, the radioactive tRNA band is cut from the gel and eluted by the crush-and-soak method as described in the transcription section.

Aminoacylation

Reaction conditions followed published procedures[44]:

^{32}P-End labeled tRNA (0.2 μ*M*, 50,000 cpm)	5 μl
E. coli aspartyl-tRNA synthetase (1 U/μl)	4 μl
RNase inhibitor (40 U/μl)	1 μl
Aspartate (1 m*M*)	6 μl
5× Aspartylation buffer	12 μl
H_2O	32 μl
	60 μl total volume

[43] F. Cobianchi and S. Wilson, *Methods Enzymol.* **152,** 94 (1987).

[44] J. R. Sampson and O. C. Uhlenbeck, *Proc. Natl. Acad. Sci. U.S.A.* **85,** 1033 (1989).

The aminoacyl synthetase stock solution (500 U/μl, kept at $-20°$) is diluted prior to use with a buffer consisting of 25 m*M* HEPES, pH 7.5, 250 m*M* KCl, 2 mg/ml bovine serum albumin (BSA), 50% (v/v) glycerol. This solution is discarded after 1 day. The 5× aspartylation buffer contained 250 m*M* HEPES–KOH, pH 7.5, 150 m*M* KCl, 100 m*M* $MgCl_2$, and 12.5 m*M* adenosine triphosphate (ATP). After 15 min incubation at 37°, 2.5 μl of bulk yeast tRNA (10 μg/μl) is added to avoid loss of labeled RNA. Following an extraction with unbuffered acidic phenol and chloroform, the RNA is precipitated by adding 1/3 of the volume of 3 *M* sodium acetate, pH 4.6. The low pH is crucial to prevent hydrolysis of the aspartyl-RNA ester. The tRNA is constantly kept on ice, handled as quickly as possible, and otherwise stored best at $-80°$. Storage overnight without loss of the amino acid is possible; however, longer storage should be avoided. Once the amino group has been reacted with the NHS ester, stability is considerably increased.[45]

Separation of Charged from Uncharged tRNAs

Separation of aminoacylated (aa) from nonaminoacylated (na) tRNAs is achieved by reacting the α amino group of the covalently bound aspartate with a *N*-hydroxysuccinimide ester linked to biotin. This facilitates immobilization of the charged tRNA to streptavidin-coated magnetic beads.[46] The precipitated, partly aminoacylated tRNA pool is taken up in 8 μl of ice-cold 10 m*M* sodium acetate, pH 4.6. The solution is kept on ice and 10 μl of 600 m*M* HEPES, pH 8.2, and 4.2 μl of 45 m*M* aqueous solution of sulfo-NHS-SS-biotin, freshly prepared before the reaction, are added. After a quick vortex and spin, the reaction is incubated for 1.5 hr on ice. Excess of biotinylation reagent is removed by applying the solution to a Microcon 10 ultrafiltration device (Millipore Corp., Bedford, MA) containing 200 μl 10 m*M* sodium acetate, pH 4.6, and reducing the volume to 5 μl by centrifugation at 4° following the manufacturer's instructions. After two additions of fresh 200 μl of 10 m*M* sodium acetate, pH 4.6, and reduction of the volume, the retentate is transferred to a second and a third Microcon and the procedure repeated. The tRNA is finally recovered by addition of 20 μl water to the last retentate. At the same time 50 μl of streptavidin magnetic beads is washed three times with 100 μl of 2.5 *M* NaCl, 100 m*M* HEPES, pH 7.5, and resuspended in 25 μl of this buffer. The tRNA is added to the beads, placed in an Eppendorf shaker for 30 min at room

[45] F. Schuber and M. Pinck, *Biochemie* **56,** 383 (1974).

[46] J. Puetz, J. Wientges, M. Sissler, R. Giege, C. Florentz, and A. Schwienhorst, *Nucleic Acids Res.* **25,** 1862 (1997).

temperature, and the supernatant removed, yielding the unbound fraction. Its volume is increased to 100 μl by addition of 55 μl of water and the tRNA is precipitated as described later. Beads are washed 3× and resuspended in 50 μl of the preceding buffer. To this solution is added 50 μl of 100 mM DTT, then shaken for 30 min at room temperature, and the supernatant recovered, which corresponds to the bound fraction. Excess DTT is removed by 3× extraction with 100 μl of ethyl acetate. The tRNA in both fractions is recovered by addition of 30 μl 3 *M* sodium acetate, pH 4.6, 325 μl ethanol, and storage at $-20°$ overnight.

Iodine Clevage

Bound and unbound tRNA fractions are recovered by centrifugation. The pellet is dissolved in 8 μl of water and the solution incubated for 5 min at 37°. This is followed by addition of 2 μl of a 10 m*M* solution of iodine in ethanol and incubation for 30 sec at 37°.[17,47] Subsequently 20 μl of loading buffer [8 *M* urea, 50 m*M* EDTA, 0.05% (w/v) of each bromphenol

[47] J. Rudinger, J. D. Puglisi, J. Puetz, D. Schatz, F. Eckstein, C. Florentz, and R. Giegé, *Proc. Natl. Acad. Sci. U.S.A.* **89,** 5882 (1992).
[48] R. Kreutzer, D. Kern, R. Giegé, and J. Rudinger, *Nucleic Acids Res.* **23,** 4598 (1995).
[49] R. B. Waring, *Nucleic Acids Res.* **17,** 10281 (1989).
[50] E. L. Christian and M. Yarus, *J. Mol. Biol.* **228,** 743 (1992).
[51] E. L. Christian and M. Yarus, *Biochemistry* **32,** 4475 (1993).
[52] D. E. Ruffner and O. C. Uhlenbeck, *Nucleic Acids Res.* **18,** 6025 (1990).
[53] B. M. Chowrira and J. M. Burke, *Nucleic Acids Res.* **20,** 2835 (1992).
[54] Y.-H. Jeoung, P. K. R. Kumar, Y.-A. Suh, K. Taira, and S. Nishikawa, *Nucleic Acids Res.* **22,** 3722 (1994).
[55] N. S. Prabhu, G. Dinter-Gottlieb, and P. A. Gottlieb, *Nucleic Acids Res.* **25,** 5119 (1997).
[56] W.-D. Hardt, J. M. Warnecke, V. A. Erdmann, and R. K. Hartmann, *EMBO J.* **14,** 2935 (1995).
[57] M. E. Harris and N. R. Pace, *RNA* **1,** 210 (1995).
[58] F. Conrad, A. Hanne, R. K. Gaur, and G. Krupp, *Nucleic Acids Res.* **23,** 1845 (1995).
[59] J. L. Jestin, E. Deme, and A. Jacquier, *EMBO J.* **16,** 2945 (1997).
[60] M. Dabrowski, C. M. T. Spahn, and K. H. Nierhaus, *EMBO J.* **14,** 4872 (1995).
[61] E. V. Alexeeva, O. V. Shpanchenko, O. A. Dontsova, A. A. Bogdanov, and K. H. Nierhaus, *Nucleic Acids Res.* **24,** 2228 (1996).
[62] O. V. Shpanchenko, M. I. Zvereva, O. A. Dontsova, K. H. Nierhaus, and A. A. Bogdanov, *FEBS Lett.* **394,** 71 (1996).
[63] W. Schnitzer and U. Von Ahsen, *Proc. Natl. Acad. Sci. U.S.A.* **94,** 12823 (1997).
[64] R. T. Batey and J. R. Williamson, *J. Mol. Biol.* **261,** 550 (1996).
[65] S. C. Blanchard, D. Fourmy, R. G. Eason, and J. D. Puglisi, *Biochemistry* **37,** 7716 (1998).
[66] P. Fabrizio and J. Abelson, *Nucleic Acids Res.* **20,** 3659 (1992).

blue and xylene xylenol] is added. To obtain optimal resolution of the 76 nucleotide (nt) full-length transcript, a 5-μl aliquot is applied to dPAGE on a $20 \times 40 \times 0.04$ cm 12% gel containing 8 *M* urea and run for 3.5 hr at 35 W. A second aliquot is applied to the same gel and the electrophoresis continued for a further 1.5 hr.

Analysis of Cleavage Pattern

After drying the gel, overnight exposure and scanning with a Fuji BAS 2500 PhosphorImager the cleavage pattern (Fig. 3) is analyzed using the TINA 1.9 software package (Raytest, Germany). The radioactivity distribution in a gel lane is determined after normalizing for gel loading by dividing the intensity of each cleavage band by the full-length tRNA band. An interference effect of a modified position becomes detectable by determination of the cleavage intensities of the aminoacylated (I_{aa}) and the non-aminoacylated fraction (I_{na}) and calculation of the ratios $F_{aa} = (I_{aa} \times 100)/(I_{aa} + I_{na})$ and $F_{na} = (I_{na} \times 100)/(I_{aa} + I_{na})$. The difference $F_{na} - F_{aa}$ is calculated for three independent experiments. Results are plotted for each nucleotide and position as exemplified for the (d)CTPαS experiments in Fig. 4. A positive value indicates overrepresentation in the uncharged fraction and thus interference with the charging reaction, whereas a negative value signifies overrepresentation in the charged fraction. If the modification has no effect, the difference is zero, since the position is equally distributed in both fractions. To allow for the experimental error, a $|F_{na} - F_{aa}|$ value of up to 5% was considered to indicate no effect, up to 10% a weak effect, up to 20% a moderate effect, and above 20% a strong effect. For those positions that showed an effect of the phosphorothioate group, an additional effect of the 2′-deoxy modification was only considered if the difference $|F_{na}-F_{aa}|$ was above 5%.

Conclusion

Critical phosphate and 2′-hydroxyl positions identified by this modification interference procedure are summarized in Fig. 5 and discussed in more detail in the original publication.[30] Generally, more critical phosphorothiate positions were identified than 2′-deoxy groups. The effect of the latter was confirmed by determination of the aminoacylation kinetics using single 2′-deoxy-modified tRNAs obtained by chemical synthesis. The strength of

[67] Y. T. Yu, P. A. Maroney, E. Darzynkiewicz, and T. W. Nielsen, *RNA* **1,** 46 (1995).
[68] R. K Gaur and G. Krupp, *Nucleic Acids Res.* **21,** 21 (1993).
[69] W. D. Hardt, V. A. Erdmann, and R. K. Hartmann, *RNA* **2,** 1189 (1996).
[70] U. Von Ahsen, R. Green, R. Schroeder, and H. F. Noller, *RNA* **3,** 49 (1997).

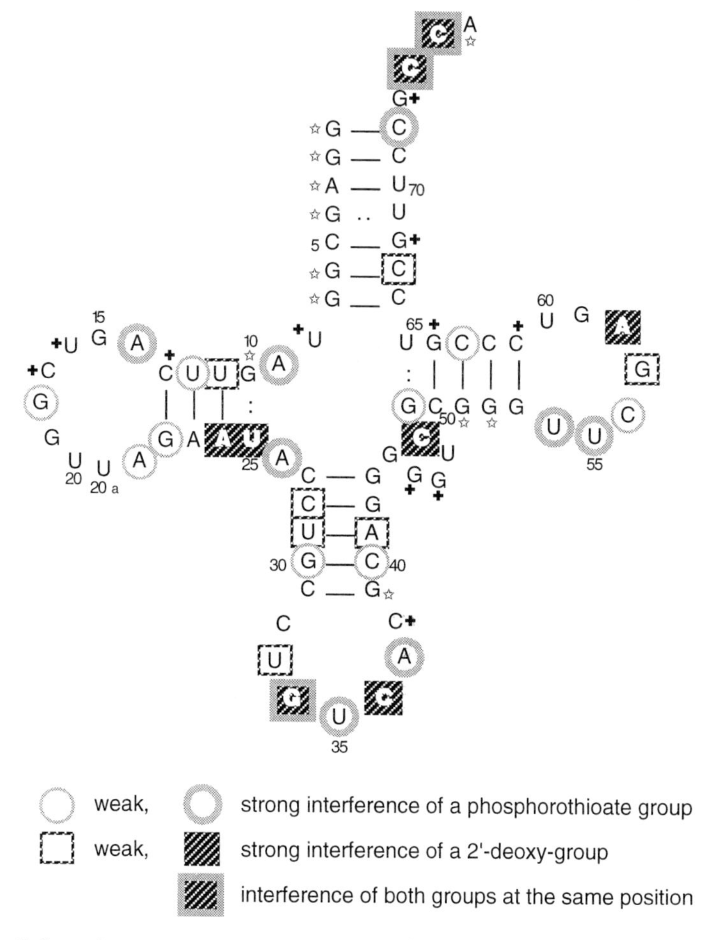

FIG. 5. Secondary structure of the *E. coli* $tRNA^{Asp}$ summarizing the interference results. I, Not gel-resolved; :, enhancement of charging. [Reprinted from C. S. Vörtler, O. Fedorova, T. Persson, and F. Eckstein, *RNA* **4,** 1444 (1998), with the permission of Cambridge University Press.]

the kinetic effects correlated very well with the intensity of the observed interference. The contribution of a single 2′-hydroxyl group was small, on the order of a three- to sixfold loss of charging efficiency. However, the presence of several hydroxyl groups reduced aminoacylation efficiency considerably.

Acknowledgment

We thank J. Morris for careful reading of the manuscript.

[6] Chemical Probing of RNA by Nucleotide Analog Interference Mapping

By SEAN P. RYDER, LORI ORTOLEVA-DONNELLY, ANNE B. KOSEK, and SCOTT A. STROBEL

Introduction

Chemical modification experiments have been widely used to address questions of RNA structure and function.[1,2] The chemistry of phosphorothioate modification interference has been expanded into a powerful method to identify the specific chemical groups that are important for RNA activity.[3–5] This approach, termed nucleotide analog interference mapping (NAIM), makes it possible to simultaneously, yet individually, test the contribution of a particular functional group at almost every position within the molecule in an assay that is as simple as RNA sequencing. In a NAIM experiment the smallest mutable unit is not the base pair, but rather the individual functional groups that comprise the nucleotides. Because the deletion or modification of particular functional groups within an RNA can severely affect its activity, this approach makes it possible to determine the chemical basis of RNA structure and function.[6]

Overview of Method

Building on the initial work of Krupp and co-workers,[4,7] NAIM utilizes α-phosphorothioate-tagged nucleotide analogs. Each analog has an incremental chemical alteration in the base or ribose sugar and a phosphorothioate substitution that acts as a chemical tag.[5,8–14] The S_p phosphorothioate

[1] S. Stern, D. Moazed, and H. F. Noller, *Methods Enzymol.* **164,** 481 (1989).
[2] L. Conway and M. Wickens, *Methods Enzymol.* **180,** 369 (1989).
[3] F. Eckstein, *Ann. Rev. Biochem.* **54,** 367 (1985).
[4] R. K. Gaur and G. Krupp, *Nucleic Acids Res.* **21,** 21 (1993).
[5] S. A. Strobel and K. Shetty, *Proc. Natl. Acad. Sci. U.S.A.* **94,** 2903 (1997).
[6] J. A. Grasby and M. J. Gait, *Biochimie* **76,** 1223 (1994).
[7] F. Conrad, A. Hanne, R. K. Gaur, and G. Krupp, *Nucleic Acids Res.* **23,** 1845 (1995).
[8] W. D. Hardt, V. A. Erdmann, and R. K. Hartmann, *RNA* **2,** 1189 (1996).
[9] L. Ortoleva-Donnelly, A. A. Szewczak, R. R. Gutell, and S. A. Strobel, *RNA* **4,** 498 (1998).
[10] L. Ortoleva-Donnelly, M. Kronman, and S. A. Strobel, *Biochemistry* **37,** 12933 (1998).
[11] S. Basu, R. P. Rambo, J. H. Cate, A. R. Ferre-D Amare, S. A. Strobel, and J. A. Doudna, *Nature Struct. Biol.* **5,** 986 (1998).
[12] C. S. Vortler, O. Fedorova, T. Persson, U. Kutzke, and F. Eckstein, *RNA* **4,** 1444 (1998).
[13] A. V. Kazanstev and N. R. Pace, *RNA* **4,** 937 (1998).
[14] A. A. Szewczak, L. Ortoleva-Donnelly, S. P. Ryder, E. Moncouer, and S. A. Strobel, *Nature Struct. Biol.* **5,** 1037 (1998).

0076-6879/00 $30.00

derivative of the nucleotide analog is randomly incorporated into the RNA by *in vitro* transcription using T7 RNA polymerase (Fig. 1, step 1).[15,16] The RNA transcripts are selected for function, such that the active and inactive variants in the population are physically separated or selectively radiolabeled (Fig. 1, step 2). The analog distribution between these populations is revealed by selectively cleaving the phosphorothioate linkage with iodine (Fig. 1, step 3), and resolving the cleavage products by polyacrylamide gel electrophoresis (PAGE) (Fig. 1, step 4).[17] Sites of analog substitution that are detrimental to function are scored as gaps in the sequencing ladder among the active RNA variants. Because every position in the sequence is a unique and independent band on the sequencing gel, a single screen defines the effect a particular analog has at every incorporated position within the RNA.

NAIM can be generalized to any analog that can be incorporated into a transcript by an RNA polymerase, because the phosphorothioate chemical tag is independent of the nucleotide analog whose location it reports. Furthermore, NAIM is applicable to any RNA that can be transcribed *in vitro* and whose function can be assayed. Such functions may include catalytic activity, protein binding, substrate or ligand binding, or folding into a specific structure. Here we outline the synthesis of phosphorothioate-tagged analogs and their *in vitro* incorporation into RNA transcripts. We also describe an interference mapping assay using a reverse splicing form of the *Tetrahymena* group I ribozyme.[18,19] This includes a procedure for the quantitation of the interference data and an interpretation of characteristic interference patterns. A modified version of this method was previously published in the journal *Methods.*[20]

Materials and Reagents

Chemicals

The four natural 5′-*O*-(1-thio)nucleoside triphosphates, ATPαS, GTPαS, UTPαS, and CTPαS, and several other analogs described in this article are available from Glen Research (Sterling, VA). Many of the

[15] M. Chamberlain, R. Kingston, M. Gilman, J. Wiggs, and A. de Vera, *Methods Enzymol.* **101,** 540 (1983).

[16] A. D. Griffiths, B. V. L. Potter, and I. C. Eperon, *Nucleic Acids Res.* **15,** 4145 (1987).

[17] G. Gish and F. Eckstein, *Science* **240,** 1520 (1988).

[18] T. R. Cech, *Ann. Rev. Biochem.* **59,** 543 (1990).

[19] A. A. Beaudry and G. F. Joyce, *Science* **257,** 635 (1992).

[20] S. P. Ryder and S. A. Strobel, *Methods,* **18,** 38 (1999).

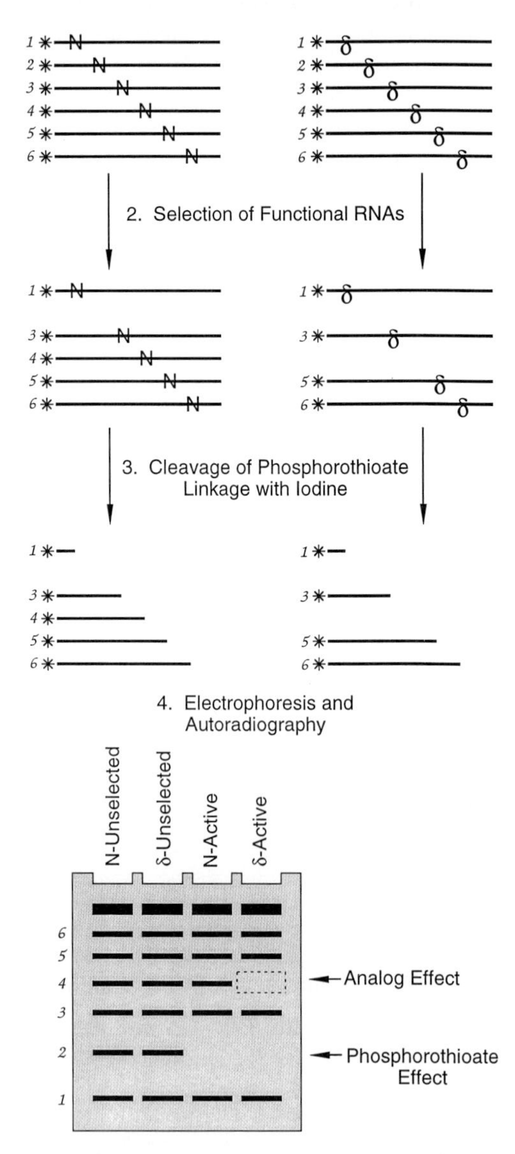

FIG. 1. Scheme for the identification of chemical groups important to RNA activity by NAIM. $\delta\alpha$S corresponds to sites of analog substitution. In this example, active molecules ligate themselves to a radiolabeled substrate. Inactive molecules do not perform this reaction, and as such are invisible to autoradiography.

unprotected nucleoside precursors are commercially available from Sigma (St. Louis, MO) or RI Chemical (Orange, CA).

[γ-^{32}P]ATP (6000 Ci/mmol) and [α-^{32}P]cordycepin triphosphate (5000 Ci/mmol): DuPont-NEN.

Tributylammonium pyrophosphate (TBAP): Sigma

All other synthetic reagents were purchased from Aldrich (Milwaukee, WI) and used in the form provided.

Equipment

All sequencing gels were visualized and quantitated on a Storm 820 using ImageQuant version 4.1 software (Molecular Dynamics, Sunnyvale, CA).

Enzymes

T7 RNA polymerase; purified from the His_6-tagged overexpression construct pT7-911Q using a Ni-agarose column (Qiagen) (T. E. Shrader, personal communication)

Y637F mutant form of T7 RNA polymerase; overexpressed from the plasmid pDPT7-Y639F and purified as described[21]

T4 polynucleotide kinase (PNK), calf intestinal alkaline phosphatase (CIP), New England Biolabs (Beverly, MA).

Yeast poly(A) polymerase, Amersham (Piscataway, NJ)

Escherichia coli inorganic pyrophosphatase, Sigma

Buffers

TEAB: triethylammonium bicarbonate, pH 8.0; prepared by bubbling CO_2 through an aqueous solution of triethylamine in water

Transcription buffer: 40 m*M* Tris-HCl, pH 7.5, 4 m*M* spermidine, 10 m*M* dithiothreitol (DTT), 15 m*M* $MgCl_2$, 0.05% Triton X-100, 0.05 μg/μl template DNA

TE: 10 m*M* Tris-HCl, pH 7.5, 0.1 m*M* EDTA

Reaction buffer: 50 m*M* HEPES, pH 7.0, 3 m*M* $MgCl_2$, and 1 m*M* manganese acetate

Loading buffer: 8 *M* urea, 50 m*M* EDTA, pH 8.0, 0.01% bromphenol blue, 0.01% xylene cyanol.

I_2–Ethanol: 100 m*M* iodine in 100% ethanol

[21] R. Sousa and R. Padilla, *EMBO J.* **14,** 4609 (1995).

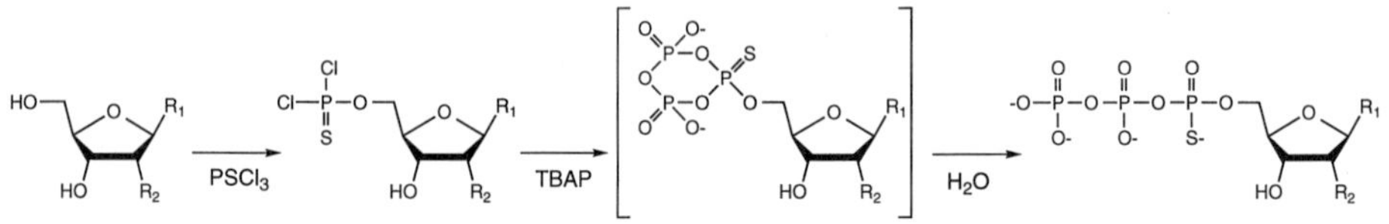

FIG. 2. Scheme for the synthesis of a 5′-*O*-(1-thio)nucleotide analog triphosphate. R_1 corresponds to base analogs, and R_2 corresponds to 2′ ribose substitutions. The unstable cyclophosphate intermediate is outlined in brackets.

Methods

Synthesis of Phosphorothioate-Tagged Nucleotide Triphosphates

The 5′-*O*-(1-thio)nucleoside triphosphates are synthesized from unprotected nucleoside precursors on a scale of approximately 100 mg of starting material (Fig. 2).[22] The reaction is carried out in two steps in a single reaction flask. In the first step, the nucleoside is converted to the 5′-*O*-(1, 1-dichloro-1-thio)phosphorylnucleoside. In the second step, the nucleotide is converted directly to the triphosphate via a cyclotriphosphate intermediate by the stepwise addition of tributylammonium pyrophosphate and water (Fig. 2). The overall yield of the reaction varies between 5% and 40%, with A analogs reacting better than G analogs, both of which react more efficiently than pyrimidine analogs. A general procedure for purine and pyrimidine analogs follows.

The nucleoside (100 mg) is transferred to a 20-ml round-bottom flask and dried overnight under vacuum at 110°. It is also possible to dry the nucleoside by coevaporating with 3 × 10 ml of anhydrous pyridine. After the nucleoside is dry, the flask is capped and purged with argon, and the nucleoside is dissolved in a minimal volume (0.5–2.0 ml) of triethyl phosphate (TEP). The volume of TEP used severely affects the reaction yield, so every effort is made to dissolve the nucleoside in less than 1 ml of TEP. If the nucleoside does not dissolve initially, the stirring solution can be heated gradually with an air gun.

For purine nucleoside analogs, 1.1 equivalents of trioctylamine (molecular weight 353.7, density 0.816 g/ml) and 1.1 equivalents of $PSCl_3$ (molecular weight 169.40, density 1.67 g/ml) are added to the nucleoside solution. This is stirred for 20–60 min under argon and the progress of the reaction monitored by quenching a small aliquot (100–200 μl) in water and resolving the products using cellulose thin-layer chromatography (TLC) in a solvent system of 0.5 *M* LiCl in water. The reaction products are visualized by

[22] A. Arabshahi and P. A. Frey, *Biochem. Biophys. Res. Com.* **204,** 150 (1994).

shadowing with a hand-held UV light. The monophosphate migrates slower on the cellulose than the unreacted starting material.

We have recently used a slightly different reaction condition to synthesize the 5′-*O*-(1-thio-1, 1-dichloro)phosphoryl nucleoside intermediate for the pyrimidine analogs. The dry nucleoside (100 mg) is dissolved in a minimum volume (0.5–2.0 ml) of TEP. The solution is cooled to 0° and trioctylamine (1.2 equivalents) and collidine (1.2 equivalents, molecular weight 121.8, density 0.917 g/ml) are added under anhydrous conditions. $PSCl_3$ (1.3–1.5 equivalents) is added dropwise to the chilled solution, and allowed to react for 30 min before warming to room temperature and reacting for an additional 45 min. The progress of the reaction and all subsequent steps are performed in the same way for purine and pyrimidine analogs.

Tributylammonium pyrophosphate (TBAP, molecular weight 451.5) is dissolved in TEP (0.1 g/ml) by gently heating the solution with an air gun. When a substantial amount of the 5′-*O*-(1-thio-1, 1-dichloro)phosphoryl nucleoside has formed, 4.0 equivalents of tributylammonium pyrophosphate (TBAP) are added per mole of $PSCl_3$, and the solution is stirred for 30 min. Adding less than 4.0 equivalents reduces the reaction yield, though we have obtained fairly good yields even when only 1.0 equivalent of TBAP is added per mole of $PSCl_3$. The progress of the reaction is monitored by quenching a small portion in a few drops of triethylamine (TEA), which causes the phosphates in the reaction to precipitate. The precipitate is collected by centrifugation, dissolved in 50 m*M* TEAB, and the products resolved by silica TLC using a solvent system of 6 : 3 : 1 *n*-propanol, ammonium hydroxide, and water. Based on TLC analysis, the reaction yields at least two distinct products, both of which migrate slower on the TLC than either the nucleoside starting material or the monophosphate. One product is the linear triphosphate, the other is the cyclotriphosphate, which will hydrolyze to the linear product after several hours at room temperature.

The nucleoside triphosphates are precipitated by adding about 60 equivalents of TEA (molecular weight 101.19, density 0.726 g/ml, calculated relative to the moles of nucleoside originally added) dropwise to the solution. The precipitate is collected by centrifugation, the solid is dissolved in 5 ml of 50 m*M* TEAB, and the reaction products hydrolyzed by incubating at room temperature overnight. The nucleoside triphosphate is purified by DEAE Sephadex chromatography using a linear gradient from 50 to 800 m*M* TEAB in a total volume of 1 liter.[23] The nucleotide typically elutes at a TEAB concentration between 0.5 and 0.6 *M*. The fractions containing the nucleoside triphosphate are pooled and lyophilized to a thin film.

[23] F. Eckstein and R. S. Goody, *Biochemistry* **15,** 1685 (1976).

The reaction products are characterized by NMR, UV, and mass spectroscopy. The correct nucleotide has three resonance peaks in the ^{31}P NMR spectra at 42–43 (α), −5–10 (γ), and −20–25 (β) ppm relative to an 85% phosphoric acid standard.[24] We have occasionally observed an uncharacterized reaction product with a resonance at 35–38 ppm, which is not incorporated into the RNA and may be inhibitory to the polymerase. The nucleotide concentration is measured by UV absorbance, and the molecular weight confirmed by mass spectrometry. In our experience, the nucleotides can be stored at −20° for at least 2 years in the freezer either as a dry solid or in TE without noticeable degradation of the triphosphates. We have used this general procedure to prepare more than 35 different nucleotide triphosphates.

Incorporation of Phosphorothioate Analogs Via in Vitro Transcription

Nucleotide analogs are randomly incorporated into RNA by *in vitro* transcription with either T7 RNA polymerase or the Y639F point mutant using the concentrations of NTP and 5′-*O*-(1-thio)nucleotide analog triphosphates shown in Table I.[21] The mutant polymerase facilitates the incorporation of analogs with functional group modifications in the minor groove.[10,21,25] Typically, transcription conditions are optimized to yield an analog incorporation level of about 5%.[26] This provides a signal that is easily detectable, but it is low enough to minimize cooperative interference due to multiple substitutions. To date, transcription conditions for 22 nucleotide analogs have been optimized (Table I), though it might be necessary to reoptimize the concentrations for specific RNA transcripts. We have found that doubling the concentrations in Table I does not necessarily result in an equivalent level of analog incorporation.

The reactions follow a standard procedure for *in vitro* transcription of RNA.[15] The DNA template is prepared by linearizing the plasmid with the appropriate restriction enzyme, the solution is extracted with an equal volume of phenol/chloroform (1 : 1) and the DNA is ethanol precipitated. The DNA pellet is resuspended in TE at a concentration of 1 μg/μl and used directly in transcription reactions. Alternatively, oligonucleotides can be used as DNA templates for transcription of smaller RNAs.[27] A typical reaction contains plasmid template DNA (0.05 μg/μl), transcription buffer, T7 RNA polymerase (5 units/μl), inorganic pyrophosphatase (0.001 units/μl), and the NTP concentrations shown in Table I. The reactions are incu-

[24] J.-T. Chen and S. J. Benkovic, *Nucleic Acids Res.* **11,** 3737 (1983).

[25] Y. Huang, A. Beaudry, J. McSwiggen, and R. Sousa, *Biochemistry* **36,** 13718 (1997).

[26] E. L. Christian and M. Yarus, *J. Mol. Biol.* **228,** 743 (1992).

[27] J. F. Milligan and O. C. Uhlenbeck, *Methods Enzymol.* **180,** 51 (1989).

TABLE I
CONDITIONS FOR INCORPORATING NUCLEOTIDE ANALOGS INTO *Tetrahymena* GROUP I INTRON[a]

Analog δTPαS (S_p isomer only)	[δTPαS] (m*M*)	[Parent NTP] (m*M*)	Polymerase (WT or Y639F)	Reference
AαS	0.05	1.0	WT	(26)
7dAαS	0.05	1.0	WT	(9)
m^6AαS	0.1	1.0	WT	(9)
DAPαS	0.025	1.0	WT	(5)
PurαS	2.0	1.0	WT	(9)
2APαS	0.5	1.0	WT	(9)
dAαS	0.8	1.0	Y639F	(9)
^{OMe}AαS[b]	2.0	0.2	Y639F	(9)
^{F}AαS	0.25	1.0	Y639F	(9)
GαS	0.05	1.0	WT	(26)
IαS	0.1	1.0	WT	(5)
m^2GαS	0.1	0.5	Y639F	(10)
dGαS	0.25	1.0	Y639F	(unpublished results)
7dGαS	0.05	1	WT	(13)
S^6GαS (4 m*M* Mn^{2+})	0.25	1.0	WT	(11)
CαS	0.05	1.0	WT	(26)
dCαS	0.75	1.0	Y639F	(unpublished results)
UαS	0.05	1.0	WT	(26)
dUαS	0.25	1.0	Y639F	(14)
^{F}UαS	0.1	1.0	Y639F	(14)
m^5UαS	0.5	1.0	WT	(unpublished results)
ψαS	0.05	1.0	WT	(unpublished results)

[a] At a level of approximately 5%.
[b] 1–2% incorporation of this analog.

bated for 2 hr at 37°. The RNA transcripts are purified on an 8 *M* urea denaturing polyacrylamide gel, visualized by UV shadowing, excised from the gel, and eluted into TE. The solubilized RNA is collected by filtration, precipitated with ethanol, and resuspended in 100–200 μl of TE. The RNA concentration is measured by UV absorbance at 260 nm.

Normalizing Incorporation Efficiency

The efficiency and distribution of analog incorporation is determined by comparing band intensities of the parental nucleotide and nucleotide analog cleavage products that result from treating ^{32}P-end labeled RNA transcripts with I_2 and resolving the cleavage products by gel electrophoresis. Ten picomoles of RNA are treated with 200 units of CIP for 1 hr at 37° to remove the 5′-terminal phosphate group. EDTA is added to a final concentration of 5 mM and heated to 80° for 15 min to inactivate the CIP. The RNA is 5′-end labeled by incubating it with 10 units of T4 polynucleotide kinase (PNK) and 30 μCi of [γ-^{32}P]ATP for 1 hr at 37°. The labeled RNA is purified by denaturing polyacrylamide gel electrophoresis, visualized by autoradiography, cut out of the gel, and eluted into 0.1% sodium dodecyl sulfate (SDS) in TE overnight. The RNA is then extracted with an equal volume of phenol/chloroform (1:1) to remove the SDS, precipitated with ethanol, and redissolved in TE. The ethanol precipitation step significantly improves the quality of resolution on the sequencing gel. The labeled RNA is mixed with two volumes of loading buffer and reacted with 1/10th volume I_2–ethanol.[17] The iodine solution selectively cleaves the RNA transcript at sites of phosphorothioate incorporation with an efficiency of about 10–15%. To enhance resolution, the reactions are heated to 90° before resolving the products on a denaturing polyacrylamide gel (Fig. 3). An equal amount of labeled RNA that has not been treated with iodine is loaded on the same gel to control for nonspecific RNA degradation (Fig. 3; lanes 6–10 and 16–20). RNAs containing the parental phosphorothioate nucleotide are also included to normalize for the efficiency and distribution of incorporation (Fig. 3; lanes 1 and 11). The gels are run for various lengths of time to maximally resolve different portions of the sequence, dried, and exposed to storage phosphor imaging plates (Kodak, Rochester, NY). The intensities of individual bands are quantitated by area integration using the software program ImageQuant version 4.1 (Molecular Dynamics).

RNA Selection Assay

NAIM requires an assay to distinguish active from inactive variants within a substituted RNA population. In general, two types of selection methods have been used. The first is to physically separate active RNAs by gel mobility shift, column chromatography, filter binding, or denaturing polyacrylamide gel electrophoresis. This approach has been used in NAIM experiments to study protein–RNA interactions, RNA–RNA interactions, RNA folding, and ribozyme cleavage activity.[4,7,8,11–13] The second approach is to use ribozyme-mediated ligation activity to selectively radiolabel active members in the population.[5,11]

Many of our NAIM experiments have utilized the 3′-exon ligation reaction performed by the L-21 G414 form of the *Tetrahymena* group I intron[18,19,28] (Fig. 4). This reaction is analogous to the reverse of the second

[28] R. Mei and D. Herschlag, *Biochemistry* **35,** 5796 (1996).

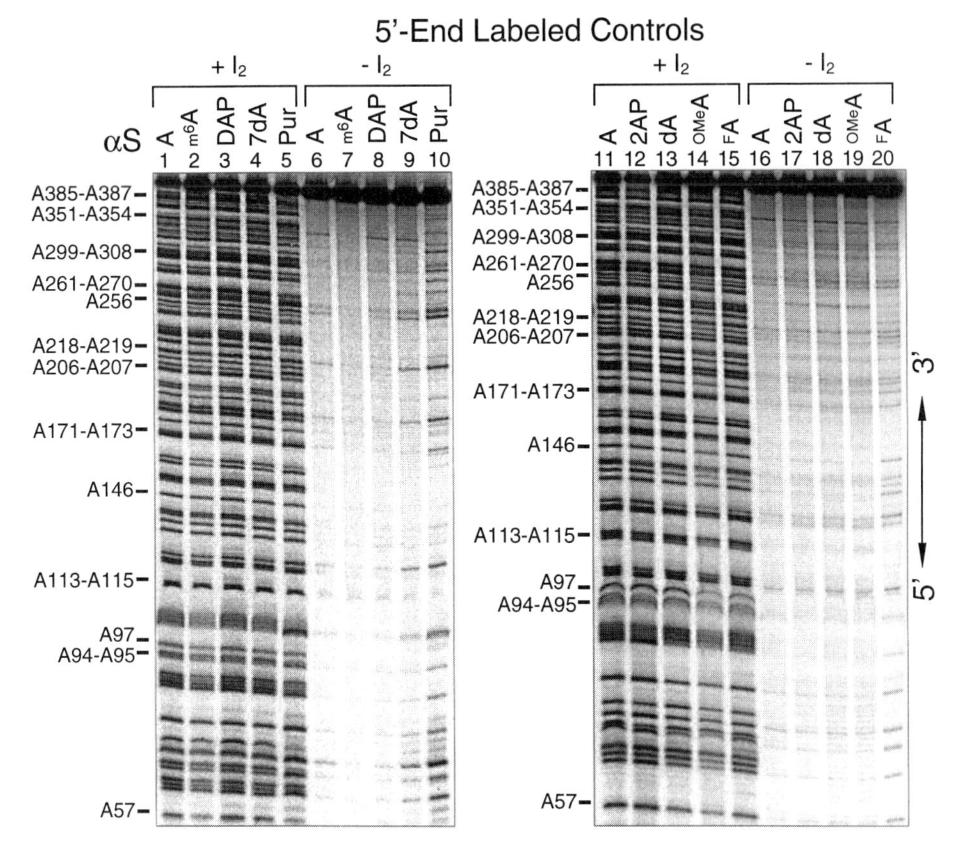

FIG. 3. Analog incorporation into the L-21 G414 group I intron.[9] The L-21 G414 RNAs containing several adenosine analogs were 5′-end labeled, gel purified, and treated with iodine to reveal the extent and positions of analog incorporation throughout the intron. The I_2-treated AαS standard is shown in lanes 1 and 11. The phosphorothioate-tagged analog incorporated into the other RNAs is listed above the lane numbers. The nucleotide numbers corresponding to several of the bands are marked to the left of each gel. The addition (lanes 1–5, 11–15) or omission (lanes 6–10, 16–20) of iodine is indicated. This particular denaturing 6% polyacrylamide gel was electrophoresed at 75 W for 1.25 hr. Longer electrophoretic times were used to improve the signal resolution of the nucleotides toward the 3′ end of the RNA (not shown).

step of intron splicing. The terminal guanosine (G414) attacks an oligonucleotide substrate comparable to 5′-3′ ligated exon and covalently transfers the 3′-exon onto the end of the intron. By 3′-end labeling the substrate oligonucleotide, the active variants in the intron population become selectively end labeled. The experimental details of the L-21 G414 selection assay are described next.

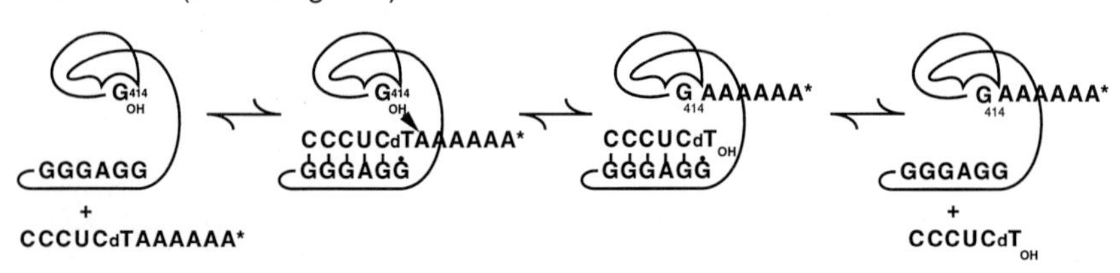

FIG. 4. The 3′-exon ligation reaction scheme for the L-21 G414 form of the group I intron.

Transcripts of L-21 G414 are incubated in reaction buffer at 50° for 10 min to prefold the RNA. An equal volume of the oligonucleotide substrate dT(-1)S [CCCUC(dT)AAAAA], radiolabeled at the 3′ end with [α-^{32}P] cordycepin, is dissolved in the same buffer and added to the ribozyme solution.[29] The reaction proceeds at 50° for 10 min, and is quenched by adding two volumes of loading buffer. The solution is split into two fractions. I_2–ethanol solution (1/10th volume) is added to one of the tubes to cleave the phosphorothioate linkages. The reactions are heated to 90° for 2 min and loaded onto a 5% denaturing polyacrylamide gel, and visualized as described for the 5′-end labeled control (Fig. 5).[9]

Data Quantitation

Quantitation of individual band intensities in the I_2 cleavage ladder is necessary to identify sites of interference due to analog substitution. Cleavage products from the 5′-end labeled unselected transcripts control for variability of analog incorporation at each position throughout the RNA (Fig. 3). These are compared to the band intensities in the selected RNA population to identify sites of phosphorothioate interference, as well as sites of nucleotide analog interference (Fig. 5). Individual band intensities for both the parental nucleotide (NαS) and the nucleotide analog ($\delta\alpha$S) are quantitated by area integration using ImageQuant version 4.1 software (Molecular Dynamics). These data are collected for both the selected and unselected RNA populations. The extent of interference at each position is calculated by substituting the individual band intensities into the equation:

$$\text{Interference} = (\text{N}\alpha\text{S selected})/(\delta\alpha\text{S selected}) \div (\text{N}\alpha\text{S unselected})/(\delta\alpha\text{S unselected})$$

[29] J. Lingner and W. Keller, *Nucleic Acids Res.* **21,** 2917 (1993).

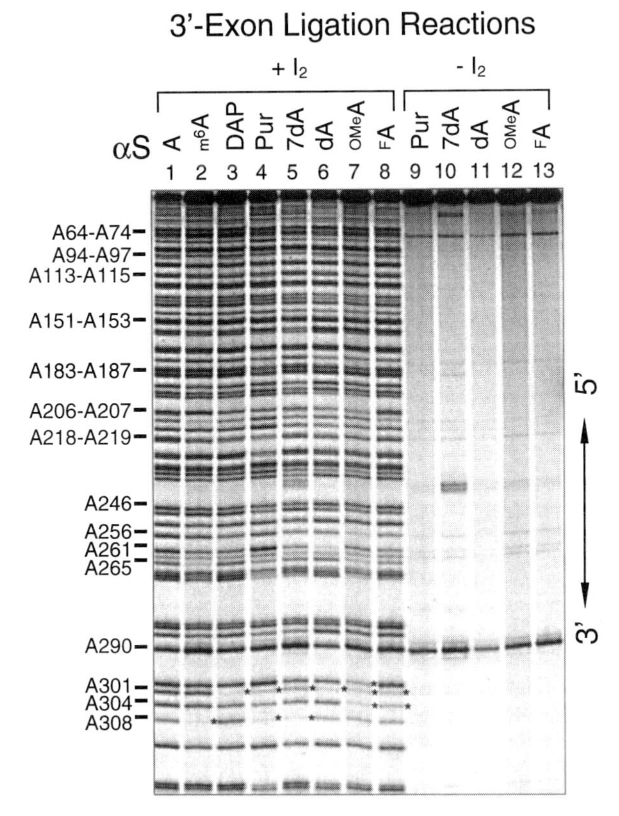

FIG. 5. The 3′-exon ligation reaction of L-21 G414 RNA with dT(-1)S.[9] The L-21 G414 RNA (50 nM) was preincubated in reaction buffer at 50° for 10 min. dT(-1)S (1 nM) radiolabeled at its 3′-end with poly(A) polymerase and [α-^{32}P]cordycepin[29] was dissolved in the same reaction buffer, added to the ribozyme solution, and the mixture then incubated at 50° for 10 min. The reaction was quenched by the addition of two volumes of loading buffer, and split into two tubes.[4] 1/10th volume of I_2–ethanol was added to one of the tubes to cleave the phosphorothioate bonds.[17] This autoradiogram reveals sites of interference throughout the intron. This denaturing 6% polyacrylamide gel was electrophoresed at 75 W for 2.25 hr. Figure legends are the same as those in Fig. 3. Lanes 1–8 are iodine cleavage reactions of analog substituted RNA pools that have reacted with dT(−1)S. Lanes 9–13 are the corresponding no I_2 controls. Some sites of interference are indicated with an asterisk. Two nucleotide positions, A302 and A306, were uninformative under these assay conditions due to a strong phosphorothioate effect at both positions. A290 is a site of significant degradation and yields no information in this assay.

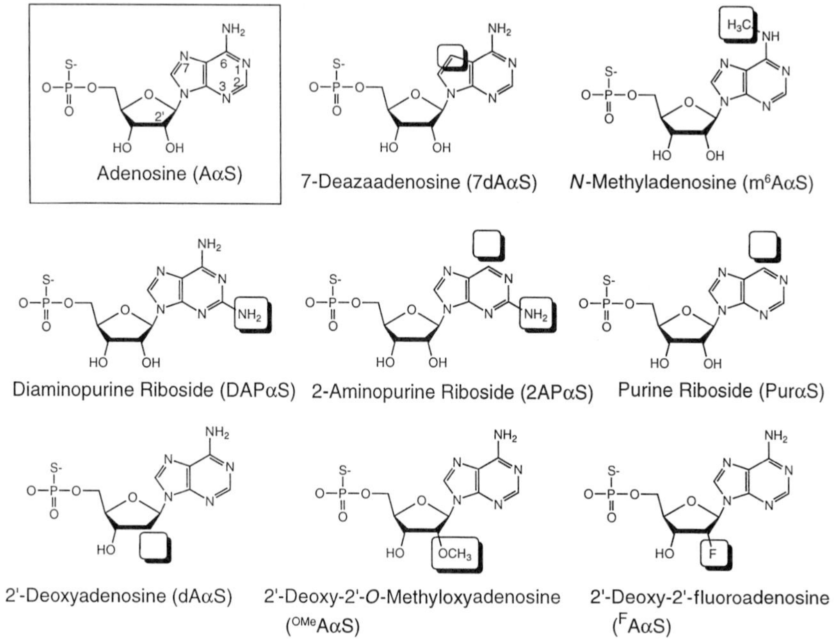

Adenosine (AαS)

7-Deazaadenosine (7dAαS)

N-Methyladenosine (m^6AαS)

Diaminopurine Riboside (DAPαS)

2-Aminopurine Riboside (2APαS)

Purine Riboside (PurαS)

2'-Deoxyadenosine (dAαS)

2'-Deoxy-2'-*O*-Methyloxyadenosine (^{OMe}AαS)

2'-Deoxy-2'-fluoroadenosine (^{F}AαS)

FIG. 6. Adenosine analogs. The chemical modification of each analog is highlighted with a box.

The interference value normalizes for any effect resulting from the phosphorothioate tag and for site-specific incorporation differences.[9] To normalize the data to a value of 1 for sites of noninterference, the average interference value for all positions that are within two standard deviations of the mean is calculated (typically between 0.8 and 1.2) and the interference value at each position divided by this average. This provides a value defined as κ that indicates the extent of interference at each site. A κ value of 1 indicates no interference, a value greater than 2 defines an interference site, and a value less than 0.5 defines a site of enhancement.

Adenosine Analogs

We and others have synthesized and tested a series of phosphorothioate-tagged nucleotide analog triphosphates for use in NAIM. Here we review the properties of these analogs in transcription reactions, and describe recurring patterns that have been observed using them in interference experiments.

Eight adenosine analogs have been utilized in NAIM (Fig. 6).[4,5,7,9,13]

Five analogs modify the nucleotide base and three modify the ribose sugar. The base analogs include purine riboside (PurαS), *N*-methyladenosine ($m^6A\alpha S$), tubercidin (7dAαS), diaminopurine riboside (DAPαS), and 2-aminopurine riboside (2APαS). The ribose sugar analogs all modify the 2′-OH group and include 2′-deoxyadenosine (dAαS), 2′-deoxy-2′-fluoroadenosine ($^{F}A\alpha S$), and 2′-*O*-methyladenosine ($^{OMe}A\alpha S$).

All of the analogs can be incorporated into an RNA transcript at approximately a 5% level of efficiency using either the wild-type T7 RNA polymerase or the Y639F mutant form of the polymerase.[9] $m^6A\alpha S$ and 7dAαS are both incorporated by the wild-type polymerase at relatively low concentrations of analog. DAPαS actually incorporates more efficiently than AαS. In contrast, PurαS incorporates poorly and only at low ATP and high PurαS concentrations. Furthermore, PurαS is incorporated rather unevenly at some, but not all, of the sites in the transcript containing two or more adjacent adenosine residues. This is also true of 2APαS, though it is not as pronounced as PurαS. dAαS and $^{F}A\alpha S$ are incorporated efficiently and uniformly by the mutant form of the polymerase. $^{OMe}A\alpha S$ can only be incorporated to about 1–2%, but the incorporation is fairly even.

Each of these analogs provides specific information about the chemical basis of RNA activity at almost every incorporated position in the transcript. PurαS, 2APαS, and $m^6A\alpha S$ measure the effect of modifications to the N-6 exocyclic amine of adenosine. PurαS and 2APαS delete the amine, and $m^6A\alpha S$ replaces one proton of the amine with a methyl group. $m^6A\alpha S$ interference indicates that either both hydrogen atoms of the amine are necessary, or that there is insufficient space in the local structure to accommodate the additional methyl group. PurαS and 2APαS interference identifies sites where the amine is important for activity.

7dAαS replaces the N-7 nitrogen with a C–H group. Interference with this analog indicates an important major groove contact to the ring nitrogen. Interference with PurαS, $m^6A\alpha S$, and 7dAαS is a strong indicator of Hoogsteen hydrogen bonding.

DAPαS and 2APαS both add an additional amine to the C-2 position of adenosine. In general, DAPαS and 2APαS show interference in areas of close packing in the minor groove of RNA.[9,30] Another characteristic effect observed with DAPαS is enhancement of activity when paired with a U in regions of the molecule where duplex stability is important for function.[5]

[30] S. A. Strobel, L. Ortoleva-Donnelly, S. P. Ryder, J. H. Cate, and E. Moncoeur, *Nature Struct. Biol.* **5,** 60 (1998).

dAαS interference identifies the 2′-OH groups important for RNA function, while ^{F}AαS delineates the role these 2′-OH groups play as either hydrogen bond donor or hydrogen bond acceptors.[9] If a 2′-OH shows interference with both analogs, it suggests that the 2′-OH is a hydrogen bond donor. If the position shows dAαS, but not ^{F}AαS interference, it argues that the 2′-OH at this site is a hydrogen bond acceptor.

In a few cases ^{F}AαS interference can also provide indirect information about the conformation of the ribose sugar for a given nucleotide.[9] ^{F}AαS interference at sites lacking dAαS interference suggests that the 2′-OH does not make a direct contribution to activity. Instead, it argues that there is an indirect effect due to the chemical nature of the fluorine substitution. The 2′-fluoro group is highly electronegative and, as such, the ribose sugar favors the C3′-endo conformation.[31] This indirect effect could result from an important C2′-endo ribose sugar conformation within the RNA. This interference pattern is observed twice within the P4–P6 domain of the group I ribozyme and both of these sites were refined as having a C2′-endo conformation within the P4–P6 crystal structure.[9,32]

Like DAPαS, ^{OMe}AαS serves as a probe for tight packing in the minor groove. Sites of ^{OMe}AαS interference that do not interfere with dAαS can be attributed to steric clash from the additional 2′-*O*-methyl group.

Guanosine Analogs

Six guanosine analogs have been utilized in NAIM[5,7,10,11,13,33] (Fig. 7). Four analogs modify the base functional groups of G, including inosine (IαS), N^2-methylguanosine (m^{2}GαS), 7-deazaguanosine (7dGαS), and 6-thioguanosine (S^{6}GαS). Two analogs, 2′-deoxyguanosine (dGαS) and 2′-*O*-methylguanosine (^{OMe}GαS), modify the 2′-OH of the ribose ring.

Transcriptional incorporation of these analogs is somewhat more complicated than the A-analog series. IαS and 7dGαS are both efficiently and evenly incorporated by the wild-type version of the polymerase,[5,13] but m^{2}GαS and dGαS are only efficiently incorporated by the Y639F mutant polymerase.[10,21] Both of these nucleotides modify functional groups in the minor groove, which is somehow tolerated by the deletion of the tyrosine hydroxyl group in the polymerase active site. S^{6}GαS is incorporated fairly well by the wild-type polymerase, but only if Mn^{2+} ion is included in the transcription reaction.[11] Because RNA containing S^{6}GαS is unusually labile,

[31] S. Uesugi, H. Miki, M. Ikehara, H. Iwahashi, and Y. Kyogoku, *Tetrahydron Lett.* **42,** 4073 (1979).

[32] J. H. Cate, A. R. Gooding, E. Podell, K. Zhou, B. L. Golden, C. E. Kundrot, T. R. Cech, and J. A. Doudna, *Science* **273,** 1678 (1996).

[33] W. D. Hardt, V. A. Erdmann, and R. L. Hartmann, *RNA* **2,** 1189 (1996).

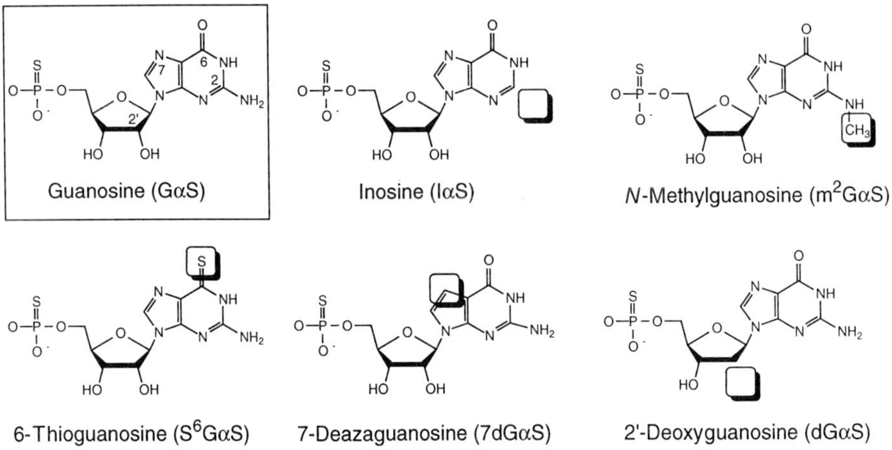

Guanosine (GαS) Inosine (IαS) *N*-Methylguanosine (m^2GαS)

6-Thioguanosine (S^6GαS) 7-Deazaguanosine (7dGαS) 2'-Deoxyguanosine (dGαS)

FIG. 7. Guanosine analogs. The chemical modification of each analog is highlighted with a box.

we found that in order to get reliable data with this analog the experiments had to be completed within 48 hr of transcription. ^{OMe}GαS incorporation has been reported using very high ratios of analog to GTP and the wild-type polymerase.[7] Unfortunately, we have not yet been able to achieve ^{OMe}GαS incorporation into the L-21 G414 RNA.

IαS and m^2GαS both modify the N-2 exocyclic amine of G. Interference from these analogs can be used to distinguish between tertiary effects within an RNA.[10] In duplex regions, IαS interference results from reduced duplex stability or loss of a tertiary contact involving the N-2 exocyclic amine.[5] By contrast, m^2GαS substitution is isoenergetic with G in the context of a G-C, a G · U, and a G · A base pair,[34] and m^2G interference has only been observed at sites of tertiary hydrogen bonding.[10]

7dGαS, like 7dAαS, replaces the N-7 nitrogen with a C–H group. Interference with this analog indicates an important major groove contact to the ring nitrogen.[13] S^6GαS also modifies a major groove functional group. It introduces an oxygen to sulfur substitution at the O-6 keto group of G. S^6GαS interference can be used to identify sites of RNA–RNA tertiary interaction, as well as sites of divalent or monovalent metal ion coordination within the RNA.[11]

dGαS, like dAαS, replaces the 2′-OH with a proton. Experiments with the 2′-fluoro derivative of G have not yet been performed, though they

[34] J. Rife, C. Cheng, P. B. Moore, and S. A. Strobel, *Nucleic Acids Res.*, **26,** 3640 (1998).

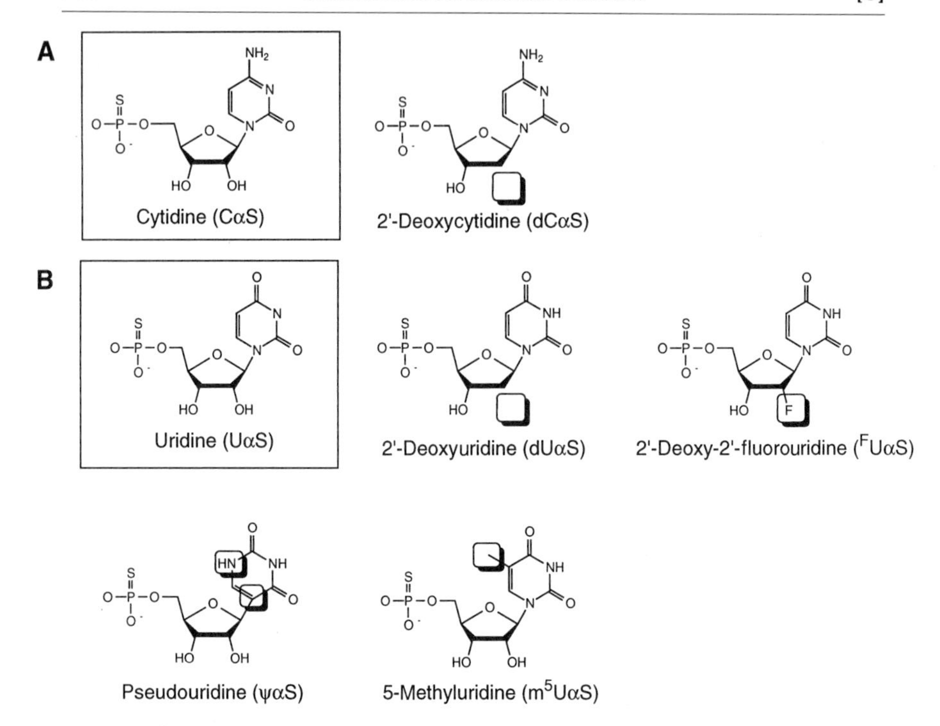

FIG. 8. (A) Cytosine analogs. (B) Uridine analogs. The chemical modification of each analog is highlighted with a box.

should provide the same kind of information obtained from the A analog series.

Pyrimidine Analogs

To date, seven pyrimidine analogs have been used in NAIM,[4,7,8,12,14] including two cytidine and five uridine analogs (Fig. 8). These include five analogs that modify the 2′-OH group of the ribose: 2′-deoxyuridine (dUαS), 2′-deoxy-2′-fluorouridine (^{F}UαS), 2′-*O*-methyluridine (^{OMe}UαS), 2′-deoxycytidine (dCαS), and 2′-*O*-methylcytidine (^{OMe}CαS). dUαS, ^{F}UαS, and dCαS are efficiently and evenly incorporated by the Y639F mutant polymerase. Incorporation experiments with the phosphorothioate-tagged 2′-*O*-methylpyrimidine analogs have only been done with the wild-type polymerase.[7] This series of analogs can be used to address the same types of questions as outlined for the A analogs.

Two additional uridine derivatives have been used in NAIM, 5-methyluridine (m^{5}UαS) and pseudouridine ($\psi\alpha$S) (L.O.D. and S.A.S., 1998, unpublished results). Both incorporate evenly and efficiently with the wild-type polymerase. $m^5U\alpha S$ introduces steric bulk in the major groove and probes for possible hydrophobic contacts to the base. $\psi\alpha S$ changes the 5 position of U to a nitrogen and is a C-linked nucleoside. Neither of these analogs show interference within the L-21 G414 ribozyme (S. Basu and S. A. Strobel, 1998, unpublished results).

Conclusion

NAIM is a generalizable chemogenetic approach that rapidly identifies functional groups that are important for the activity of an RNA. This method uses a series of 5′-*O*-(1-thio)nucleoside analogs to probe the contribution specific functional groups make to RNA activity. NAIM utilizes a pool of randomly substituted RNAs generated by *in vitro* transcription, and the active RNAs are identified through a selection experiment. In this way, a particular RNA can be screened with multiple analogs in a time-efficient manner. NAIM is applicable to any RNA with a known function. It can be used to study RNA catalysis through cleavage or ligation, RNA interactions with protein and other ligands, and the steps of RNA folding. This approach should make it possible to identify the chemical groups important for a wide variety of RNA and potentially DNA activities.

Acknowledgments

We thank R. Sousa for the gift of the Y639F polymerase, and L. Ortoleva-Donnelly, A. Oyelere, L. Beigelman, and J. Adamic for assistance with phosphorothioate synthesis. This work was supported by NSF grant CHE-9701787, a Beckman Young Investigator Award, and a Searle Foundation Scholarship.

[7] Joining of RNAs by Splinted Ligation

By MELISSA J. MOORE and CHARLES C. QUERY

Introduction

Recent improvements in techniques for the synthesis of site-specifically modified RNAs have opened numerous new experimental approaches for RNA structure/function studies (for example, see other chapters in this volume). For example, it is now possible to determine what proteins or other RNAs interact with a single internal nucleotide of interest by purposely incoporating a photo-cross-linkable group and associated radio-

0076-6879/00 $30.00

active tag at that position. The exact role of each atom or functional group on an important nucleotide can also be determined by selectively replacing each in turn, be it on the base, sugar, or phosphate. Similarly, the functional role(s) of naturally modified bases in a purified cellular RNA may be determined by cutting regions out of the natural RNA and pasting them into unmodified *in vitro* transcripts for subsequent functional assays. Introduction of heavy atoms or paramagnetic isotopes at selected positions can significantly facilitate solution of X-ray and nuclear magnetic resonance (NMR) structures, and the folding pathway of an RNA can be probed by fluorescence energy transfer between site-specifically positioned fluorescence donor and acceptor groups.

Most of the above experiments have been made possible by advances in two areas: (1) chemical synthesis of RNA and (2) RNA joining reactions. Chemical synthesis, while still expensive for RNA, is now almost as efficient as DNA synthesis and a wide range of modified precursors is commercially available. Thus, large quantities of RNAs even as long as 50 nucleotides containing single or multiple modifications can now be made synthetically. Synthesis of longer RNAs or incorporation of internal radioactive tags, however, requires the ability to piece shorter fragments together. Such RNA ligation reactions are the main subject of this article.

Methods for Joining RNA

Chemical Ligation

Both RNA and DNA fragments can be joined chemically with either cyanogen bromide or water-soluble carbodiimides.[1,2] The segments to be joined are usually hybridized to a complementary template to create a nicked duplex in which one substrate supplies the terminal monophosphate (e.g., a 5′-phosphate) and the other a nucleophilic oxygen (e.g., a 3′-hydroxyl). Unfortunately, however, chemical coupling reactions can be quite slow (0.2–6 days) and are prone to side reactions between base functional groups in single-stranded regions and the coupling reagents. For these reasons, chemical ligation is most useful for creating unusual internucleotide linkages that cannot be made enzymatically by T4 RNA or DNA ligase (see later discussion). Examples of such unnatural linkages include 2′-5′ phosphodiesters, 5′-N-P or 3′-N-P phosphoramides, and tri-substituted pyrophosphates. The latter in particular are very useful for

[1] Z. A. Shabarova, *Biochimie* **70,** 1323 (1988).

[2] N. G. Dolinnaya, N. I. Sokolova, D. T. Ashirbekova, and Z. A. Shabarova, *Nucleic Acids Res.* **19,** 3067 (1991).

probing protein–nucleic acid interactions as they are relatively stable to hydrolysis, but do react readily with primary amines (such as lysine side chains in proteins) to form stable phosphoramide derivatives.[3,4]

T4 RNA Ligase

T4 RNA ligase catalyzes the ATP-dependent joining of a 3′ "donor" RNA having a 5′-terminal monophosphate and a 5′ "acceptor" RNA terminating in a 3′-hydroxyl. Both substrates must be single stranded around the ligation joint and the product is a standard 3′-5′ phosphodiester linkage.[5,6] This enzyme is well suited for many RNA joining applications, including ligation of very short oligomers,[7,8] construction of tRNAs with altered anticodon loops,[9,10] and 3′-end labeling of RNA with [5′-^{32}P]pCp.[6,11,12] However, several features of T4 RNA ligase limit its usefulness for reactions containing multiple or long RNAs. First, because the enzyme strongly prefers single-stranded substrates, ligation reactions using T4 RNA ligase cannot be readily templated. This means that side products such as circles and homodimers can predominate unless the 3′-OH of the donor RNA is protected or destroyed.[5,6] Second, in cases where the acceptor RNA is contaminated with $N + 1$ transcripts, extra unwanted nucleotides will be inadvertantly incorporated at the ligation junction.[13] Third, ligation yields can be quite dependent on the sequence at the 3′ end of the acceptor (uridine at either the penultimate or 3′-terminal position is especially bad) and, to some extent also on the donor sequence.[7,14] Fourth, at the concentrations achievable for long RNAs (often submicromolar), the high K_m of the

[3] N. A. Naryshkin, M. A. Farrow, M. G. Ivanovskaya, T. S. Orestkaya, Z. A. Shabarova, and M. J. Gait, *Biochemistry* **36,** 3496 (1997).

[4] N. A. Naryshkin, M. G. Ivanovskaya, T. S. Orestkaya, E. M. Volkov, M. J. Gait, and Z. A. Shabarova, *Bioorg. Khim.* **22,** 592 (1996).

[5] O. C. Uhlenbeck and R. I. Gumport, *in* " The Enzymes" (P. D. Boyer, ed.), p. 31. Academic Press, New York, 1982.

[6] P. J. Romaniuk and O. C. Uhlenbeck, *Methods Enzymol.* **100,** 52 (1983).

[7] T. E. England and O. C. Uhlenbeck, *Biochemistry* **17,** 2069 (1978).

[8] M. Krug, P. L. D. Haseth, and O. C. Uhlenbeck, *Biochemistry* **21,** 4713 (1982).

[9] L. Bare, A. G. Bruce, R. Gestland, and O. C. Uhlenbeck, *Nature (Lond.)* **305,** 554 (1983).

[10] W. L. Wittenberg and O. C. Uhlenbeck, *Biochemistry* **24,** 2705 (1985).

[11] J. R. Barrio, M. D. C. G. Barrio, N. J. Leonard, T. E. England, and O. C. Uhlenbeck, *Biochemistry* **17,** 2077 (1978).

[12] C. Enright and B. Sollner-Webb, *in* "RNA Processing, Vol. II: A Practical Approach" (S. J. Higgins and B. D. Hames, eds.), p. 135. IRL Press, Oxford, 1994.

[13] M. J. Moore and P. A. Sharp, *Science* **256,** 992 (1992).

[14] E. Romaniuk, L. W. McLaughlin, T. Neilson, and P. J. Romaniuk, *Eur. J. Biochem.* **125,** 639 (1982).

enzyme for polynucleotides (>1 mM)[5,15] means the reaction can proceed quite slowly.

T4 DNA Ligase

Like T4 RNA ligase, T4 DNA ligase catalyzes the ATP-dependent joining of a 5′-acceptor substrate to a 3′-phosphate donor substrate to form a natural 3′-5′ phosphodiester linkage between them.[16,17] A crucial difference, however, is that T4 DNA ligase is specific for double-stranded substrates—once known as T4 polynucleotide ligase, this enzyme readily joins nicks in double-stranded DNA as well as a number of hybrid duplexes.[18,19] A major advantage is that the ligation reactions can be targeted to form specific products by including bridging deoxyoligonucleotide templates to create nicked (RNA:RNA)/DNA duplexes at only the desired sites of ligation.[13,20] Therefore, circularization or oligomerization of the phosphate donor substrate is not particularly problematic, and three- and even four-way ligations are quite doable.[20,21] Another advantage is that the activity of T4 DNA ligase is not highly dependent on the exact sequences at the ligation junction—it requires only that the nucleotides on either side of the nick be base paired. Because the enzyme does not ligate nicks containing extra, nonpaired nucleotides, $N + 1$ acceptors are effectively excluded from the ligation products.[13] Lastly, because this enzyme has a K_m for polynucleotide duplexes in the submicromolar range,[16,17] efficient ligations can be performed even with pico- or femtomolar amounts of substrates. These quantities are easily attainable from small-scale transcription reactions of even quite long RNAs and are ideal for the preparation of RNAs containing a single, high specific activity ^{32}P label.

Procedures for Splinted RNA Ligation with T4 DNA Ligase

For the reasons just outlined, T4 DNA ligase is often the method of choice for joining two or more RNA molecules. The following sections contain specific procedures for carrying out such ligation reactions. Proce-

[15] O. C. Uhlenbeck and V. Cameron, *Nucleic Acids Res.* **4,** 85 (1977).

[16] N. P. Higgins and N. C. Cozzarelli, *Methods Enzymol.* **68,** 50 (1979).

[17] M. J. Engler and C. C. Richardson, *in* "The Enzymes" (P. D. Boyer, ed.), p. 3. Academic Press, New York, 1982.

[18] K. Kleppe, J. H. v. D. Sande, and H. G. Khorana, *Proc. Nat. Acad. Sci. U.S.A.* **67,** 68 (1970).

[19] G. C. Fareed, E. M. Wilt, and C. C. Richardson, *J. Biol. Chem.* **246,** 925 (1971).

[20] M. J. Moore and C. C. Query, *in* "RNA-Protein Interactions: A Practical Approach" (C. W. J. Smith, ed.), p. 75. IRL Press, Oxford, 1998.

[21] M. J. Moore and P. A. Sharp, *Nature (Lond.)* **365,** 364 (1993).

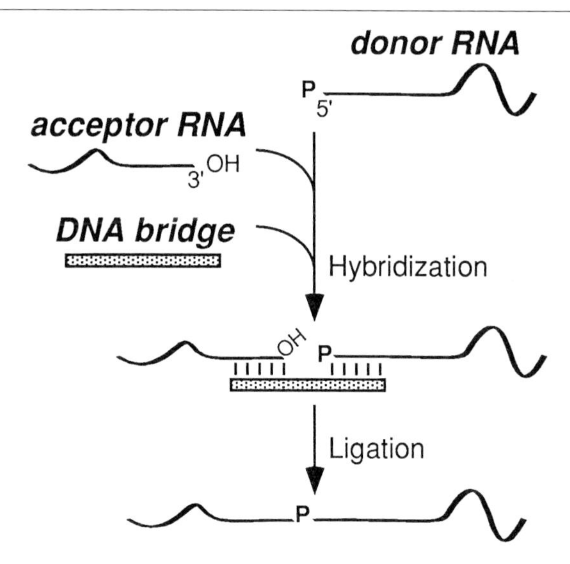

FIG. 1. Generic scheme for templated RNA–RNA ligations. Two RNAs—the donor having a 5′-monophosphate and the acceptor with a 3′-OH—are hybridized to a complementary DNA bridge or template. The two RNAs are then joined using T4 DNA Ligase. See procedure 1 for detailed description.

dure 1 gives generic conditions for RNA–RNA ligations using T4 DNA ligase. The two basic steps are hybridization of the RNA substrates to a bridging cDNA oligonucleotide template and then incubation with T4 DNA ligase (Fig. 1). This protocol is readily adaptable and can be used to make pico- to nanomolar amounts of linear ligated products.

Occasionally it may be of interest to generate circular or linear dimer RNAs (Fig. 2). Circular RNAs can be used to determine whether the end(s) of an RNA are required for biologic function. For example, ultimate proof that ribosomes can initiate translation from IRES elements without involvement of a cap structure came from translation of a circular mRNA.[22] Procedure 2 gives a detailed protocol for synthesis and confirmation of circular RNAs. The key to generating circles is to carry out the hybridization at low RNA and bridging cDNA oligonucleotide concentrations to favor formation of circular monomers over linear dimers.

Procedures 3 and 4 are for creating RNAs that contain a single internal radioactive label of high specific activity. Such RNAs are extremely useful for short-wave UV cross-linking studies to identify proteins that interact with the position labeled (Ref. 20 and references therein). The main differ-

[22] C. Y. Chen and P. Sarnow, *Science* **268,** 415 (1995).

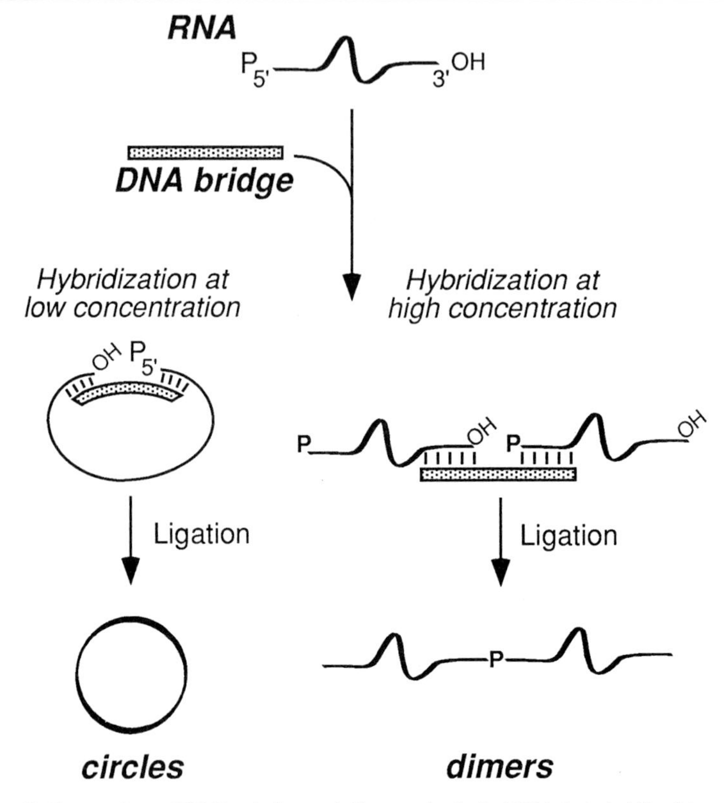

FIG. 2. Generation of RNA circles and dimers. A single RNA is hybridized to a DNA bridge complementary to both 5′ and 3′ ends of the RNA. Low concentration (*left*) favors hybridization of only one RNA molecule to the bridge to produce circles, whereas higher concentrations (*right*) allow two or more RNAs to hybridize resulting in dimers (and higher order multimers). See procedure 2 for a detailed protocol.

ences between these protocols are the amounts of radioactivity used and whether the ligase reaction immediately follows the phosphorylation reaction without changing buffers and tubes. Procedure 3 is designed to incorporate as much of the label as possible. To do so, the phosphorylation reaction is performed with the donor RNA in about fourfold molar excess over the labeled ATP. Thus, while the bulk of the label becomes incorporated, the majority of the donor RNA remains unlabeled. Because it lacks a 5′-phosphate, however, this unlabeled RNA will not participate in or interfere with the subsequent ligation. On the other hand, in some situations it is better to maximize the molar quantity of donor RNA that becomes phosphorylated. This can easily be accomplished by just reversing the ratios of adenosine triphosphate (ATP) and donor RNA; i.e., [ATP] $\geq 4 \times$ [donor RNA]. Procedure 4 is one such protocol. Notably, it uses a different source of ATP that is less pure, but is both more concentrated and less expensive than that used in Procedure 3. For both protocols it is assumed that the donor initially has a free 5′-OH to be phosphorylated. If this RNA is made by transcription, a terminal 5′-OH can be generated either by initiating the transcript with an X(3′)p(5′)G dinucleotide or by treatment with phosphatase to remove any terminal phosphates (see Fig. 3 and later discussion). With the latter, it is key that the phosphatase be completely inactivated or otherwise removed prior to attempting the kinase reaction. Because both Procedures 3 and 4 use high quantities of radioactivity, it is crucial to take extra precautions not to contaminate work areas and microcentrifuges.

Reagents

General Reagents for Splinted Ligation (Procedure 1)

RNAs to be ligated (in water)
cDNA oligonucleotide template (in water or TE)
10× Ligase buffer [e.g., 500 m*M* Tris-HCl (pH 7.5), 100 m*M* $MgCl_2$, 200 m*M* dithiothreitol (DTT), 0.5 mg/ml bovine serum albumin (BSA) (per New England Biolabs, Beverley, MA); or 660 m*M* Tris-HCl (pH 7.6), 66 m*M* $MgCl_2$, 100 m*M* DTT (per USB, Cleveland, OH/Amersham, Piscataway, NJ)]
RNase-free water [either diethyl pyrocarbonate (DEPC)-treated or, preferably, from a Milli-Q (Millipore, Bedford, MA) or similar deionization system]
10 m*M* ATP
T4 polynucleotide kinase
T4 DNA ligase, ≥10 Weiss (USB) units/μl (multiple sources; note that some suppliers use a different unit definition)
RNasin (Promega, Madison, WI)
RQ1 RNase-free DNase (Promega, Madison, WI)

Reagents Specific to Procedure 2

10× Hybridization buffer [100 m*M* Tris-HCl (pH 7.5), 1 *M* NaCl, 1 m*M* EDTA]
250 m*M* $NaHCO_3$ (pH 9.0)

Reagents Specific to Procedure 3

[γ-^{32}P]ATP (10 mCi/ml, 3000 Ci/mmol = 3.33 μM; multiple suppliers)
RNAs to be ligated, 20 μM stocks each (in water)
cDNA template, 20 μM stock (in water or TE)

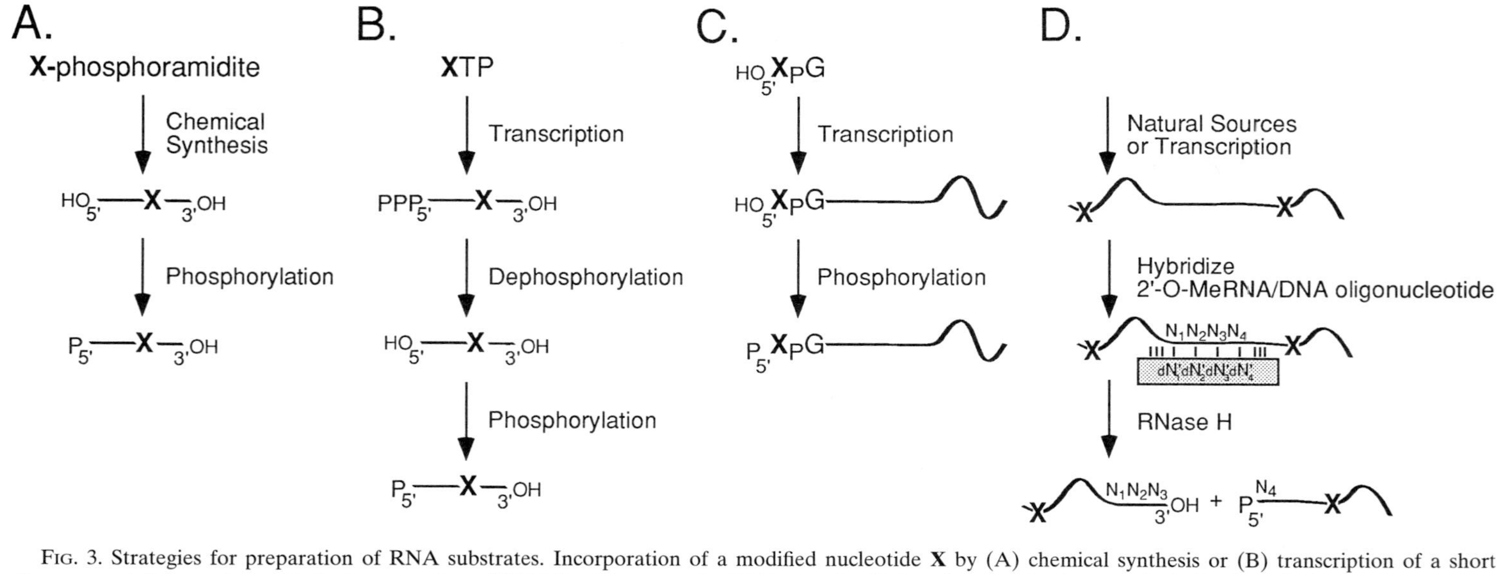

FIG. 3. Strategies for preparation of RNA substrates. Incorporation of a modified nucleotide **X** by (A) chemical synthesis or (B) transcription of a short oligomer containing a single site for incorporation of **XTP.** In either case, the RNA must be 5′ phosphorylated prior to its inclusion in an RNA ligation reaction, if it is to be the donor RNA. (C) A modified nucleotide may also be included by initially incorporating **X** into a dinucleotide, which is then used to prime transcription of the donor RNA. The donor is subsequently phosphorylated and joined to an upstream acceptor. (D) A naturally occurring RNA containing modifications may be cleaved at particular locations using RNase H and a chimeric 2′-*O*-methyl-RNA/DNA oligonucleotide complementary to the site of desired cleavage. The chimeric oligonucleotide contains four deoxy positions (dN'_{1-4}) that are flanked by 2′-*O*-methyl residues. Cleavage by RNase H occurs between complementary positions 3 and 4 the target RNA.[25–28]

Reagents Specific to Procedure 4

- [γ-^{32}P]ATP (150 mCi/ml, 6000 Ci/mmol = 25 μM; DuPont-NEN, Boston, MA)
- RNAs to be joined (50 μM each in water)
- DNA template (50 μM in water)
- Polyvinylpyrrolidone-40 (PVP-40) (10% in water, w/v) (optional)
- Phenol : chloroform (phenol : chloroform : isoamyl alcohol 25 : 24 : 1, v/v/v)
- Chloroform (chloroform : isoamyl alcohol 24 : 1, v/v)
- Ethanol (absolute)
- 0.2 *M* sodium acetate, 1 *M* $MgCl_2$

Preparation of RNA Substrates

The RNA substrates to be used for ligation by T4 DNA ligase can be prepared in any number of ways (Fig. 3), as long as the acceptor substrate terminates in a free 3′-OH and the donor substrate begins with a 5′ phosphate. If the donor RNA is synthesized chemically, a 5′ monophosphate can be either incorporated during the synthesis or added later using hot or cold ATP and T4 polynucleotide kinase (Fig. 3A). If the donor RNA is transcribed *in vitro*,[20,23,24] it can be dephosphorylated and then kinased (Fig. 3B), initiated with a 5′-XpG-3′ dinucleotide primer and then kinased (Fig. 3C), or simply initiated with GMP. If one or more *in vivo* modified bases are required for biologic activity of the ligated product, RNAs purified from natural sources can also be used as ligation substrates. In this case, individual fragments of the natural RNA can be prepared by cleaving it at particular locations with complementary 2′-*O*-methyl-RNA/DNA chimeric oligonucleotides and RNase H (Fig. 3D).[25–28]

When using RNAs prepared by runoff transcription, it is also important to consider the following. First, such transcriptions yield significant amounts of short, abortive initiation products in addition to full-length RNA.[29,30]

[23] J. K. Yisraeli and D. A. Melton, *Methods Enzymol.* **180,** 42 (1989).

[24] B. Chabot, *in* "RNA Processing, Vol. I: A Practical Approach" (S. J. Higgins and B. D. Hames, eds.), pp. 1–30. IRL Press, Oxford, 1994.

[25] H. Inoue, Y. Hayase, S. Iwai, and E. Ohtsuka, *FEBS Lett.* **215,** 327 (1987).

[26] J. Lapham and D. M. Crothers, *RNA* **2,** 289 (1996).

[27] Y.-T. Yu and J. A. Steitz, *RNA* **3,** 807 (1997).

[28] Y.-T. Yu, M.-D. Shu, and J. A. Steitz, *EMBO J.* **17,** 5783 (1998).

[29] J. F. Milligan, D. R. Groebe, G. W. Witherell, and O. C. Uhlenbeck, *Nucleic Acids Res.* **15,** 8783 (1987).

[30] M. L. Ling, S. S. Risman, J. F. Klement, N. McGraw, and W. T. McAllister, *Nucleic Acids Res.* **17,** 1605 (1989).

Because these aborted RNAs have the same 5′-end sequence as the full-length donor, if they are not removed prior to the hybridization step, they can compete with the donor for the cDNA template and thereby significantly decrease the yield of the desired product. Since standard desalting columns (e.g., Sephadex G-50) often fail to efficiently eliminate such abortive transcripts, it is advisable to gel purify donor transcripts prior to ligation. On the other hand, if the acceptor substrate is a runoff transcript, it need only be gel filtered prior to use—any abortive initiation products would be too short to interfere with the ligation (unless, of course, the sequences at the 5′ ends of both substrates are identical or the acceptor is very short). In our hands, using a desalted acceptor and gel purified donor often results in much better ligation yields than when both are gel purified.

Many runoff transcripts also have significant heterogeneity at their 3′ ends due to the production of so-called $N + 1$ products.[23,31] In most cases, little can be done about this (with long transcripts it is all but impossible to purify the desired RNA away from the $N + 1$ RNAs), although different 3′ ends seem to elicit more or less heterogeneity. Luckily, T4 DNA ligase will only ligate the N length acceptors.[13] However, a significant amount of $N + 1$ transcript can inhibit ligation efficiency by hybridizing to the DNA template and effectively sequestering some of the donor substrate in unreactive hybrids. Although this effect can never be completely overcome, one remedy is to provide the acceptor in excess and perform two or more cycles of hybridization and ligation. Another option is to transcribe the acceptor as a longer precursor and then cleave it to the desired length with RNase H and a complementary 2′-*O*-methyl-RNA/DNA oligonucleotide.[25,26] Of course, the best way to produce homogeneous RNA is by chemical synthesis; these RNAs give near quantitative yields in RNA ligations.

Procedure 1: Generic Conditions for Joining RNAs with T4 DNA Ligase

Hybridization. In an Eppendorf tube, mix the RNAs, cDNA template, water, and the amount of 10× ligase buffer dictated by the desired volume of the reaction in the ligation step. Hybridize the RNA and DNA pieces by heating at 75° for 2 min, followed by room temperature for 5 min. Centrifuge briefly to return any condensate to the bottom of the tube.

Ligation. To the cooled tube, add ATP to a final concentration between

[31] J. F. Milligan and O. C. Uhlenbeck, *Methods Enzymol.* **180,** 51 (1989).

0.2 and 1 mM, T4 DNA ligase (see comment later about enzyme concentration), and RNasin (if desired). After all components have been added, the reaction should contain 1× ligase buffer. Incubate at 30° for 2–4 hr, then add at least 1 unit of RNase-free DNase for every 1 μg of cDNA template and incubate at 37° for 15–30 min.

Product Purification. Purify the ligated product by denaturing polyacrylamide gel electrophoresis (PAGE). If the reaction volume is small (<20 μl), an equal volume of 2× gel loading buffer can be added and the sample loaded directly. For larger volumes it is advisable to phenol : $CHCl_3$ extract and ethanol precipitate the sample first, because such large reactions contain more protein, which can lead to smearing of the RNA in the gel.

Procedure 2: Generation of Circular RNAs

Hybridization (for 120-μl Reaction, Using 60 pmol 5'-Phosphorylated RNA). To 102 μl RNase-free water, add 12 μl 10× hybridization buffer, 3.0 μl 20 μM RNA and 3.0 μl 20 μM cDNA bridge spanning the ends of the RNA (total volume 120 μl). Hybridize by heating at 75° for 2 min, followed by slow cooling at room temperature. Add 12 μl 3 M sodium acetate, 1.2 μl 1 M $MgCl_2$, and 3 volumes cold ethanol to precipitate. Incubate on dry ice for 10 min, or at −20° for at least 3 hr. Collect the precipitated RNA · DNA hybrid by microcentrifugation at top speed for 30 min at 4°. Remove supernatant, rinse with 0.5 ml 70% ethanol and recentrifuge as above for 5–10 min. Again, remove supernatant and dry briefly in a Speed-Vac.

Ligation. To the dry RNA · DNA hybrid, add 10.5 μl water, 1.5 μl 10× ligation buffer, 1.0 μl 10 mM cold ATP, and 2.0 μl T4 DNA ligase (total volume 15 μl). Incubate at 30° for 2–4 hr. Add 1 μl of RQ1 RNase-free DNase and incubate at 37° for 15–30 min.

Product Purification. Add 16 μl 2× gel loading buffer and load into a large well on a denaturing polyacrylamide gel. Labeled linear and dimer RNA should be run as markers. Circular RNAs can be identified by running the ligation reaction on both high and low percent polyacrylamide gels; in high percent gels, the circular RNA will migrate aberrantly relative to size in comparison to linear RNA.

Confirmation of Circular Topology. Subject a sample of the circular RNA to limited alkaline cleavage by incubation in 250 mM $NaHCO_3$ (pH 9.0) at 90° for 3 min, and subsequent analysis on a high percentage denaturing polyacrylamide gel. If the RNA is circular, it should be first converted to a form that comigrates with linear RNA followed by a hydrolysis ladder migrating below the linear RNA.

Procedure 3: Radioactive Labeling and RNA Ligation in One Tube

Phosphorylation of Donor RNA (for 5-μl Reaction, Using 60 pmol RNA). In a Speed-Vac, dry down 5 μl of [γ-^{32}P]ATP (50 μCi, 16 pmol) in a 0.5- or 1.5-ml tube. This is easily done in a closed microcentrifuge tube with a hole punched in the top using a 25-gauge needle to minimize the chance of accidental radioactive contamination. To the dried ATP, add in the following order: 0.5 μl 10× ligation buffer, 3.0 μl 20 μ*M* donor RNA, 1.0 μl water, and 0.5 μl T4 polynucleotide kinase (final volume 5.0 μl). Incubate at 37° for 60 min. Inactivate the kinase by incubating at 75° for 10 min or 92–95° for 2 min. Store on ice.

Hybridization and Ligation of Donor and Acceptor RNAs (for 15-μl Reaction, Using 60 pmol RNA). To the inactivated kinase reaction, add 1.0 μl 10× ligation buffer, 3.0 μl 20 μ*M* acceptor RNA, and 3.0 μl 20 μ*M* cDNA bridge. Hybridize by heating at 75° for 2 min, followed by 5 min at room temperature. To the hybridized reaction, add 1.0 μl 10 m*M* cold ATP and 2.0 μl T4 DNA ligase (total volume 15 μl). Incubate at 30° for 2–4 hr, then add 1 μl of RQ1 RNase-free DNase and incubate at 37° for 15–30 min.

Product Purification. Add 16 μl 2× gel loading buffer and load into a large well on a denaturing polyacrylamide gel. Labeled acceptor and donor RNAs and full-length RNA that correspond to the ligated product should be run as markers.

Procedure 4: Radioactive Labeling and RNA Ligation in Different Tubes

Phosphorylation of Donor RNA (for 5-μl Reaction, Using 50 pmol RNA). Dry down 10 μl of [γ-^{32}P]ATP as in procedure 3. To this tube add 0.5 μl 10× buffer, 1 μl 50 μ*M* donor RNA, 0.5 μl RNasin, 0.5 μl T4 polynucleotide kinase, and 2.5 μl water (total volume 5.0 μl). Incubate at 37° for 15 min. Add 50 μl 0.3 *M* sodium acetate, then phenol : chloroform and chloroform extract. Add 0.5 μl 1 *M* $MgCl_2$ and 3 volumes cold ethanol to precipitate. Incubate on dry ice for 10 min, or at −20° for at least 3 hr. Collect the precipitated RNA by microcentrifugation at top speed for 30 min at 4°. Carefully remove the supernatant containing free [γ-^{32}P]ATP. Rinse with 0.5 ml 70% ethanol and recentrifuge for 5–10 min. Again, remove the supernatant, and dry briefly in a Speed-Vac.

Hybridization and Ligation (for 10 μl, 50 pmol). To the dry, labeled donor RNA add 7.0 μl water, 1.0 μl 10× buffer, 1.0 μl 50 μ*M* acceptor RNA (50 pmol), and 1.0 μl 50 μ*M* cDNA template (50 pmol) (total volume 10 μl). Hybridize by incubating at 65° for 5 min and then slowly cooling to room temperature (about 5–10 min). Add 10.0 μl 2× ligation mix, comprised of premixed 2.0 μl water, 1.0 μl 10× buffer, 2.0 μl 10 m*M* cold

ATP, 0.5 μl RNasin, 1.5 μl PVP-40 (optional; PVP is typically used to increase the effective concentration of reactants and may increase the efficiency of some ligations), and 3.0 μl T4 DNA ligase (total volume 20.0 μl). Incubate at 30° for 2–4 hr, then add 1 μl of RQ1 RNase-free DNase and incubate at 37° for 15–30 min.

Product Purification. Purify ligated product as in procedure 3.

General Comments

Enzyme Concentration

Efficient ligation of RNA substrates requires stoichiometric or greater concentrations of T4 DNA ligase because this enzyme cannot turnover effectively on RNA-containing duplexes.[32] Therefore, only concentrated enzyme preparations with high specific activities can be used for RNA ligations. As a general rule, one Weiss unit of activity corresponds to 1 pmol of enzyme, so use at least one Weiss unit of enzyme for every picomole of phosphorylated donor substrate to be ligated. If the amount of ligase required becomes cost inhibitory, a $(His)_6$-tagged version of the enzyme is available.[33]

RNA Substrate and Bridging cDNA Template Concentrations

The final concentrations of RNAs and cDNA template generally used are between 0.1 and 5 μ*M*. If one or more of the stocks is too dilute, that component can be added to the reaction tube first, dried down, and then redissolved in a mixture of 10× ligase buffer and water. It is imperative, however, that the final concentration of the bridging cDNA template be equal to or intermediate between those of the RNAs to be ligated. If excess template is present, then individual acceptor and donor RNA molecules will hydridize to different template molecules and therefore have no ligation partner. Thus it is counterproductive to try to "drive" the hybridization by adding excess cDNA template. However, one can use higher concentrations of one of the substrates to increase the possibility that the limiting substrate "sees" a potential ligation partner hybridized to the same cDNA template. In this case, the cDNA concentration would be somewhere between that of the two substrates.

[32] M. Suntharalingam, E. Dulude, and M. J. Moore, unpublished data.

[33] S. A. Strobel and T. R. Cech, *Science* **267,** 675 (1995).

Substrate and Template Lengths

There is apparently no upper limit to the lengths of RNA substrates and cDNA templates for exon ligation. At the other extreme, RNA substrates as short as 6 nucleotides can be ligated quantitatively in the standard incubation time of 2 hr, whereas longer incubation times are required for 4- and 5-nucleotide substrates. We have not been able to detect ligation products for substrates of 3 nucleotides or fewer.[32] Lower ligation temperatures (e.g., room temperature or 16°) also aid in the joining of short RNAs. Generally, a cDNA template that extends 8–10 nucleotides on either side of the ligation junction is sufficient for quantitative ligations, but ligation efficiencies fall off dramatically with shorter templates.[32] However, if other factors decrease the stability of the duplex (e.g., a high percentage of U's in the duplex region or significant RNA secondary structure in the area), longer cDNA templates are advised. Other tactics for dealing with poor hybridization efficiencies because of internal RNA structure are to include a second "disrupter" oligonucleotide to break up tertiary interactions[34] or, if possible, change the sequence of the RNA at or near the ligation junction.[20]

Multipart Ligations

Some synthesis projects require three-way (or even four-way!) ligations.[20,21] If the middle RNA is short, then a single cDNA template spanning all three RNA pieces can be used. Alternatively, individual ligation junctions can be spanned by separate cDNA oligonucleotides. Such multiple ligations can often be carried out simultaneously by simply including all of the substrates and templates in the same reaction. However, since the overall efficiency of any multipart ligation is the product of the efficiencies of the individual ligations, it can be imperative to make sure each reaction proceeds efficiently on its own before attempting any multipart ligation. Also for simultaneous reactions, the sequences at all the ligation junctions must vary enough so that the substrates do not cross-hybridize to the wrong place on the cDNA. Additionally, if a simultaneous multipart ligation will use more than one cDNA template, the base-pairing regions of the templates with the substrates should not overlap or they will interfere with each other.

Ligation of Other Duplexes and Effects of Mutations

In addition to (RNA/RNA) · DNA duplexes, (RNA/DNA) · DNA and (DNA/RNA) · DNA hybrids are good substrates with the same enzyme

[34] S. A. Strobel and T. R. Cech, *Biochemistry* **32,** 13593 (1993).

concentration requirements as (RNA/RNA) · DNA hybrids. Ligations can also be performed with an RNA bridge, but these reactions are markedly less efficient.[32] Interestingly, perfect base-pairing complementarity is not necessary at every position of the duplex.[35,36] We have found that RNAs containing single mutations close to the ligation junction can often be joined to a ligation partner using just one cDNA template.[37] This property is quite useful, as one may often desire to change a single position in the RNA of interest, and it is not necessary to synthesize a new cDNA template for each variant.

DNase Treatment

Incubation with RNase-free DNase after the ligation reaction degrades the bridging cDNA template. This step is not strictly necessary because the cDNA oligonucleotide is usually separated from the ligated RNA on denaturing gel purification. However, if the ligated product is to be incubated with any crude cellular extract in a later step, DNase treatment is advisable since any template that remains in the sample could activate RNase H cleavage of the ligated RNA. On the other hand, if the ligated product will contain two or more contiguous deoxynucleotides, do not treat with DNase.

Side Reactions

A number of other activities of T4 DNA ligase have been reported for DNA substrates, including partially nontemplated joining and reactions in which the optimal substrate is not perfectly basepaired.[35,38–40] In some of these cases, the substrate is a hairpin DNA in which the ligation site is presumably somewhat strained; in contrast, it is clear from selection experiments that perfect base complementarity is optimal for a typical DNA : DNA/DNA substrate.[35] These other actions may also occur on RNA, although this has not yet been specifically investigated, and could account for a number of additional minor product RNAs that are sometimes detectable in RNA ligations. In general, however, the templated reactions discussed in this article should predominate over any of the described "side reactions" and be easily separable by gel purification from minor products.

[35] K. Harada and L. E. Orgel, *Nucleic Acids Res.* **21,** 2287 (1993).
[36] D. Y. Wu and R. B. Wallace, *Gene* **76,** 245 (1989).
[37] C. C. Query, M. J. Moore, and P. A. Sharp, *Genes Develop.* **8,** 587 (1994).
[38] L. M. Western and S. J. Rose, *Nucleic Acids Res.* **19,** 809 (1991).
[39] D. F. Bogenhagen and K. G. Pinz, *J. Biol. Chem.* **273,** 7888 (1998).
[40] O. Madrid, D. Martín, E. A. Atencia, A. Sillero, and M. A. G. Sillero, *FEBS Lett.* **433,** 283 (1998).

[8] Heavy Atom Derivatives of RNA

By BARBARA L. GOLDEN

Introduction

Recently, crystallographic studies of RNA have enjoyed a rebirth. This is due partly to technological advances that allow milligram quantities of RNA to be transcribed *in vitro* or smaller RNAs to be synthesized. A second advance is the development of sparse matrix approaches,[1–4] which allow rapid identification of crystallization conditions for RNA molecules. As with proteins, there is a second major stumbling block in RNA crystallography: the preparation of suitable heavy atom derivatives of the molecule. These derivatives are essential to the calculation of crystallographic phases and generation of an electron density map when solving new structures.

Heavy atom derivatives may be obtained in several ways. Compounds can be soaked into preformed crystals or cocrystallized with the molecule. If there are sites within the molecule that tightly bind to the compound, a useful derivative may be obtained. This is the classical method of heavy atom derivatization and it has been used to solve the majority of the new protein structures. It is not necessarily trivial, however, to obtain derivatives of RNA in this manner. RNAs are polyanionic and therefore can frequently interact with cationic metal compounds in a nonspecific manner. The structure of RNAs is often dependent on the presence of divalent metal ions, usually magnesium, and the magnesium in the crystallization conditions can compete with heavy atom binding. RNA also lacks the library of functional groups present in protein molecules (i.e., the sulfhydryl group of cysteine), thus the type of interactions often involved in heavy atoms binding to proteins are lacking in RNA. Additionally, soaking crystals in heavy atom compounds can be detrimental to the diffraction of the crystals, or alter the packing of the RNA within the crystals, rendering the derivative nonisomorphous with native crystals (although this is less of an issue in the age of MAD phasing). This being said, specific binding sites for soaked-in heavy atom compounds have been found in the P4–P6 domain of the *Tetrahymena*

[1] J. A. Doudna, C. Grosshans, A. Gooding, and C. E. Kundrot, *Proc. Natl. Acad. Sci. U.S.A.* **90,** 7829 (1993).

[2] W. G. Scott, J. T. Finch, R. Grenfell, J. Fogg, T. Smith, M. J. Gait, and A. Klug, *J. Mol. Biol.* **250,** 327 (1995).

[3] A. C. Anderson, B. E. Earp, and C. A. Frederick, *J. Mol. Biol.* **259,** 696 (1996).

[4] B. L. Golden, E. R. Podell, A. R. Gooding, and T. R. Cech, *J. Mol. Biol.* **270,** 711 (1997).

0076-6879/00 $30.00

group I,[5] a 247-nucleotide ribozyme derived from the same intron,[6] the hammerhead ribozyme,[7] the fragment I domain of 5S rRNA,[8] and various tRNAs.[9]

The alternative to this "soak and pray" methodology is engineering heavy atoms or heavy atom binding sites into the covalent structure of the RNA to create derivatives. There are advantages to using engineered derivatives. First, the modification introduced is usually minimal. Therefore the macromolecule often behaves like the native, underivatized, species, and crystals of the derivatized macromolecule often diffract as well as the native crystals. Second, since the heavy atom is present in most of the molecules within the crystal, engineered derivatives often result in tightly bound heavy atoms with high occupancy. Third, the site of the modification within the primary sequence is known. This provides a landmark in the initial interpretation of the electron density maps, a distinct advantage, especially when dealing with lower resolution maps. Engineering of heavy atom binding sites into proteins (i.e., by selenomethionine incorporation[10]) or introduction of cysteine residues to bind mercury[11] is one of the reasons for the explosion in the number of protein crystal structures solved in recent years. RNA crystallography does not yet have such a simple methodology for production of heavy atom derivatives. Much more effort is involved in preparing large RNAs with specific modifications in quantities sufficient for a crystallization experiment. This article explores methods that can be used to incorporate heavy atoms or heavy atom binding sites into RNAs for use as heavy atom derivatives. Included are a review of the modifications suitable for creating useful heavy atom derivatives, strategies available for incorporating the modification into large RNAs, and methods for probing crystalline RNA to help identify sites within the RNA crystal that remain accessible for modification.

Bromine or Iodine Derivatives of RNAs

The simplest way to create a heavy atom derivative of RNA is to incorporate one of the halogenated pyrimidines in place of its unmodified

[5] J. H. Cate, A. R. Gooding, E. Podell, K. Zhou, B. L. Golden, C. E. Kundrot, T. R. Cech, and J. A. Doudna, *Science* **273,** 1678 (1996).

[6] B. L. Golden, A. R. Gooding, E. R. Podell, and T. R. Cech, *Science* **282,** 259 (1998).

[7] H. W. Pley, K. M. Flaherty, and D. B. McKay, *Nature* (*Lond.*) **372,** 68 (1994).

[8] C. C. Correll, B. Freeborn, P. B. Moore, and T. A. Steitz, *Cell* **91,** 705 (1997).

[9] S.-H. Kim, W.-C. Shin, and R. W. Warrant, *Methods Enzymol.* **114,** 156 (1985).

[10] W. A. Hendrickson, J. R. Horton, and D. M. LeMaster, *EMBO J.* **9,** 1665 (1990).

[11] G. F. Hatfull, M. R. Sanderson, P. S. Freemont, P. R. Raccuia, N. D. F. Grindley, and T. A. Steitz, *J. Mol. Biol.* **208,** 661 (1989).

counterpart. Bromine or iodine substitution at the 5 position of uracil or cytosine residues is an often used modification. Halogenation of RNA for use as heavy atom derivatives is a classical method used in the investigation of tRNA structures.[12,13] Bromine derivatives have the advantage of an anomalous diffraction signal that can be harnessed, using a synchrotron radiation source, to provide additional phase information. The higher electron density of an iodine atom, however, may make initial localization of heavy atom positions within the unit cell easier, especially when using a rotating enode X-ray source. This may be the more important consideration when working with large RNAs.

The most common means of incorporating bromine or iodine is to chemically synthesize the RNAs and incorporate them in the synthesis. Reagents for making all of these modifications are available commercially. Chemically synthesized, modified RNAs have provided useful halogenated derivatives of the hammerhead ribozyme,[7,14] a pseudoknot,[15] and numerous RNA duplexes.[16,17] If carefully designed, the modification may also be introduced by transcription. For example, if there are only one or two uridines in this fragment, transcription in the presence of bromouridine triphosphate in place of uridine triphosphate may provide a suitable modification.[18] However, as multiple modifications are introduced, each modification has the potential to interfere with RNA structure or crystal contacts. The all-or-nothing nature of this method may make it less useful in the long run. When working with halogenated RNAs, it is important to protect the sample from light during purification, crystallization, and, if possible, even during data collection to maximize the occupancy of the heavy atom.

Mercury Derivatives of RNA

An alternate method of derivatization may be accomplished by incorporating a mercury-binding site into a RNA via a phosphorothioate in the

[12] M. Pasek, M. P. Venkatappa, and P. B. Sigler, *Biochemistry* **12,** 4834 (1973).

[13] J. Tropp and P. B. Sigler, *Biochemistry* **18,** 5489 (1979).

[14] W. G. Scott, J. T. Finch, and A. Klug, *Cell* **81,** 991 (1995).

[15] S. E. Lietzke, C. L. Barnes, V. F. Malone, J. T. Jones, and C. E. Kundrot, *in* "Structure, Motion, Interaction and Expression of Biological Macromolecules, Proceedings of the Tenth Conversation" (R. H. Sarma and M. H. Sarma, eds.), p. 91. State University of New York, Albany, 1997.

[16] S. E. Lietzke, C. L. Barnes, J. A. Berglund, and C. E. Kundrot, *Structure* **4,** 917 (1996).

[17] C. C. Correll, A. Munishkin, Y. L. Chan, Z. Ren, I. G. Wool, and T. A. Steitz, *Proc. Natl. Acad. Sci. U.S.A.* **95,** 13436 (1998).

[18] L. Su, L. Chen, M. Egli, J. M. Berger, and A. Rich, *Nature Struct. Biol.* **6,** 285 (1999).

RNA backbone. This technique has been used to solve crystal structures of DNA–protein complexes (see, for example, Refs. 19 and 20) and, in theory, it should prove equally powerful in solving RNA structures. However it has not yet been used to successfully derivatize a RNA molecule (see Ref. 21). There are distinct advantages to mercury as a derivative. Mercury is heavy enough to be easily located within the unit cell using Patterson maps, and it has an anomalous scattering signal that can be harnessed to provide additional phasing information. Thus, it may be very helpful in solving the structure of larger RNAs. Technically this method is more complicated than use of halogenated pyrimidines in that mercuration is an additional step in the process, and conditions for mercurating the crystal may need to be explored to optimize the usefulness of the heavy atom derivative. Both R_p and S_p isomers are made during chemical synthesis of phosphorothioate-containing RNAs, and these must often be separated prior to crystallization (see, for examples, Refs. 21 and 22). To minimize interference by the modification and to maximize the occupancy of the heavy atom, independent crystallization and derivatization trials need to be performed on each isomer. Additionally, the nucleotide 5′ to the modification should be synthesized with a deoxyribose to prevent loss of the sulfur atom during mercury treatment.

Mercury can be bound to the sulfur either prior to crystallization or soaked into preformed crystals, and both should be tried. The mercury atom can be provided as an inorganic compound such as mercury chloride or mercury acetate, or as an organic mercurial compound such as methylmercury chloride or ethylmercury phosphate. The latter compounds have the advantage that they have only one labile ligand and therefore cannot cross-link two modified RNA strands. If the RNA is to be mercurated after crystallization, the mercury compound is introduced into crystallization solutions, typically at concentrations between 0.1 and 5 m*M*.

Another rarely explored option is direct mercuration at the C-5 position of pyrimidines.[21,23] This can be done by reaction for 3–4 hr at 50° with a molar excess of mercury acetate in acetate buffer at pH 5–6. All of the pyrimidines within an oligonucleotide are going to be subject to mercuration by this methodology. Thus, the usefulness of this approach may be limited.

[19] W. Yang and T. A. Steitz, *Cell* **82,** 193 (1995).

[20] M. P. Horvath, V. L. Schweiker, J. M. Bevilacqua, J. A. Ruggles, and S. C. Schultz, *Cell* **95,** 963 (1998).

[21] C. C. Correll, B. Freeborn, P. B. Moore, and T. A. Steitz, *J. Biomol. Struct. Dynamics* **15,** 165 (1997).

[22] E. C. Scott and O. C. Uhlenbeck, *Nucleic Acids Res.* **27,** 479 (1999).

[23] R. M. K. Dale, D. C. Livingston, and D. C. Ward, *Proc. Natl. Acad. Sci. U.S.A.* **70,** 2238 (1973).

Site Selection

The first consideration when engineering a heavy atom binding site into the RNA is to refrain from disturbing interactions required for proper folding of the RNA and for formation of crystal contacts. Halogenated pyrimidines alter the functional groups in the major groove and modification of the phosphate oxygen atoms is largely going to affect the major groove as well. Thus, both modifications steer clear of the shallow minor groove crucial for many tertiary interactions within RNA structures. Within the P4–P6 domain, the best derivatives were obtained by modification of uridines involved in G · U wobble pairs.[24] It will be interesting to see if this continues to be a trend. While some thought should go into deciding where the effects of the modification will be minimized, this is always going to be somewhat of a trial-and-error proposition. Where biochemical assays are available to report on the structure, it may be worthwhile to make the modified RNA in biochemical quantities first, and scale up those modifications that pass this preliminary test.

Engineering Large RNAs

If the RNA under investigation is too long to be made in quantity by chemical synthesis, than engineering a heavy atom derivative becomes more difficult. Modification of RNAs greater than ~30 nucleotides requires that the RNA be made in two pieces, one carries the modification, and the second, which can be made by *in vitro* transcription, makes up the remainder of the molecule. These two pieces must then be reconstituted somehow prior to crystallization to make the full-length product. One means of reconstitution is to covalently join the two pieces using T4 DNA ligase and a DNA splint.[25] This method requires that the RNA corresponding to the 5′ end has a free 3′-hydroxyl group, and the RNA corresponding to the 3′ end possess a 5′ phosphate. Phosphorylation at the 5′ end can be accomplished by transcription in the presence of 10-fold molar excess of GMP over GTP or by T4 polynucleotide kinase. Synthetic RNAs can be phosphorylated enzymatically using T4 polynucleotide kinase or chemically during the synthesis. The two RNAs are then mixed with a DNA oligonucleotide which is complementary to the junction to be joined, and usually heated and cooled to anneal the three oligonucleotides. This forms a DNA–RNA heteroduplex with a nick at the junction of the two RNAs. If this structure is correctly formed, T4 ligase is capable of covalently linking the two RNAs.

[24] B. L. Golden, A. R. Gooding, E. R. Podell, and T. R. Cech, *RNA* **2,** 1295 (1996).
[25] M. J. Moore and P. A. Sharp, *Science* **256,** 992 (1992).

A typical ligation reaction might contain the following:

60–70 mM Tris-HCl, pH 7.8–8.2
10 mM $MgCl_2$
10 mM Dithiothreitol (DTT)
0.5–1.0 mM adenosine triphosphate (ATP)
20 μM each RNA
20 μM DNA splint
50–250 units/μl T4 DNA ligase

This reaction can be performed at room temperature or at 16°, which is the temperature at which the ligase possesses maximal activity. To maximize the efficiency of the reaction with a minimum of ligase, the reaction can be run over the course of several hours or even overnight.

It is important to design the system such that the ligation reaction is efficient. Poor yields in the ligation reaction can result if a strong secondary or tertiary structure interferes with proper annealing of the splint. Ligation efficiency should be checked by performing small-scale reactions with unmodified RNAs prior to undertaking large-scale reactions. It may be necessary to explore several variations to find a system that allows efficient ligation. Several steps may be taken to increase a low yield. Most of these involve changing the annealing protocol. Band shift analysis may be used to monitor annealing of the three oligonucleotides independent of the ligation reaction. Annealing can be accomplished by heating the oligos to 95° and snap cooling on ice. Alternately the molecules may be slowly cooled from 95° to room temperature in a heating block or a using a thermal cycle. Other systems have benefited from heating for longer time periods at lower temperatures. The order of addition of the three components can be crucial; it may be necessary to anneal the DNA to one RNA molecule, then reanneal in the presence of the second RNA. Occasionally redesigning the DNA splint has had dramatic effects. Addition of one or two extra complementary nucleotides to the end of the splint can help in formation of the ternary complex, perhaps due to disruption of residual secondary structure in the RNA. Also along these lines, a second DNA oligonucleotide complementary to the RNA at a distal site may be added to the annealing reaction to disrupt secondary structure within the RNA during ligation reaction.[26] Sodium chloride (up to 20 mM) may be added to the buffer to increase the efficiency of the annealing reaction, however higher concentrations (~50 mM) will inhibit T4 ligase.

As an alternative to ligation, the two pieces can be simply be annealed by base pairing and tertiary interactions without covalent ligation, creating a nicked version of the crystallized molecule.[24] This method has the benefit that optimization of the ligation reaction is not necessary, however, the

[26] S. A. Strobel and T. R. Cech, *Comments* **19,** 89 (1992).

nick introduced cannot profoundly destabilize the overall structure or else crystallization will be effected. The two options are somewhat complementary. If the RNA has strong secondary and tertiary structure, residual structure in the RNA can interfere with the ligation reaction. The same strong tertiary structure often results in stable folding of a nicked species of the RNA. This may allow assembly of the two pieces without covalent ligation. Assembly can be assayed by native gel-shift analysis or by biochemical assays. The trick to using either of these approaches is to find a site within 30 nucleotides of either the 5′ or 3′ terminus that will allow efficient ligation or, in the case of the latter method, will tolerate the introduction of a break in the backbone.

The use of *cis*- or *trans*-acting ribozymes (often hammerhead sequences) may be very useful in the design of these two-piece systems.[24] Posttranscriptional processing of RNAs has several advantages. First, if a deletion is to be made at the 5′ end of a transcript, the new 5′ end must retain a sequence that allows efficient T7 transcription, usually this sequence begins with a guanosine and is purine rich.[27] This is an additional constraint that may be difficult to work around. If a hammerhead sequence is incorporated at the 5′ end of the transcript, however, the efficient T7 start site precedes the hammerhead and thus is cleaved off during posttranscriptional processing. The sequence at the 5′ end of the final product has no sequence constraints, and possesses a free 5′-hydroxyl group that is readily phosphorylated using T4 polynucleotide kinase. This strategy allows more flexibility in situations where the smaller oligonucleotide is to be added at the 5′ end of the molecule.

If the smaller oligonucleotide is to be added to the 3′ end of the molecule, hammerhead processing of the 3′ end may or may not be advisable. On one hand, this technique eliminates heterogeneity at the 3′ end characteristic of T7 polymerase transcripts. This is important if this RNA is to be ligated because the uncoded nucleotides added to a portion of the T7 transcripts will interfere with the ligation reaction. However, hammerhead processing imposes some restriction on the sequence—specifically the 3′-terminal residue may not be a guanosine. Additionally, the cleavage product of the hammerhead results in a 3′ end that is blocked with a 2′,3′-cyclic phosphate which must be removed if the product is to be covalently joined to another RNA by T4 ligase.

Analysis of Crystallized RNA

Biochemical characterization of crystalline RNA can provide strategies for the introduction of modifications. During the crystallization process,

[27] J. F. Milligan and O. C. Uhlenbeck, *Methods Enzymol.* **180,** 51 (1989).

the RNA within the crystals often accumulates nicks. RNA obtained from redissolved crystals can be reverse transcribed or labeled at the 5′ end with [γ-^{32}P]ATP using T4 polynucleotide kinase. Breaks in the RNA are revealed as strong stops not present in freshly prepared RNA. If nicks are present within 30 nucleotides of either terminus, this provides strong evidence that it will be possible to cocrystallize two appropriately designed RNAs without prior ligation.

It may also be possible to identify and avoid sites that are involved in crystal contacts by chemical modification of RNA within the crystal. Dimethylsulfate (DMS) mapping of crystallized RNA was used to successfully predict a crystal contact in the P5c loop of the P4–P6 domain.[28] Comparison of the modification pattern in the crystal and in solution may also reveal formation of nonnative structures resulting from crystal packing prior to actually solving the structure. Crystallization of nonnative structures is a continuing problem in RNA crystallography.[15,29,30]

To perform analyses on crystallized RNA, a suitable stabilizing buffer must first be identified that will not dissolve or crack the crystals. This stabilizing buffer usually contains precipitant, buffer, magnesium, polyamines, and other salts, but no RNA, and it will be different for every crystal form. By carefully washing crystals with the buffer prior to dissolving them, crystallized RNA can be selectively isolated from the drop. Thus, the analysis will correspond only to the RNA incorporated in the crystal and not the molecules that remain soluble during the crystallization process. The *Tetrahymena* ribozyme crystals,[4] which are ~70% solvent and ~200–400 μm in each dimension, are calculated to have ~4 m*M* RNA within the crystal (1 *M* nucleotides), and to contain ~1.0 μg of RNA each. This is going to vary somewhat depending on the solvent content of the crystal and the size of the RNA, but in general, one medium-sized crystal will be more than sufficient to allow optimization of reverse transcription experiments. Crystals that are poorly formed and therefore unsuitable for a diffraction experiment are fine for these analyses.

To examine the DMS reaction pattern of the RNA within the crystal, a procedure similar to that used by Zaug and Cech[31] is used. Prior to modification, crystals are washed several times in stabilization buffer. The crystals are then modified by treatment with stabilization buffer that contains 10% DMS stock solution (2 μl DMS, 7 μl absolute ethanol). Using a different crystal for each experiment, several time points (15 min to several

[28] B. L. Golden and T. R. Cech, unpublished results.

[29] S. R. Holbrook, C. Cheong, I. Tinoco, Jr., and S. H. Kim, *Nature* (*Lond.*) **353,** 579 (1991).

[30] J. Nowakowski, P. J. Shim, G. S. Prasad, C. D. Stout, and G. F. Joyce, *Nature Struct. Biol.* **6,** 151 (1999).

[31] A. J. Zaug and T. R. Cech, *RNA* **1,** 363 (1995).

hours) are observed. The reaction is then quenched by dissolving the crystals in 0.5 volumes of stop solution (0.5 *M* 2-mercaptoethanol, 0.75 *M* sodium acetate, pH 5.5). The redissolved RNA can then be ethanol precipitated by addition of 3 volumes of ethanol, and resuspended in 100 μl volume. Under ideal conditions, modification of the RNA is readily apparent in the analysis that followed, but the crystal remains visually intact with no cracking or surface striations.

Modification patterns are revealed by reverse transcription of the RNA using avian myeloblastosis virus (AMV) reverse transcriptase.[31] Adenosine and cytosine residues accessible to modification are identified as stops in a reverse transcription reaction. These nucleotides can then be identified by comparison with a sequencing ladder. The amount of RNA in a crystal with dimensions of ~200 μm is sufficient for ~100 reverse transcriptase reactions. It is important to compare the modified RNA to unmodified crystalline RNA because nicks in the backbone accumulate during crystallization and these too will appear as stops in the reverse transcription reaction.

Fe · EDTA protection analysis might seem to be ideal for this type of analysis, since this reagent reveals regions of the RNA that are inaccessible to the solvent. Presumably regions involved in crystal contacts will be protected from the solvent. However MPD and polyethylene glycol which are often used as precipitants are capable of quenching this free radical formation. It will be interesting to see if crystals grown under high salt conditions can be probed using free-radical reagents.

Acknowledgments

I am grateful to Tom Cech, Anne Gooding, and Elaine Podell for help with many of the technical details outlined here. This is paper 15955 from the Purdue Agriculture Experiment station.

[9] Site-Specific Cleavage of Transcript RNA

By JON LAPHAM and DONALD M. CROTHERS

Introduction

We describe a method by which transcript RNA is cleaved site specifically using ribonuclease H (RNase H) and a targeting 2′-*O*-methyl-RNA/DNA chimera. The reaction is high yielding, has no sequence specificity requirements, and has been adapted to the cleavage of large quantities

0076-6879/00 $30.00

(milligram) of RNA suitable for the sample preparation needs of biophysical techniques requiring large amounts of RNA, such as NMR spectroscopy and X-ray crystallography.

RNase H is an endonuclease that hydrolyzes the phosphodiester backbone of RNA in an RNA/DNA hybrid.[1] The product of the reaction contains 5′-phosphate and 3′-hydroxyl termini at the site of hydrolysis.[2] The biological role of RNase H is to remove RNA during processes that generate RNA/DNA duplexes, such as during reverse transcription of a retroviral genome.[3,4] Methods have been developed for overexpressing and purifying RNase H (*Escherichia coli*),[5] allowing for synthesis of the enzyme in large quantities at relatively low cost.

RNase H cleavage of RNA is not, however, site specific by nature. If a long stretch of DNA is bound to a target RNA molecule, RNase H will cleave the RNA strand in random positions shared by the DNA/RNA duplex. Inoue *et al.*[6] recognized that the activity of RNase H could be exploited for site-specific cleavage if the number of DNA nucleotides bound to the RNA could be reduced, the reasoning being that as the size of the DNA became smaller, the cleavage position on the target RNA molecule would become more site specific. In fact, they demonstrated that if the number of DNA nucleotides was reduced to four, the target RNA was cleaved in a single, predictable position. To increase the thermodynamic stability of this DNA/RNA duplex, the four DNA nucleotides are surrounded by 2′-*O*-methyl-RNA nucleotides. This chimeric 2′-*O*-methyl-RNA/DNA construct binds to the opposite strand target RNA, but only ribonucleotides opposite the DNA portion of the chimera are recognized as possible cleavage sites. Using this construct, the hydrolysis of the RNA occurs between the nucleotides positioned opposite of the 5′-most DNA nucleotide as shown in Fig. 1. We have found that 2′-*O*-methyl-RNA nucleotides 5′ to the DNA segment are not required.[7] This allows a single 2′-*O*-methyl-RNA/DNA chimera sequence to be used to remove a common leader sequence from a variety of RNA sequences 3′ to the cleavage site (see Applications section).

In addition to this RNase H method, another technique has been devel-

[1] I. Berkower, J. Leis, and J. Hurwitz, *J. Biol. Chem.* **248,** 5914 (1973).

[2] R. J. Crouch and M.-L. Dirksen, *in* "Nuclease" (S. M. Linn and R. J. Roberts, eds.), p. 211. Cold Spring Harbor Laboratory Press, Cold Spring Harbor, New York, 1982.

[3] A. Hizi, R. Tal, M. Shaharabany, and S. Loya, *J. Biol. Chem.* **266,** 6230 (1991).

[4] S. P. Goff, *J. Acquired Immune Defic. Syndr.* **3,** 817 (1990).

[5] S. Kanaya, A. Kohara, M. Miyagawa, T. Matsuzaki, K. Morikawa, and M. Ikehara, *J. Biol. Chem.* **264,** 11546 (1989).

[6] H. Inoue, Y. Hayase, S. Iwai, and E. Ohtsuka, *FEBS Lett.* **215,** 327 (1987).

[7] J. Lapham and D. M. Crothers, *RNA* **2,** 289 (1996).

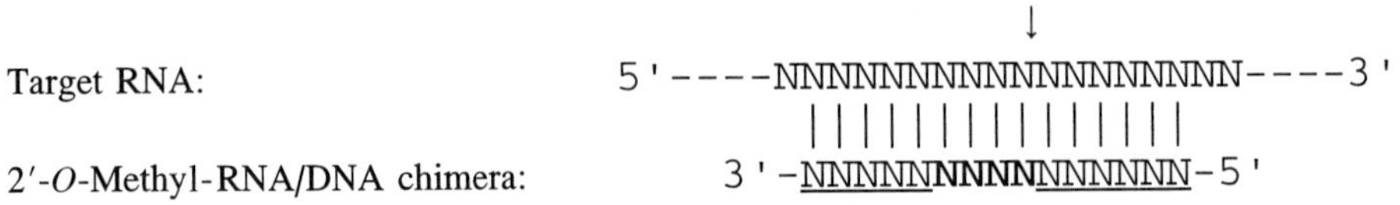

```
                                                    ↓
Target RNA:                      5'----NNNNNNNNNNNNNNNNNNNN----3'
                                         |||||||||||||||
2'-O-Methyl-RNA/DNA chimera:          3'-NNNNNNNNNNNNNNNNNN-5'
```

FIG. 1. RNase H cleavage position. Underline characters, N, represent 2′-*O*-methyl-RNA. Bold characters, **N**, represent DNA. Regular characters, N, represent RNA. The arrow, ↓, indicates the position of cleavage.

oped to create a site-specific RNA endonuclease. The hammerhead ribozyme[8] is an RNA found in some plant viroids that, under the proper conditions, cleaves itself spontaneously, producing a 5′-hydroxyl and 2′,3′-cyclic phosphate termini.[7,9] The self-cleaving domain from this RNA can be constructed with as few as 50 nucleotides comprising three helices and a conserved central core of 11 nucleotides. This ribozyme has been used to construct an *in vitro* sequence-specific RNA cleaving restriction enzyme that can be added to the target RNA molecule *in trans.*[10,11] However, the hammerhead reaction does have some sequence requirements at the three nucleotides around the site of cleavage,[12] limiting the possible sites that may be targeted.

A disadvantage of the RNase H method is that it requires a 1 : 1 molar equivalent of the targeting chimera, whereas the *trans* hammerhead reaction is catalytic with the same ribozyme cleaving many target RNAs.[10] The advantages of the RNase H method are that it should work for any given sequence of RNA with little or no sequence requirements, the cleavage position is very well defined, and it produces termini that make the product RNAs ready for subsequent reactions, such as ligations.

Materials and Methods

The RNase H used in the cleavage reactions is obtained from Pharmacia (Piscataway, NJ) or Sigma (St. Louis, MO), where 1 unit is defined as able to catalyze the production of 1 nmol of acid-soluble RNA nucleotide in 20 min at 37°. The T4 DNA ligase used in the ligation reactions is obtained from New England Biolabs (Beverly, MA) at 400 units/μl.

The T7 RNA polymerase used to transcribe the target RNAs is overex-

[8] A. C. Forster and R. H. Symons, *Cell* **49,** 211 (1987).

[9] C. J. Hutchins, P. D. Rathjen, A. C. Forster, and R. H. Symons, *Nucleic Acids Res.* **24,** 3627 (1986).

[10] O. C. Uhlenbeck, *Nature* **328,** 596 (1987).

[11] J. Haseloff and W. L. Gerlach, *Nature* **334,** 585 (1988).

[12] D. E. Ruffner, G. D. Stormo, and O. C. Uhlenbeck, *Biochemistry* **29,** 10695 (1990).

pressed and purified following previously published procedures.[13–15] The buffer is 5× RNase H reaction buffer:

100 mM HEPES–KOH, pH 8.0
250 mM KCl
50 mM $MgCl_2$
5 mM dithiothreitol (DTT)

The chimera is annealed to the target RNA in the reaction buffer by heating to a temperature above the melting temperature of any secondary structure elements in either the target RNA or the 2′-*O*-methyl chimera, typically from 70 to 95°, and slowly cooling to room temperature. This annealing process works best if the chimera is at a higher concentration (1.2×) than the target RNA and the overall concentration of the complex is high (millimolar) to drive the bimolecular complex formation.

The RNase H enzyme is then added to a final concentration of 20 units per 100 μl of reaction. The reaction typically takes from 30 min to 3 hr for completion. The reaction is stopped by addition of enough ethylenediaminetetraacetic acid (EDTA) to chelate the magnesium (the divalent ion required by RNase H for cleavage). The final products can be purified by using standard denaturing polyacrylamide gel electrophoresis (PAGE) techniques.

Applications

Increased Yield on T7 RNA Polymerase Transcription

The bacteriophage protein T7 RNA polymerase is often used in the production of *in vitro* transcribed RNA.[16,17] Although the RNA polymerase from other sources has been used to synthesize RNA, this polymerase from T7 has been found to be the most amenable to large-scale RNA preparation and can be readily obtained in large quantities by overexpression and purification techniques.[13–15] The overall transcription yield is very sensitive to the 5′ sequence of the product RNA, with the highest yield coming from 5′-end sequences largely composed of guanines. For this reason, RNAs used in biophysical studies are often designed with sequence modifications at their 5′ ends to increase the transcription yield. These changes in se-

[13] J. Grodberg and J. J. Dunn, *J. Bacteriol.* **170,** 1245 (1988).

[14] P. Davenloo, A. H. Rosenberg, J. J. Dunn, and F. W. Studier, *Proc. Natl. Acad. Sci. U.S.A.* **81,** 2035 (1984).

[15] V. Zawadzki and H. J. Gross, *Nucleic Acids Res.* **19,** 1948 (1991).

[16] J. F. Milligan, D. R. Groebe, G. W. Witherell, and O. C. Uhlenbeck, *Nucleic Acids Res.* **15,** 8783 (1987).

[17] J. F. Milligan and O. C. Uhlenbeck, *Methods Enzymol.* **180,** 51 (1989).

quence can sometimes lead to undesirable physical properties in the RNA. For instance, it is known that non-base-paired guanines can form intermolecular aggregations at high concentrations, a problem that must be avoided in NMR and X-ray crystallography studies. Thus, a site-specific RNA endonuclease would be helpful to overcome this problem by allowing one to transcribe an RNA with a high-yield 5′-end sequence, which could be selectively cleaved, leaving the desired RNA sequence. In a previous study,[7] we reported an approximate 13-fold increase in the overall production yield of a poorly transcribing 30-nucleotide RNA sequence compared to the same RNA with a additional 15-nucleotide leader sequence (LDR) added to the 5′ end (Fig. 2). After the cleavage reaction and final purification gel (Fig. 3), the overall yield increase was approximately 5-fold as compared to the transcription of the original 30-nucleotide RNA.

This reaction can also be performed on a solid matrix by complexing a chimera biotinylated at its 3′ termini with a streptavidin/agarose bead.[7] This allows for quick purification of the product RNA by simple extraction from the supernatant, and the solid-phase media may be used repeatedly.

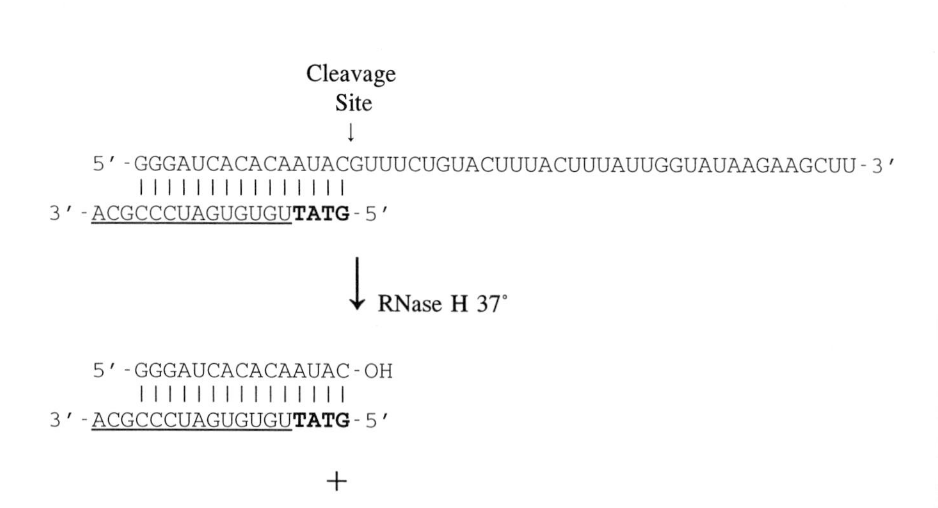

FIG. 2. The 45-nucleotide target RNA is the top strand, the first 15 nucleotides are the LDR high-yield T7 RNA polymerase sequence and the bottom strand is the 2′-*O*-methyl-RNA/DNA chimera used to target the cleavage at the position shown, between C_{15} and G_{16}. The underlined nucleotides are 2′-*O*-methyl-RNA, the bold nucleotides are DNA, and the remainder are normal RNA. After the RNase H cleavage reaction, the desired 30-nucleotide RNA can be purified from the LDR RNA and the 2′-*O*-methyl chimera. [Adapted from J. Lapham and D. M. Crothers, *RNA* **2,** 289 (1996).]

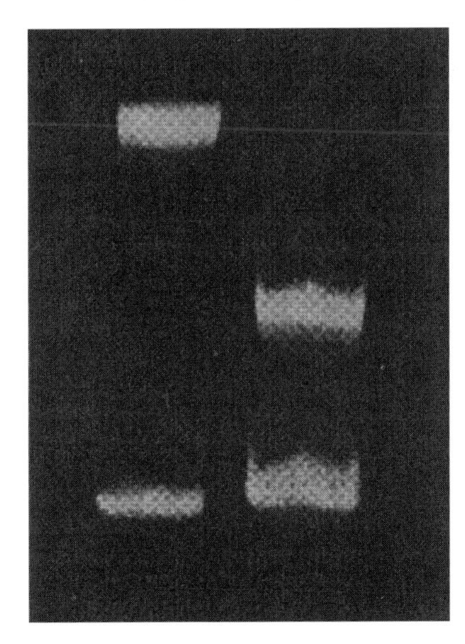

FIG. 3. Large-scale cleavage of a 45-nucleotide RNA. Ethidium-stained 20% PAGE before and after the RNase H reaction. This demonstrates that the reaction works to nearly complete conversion of the 45-nucleotide RNA into the 30-nucleotide RNA product [Reprinted with permission from J. Lapham and D. M. Crothers, *RNA* **2,** 289 (1996).]

Religation of RNA Cleavage Products

As mentioned previously, the product of the RNase H cleavage reaction is a 3′-hydroxyl and a 5′-phosphate. This is the required end chemistry needed for ligating two pieces of RNA together using the DNA guide and T4 DNA ligase method.[18] This fact has been taken advantage of in a number of ways. We have previously reported a method of constructing a "segmentally" isotopic-labeled RNA using this technique.[19] Pulse sequences in NMR spectroscopy have been developed for selectively observing the NMR signal arising from protons directly attached to $^{13}C/^{15}N$ ("isotope labeled protons") or $^{12}C/^{14}N$ ("unlabeled protons") using heteronuclear filters.[20] This is powerful in that it offers a method of selectively studying a small segment of a larger complex, thus simplifying the interpre-

[18] M. J. Moore and P. A. Sharp, *Science* **256,** 992 (1992).
[19] J. Xu, J. Lapham, and D. M. Crothers, *Proc. Natl. Acad. Sci. U.S.A.* **93,** 44 (1996).
[20] G. Otting and K. Wuthrich, *Quart. Rev. Biophys.* **23,** 39 (1990).

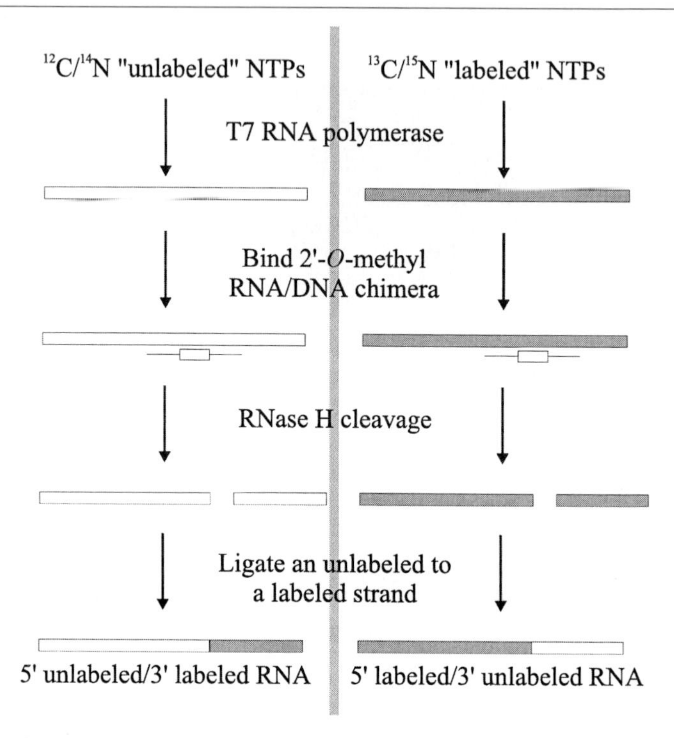

FIG. 4. Flowchart for synthesizing segmentally labeled RNA samples for study with isotope filtered nuclear magnetic resonance techniques. [Adapted from J. Xu, J. Lapham, and D. M. Crothers, *Proc. Natl. Acad. Sci. U.S.A.* **93,** 44 (1996).]

tation of the results. First, a full-length RNA is transcribed using $^{12}C/^{14}N$ isotope nucleotide triphosphates (NTPs) and in a separate reaction this same RNA is transcribed using $^{13}C/^{15}N$ isotope NTPs.[21,22] Both RNAs are then cleaved by RNase H and 2′-*O*-methyl-RNA/DNA chimera at the same position in the sequence. The halves of the RNAs are then purified and an isotope labeled half is joined to an unlabeled half, forming two segmentally labeled samples as shown schematically in Fig. 4.

Religation of postcleaved RNA has also been used to introduce a 4-thiouridine (^{4S}U) nucleotide in a pre-mRNA substrate.[23] The ^{4S}U nucleo-

[21] E. P. Nikonowicz, A. Sirr, P. Legault, F. M. Jucker, L. M. Baer, and A. Pardi, *Nucleic Acids Res.* **20,** 4507 (1992).

[22] R. T. Batey, M. Inada, E. Kujawinski, J. D. Puglisi, and J. R. Williamson, *Nucleic Acids Res.* **20,** 4515 (1992).

[23] Y.-T. Yu and J. A. Steitz, *RNA* **3,** 807 (1997).

tide can be used as a structural probe because of its propensity to cross-link when exposed to UV light. The pre-mRNA molecule is first cleaved specifically using the RNase H method, then the ^{4S}U nucleotide is added to the 5′ half of the RNA with T4 RNA ligase (a template-free reaction). The 5′ and 3′ halves of the RNA are then ligated using the DNA guide template and T4 DNA ligase method.[18] The final product contains a ^{4S}U nucleotide at any desired position within the pre-mRNA molecule.

Detection of 2′-O-Methyl Positions in RNA

Yu *et al.*[24] used the RNase H cleavage reaction to probe a target RNA for sites of 2′-*O*-methyl modification. Since the cleavage of RNA by RNase H is presumed to go through a 2′-O-P-O-3′ intermediate, they assumed that if the 2′-hydroxyl of the target RNA were blocked with a methyl group, the reaction would be inhibited. They were able to detect sites of methylation in both chemically synthesized RNA with known methylation positions and *in vivo* purified RNA with unknown sites.

Discussion

The most important step in successfully cleaving RNA with this method is ensuring that all the target RNA molecules are bound by the 2′-*O*-methyl-RNA/DNA chimera. This can be accomplished by examining a native gel band shift of the target RNA before and after addition of the chimera. The chimera will have to compete with any self-structure the target RNA may form, so the length of the construct may have to be lengthened to favor complete complex formation.[25]

Note that it has been reported that the position of cleavage may depend on the source of RNase H used.[26] In our studies, the RNase H was always purchased from either Pharmacia or Sigma and the cleavage patterns were as discussed. However, it has been reported[24] that RNase H purchased from Boehringer Mannheim (Indianapolis, IN) may cleave the target RNA one nucleotide in the 5′ direction, so care should be taken in standardizing the enzyme source and verifying the cleavage position.

We describe a simple, effective method of cleavage transcript RNA. The reaction is general, having no sequence requirements, can be scaled up to synthesize large quantities of RNA, and the products of the reaction are capable of being utilized in a subsequent ligation reaction without further preparation.

[24] Y.-T. Yu, M.-D. Shu, and J. A. Steitz, *RNA* **3,** 324 (1997).
[25] Y. Hayase, H. Inoue, and E. Ohtsuka, *Biochemistry* **29,** 8793 (1990).
[26] J. Lapham, Y.-T. Yu, M.-D. Shu, J. A. Steitz, and D. M. Crothers, *RNA* **3,** 950 (1997).

[10] Using DNAzymes to Cut, Process, and Map RNA Molecules for Structural Studies or Modification

By ANNA MARIE PYLE, VI T. CHU, ECKHARD JANKOWSKY, and MARC BOUDVILLAIN

Introduction

Once an RNA transcript has been made, it can be difficult to perform manipulations on it. Unlike DNA, which can be cut with sequence-specific restriction endonucleases, there are few practical ways to cleave an existing transcript into pieces of desired sequence and length. This makes it difficult to rapidly incorporate modified nucleotides into long transcripts, map sites of modification on an RNA molecule, or produce termini of homogeneous length on RNA transcripts. Although ribozymes can be designed to process RNA into pieces of defined length,[1,2] and targeted RNase H methods have proven to be highly effective[3] we have found that the use of simple DNAzyme molecules represents a simple and inexpensive way to stoichiometrically cleave large transcripts for a variety of different applications.

The type of DNAzyme described here was developed by Santoro and Joyce through *in vitro* selection.[4] These molecules are DNA oligonucleotides containing a small catalytic core that anneals to RNA targets through hybridization of two flanking arms. Upon addition of Mg^{2+} or other divalent ions, the DNAzyme cleaves the RNA specifically at a single designated site. A highly reactive catalytic core has been optimized and kinetically characterized in elegant studies that have been previously published.[5,6] DNAzymes can be targeted to cleave many different RNA sequences efficiently under multiple-turnover conditions.[5] However, we find that the greatest variety of target RNA substrates can be cleaved to completion by using single-turnover protocols in which the DNAzyme is in excess.[7,8]

Among the methods that are available for sequence-specific cleaving and tailoring of RNA molecules, an advantage of DNAzymes is that one

[1] C. A. Grosshans and T. R. Cech, *Nucleic Acids Res.* **19,** 3875 (1991).
[2] A. R. Ferre-D'Amare and J. A. Doudna, *Nucleic Acids Res.* **24,** 977 (1996).
[3] J. Lapham and D. M. Crothers, *RNA* **2,** 289 (1996).
[4] S. W. Santoro and G. F. Joyce, *Proc. Natl. Acad. Sci. U.S.A.* **94,** 4262 (1997).
[5] S. W. Santoro and G. F. Joyce, *Biochemistry* **37,** 13330 (1998).
[6] N. Ota, M. Warashina, K. Hirano, K. Hatanaka and K. Taira, *Nucleic Acids Res.* **15,** 3385 (1998).
[7] M. Boudvillain and A. M. Pyle, *EMBO J.* **17,** 7091 (1998).
[8] V.-T. Chu, Q. Liu, M. Podar, P. S. Perlman and A. M. Pyle, *RNA* **4,** 1186 (1998).

0076-6879/00 $30.00

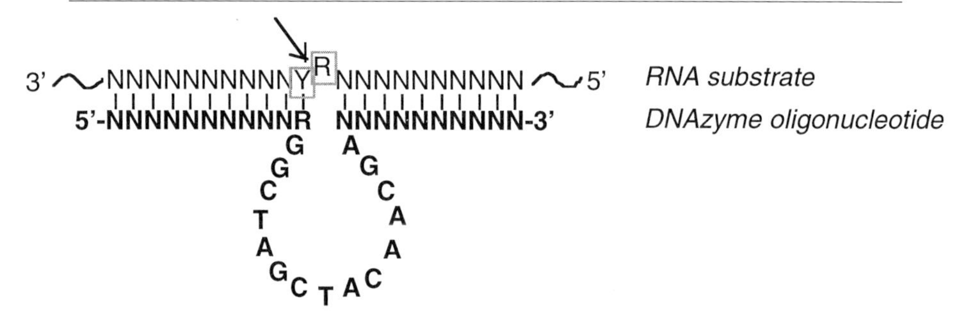

FIG. 1. Schematic of a DNAzyme (bold) from the "10-23" family, showing the looped-out sequence of the catalytic core. The binding arms are shown base paired to the RNA substrate (plain text). The arrow indicates the cleavage site between the boxed purine (R) and pyrimidine (Y) nucleotides on the substrate RNA, which are the only sequence requirements.

needs only to add a simple DNA molecule to the RNA transcript, and one can order a synthetic DNAzyme much as one would routinely order a primer. Several protocols are presented here, having been optimized for three separate applications involving the preparative and analytical scale cleavage of RNA transcripts, and the mapping of branch points or other modifications on an RNA molecule.

General Method

DNAzymes in the "10-23" family[4] cleave at any 5′-purine-pyrimidine-3′ linkage (5′-RY-3′) in an RNA molecule, provided that their "binding arms" anneal on either side of the target RNA purine (Fig. 1). In general, 5′-RU-3′ sequences are cleaved with higher efficiency than 5′-RC-3′. In this arrangement, the target purine is unpaired. The catalytic portion of the DNAzyme is the 15-nucleotide sequence 5′-GGCTAGCTACAACGA-3′, which loops out from between the binding arms.[4,5]

Binding arms are typically chosen so that each arm has a duplex stability of 10–12 kcal/mol (9–14 nucleotides depending on sequence) when paired with target RNA.[5] To calculate the free energy of each arm in its hybridized form, it is useful to sum published values for the free energies of nearest-neighbor pairs in RNA–DNA duplexes.[9]

The cleavage reaction results in 5′-hydroxyl and 2′,3′-cyclic phosphate termini on the RNA.[5] In the event that the RNA reaction product will be used in a subsequent ligation reaction, the termini must be modified. The

[9] N. Sugimoto, S. I. Nakano, M. Katoh, A. Matsumura, H. Nakamuta, T. Ohmichi, M. Yoneyoma and M. Sasaki, *Biochemistry* **34,** 11211 (1995).

2′,3′-cyclic phosphate terminus can be converted to a mix of 3′- and 2′-phosphates by treatment with maleic acid.[10] Treatment with calf intestinal phosphatase (CIP) then removes the terminal phosphates, resulting in a uniform 3′-hydroxyl terminus.[11] The 5′-hydroxyl terminus can be phosphorylated with T4 polynucleotide kinase.

Materials

DNAzyme oligonucleotide (typically 35–40 nucleotides in length)
Unlabeled or ^{32}P-end-labeled RNA
5× Reaction buffer (750 m*M* NaCl, 200 m*M* Tris-HCl, pH 8.0)
5× Magnesium acetate or $MgCl_2$ (300 m*M*)
RQ RNase-free DNase (Promega, Madison, WI)

Protocols

Preparative-Scale DNAzyme Cleavage of RNA

This procedure can often eliminate the need for recloning plasmid transcription templates when one requires smaller pieces of a parent RNA transcript. For example, transcripts can be cut into one or more pieces at any desired purine-pyrimidine linkage to create deletion mutants. Alternatively, the approach can be used to trim T7 RNA polymerase transcripts of transcriptionally ragged 3′ and 5′ ends without cloning a ribozyme at the terminus.

Protocol. In this procedure, RNA (~1–10 μM final concentration) is combined with DNAzyme (~30 μM final concentration) in 0.1× reaction buffer (i.e., 15 m*M* NaCl, 4 m*M* Tris-HCl). The low-salt buffer promotes proper annealing and prevents extensive nonspecific binding between RNA and DNAzyme. The solution is heated at 95° for 3–4 min to denature the RNA and DNA, and then placed on ice for 5 min. The reaction is then incubated at 25° for 10 min. The buffer concentration is brought to 1× by addition of 5× reaction buffer, and the temperature is raised to 37–42°. Reaction is initiated by addition of 5× $MgCl_2$ solution to a final concentration of 60m*M*. Reaction times can vary from 30 min to 4 hr, depending on the efficiency of individual DNAzyme/RNA sequence pairs. The reaction tube is then placed on ice and the nucleic acids are ethanol precipitated. If the DNAzyme oligonucleotide is likely to interfere with subsequent

[10] S. L. Alam, C. C. Nelson, B. Felden, J. F. Atkins and R. F. Gesteland, *RNA* **4,** 607 (1998).
[11] M. J. Maunders, ed., "Alkaline Phosphatase," Vol. 16. Humana Press, Totowa, New Jersey, 1993.

purification of the RNA (for example, if RNA and the DNAzyme are of similar lengths), the DNAzyme can be eliminated by digestion with RNase-free DNase (Promega). After ethanol precipitation, the sample is diluted in the RQ DNase buffer (Promega) in a volume of 100 μl, 15 units of DNase are added, and the reaction is incubated at 37° for 1 hr. The solution is then phenol extracted and ethanol precipitated again. The processed RNA is then purified by denaturing polyacrylamide gel electrophoresis (PAGE).

Optimization of Conditions. Prior to using a DNAzyme, it is often helpful to run an analytical-scale reaction in order to optimize reaction conditions, because each DNAzyme/target RNA context is slightly different. Conditions that can be useful to vary include the concentration of Mg^{2+}, RNA/DNA ratio, concentration of DNAzyme, reaction temperature, time of reaction, or low concentrations (~0.5 M) of urea. It has also been reported that other cations, such as Ca^{2+} or Mn^{2+} in the presence of NaCl and N-(2-hydroxyl)piperazine-N'-3-propanesulfonic acid (EPPS) buffer can have a highly stimulatory effect on reaction.[5] Depending on the sequence of the flanking arms, it appears that there are, in general, two classes of DNAzyme: one fast and one slow. Whether a DNAzyme is in one class or another has, up to this point, not been apparent from the sequence or structure of the DNAzyme or its target RNA. In general, fast DNAzymes require only 5 min at 37° to completely cleave an RNA substrate, while slow ones may yield ~40% cleavage after 4 hr at 42° under the conditions described here.

It is important to emphasize that DNAzymes have been found to function with high efficiency under conditions of RNA excess in certain contexts, so it is advisable to try published multiple-turnover protocols during optimization of the reaction.[5] In particular, it has been found that thermocycling during reaction can cause a catalytic amount of DNAzyme to dissociate from RNA products and catalyze new rounds of RNA cleavage (John Burke, personal communication, 1997). Even under single-turnover conditions, it should be possible to use lower concentrations of DNAzyme than those reported here. The fact that we observe greater extents of reaction with higher DNAzyme concentration may simply be the result of the sequences or systems we have studied.

Analytical-Scale DNAzyme Cleavage of Labeled RNA

This procedure has been adapted in our laboratory for creating homogeneous 3′ ends on ^{32}P-labeled transcripts used in group II intron structural studies.[7] One begins this procedure by first cutting the transcription template with a restriction enzyme that cleaves at least 20 nucleotides further downstream from the desired 3′ terminus of the product RNA transcript.

Protocol. Approximately 20 pmol of transcript (containing an elongated 3′ end) is 5′-end-labeled with [α-^{32}P]ATP and polynucleotide kinase. After purifying the labeled RNA by phenol extraction and ethanol precipitation, the pellet is resuspended in 4 μl of H_2O. Two microliters of DNAzyme (from a 300–400 μM stock solution, in water) is added, the mixture is heated for 2 min at 95°, and then placed on ice for ~1 min. In our hands, the fast-cooling step described here and in the previous protocol works best, although slow-cooling steps have been successful in other contexts.[5] The chilled sample is then quickly combined with 2 μl of 5× DNAzyme buffer and 2 μl of 5× $MgCl_2$ (for a total volume of 10 μl), the solution is vortexed, and the reaction is incubated for 2 hr at 37°. The cleaved and uncleaved products are resolved by denaturing PAGE. Radioanalytic quantitation shows that, for the substrates we have studied,[7] ≥80% of precursor RNA molecules are cleaved by this procedure. The procedure can be optimized by changing the same experimental variables described in the previous section.

Mapping Branch Points and Other RNA Modifications Using DNAzymes

In this procedure, two DNAzymes are used to excise a small piece of RNA from a much larger parent RNA in order to study the fragment in fine detail.[8] This allows high-resolution mapping of the position and composition of base modifications, or as in this case, 2′-5′-3′ branched nucleotides common in processed group II or spliceosomal introns (Fig. 2). After excision from the surrounding RNA, the fragment can be 5′- or 3′-end labeled for sequencing, or it can be subjected to fingerprinting techniques to study specific nucleotide modifications. The procedure described here has been optimized specifically for studying the branch point of group II intron ai5γ,[8] but alterations can be made for the study of many other modified nucleotides.

Protocol. Lariat RNA is isolated from a splicing reaction in which the precursor RNA was lightly doped with [α-^{32}P] UTP to permit visualization of reaction products.[8] After isolation and purification, the lariat RNA (45 pmol) is resuspended in an RNA storage buffer (60 μl of ME, containing 10 mM MOPS, pH 6.5, 1 mM EDTA), heated to 90° for 2 min, briefly spun, and placed on ice. The mild, low-pH buffer used in this step is relatively stable to temperature fluctuations and helps to reduce degradation when RNA samples are subjected to heat denaturation prior to reaction. At the same time, the two DNAzymes (9 nmol of each) are each suspended in 15 μl of TE (10 mM Tris, pH 8.0, 1 mM EDTA), incubated separately at 90° for 1 min, briefly spun, and placed on ice. Then the RNA, the two DNAzymes, 30 μl of the DNAzyme buffer, and 30 μl of the 5×

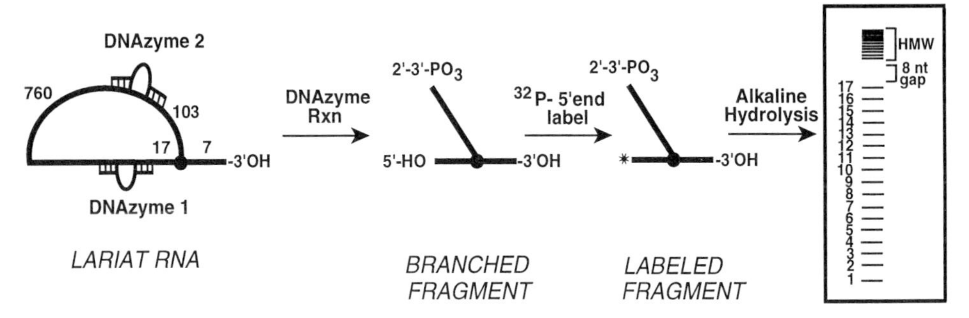

FIG. 2. Simultaneous use of two DNAzymes to isolate a group II intron branch point for high-resolution mapping studies. The large black dot indicates the site of the 2′-5′-3′ branch-point linkage. DNAzyme 1 was designed to cleave the sequence 17 nucleotides upstream of the branch point. DNAzyme 2 targets the sequence 103 nucleotides from the intron 5′ end. In the case shown, 5′-end labeling of the product fragment results in a 17-nucleotide span between the 5′ label and the branch point. Alkaline hydrolysis therefore results in a 17-nucleotide ladder, followed by a gap separating the branch point from higher molecular weight species. This pinpoints the site of the branch point, and the result is confirmed by 3′-end labeling and mapping the other side of the fragment.

$MgCl_2$ stock are combined, mixed, and incubated at 37° for 5 hr. The total volume of this reaction is 150 μl. After reaction, the sample is purified by PAGE on a 6% denaturing gel. For branch-point characterization,[8] the sample is either 5′- or 3′-end labeled with ^{32}P and then mapped by alkaline hydrolysis.[12]

Optimization of Conditions. Note that in this particular procedure, there is no preannealing step in which DNAzymes are combined with lariat RNA and incubated in the absence of salt (or low salt). This is because annealing steps of the type described in the previous protocols had no effect on the DNAzyme reactions in this case. In optimizing this reaction, increasing the $MgCl_2$ concentration and the incubation time were found to have the greatest stimulatory effects on reaction extent. In addition, longer DNAzyme arms ($\Delta G^\circ > 14.5$ kcal/mol for each arm rather than 12 kcal/mol) were found to increase the extent of product formation, presumably because their slower off-rate led to increased cleavage under these single-turnover conditions.

Applications

In the preceding protocols, variations on the DNAzyme reaction have been used to generate short, uniform RNA fragments from long RNAs,

[12] Y. Kuchino and S. Nishimura, *Methods Enzymol.* **180,** 154 (1989).

on both the preparative and analytical scale. The technique has been applied to generate homogeneous 3′ ends on an RNA transcript,[7] which is useful for crystallographic applications, as well as other types of structural studies. In addition, a double DNAzyme protocol is presented for mapping modifications within a large piece of RNA.[8] DNAzymes are likely to be particularly useful for "cut-and-paste" approaches to incorporating modified nucleotides within a long piece of RNA.[13,14] Typically, incorporation of a single modified residue within a long RNA has been performed through template-directed ligation. This technique often involves building new plasmid constructs that encode the RNA transcripts that flank each side of a synthetic RNA to be inserted. Using DNAzyme reactions, one can conceivably exploit existing RNA transcripts, simply snipping out an internal region and replacing it with a synthetic RNA through template-directed ligation with T4 DNA ligase. (Note that the RNA products of DNAzyme reactions have termini that must be modified prior to ligation reactions, as described earlier.) Similarly, DNAzyme reactions could be used to clip existing transcripts for internal incorporation of single modified pNp nucleotides such as 4-thiouridine cross-linking agents.[15]

[13] R. A. Zimmerman, M. J. Gait and M. J. Moore, *in* "Modification and Editing of RNA" (H. Grosjean and R. Benne, eds.), pp. 59–84. ASM Press, Washington, DC, 1998.

[14] M. J. Moore and C. C. Query, in "RNA Protein Interactions: A Practical Approach" (C. W. J. Smith, ed.), pp. 75–108. Oxford University Press, 1998.

[15] Y. T. Yu and J. A. Steitz, *RNA* **3,** 807 (1997).

Section II

RNA Structure Determination

A. X-Ray Crystallography
Articles 11 through 13

B. Nuclear Magnetic Resonance Spectroscopy
Articles 14 through 16

C. Electron Microscopy
Articles 17 through 19

[11] Purification, Crystallization, and X-Ray Diffraction Analysis of Small Ribozymes

By JOSEPH E. WEDEKIND and DAVID B. MCKAY

Introduction

Small ribozymes[1] are a family of molecular catalysts composed solely of RNA. These enzymes are derived from the genomic or antigenomic RNA of small satellite RNA viruses that infect plants, in the case of the hammerhead and hairpin ribozymes (HP), or from human pathogens, such as the hepatitis delta virus (HDV). Another ribozyme source comes from the mitochondrial Varkud satellite plasmid of *Neurospora,* referred to as the VS ribozyme. Small ribozymes range in size from 24 to 150 nucleotides, each with a different sequence and fold. However, all cleave a specific phosphodiester bond on a target "substrate" RNA strand that is at least partially complementary, through Watson–Crick base pairing, to the ribozyme strand. The cleavage products are a 2′,3′-cyclic phosphodiester and a free 5′-hydroxyl group. In addition, the catalytic activities of these ribozymes are enhanced or greatly stimulated by low concentrations of metal ions such as Mg^{2+}, Co^{2+}, Mn^{2+}, or Zn^{2+}. Other small ribozymes have been derived artificially for the capacity to catalyze a specific chemical reaction. The leadzyme is one example of a ribozyme produced by *in vitro* selection for its ability to cleave a specific phosphodiester bond in the presence of Pb^{2+}.

Ribozymes are interesting not only for their unusual tertiary structures, but also for their chemistry. These enzymes have been exploited in a variety of areas ranging from the simplest molecular biological applications to the combatants of human disease. It is logical that a major step to understanding these molecules comes from the knowledge of their three-dimensional structures. This article provides an overview of some practical strategies used in the purification, crystallization and X-ray crystallographic structure determinations of small ribozymes. Because we have solved the structures of two ribozymes in our laboratory by multiple isomorphous replacement—the hammerhead[2] and leadzyme[3]—they will be cited most often as examples in our case studies.

[1] D. B. McKay and J. E. Wedekind, *in* "The RNA World. The Nature of Modern RNA Suggests a Prebiotic RNA" (R. F. Gesteland, T. R. Cech, and J. F. Atkins, eds.), pp. 265–286. Cold Spring Harbor Laboratory Press, Cold Spring Harbor, New York, 1999.

[2] H. W. Pley, K. M. Flaherty, and D. B. McKay, *Nature* (*Lond.*), **372,** 68 (1994).

[3] J. E. Wedekind and D. B. McKay, *Nature Struct. Biol.* **6,** 261 (1999).

0076-6879/00 $30.00

Strategies to Promote Crystal Packing

One of the rate-limiting steps in any structure determination is the preparation of crystals suited to X-ray diffraction analysis. For this task it may be possible to induce crystal growth by selectively modifying the sequence of the target RNA construct. One method involves the incorporation of "sticky-ended" stems. In the event that the RNA is expected to form a duplex with blunt helical stems [Eq. (1)], a single G overhang at one

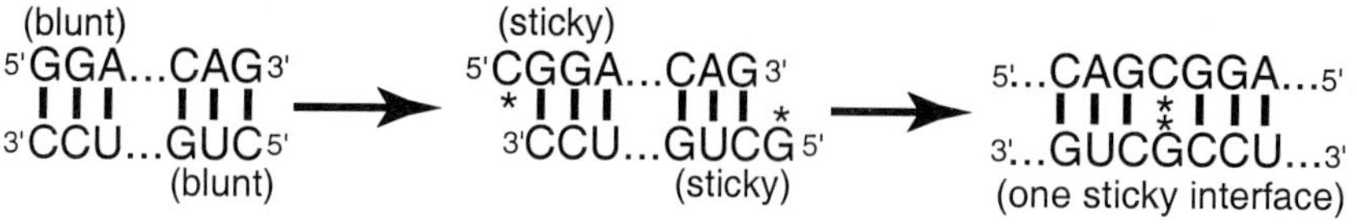

```
(blunt)                     (sticky)
5'GGA...CAG3'               5'CGGA...CAG3'                 5'...CAGCGGA...5'
  III   III     ──────►       *III   III *     ──────►         III*III          (1)
3'CCU...GUC5'                3'CCU...GUCG5'                3'...GUCGCCU...3'
         (blunt)                      (sticky)             (one sticky interface)
```

5′ end, with a C overhang at the other, may promote sticky packing end to end [Eq. (1)]. This strategy proved useful in the leadzyme crystallization, in which molecules are observed to pack end to end in a pseudocontinuous A-form helix. The G* and C* overhangs form a Watson–Crick base pair at the intermolecular interface [Eq. (1)] such that respective sticky ends of the molecule effectively glue the crystal together along one axis of the lattice. In order for pseudocontinuous packing to obey crystallographic symmetry, individual molecules must exist as an integral number of complete helical turns. For an ideal A-form RNA helix, this requires 11 nucleotides. The leadzyme construct utilized an asymmetric 11-mer complexed with a 13-mer; this construct was anticipated to form an asymmetric internal loop flanked by A-form helical stems.

A second strategy incorporates a protein binding site into the ribozyme. This method utilizes a well-characterized protein–RNA interaction placed at an innocuous location in the nucleic acid sequence. Such a strategy was employed for the HP ribozyme, although not in the context of crystallization. This work demonstrated that a nonconserved region of helix 4 of the HP could be extended to include a binding site for phage R17 coat protein.[4] Such a modification appears to stabilize the ribozyme, and the HP remains fully active while R17 is bound. A second example has been utilized successfully in the crystallization[5] and structure determination[6] of a construct of the HDV genomic ribozyme. In this case, the U1A–RNA binding domain was included as a modification of HDV stem loop P4. U1A was overexpressed as the selenomethionine derivatized protein and the HDV was cocrystallized as the U1A–RNA complex. The selenium derivatized crystals

[4] B. Sargueil, D. B. Pecchia, and J. M. Burke, *Biochemistry* **34,** 7739 (1995).

[5] A. R. Ferré-D'Amaré, K. Zhou, and J. A. Doudna, *J. Mol. Biol.* **279,** 621 (1998).

[6] A. R. Ferré-D'Amaré, K. Zhou, and J. A. Doudna, *Nature* **395,** 567 (1998).

were used in a multiwavelength anomalous diffraction (MAD) experiment, which provided a source of experimental phases.

Chemical Synthesis as RNA Source

A traditional source of RNA for structural studies has been *in vitro* transcription by T7 RNA Polymerase.[7] Subsequently, the RNA is purified to remove failure sequences and $N + 1$ extensions, by use of denaturing polyacrylamide gels.[8,9] Because these areas have been documented previously, we will describe an alternative approach in which crystallization quality RNA is prepared from solid-phase chemical synthesis (CPG support) by use of commercially available phosphoramidites; also see Kundrot[9] and Wincott *et al.*[10] The use of this approach can be very powerful because it allows the explicit incorporation of nonstandard nucleotides into the target RNA sequence. For crystallography, perhaps the most useful modification is halogenation of the nucleotide base; this includes the 5-bromo- and 5-iodo adducts of uridine (U) and cytidine (C), as well as the 8-bromo adduct of adenosine (A). Standard and modified phosphoramidites can be purchased from a variety of companies such as Glen Research (Sterling, VA) or ChemGenes, Inc. (Waltham, MA) as the 5′-dimethoxytriphenylmethyl-2′-*tert*-butyldimethylsilylribonucleoside 3′-cyanoethyl (5′-DMT-2′-tBDSilyl-ribonucleoside 3′-CNEt) form.

Preparation of Synthetic Oligonucleotides

Strands synthesized at the 1–10 μM scale can be purchased, linked to the CPG support, from Oligo Etc, Inc. (Wilsonville, OR), CruaChem, Inc. (Sterling, VA), the Keck Center for RNA synthesis at Yale University (New Haven, CT), or the PAN Facility at Stanford University (Stanford, CA). If necessary, residual acetonitrile from synthesis is removed from the synthesis cartridge *in vacuo*. For standard ribonucleotides, the base protection groups are removed in ethanolic ammonium hydroxide (1 : 3; specific gravity 0.896) at 55° for 17 hr. Halogenated strands should be protected from light, and deblocked at room temperature with mild orbital shaking for 40 hr. Ethanolic ammonium hydroxide is removed expeditiously

[7] J. F. Milligan, D. R. Groebe, G. W. Witherell, and O. C. Uhlenbeck, *Nucleic Acids Res.* **15,** 8783 (1987).

[8] H. W. Pley, D. S. Lindes, C. DeLuca-Flaherty, and D. B. McKay, *J. Biol. Chem.* **268,** 19656 (1993).

[9] C. Kundrot, *Methods Enzymol.* **276,** 143 (1997).

[10] F. Wincott, A. DiRenzo, C. Shaffer, S. Grimm, D. Tracz, C. Workman, D. Sweedler, C. Gonzalez, S. Scaringe, and N. Usman, *Nucleic Acids Res.* **23,** 2677 (1995).

in a Speed-Vac. The remaining aqueous solution is flash frozen in liquid nitrogen (−198°) and lyophilized to form a fluffy white powder.

The DMT group can be removed by addition of 0.25–0.50 ml of concentrated acetic acid to the lyophilized powder. The reaction is allowed to proceed at 23° for 30 min with gentle mixing. Hydrolysis of the DMT group may be accompanied by a pink or orange color change. The reaction is stopped by flash freezing the reactants in liquid nitrogen. The volatile acid is drawn off by a lyophilizer equipped with an acid trap. The resulting white powder may appear granular or clumpy, but should not be discolored.

The 2′-tBDSilyl group can be removed through the addition of 0.25–0.50 ml of dry premixed 1 *M* tetrabutylammonium fluoride (TBAF) in THF (Aldrich, Milwaukee, WI) to the lyophilized DMT-off RNA. The reaction proceeds at 23° for approximately 12 hr with gentle mixing. Desilylation is quenched by addition of an equal volume of 1 *M* triethylammonium acetate (TEAA) buffer, pH 7. The solution is loaded onto a DNA grade superfine Sephadex G-25 (Sigma, St. Louis, MO) size-exclusion column, approximately, 10 × 300 mm, using Milli-Q grade (Millipore Corp., Bedford, MA) water as the solvent; RNA should elute in the void volume (about 5 ml). Fractions of 1 ml are collected and analyzed for RNA by spectrophotometric absorption at 260 nm. Fractions free of TBAF are pooled, flash frozen in liquid nitrogen or a dry ice/ethanol bath, and lyophilized. The RNA is now susceptible to cleavage from the addition of metal cofactors or by contamination with ribonucleases. Special care should be taken to avoid contamination from this point onward. All ensuing procedures should be performed under aseptic conditions using the highest quality reagents and water.

HPLC Purification

In our studies, chemically synthesized oligonucleotides were utilized if sequences were ≤13-mers in length. Longer strands can be problematic since they tend to form undesirable secondary structures that make purification unreliable under nondenaturing conditions; yields from chemical synthesis become increasingly poorer with longer sequences, as well, due to coupling failures. RNA strands are purified at 23° through modification of a previously reported method, which was designed to separate RNAs containing racemic mixtures of phosphorothioates.[11] The chromatography medium is μBondapak C_{18} (Waters, Milford, MA) with a 125-Å pore size. Semianalytical (7.8 × 300 mm) or preparative (19 ×

[11] G. Slim and M. J. Gait, *Nucleic Acids Res.* **19,** 1183 (1991).

300 mm) scale columns, equipped with guard columns, are utilized for our applications. The buffer system is 0.10 *M* TEAA, pH 7 (buffer A), and 0.10 *M* TEAA with 50% acetonitrile (buffer B). Individual lyophilized strands (DMT-off and tBDSilyl-off) are dissolved in buffer A and injected as 50-μl to 1-ml volumes onto the C_{18} column. Typical preparative scale purifications utilize a gradient of 15–24% buffer B over 60 min with a flow rate of 7 ml/min. The elution of the RNA is monitored by UV absorption at 260 nm. A majority of the synthesis product elutes as a single sharp peak of the desired oligonucleotide, which can be collected in a single vessel. The RNA is flash frozen and lyophilized. This method allows the purification of a 1-μmol scale synthesis in a single high-performance liquid chromotography (HPLC) injection.

Desalting RNA

Prior to crystallization, RNA should be desalted to remove residual TEAA from the HPLC purification. Samples can be desalted by use of a 12-ml (2-g) Sep-Pak Vac C_{18} cartridge (Waters, Milford, MA). Such cartridges may be treated as small gravity flow columns. To activate the resin, 40 ml of acetonitrile is passed through the cartridge, followed by washing with an equal volume of water. The lyophilized RNA is resuspended in $\leq$2 ml of water and loaded onto the column. When loaded, the cartridge is washed with 25 ml of water. The desalted RNA is eluted in bulk by application of 20 ml of 40% aqueous acetonitrile. The RNA is flash frozen and lyophilized to give a white fluffy powder, which is stored at $-70°$. A small amount of RNA can be resuspended in water, and submitted for mass spectroscopy, which should reveal a single predominant species of the expected molecular weight. In addition, the RNA may be subjected to analytical denaturing polyacrylamide gel electrophoresis (PAGE)[12] by use of a minigel (10×8 cm^2) apparatus. One example is the Hoefer SE 200 series (Amersham Pharmacia Biotech, Piscataway, NJ). For detection of nanogram quantities of RNA, the gel can be silver stained.[13]

This purification method yields approximately 0.5–1.5 mg of pure RNA per 1 μmol synthesis. Halogenated oligonucleotides prepared by this procedure retain nearly 100% of their iodo or bromo substituents. This method is suited to the separation of racemic phosphorothioates as well.

[12] J. Sambrook, E. F. Fritsch, and T. Maniatis, "Molecular Cloning: A Laboratory Manual." Cold Spring Harbor Laboratory Press, New York, 1989.

[13] H. Blum, H. Beier, and H. J. Gross, *Electrophoresis* **8,** 93 (1987).

Crystallization

Several brief protocols have been devised to screen for nucleic acid crystallization conditions.[9,14–17] Such searches can be conducted most effectively by use of the hanging drop vapor diffusion method.[18] This is conducive to the use of 24-well Linbro or VDX (Hampton Research, Laguna Niguel, CA) style plates containing 1 ml of each precipitant mixture per well. First, the pure lyophilized RNA strands are dissolved in a suitable buffer at low ionic strength in an appropriate molar ratio. The total RNA concentration may vary according to the RNA solubility. The RNA, 2–7 μl, is pipetted onto the center of a 22-mm siliconized autoclaved coverslip; AquaSil (Pierce, Rockford, IL) provides a suitable siliconization medium. An equal volume of precipitant solution from the well is added to the RNA drop without manual mixing. The coverslip is inverted and placed over the precipitant well, and is sealed by preapplication of a thin layer of grease or heavy paraffin oil (J. T. Baker, Phillipsburg, NJ) on the rim of the plate. Initial screening experiments should be performed in duplicate at 4° and 20°. Precipitation and other solubility properties should be noted immediately upon setup, and each time the plates are examined for crystals. Typically, plates are stored for 3–5 days before their first inspection for crystals.

Precipitants in RNA Crystallization

If rapid screening protocols fail to provide diffraction quality crystals, a more extensive search must be conducted. This may require the exhaustive combination of multiple cations, anions, polymers, organic agents, metals, or polyamines. In this manner, the effect of a single variable is changed from one experiment to the next; each experiment consists of a single chamber of the 24-well plate. A list of these variable components is provided in Table I, which should be made up as concentrated stock solutions (sterile filtered and autoclaved if possible). For example, there are 18 possible cation–anion combinations alone. These can be combined with numerous polymers, such as poly(ethylene) glycol (PEG), or organic compounds, such as 2-methyl-2,4-pentane diol (MPD) at variable concentrations (for example, 5–35% quantities augmented in 5% increments). To reduce the

[14] J. A. Dounda, C. Grosshans, A. Gooding, and C. E. Kundrot, *Proc. Natl. Acad. Sci. U.S.A.* **90,** 7829 (1993).

[15] W. G. Scott, J. T. Finch, R. Grenfell, J. Fogg, T. Smith, M. J. Gait, and A. Klug, *J. Mol. Biol.* **250,** 327 (1995).

[16] I. Berger, C.-H. Kang, N. Sinha, M. Wolters, and A. Rich, *Acta Crystallogr.* **D52,** 465 (1996).

[17] J. H. Cate and J. A. Doudna, *Methods Mol. Biol.* **74,** 379 (1997).

[18] A. McPherson, "Preparation and Analysis of Protein Crystals." Krieger, Melbourne, Florida, 1989.

TABLE I
PRECIPITANTS FOR RNA CRYSTALLIZATION

Precipitants					Metals		
Salts							
Cations	Anions	Polymers	Organics	Buffers	Divalent	Trivalent	Polyamines
		PEG[a]	EG		Mg^{2+}	Lu^{3+}	Cadaverine
Li^+	Cl^-	200, 400	Glycerol	Cacodylate	Ca^{2+}	Sm^{3+}	Putrescine
Na^+	SO_4^{2-}	1k, 2k	MPD[b]	Bis–Tris	Sr^{2+}	Nd_{3+}	Spermidine
K+	$C_2H_3O_2^-$	3350	Ethanol		Ba^{2+}	$Co(NH_3)_6^{3+}$	Spermine
NH_4^+		5k	Dioxane		Mn^{2+}	$Os(NH_3)_5^{3+}$	
$N(CH_3)_4^+$		8k, 10k	DMSO[d]		Zn^{2+}		
$N(CH_3CH_2)_4^+$		MME[c] 2k	1,3-Butanediol		Cd^{2+}		
		MME 5k	1,4-Butanediol		Co^{2+}		
		Jeffamine[e] M600	1,6-Hexanediol		Ni^{2+}		

[a] PEG, poly(ethylene) glycol; EG, ethylene glycol; PEG 1k, PEG 1000.
[b] MPD, 2-methyl-2,4-pentanediol.
[c] MME, monomethyl ether.
[d] DMSO, dimethyl sulfoxide.
[e] Jeffamine M600, *O*-(2-aminopropyl)-*O'*-(2-methoxyethyl)polypropylene glycol 500.

number of variables to screen, the pH of each experiment can be maintained near neutral, since pH does not appear to influence RNA crystallization; typically one can use 0.05 *M* cacodylate buffer (a bacteriostatic) at pH 6. However, since cacodylate may form insoluble complexes with di- and trivalent metals, it should be substituted with a less reactive buffer, such as Bis–Tris, while searching for heavy atom derivatives. In addition, small amounts of polyamine, such as 1–20 m*M* spermine hydrochloride, should be included in screening experiments.

Choice of Cation

One of the most important aspects in small ribozyme crystallization is the selection of an appropriate divalent or trivalent cation. Several cases presented here demonstrate that the inclusion of such ions in the crystallization medium can have profound effects on crystal growth and habit. Because all small ribozymes require or are stabilized by metals, it is sensible to include 1–20 m*M* of such ions in crystallization screens, as long as they do not activate the ribozyme for *cis* or *trans* cleavage reactions. Several multivalent cations are listed in Table I. Most of these are likely to make direct, first sphere contacts, to the RNA through interactions with nitrogen or oxygen ligands. As such, $Co(NH_3)_6^{3+}$ appears to bind in a unique manner, since it can mimic a hydrated magnesium ion, $Mg(H_2O)_6^{2+}$. Often, this

TABLE II
HAMMERHEAD X-RAY DIFFRACTION[a]

Group IIA cation (radius, Å)[b]	Diffraction limit (Å)
Mg^{2+} (0.86)	2.3
Ca^{2+} (1.14)	2.5
Sr^{2+} (1.32)	3.4
Ba^{2+} (1.49)	6.0

From F. A. Cotton and G. Wilkinson, "Advanced Inorganic Chemistry," 5th Ed. Wiley and Sons, New York, 1988.

[a] X-ray diffraction data were collected on an R-axis IIC image plate detector using graphite monochromatized CuKα X rays from a Rigaku RU200 rotating anode operated at 4.5 kW with a 0.3- × 3-mm focus cup.

[b] Assuming a coordination number of six.

relatively inert metal complex serves a purely structural role, and is known to coordinate the bases of tandem GU mismatches in RNA.[19]

An extreme example of how multivalent cations influence RNA crystals has been observed with the hammerhead ribozyme. In this case, the ribozyme 34-mer, complexed with a 13-mer substrate analog, was crystallized in the presence of either Mg^{2+}, Ca^{2+}, Sr^{2+}, or Ba^{2+} from high concentrations of $K(C_2H_3O_2)$. Regardless of the group IIA ion used, crystals grew with the identical hexagonal habit and were indistinguishable by all other optical and growth properties. However, crystals grown in the presence of ions larger than Mg^{2+} diffracted exceedingly worse as a function of ionic radius (Table II). In addition, Co^{2+} appeared to increase the mosaic spread[20] of these crystals to values of 3° to 5°; crystals grown from Mg^{2+} have an anisotropic mosaicity with values ranging from 0.6 to 1.2°.

In the context of multivalent ions, it is interesting to note the properties of some typical crystals grown in our laboratory (Table III). In the case of the hammerhead, it is possible to substitute Mg^{2+} for Lu^{3+} keeping all other crystallization conditions constant. This single change of ion coincides with a change in crystal habit and space group; in the presence of Mg^{2+}, crystals appear as hexagonal plates, space group *C2;* whereas Lu^{3+} produces well-diffracting hexagonal rods, space group $P6_322$ (Table III). Due to the highly mosaic and anisotropic characteristics of X-ray diffraction from the group IIA sensitive *C2* crystal form, the $P6_322$ crystal form was pursued instead.

[19] J. H. Cate and J. A. Doudna, *Structure* **4,** 1221 (1996).

[20] Mosaic spread is defined as the full-width half-maximum of the rocking-curve profile.

TABLE III
CRYSTALLIZATION AND DIFFRACTION PROPERTIES OF RIBOZYME CRYSTALS[a]

Ribozyme construct (m*M*)	Crystallization conditions	Cell constants (Å), space group, number per asu[b]	Resolution limit (Å)	R_{sym} (%)[c] (last shell)
Hammerhead 34-mer + 13-mer (0.30 m*M*)	1.9–2.2 *M* $(NH_3)_2SO_4$, 10 m*M* buffer, pH 6, 0–100 m*M* $MgCL_2$, 0–2 m*M* spermine hydrochloride, 4°	a = b = 89.7, c = 185.8 *P3₂21*	2.60	6.9 (NA[d])
Hammerhead 34-mer + 13-mer (0.46 m*M*)	3.2 *M* $K(C_2H_3O_2)$, 50 m*M* Bis–Tris-Cl, pH 6, 1 m*M* spermidine hydrochloride, 1–20 m*M* $Mg(C_2H_3O_2)_2$, 4°	a = 77.9, b = 134.4, c = 84.6, β = 104.5° *C2* 6 per asu	2.25	5.6 (11.0)
Hammerhead 34-mer + 13-mer (0.46 m*M*)	3.2 *M* $K(C_2H_3O_2)$, 50 m*M* Bis-Tris-Cl, pH 6, 1 m*M* spermidine hydrochloride, 1 m*M* $Lu_2(SO_4)_3$, 4°	a = b = 70.0, c = 163.9 $P6_322$ 1–2 per asu	1.65	5.8 (26.5)
Hepatitis δ virus (RC1) 61-mer circle + 8-mer (0.37 m*M*)	2.7 *M* $Li(C_2H_3O_2)$, 50 m*M* Bis-Tris-Cl, pH 6, 1 m*M* spermine-hydrochloride, 20 m*M* $Mg(C_2H_3O_2)_2$, 1 m*M* $Co(NH_3)_6Cl_3$, 20°	a = 75.4, b = 85.4, c = 84.7, β = 96.6° $P2_1$ 4 per asu	3.20	8.0 (31.4)
Leadzyme (LZ4) 13-mer circle + 11-mer (0.84 m*M*)	20–25% MPD, 50 m*M* sodium-cacodylate, pH 6, 1 m*M* spermine-hydrochloride, 20 m*M* $Mg(C_2H_3O_2)_2$, 20°	a = b = 60.4, c = 133.1 $P6_122$ 2 per asu	2.70	6.4 (21.4)

[a] All X-ray data were collected at the Stanford Synchroton Radiation Laboratory on beamline 7-1 or 9-1, except the C2 form of the hammerhead, data for which were collected as described in Table II. X-ray data were collected at temperatures close to −173°.
[b] Number per asymmetric unit (asu); determined experimentally or derived from estimates of the solvent content of crystals and/or symmetry elements of self-rotation functions.
[c] $R_{sym} = \Sigma_{hkl}|I_j - \langle I_j\rangle|/\Sigma_{hkl}|I_j|$, where I_j is the jth intensity measurement of a reflection, and the sum is over all reflections.
[d] NA, not available.

The mosaicity of these Lu^{3+} derived crystals is isotropically 0.45°. Such a dependence on monovalent cations was documented previously for the $P3_221$ crystal form of the hammerhead, as well, in which crystals grown from solutions of ammonium sulfate were adapted to lithium sulfate. The use of Li^+ brought about an increase in X-ray diffraction resolution, and provided a more suitable medium for heavy atom derivative searches.[2]

Although the dependence on cations appears to be extreme for the hammerhead crystallization, it is apparent that cation valency and radius can play a decisive role in RNA crystallization and X-ray diffraction properties.

X-Ray Diffraction Analysis

Ideally, RNA crystals should be as large as possible. In practice, one can examine crystals ranging in size from $1.0 \times 1.0 \times 1.0\ mm^3$ to $0.015 \times 0.015 \times 1.0\ mm^3$. These are typical dimensions for the hammerhead ($P3_221$) and leadzyme crystal forms of Table III. Crystals should be single and free from imperfections, as well; however, in the case of the leadzyme, crystals nucleate from a central locus with multiple needles growing side by side, much like a pin cushion. Such crystals are operative if single crystals can be cut away from the nucleation site with a thin knife blade or glass fiber. These needles can be transferred on the tip of a keratin bristle into synthetic mother liquor, where they are ready for X-ray diffraction analysis. In the case of the HDV (Table III), crystals grow as individual thin plates, with dimensions of $0.01 \times 0.3 \times 0.4\ mm^3$. In such cases, the crystal size must be increased by multiple rounds of macroseeding,[21] which can increase the thickness by five- to sevenfold; this is necessary for suitable X-ray diffraction.

Due to their radiation sensitivity, RNA crystals must be flash cooled at low temperatures (−173°) prior to the initiation of an extensive X-ray diffraction study. However, it is recommended that crystals be mounted initially in thin-walled quartz capillaries[22] (Charles Supper, Inc., Natick, MA) in order to assess diffraction properties, such as resolution and mosaic spread. This serves as a control procedure since it eliminates the time and difficulties spent in the development of a new cryoprotection procedure.[23] Capillary mounts avoid the uncertainty of whether poor diffraction is a property of the crystal, or the consequence of a failed freezing attempt.

It has been our experience that most RNA crystals, whether frozen or at 23°, rarely diffract X rays to high resolutions by use of conventional X-ray sources. As shown in Table III, the larger ribozyme construct of the HDV (69 nucleotides) diffracts to 3.2-Å resolution at a synchrotron radiation (SR) source, with a typical R_{sym} of 8.0%; in-house the same crystals diffract weakly to 4.5-Å resolution. Similarly, the ammonium sulfate crystal form of the hammerhead diffracts significantly better (by approximately

[21] C. Thaller, G. Eichele, L. H. Weaver, E. Wilson, R. Karlsson, and J. N. Jansonius, *Methods Enzymol.* **114,** 132 (1985).

[22] I. Rayment, *Methods Enzymol.* **114,** 136 (1985).

[23] We estimate that fewer than 1 in 5 new RNA crystal forms diffract X rays well enough to pursue for structural studies.

0.5-Å resolution) at a SR source, compared to in-house. Most likely, RNA crystals suffer less radiation damage at SR sources, since there is less absorption and X-ray data are collected over a much shorter period of time (hours rather than days). It is recommended that diffraction analyses of RNAs be conducted using SR, which is suited to especially difficult cases. Helliwell[24] provides a gazetteer of SR sources that can be useful for planning X-ray diffraction experiments.

Synthetic Mother Liquors and Cryoprotection

A good synthetic mother liquor is essential for the successful X-ray diffraction analysis of any macromolecular crystal. This should be a stabilizing solution, usually derived from the precipitating agent, in which crystals can be stored indefinitely (if possible) without deterioration of their diffraction properties. Often this solution is the solvent for heavy atom compounds used in derivatization, and serves as the basis of cryoprotectant solutions used to adapt the crystal for flash cooling at liquid nitrogen temperatures. As a starting point, one should attempt to increase the concentration of any organic or polymeric precipitant used in the crystallization (Table III) by as much as 5–10%. This reduces the bulk solvent in the crystal and may make the RNA less reactive to heavy atom compounds, and closer to the final freezing conditions. If possible, the ionic strength should be increased by as much as 30%, as well. This is important for alleviating osmotic shock that may occur during the adaptation into cryoprotectant. Methods for preserving crystals by flash-cooling at low temperatures have been reviewed elsewhere for proteins,[25] but are applicable here as well.

In our case study of small ribozyme crystals, it is interesting to note how our RNA crystals are cryoprotected prior to X-ray diffraction experiments. Typical cryoprotectants include glycerol, PEG 400, MPD, ethylene glycol (EG), or high concentrations of salt. The ammonium sulfate crystal form of the hammerhead (Table III) can be cryoprotected in 25% glycerol[2] or EG.[26] The *C2* and $P6_322$ crystal forms of the hammerhead are prepared from high concentrations of $K(C_2H_3O_2)$. Such high ionic strength allows these crystals to be frozen directly, after a 30-sec exposure to the 3.2 *M* $K(C_2H_3O_2)$ mother liquor (Table III). A similar method has been successful for cryoprotecting crystals of the HDV, which grows from 2.7 *M* $Li(C_2H_3O_2)$ (Table III). The leadzyme, which is grown from 20 to 25% MPD (Table III), can be transferred to solutions of 35% MPD for less than 30 sec or stored indefinitely—either timescale for cryoprotection is suitable for direct

[24] J. R. Helliwell, "Macromolecular Crystallography with Synchrotron Radiation." Cambridge University Press, Cambridge, Massachusetts, 1992.

[25] D. W. Rodgers, *Methods Enzymol.* **276,** 183 (1997).

[26] J. E. Wedekind and D. B. McKay, unpublished work (1997).

flash freezing. Crystals of the leadzyme soaked in 35% MPD are suited to screen for heavy atoms, as well. This has the advantage that no additional cryoprotection steps on the heavy atom sensitized crystals are necessary prior to freezing.

Strategies to Prepare Heavy Atom Derivatives

Several strategies have been designed to incorporate heavy atoms into RNA crystals as a means of obtaining experimental phase information. The traditional method, known locally as "soak and pray" (see Rould[27]), has not been particularly successful for RNAs, most likely due to their polyanionic nature and general dearth of reactive chemical groups. However, under crystallization conditions of low ionic strength, heavy atoms can be soaked successfully into ribozyme crystals to provide isomorphous derivatives. In the case of the leadzyme,[3] five heavy atom derivatives were prepared in this manner by use of divalent or trivalent ions (Table IV). This approach was a last resort, after bromouracil and iodouracil strands failed to crystallize. In high-salt conditions of 2.4 *M* Li_2SO_4, the hammerhead ribozyme will bind specifically to Cd^{2+} (Table IV), when soaked in 0.10 *M* $CdSO_4$. However, such Cd^{2+} sites refine with high-temperature factors in all three independent copies, and are probably not suited as heavy atom derivatives.

Another strategy to obtain derivatives is to cocrystallize a heavy atom with the ribozyme. In the case of the HDV RC1, it has been possible to prepare crystals in the presence of $Os(NH_3)_5{}^{3+}$ (Aldrich, Milwaukee, WI) in lieu of the $Co(NH_3)_6{}^{3+}$ used normally for crystallization (Table III). This strategy was inspired by the success of a similar method used in the structure determination of the P4–P6 domain of the group I intron.[19,28] Cocrystallization in the presence of a lanthanide provides another alternative for RNA derivatization. This approach has been successful for the hammerhead ribozyme in which Lu^{+3} was used to replace Mg^{2+} in the crystallization medium. This minor change resulted in the growth of a new crystal form, changing the space group from *C2* to $P6_322$ (Table III). An anomalous difference Patterson synthesis, derived from X-ray data collected at the lutetium L_{III} absorption edge, revealed that the heavy atom binds on the crystallographic six-fold axis. This condition limited the phase information collected in a MAD experiment to a set of phases for only the $l = 2n$ reflections. The high concentration of $K(C_2H_3O_2)$ in the mother liquor has precluded attempts to prepare additional heavy atom derivatives.

A more rational approach to obtaining isomorphous derivatives is

[27] M. A. Rould, *Methods Enzymol.* **276,** 461 (1997).

[28] J. H. Cate, A. R. Gooding, E. Podell, K. Zhou, B. L. Golden, C. E. Kundrot, T. R. Cech, and J. A. Doudna, *Science* **273,** 1678 (1996).

TABLE IV
HEAVY ATOM BINDING SITES

Site[a]	Heavy atom ion	Functional group	Distance (Å) (Copies 1 to n)
Hammerhead[b]			
1	Cd^{2+}	A9.0 O1P	2.6, 2.4, 2.5
		G10.1 N7	2.3, 2.4, 2.5
Leadzyme			
1a	Ba^{2+}	G42 O6	nb[c], 3.5
		U41 O4	nb, 3.3
2	Ba^{2+}	C23 O2′	nb, 3.3
		C23 O2	nb, 3.5
		A25 O2P	nb, 2.9
		A45 N1	nb, 2.9
3	Ba^{2+}	G46 O6	3.4, 3.0
		U47 O4	nb, 3.5
1b	Pb^{2+}	G42 N7	nb, 3.1
		G42 O6	nb, 3.5
		U41 O4	nb, 3.5
4	Lu^{3+}	C40 O2P	2.4, nb
5		C49 O2P	nb, 3.5
6	UO_2^{2+}	G43 O1P	nb, 3.3
7		site 6[d]	nb, 2.9
8		G46 O1P	nb, 3.7
9		site 8[d]	nb, 3.3

[a] Site numbers with letter suffixes bind more than one ion type, such as Pb^{2+} or Ba^{2+}.
[b] Wedekind and McKay, unpublished (1997).
[c] nb, did not bind in this copy.
[d] Site binds close to another U atom; sites 6 and 7 and 8 and 9 are too close (2.8–3.2 Å) to distinguish if they are distinct.

through the incorporation of iodo- or bromo-substituted nucleotides into the nucleic acid strands. This strategy was effective for the hammerhead ribozyme structure, which utilized an all-DNA substrate analog bound to an RNA ribozyme 34-mer. In this case, 5-iodouracil was substituted for each of the respective thymine positions in the DNA strand. The superior quality of iodine as a derivative and its $7e^-$ anomalous signal at 1.5-Å wavelength made it possible to solve the hammerhead structure by use of an in-house X-ray source. A second derivatization approach incorporates phosphorothioates at specific nonbridging oxygen positions of the phosphodiester backbone. Crystals grown with the sulfur substitution can be reacted with a mercurial derivative, due to the extensive thiolate character of the P–S bond.[29] In practice, phosphorothioates have been used directly as the

[29] P. A. Frey, *Adv. Enzymol. Relat. Areas. Mol. Biol.* **62,** 119 (1989).

racemic mixture from chemical synthesis[30] or in conformationally pure forms.[31]

Heavy Atom Positions—A Case Study

The most common method for obtaining experimental phase information is by use of Patterson methods. In this manner, differences in the structure factor amplitudes between native, F_P, and derivatized, F_{PH}, RNA crystals are used as coefficients in a Patterson synthesis. This map of interatomic vector space (u, v, w) can yield the positions of heavy atoms in the derivatized RNA crystals. In the general case, the number of nonorigin interatomic vectors increases as $N(N-1)$, where N is the total number of heavy atom sites in the unit cell. It is obvious that such an analysis becomes a difficult prospect when the number of sites in the asymmetric unit (asu) is large and the space group symmetry is high. For the case of the leadzyme (Table IV), there are 4 Ba^{2+} sites in 12 asymmetric units; this produces 188 nonorigin Patterson vectors per asymmetric unit. However, for the single-site Pb^{2+} derivative, there are only 11 nonorigin vectors per asymmetric unit. Thus, the Pb^{2+} site is best suited (in retrospect) to a manual solution of the unique site.[32] Representative Harker sections of the Patterson can be used to calculate the fractional coordinates of a particular lead site in the asu (Fig. 1). Although these maps are interpretable *a priori,* they are somewhat noisy (contours begin at 1.25σ and increase in 0.5σ increments). This is probably due to the low level of lead substitution, as well as some degree of nonisomorphism. This latter effect may be due to the fact that lead is the cofactor necessary for lead-dependent cleavage, and its binding and chemical reactivity may be disruptive to the crystal lattice. In any event, any detectable signal above noise on a Harker section should be scrutinized when considering possible solutions. In practice, most Patterson maps derived from small ribozymes have been less than optimal. One method to maximize the difference Patterson signal is to local scale F_{PH} to F_P. This procedure can be implemented in SOLVE[33] and has been reviewed elsewhere.[27]

Ultimately, in the leadzyme structure determination, the heavy atom positions for Ba^{2+}, Pb^{2+}, and UO_2^{2+} were solved by use of an automated

[30] W. Yang and T. A. Steitz, *Cell* **82,** 193 (1995).

[31] C. C. Correll, B. Freeborn, P. B. Moore, and T. A. Steitz, *J. Biomol. Struct. Dyn.* **15,** 165 (1997).

[32] Once solved, the lead site can be used to locate additional heavy atom sites by use of difference Fouriers.

[33] T. C. Terwilliger, "SOLVE: An Automated Crystallographic Structure Solution Program for MIR and MAD." www.solve.lanl.gov, Los Alamos National Laboratory, New Mexico, 1997.

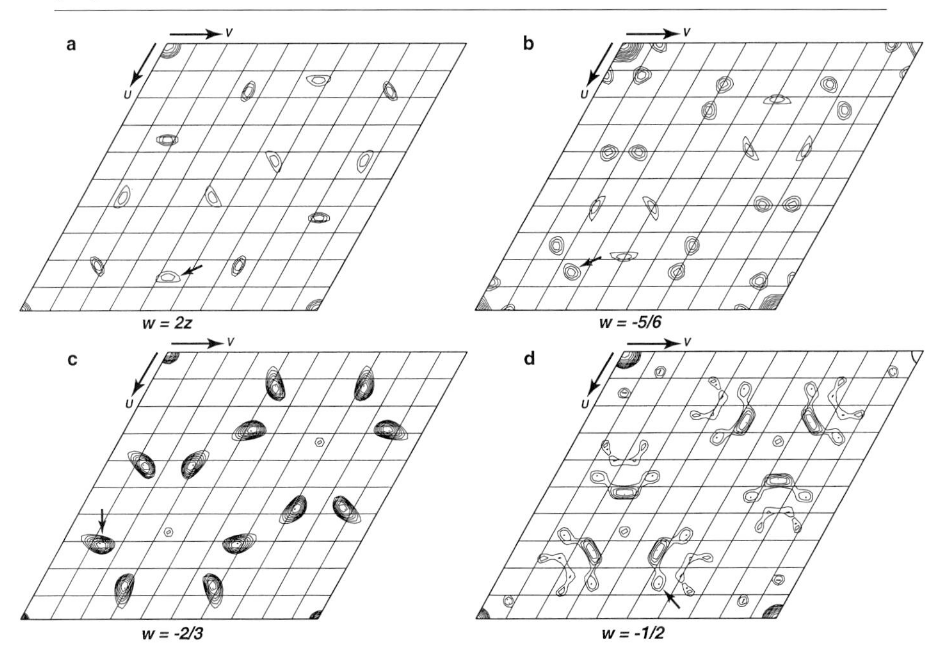

FIG. 1. Representative Harker sections for a lead acetate derivative of the leadzyme. Crystals were soaked for 19 hr in 50 μM lead acetate. Isomorphous difference Patterson maps were calculated from 5- to 25-Å resolution with coefficients $|F_{PH} - F_P|^2$, where F_{PH} is the heavy atom derivative structure factor amplitude and F_P is the native structure factor amplitude. The maps are contoured at a starting height of 1.25σ and increase in 0.5 σ increments. Sections of the entire unit cell in u and v are shown. An arrow inside each section indicates the Harker peak arising from lead binding at site: $x = 0.28$, $y = 0.43$, $z = 0.063$. (a) The $u = y$, $v = 2y$, $w = 2z$ section. (b) The $u = x - y$, $v = x$, $w = -5/6$ section. (c) The $u = 2x - y$, $v = x + y$, $w = -2/3$ section. (d) The $u = 2x$, $v = 2y$, $w = -1/2$ section.

difference Patterson search method, as implemented in SOLVE.[34] Solutions were validated by use of cross-difference and residual difference Fouriers. Heavy atom sites for the Lu^{3+} derivative were found subsequently by use of difference Fouriers. The ambiguity in space group enantiomorph (*P6₁22* or *P6₅22*) was resolved by phasing the anomalous differences for the Ba^{2+} derivative. Although phases were calculated by use of SOLVE and MLPHARE[35] (25- to 3-Å resolution), the degree of nonisomorphism for the UO_2^{2+} and Pb^{2+} derivatives[3] appeared to be modeled and weighted most effectively by use of SHARP.[36] The latter method gave the best quality

[34] T. C. Terwilliger, S.-H. Kim, and D. Eisenberg, *Acta Crystallogr.* **A43,** 1 (1987).
[35] Z. Otwinowski, "MLPHARE." Daresbury Laboratory, Warrington, U.K., 1991.
[36] E. de La Fortelle and G. Bricogne, *Methods Enzymol.* **276,** 472 (1997).

experimentally phased electron density maps, which were interpretable and used for model building, after successive cycles of density modification and phase combination by use of SOLOMON.[37,38] Density modification of the MIR phases was performed iteratively from 3- to 25-Å resolution in order to choose the most suitable solvent content; this was necessary to produce a map with the most contiguous and recognizable features. Inspection of maps and model building of nucleic acids can be performed readily by use of the graphics program O.[39]

Initial Electron Density Maps

The initial electron density map to be used in model building should be calculated to the limit of useful experimental phase information; several maps should be inspected at different resolutions if necessary. Those maps in which the solvent boundary is recognizable, with a well-connected polyphosphate backbone (phosphorous atoms $\geq 2\sigma$ contour level), may be suited to model build. Examples of high-quality experimental electron density maps have been published for the hammerhead[2] and P4–P6[28] structures. Experimental electron density for the leadzyme 24-mer, calculated using data between 3- and 25-Å resolution, is shown in Fig. 2. Although the leadzyme molecule is small, the initial map was not high quality, making an interpretation of the central 6-nucleotide internal loop most difficult. As in most ribozyme structure determinations, modeling of regions that do not conform to canonical A-form helical duplexes (the interesting parts) can be most confounding. In the leadzyme, the bulged region contains three nucleotides that project their bases away from the helical core (Fig. 2). With such irregular geometry, it was not possible to build this region without first constructing the helical stems that flank the internal loop and to correctly assign the sequence to the electron density; fortunately, the existence of the stems was predicted from the primary structure. Based on the asymmetry of the respective nucleotide strands (Table III) the sequence could be assigned; this took into consideration the absence of electron dense 3′- or 5′-phosphoryl groups at the molecular boundary, which was difficult to assess since the nucleotide stems packed, by design [Eq. (1)], into a pseudocontinuous helix. Ultimately, it was possible to build both 24-nucleotide structures in the asymmetric unit independently, prior to refinement.

Although a powerful constraint on phase refinement, noncrystallographic symmetry averaging of experimental electron density maps is not usually an option in RNA crystallography. This is due to large conforma-

[37] J. P. Abrahams and A. G. W. Leslie, *Acta Crystallogr.* **D42,** 30 (1996).

[38] J. P. Abrahams, *Acta Crystallogr.* **D53,** 371 (1997).

[39] T. A. Jones, J. Y. Zhou, S. W. Cowan, and M. Kjeldgaard, *Acta Crystallogr.* **A47,** 110 (1991).

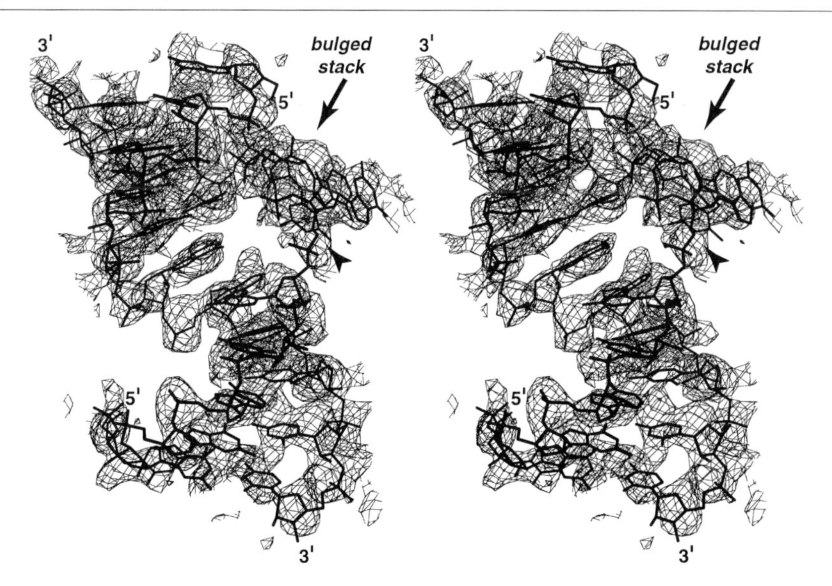

FIG. 2. Stereo representation of the initial experimental electron density map of leadzyme molecule 1 used for model building. The map was calculated from 3- to 25-Å resolution using the native structure factor amplitudes, F_N; multiple isomorphous replacement phases used in the calculation have been density modified by use of SOLOMON.[37,38] The final refined model is indicated as a stick model. The major groove appears as a prominent central cleft. Bulged and stacked nucleotide bases (G24, A25, and G26) pointing away from the helical core are labeled. The scissile bond at the leadzyme active site is indicated by an arrowhead.

tional differences in the RNA backbone between independent molecules in the crystallographic asu. This was true for both the hammerhead and the leadzyme crystal structures, which have multiple independent copies (Table III). Typical root mean square (rms) deviation among pairs of hammerhead molecules is 3.0 Å. For the leadzyme molecules, the rms deviation from a pairwise superposition is 2.0 Å; individual stems give rms superposition values of 0.50 and 0.70 Å. These differences indicate a high degree of flexibility in RNA structures. This may be one explanation for the ineffectiveness of density averaging, and is likely to be the reason why molecular replacement has not been especially effective in ribozyme structural studies.

Refinement

Once the initial ribozyme model has been built, it is often possible to improve the fit of the model to the experimental electron density by use of rigid body refinement. In a typical case, the entire molecule is refined

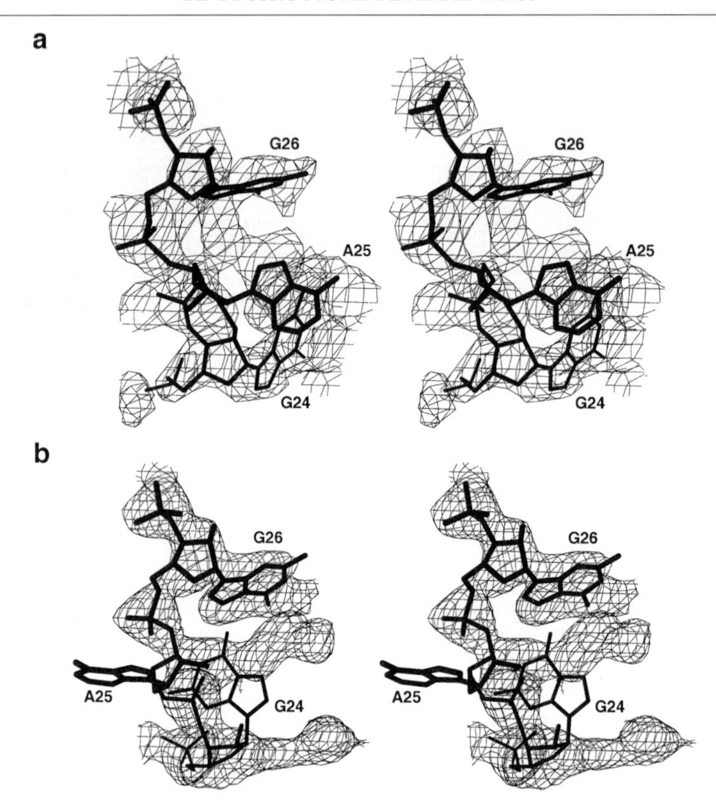

FIG. 3. Representative electron density maps for the internal loop region of the leadzyme (molecule 2) at various stages of the structure determination; images are stereo pairs. Bulged nucleotides are labeled G24, A25, and G26. The bonds for the RNA model are shown as black sticks. (a) An experimentally phased and density modified map (3- to 25-Å resolution) contoured at 1σ. The model was built manually with additional refinement by rigid body minimization. (b) Sigma A-weighted ($2F_o - F_c$) map after simulated annealing of the model in (a); phases for maps in (b), (c), and (d) were calculated from the respective RNA models (2.7- to 25-Å resolution). (c) Sigma A-weighted ($F_o - F_c$) map for the model shown in (b) contoured at $+2.5\sigma$ (black) and -2.5σ (gray). (d) SigmaA weighted ($2F_o - F_c$) map for the final refined model at 2.7-Å resolution.

as a single rigid unit, using data between 3- and 10-Å resolution. At this point it is desirable to minimize against the experimental phase probability distribution as implemented in CNS[40] or REFMAC.[41] Next, the model may be divided into subdomains, if appropriate. Last, in cases where the

[40] A. T. Brünger, P. D. Adams, G. M. Clore, P. Gros, R. W. Grosse-Kunstleve, J.-S. Jiang, J. Kuszewski, M. Nilges, N. S. Pannu, R. J. Read, L. M. Rice, T. Simonson, and G. L. Warren, *Acta Crystallogr.* **D54,** 905 (1998).

[41] G. N. Murshudov, A. A. Vagin, and E. J. Dodson, *Acta Crystallogr.* **D53,** 240 (1997).

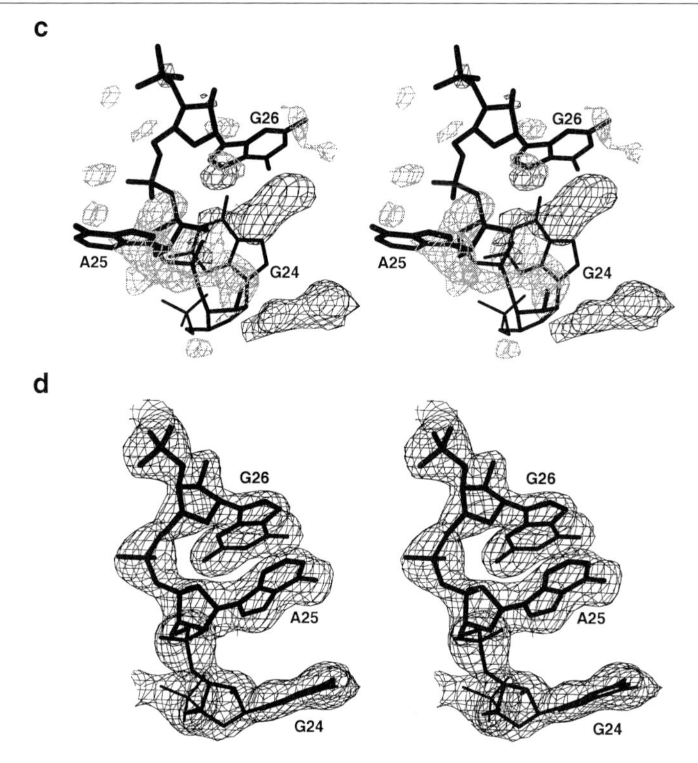

FIG. 3. (*continued*)

nucleotide electron density is exceptionally poor, individual nucleotides can be partitioned into three groups consisting of base, phosphate, and ribose. Rigid body refinement should be performed simultaneously on all three components, while paying close attention to the R_{free} value.[42] Any step that increases the R_{free} can be considered overfitting and should be eliminated from refinement. This strategy was employed for the leadzyme, treating each nucleotide separately (as three components) over 48 nucleotides. This yielded a crystallographic R_{factor} of 46.2% (R_{free} = 40.2%) as compared to the initial R_{factor} of 49.2% (R_{free} = 46.2%). To demonstrate that this procedure is generally nondisruptive to the model, a representative region of the experimental electron density with the final rigid body refined model is shown in Fig. 3a. Although the geometry for the bulged nucleotides is poor, the model appears to fit the electron density better than that produced by manual building (not shown).

The model can be improved further by use of Cartesian or torsional

[42] A. T. Brünger, *Nature* (*Lond.*) **355,** 472 (1992).

simulated annealing[43,44] with a maximum likelihood target for the residual structure factor amplitudes.[45,46] This procedure significantly improves model geometry. For the leadzyme, the application of Cartesian annealing reduced the R_{factor} to 32.9% with R_{free} = 35.5% (2.7- to 25-Å resolution); compare Fig. 3a and 3b. However, in an attempt to refine the model geometry, molecular dynamics positioned the bulged nucleotides at G24 and A25 incorrectly, toward the core of leadzyme. This was observed subsequently in a reduced bias sigmaA-weighted[45,47] $(2mF_o - DF_c)$ electron density map (Fig. 3b). In addition to $(2F_o - F_c)$ maps, incorrect features of a model can be detected by use of residual difference $(F_o - F_c)$ maps contoured at both positive and negative electron density levels. For example, Fig. 3c shows the sigmaA-weighted residual map for the annealed model in Fig. 3b. Although the backbone is positioned correctly (no residual density), the bases for G24 and A25 are surrounded by negative density, contoured at $-2.5\ \sigma$; in addition, there are large portions of positive density at the $+2.5\ \sigma$ level that correspond to the correct positions of the misplaced bases (Fig. 3c). Additional cycles of manual rebuilding and positional refinement are required to correct these errors.

Ultimately, the leadzyme model was refined to an R_{factor} of 25.4% with R_{free} = 35.5%, using all data between 2.7- and 25-Å resolution.[3] Representative electron density for the refined structure is shown in Fig. 3d. In general, the strategies described here for our case studies are generally applicable to other small ribozymes. This work is intended to provide a practical guide to those interested in initiating structure determinations of small RNAs. Methods for solving larger RNA structures are presented in this volume, as well.[48]

Acknowledgments

We thank M. Sousa and C. Kielkopf for advice and assistance in the development of RNA purification methods, and A. R. Kolatkar for comments on this manuscript. Pure HDV RC1 used in this study was a generous gift from M. Been, Duke University; hammerhead substrate analogs were the gift of N. Usman and L. Beigelman, Ribozyme Pharmaceuticals, Inc. J.E.W. is a Burroughs Wellcome Fund Fellow of the Life Sciences Research Foundation (LSRF). This work was supported by funds from the LSRF and an NIH grant to D.B.M. SSRL is funded by the DOE, Office of Basic Energy Sciences. The Biotechnology Program is supported by the NIH, National Center for Research Resources Biomedical Technology Program, and the DOE, Office of Biological and Environmental Research.

[43] A. T. Brünger, J. M. Kuriyan, and M. Karplus, *Science* **235,** 458 (1987).
[44] L. M. Rice and A. T. Brünger, *Proteins* **19,** 277 (1994).
[45] P. D. Adams, N. S. Pannu, R. J. Read, and A. T. Brünger, *Proc. Natl. Acad. Sci. U.S.A.* **94,** 5018 (1997).
[46] N. S. Pannu and R. J. Read, *Acta Crystallogr.* **A52,** 659 (1996).
[47] R. J. Read, *Acta Crystallogr.* **A42,** 140 (1986).
[48] J. H. Cate and J. A. Doudna *Methods Enzymol.* **317** [12], 2000 (this volume).

[12] Solving Large RNA Structures by X-Ray Crystallography

By JAMIE H. CATE and JENNIFER A. DOUDNA

Introduction

Structured RNAs play key roles in many aspects of biology, from RNA processing to protein synthesis and transport. Advances in the large-scale synthesis and purification of RNA have fostered structural studies by X-ray crystallography. In this article, we discuss those aspects of solving large RNA crystal structures that differ from the approaches developed for protein crystallography. Common strategies in crystallography are found in Volumes 276 and 277 of this series.

Preparation of Well-Diffracting Crystals

The first hurdle to solving an RNA structure by X-ray crystallography is obtaining crystals that diffract X rays to biochemically useful resolution. The electron-dense phosphate backbone of RNA is readily detectable at 5-Å resolution,[1] but to identify individual bases and registration of the sequence, 3-Å resolution or better is generally necessary. As for proteins, high-quality RNA crystals form by utilizing exposed tertiary contacts. These contacts are sometimes inherent in the RNA, as for the tRNA anticodon,[2,3] the three major crystal contacts in the *Tetrahymena* ribozyme P4–P6 domain crystals,[4] the GAAA tetraloop in the hammerhead ribozyme structures,[5,6] and loop E in the 5S rRNA fragment structure.[7] Recent work on engineering RNA–RNA crystal contacts showed that RNAs with GAAA tetraloop–tetraloop receptor motifs included for intermolecular contacts yielded many more crystal forms than controls.[8] However, the most successful design

[1] B. L. Golden, A. R. Gooding, E. R. Podell, and T. R. Cech, *Science* **282,** 259 (1989).
[2] F. L. Suddath, G. J. Quigley, A. McPherson, D. Sneden, J. J. Kim, S. H. Kim, and A. Rich, *Nature* **248,** 20 (1974).
[3] J. D. Robertus, J. E. Ladner, J. T. Finch, D. Rhodes, R. S. Brown, B. F. Clark, and A. Klug, *Nature* **250,** 546 (1974).
[4] J. H. Cate, A. R. Gooding, E. Podell, K. Zhou, B. L. Golden, C. E. Kundrot, T. R. Cech, and J. A. Doudna, *Science* **273,** 1678 (1996).
[5] H. W. Pley, K. M. Flaherty, and D. B. McKay, *Nature* (*Lond.*) **372,** 68 (1994).
[6] W. G. Scott, J. T. Finch, and A. Klug, *Cell* **81,** 991 (1995).
[7] C. C. Correll, B. Freeborn, P. B. Moore, and T. A. Steitz, *Cell* **91,** 705 (1997).
[8] A. R. Ferre-D'Amare, K. Zhou, and J. A. Doudna, *J. Mol. Biol.* **279,** 621 (1998).

METHODS IN ENZYMOLOGY, VOL. 317
0076-6879/00 $30.00

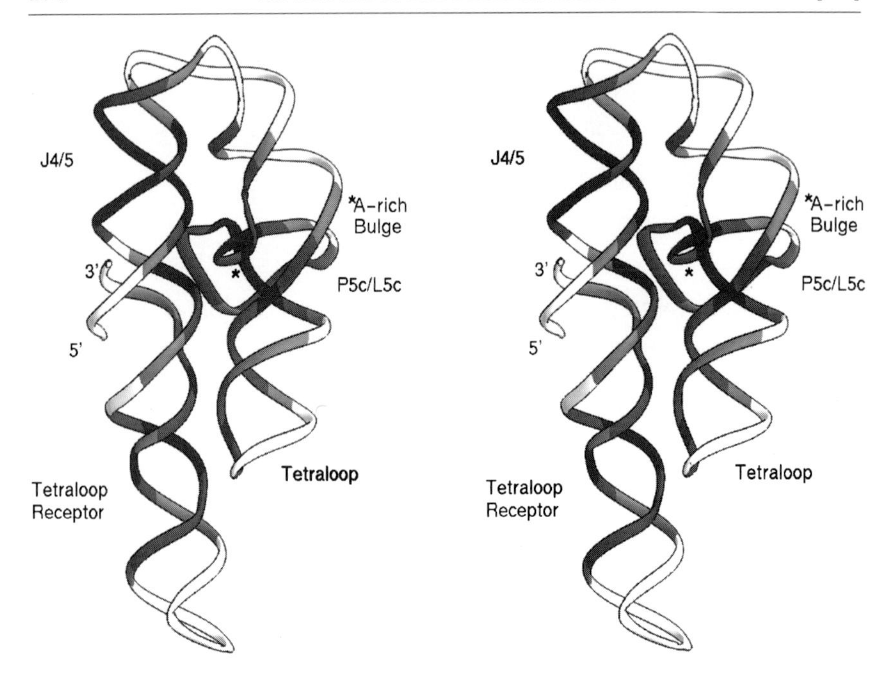

FIG. 1. Temperature factor (B factor) distribution in the P4–P4 domain crystal structure. A stereo view of one of the two molecules in the asymmetric unit of the crystal is shown, shaded according to B factor ranges as follows: black, $B \leq 24$ Å^2; dark gray, $24 < B \leq 36$ Å^2; light gray, $36 < B \leq 48$ Å^2; white, $B > 48$ Å^2. The other molecule in the asymmetric unit has similar temperature factor distributions, due to noncrystallographic symmetry (NCS) restraints. Note that the darker regions correspond to intramolecular tertiary structure and/or crystal packing contacts. The B factors can change dramatically within a few residues, for example, in the tetraloop strand and above J4/5.

incorporated a binding site for the well-characterized RNA binding domain of the U1A protein[9,10] into the hepatitis delta virus (HDV) ribozyme, thus providing stable protein–protein contacts in the crystal lattice.[11]

One problem with RNA is that regions of the molecule not involved in crystal contacts tend to have very high temperature factors (B factors), indicating motion or disorder (Fig. 1). This may explain in part why most RNA crystals do not diffract X rays well. If an RNA molecule of interest fails to crystallize or yields poorly ordered crystals, the introduction of crystallization modules into solvent-exposed regions of the RNA may be helpful.[8] The key is to insert the modules so that they do not disrupt or change the structure of the RNA. Phylogenetic covariation analysis, when

[9] K. Nagai, C. Oubridge, T. H. Jessen, J. Li, and P. R. Evans, *Nature* **348,** 515 (1990).
[10] C. Oubridge, N. Ito, P. R. Evans, C. H. Teo, and K. Nagai, *Nature* **372,** 432 (1994).
[11] A. R. Ferre-D'Amare, K. Zhou, and J. A. Doudna, *Nature* (*Lond.*) **395,** 567 (1998).

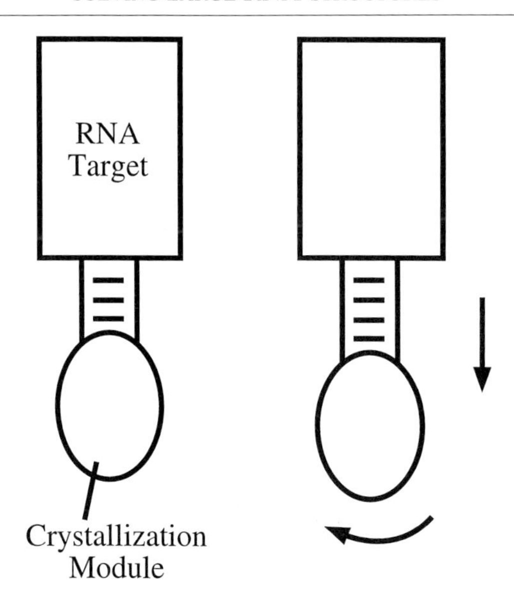

FIG. 2. Design of crystal packing interactions in RNA. The position of the crystallization module can be varied by inserting different length helical segments between the RNA of interest and the module. Each additional base pair increases the distance between the two as well as rotating their relative orientations by ~30°.

it is available, often reveals variable regions of the RNA where insertions, deletions, and sequence changes are allowed while maintaining function. Chemical probes of RNA structure, including Fe(II)-EDTA[12,13] and base modification reagents such as DMS, kethoxal, and diethyl pyrocarbonate (DEPC)[14] are also useful for identifying exposed regions of the RNA that are not critical for function. These sites can then be altered to include, for example, a binding site for the U1A–RNA binding domain or recognition motifs for tetraloops. It is important to design multiple constructs to vary the length of the stem connecting the inserted element to the core of the RNA, which will alter their relative orientations (Fig. 2). It is also essential to test that the biochemical activity of the designed molecules is not significantly different from that of the wild-type RNA.

Once suitable crystals are in hand, two primary factors determine how rapidly the structure can be solved. First, highly reproducible conditions for growing the crystals must be found. It may take hundreds of crystals

[12] D. W. Celander and T. R. Cech, *Biochemistry* **29,** 1355 (1990).
[13] D. W. Celander and T. R. Cech, *Science* **251,** 401 (1991).
[14] S. Stern, D. Moazed, and H. F. Noller, *Methods Enzymol.* **164,** 481 (1988).

to find appropriate stabilizing conditions and the heavy atom derivatives needed to solve the structure. Several approaches may increase the yield of usable crystals. The length and conformational purity of the sample should be verified by denaturing and native gel electrophoresis, and by dynamic light scattering.[15] Although 2- to 4-μl crystallization drop volumes are useful for initially screening conditions, drop volumes as large as reasonably possible should be used to obtain large crystals for heavy atom derivatization. The kinetics of equilibration are retarded in sitting drop volumes of 10–30 μl, resulting in more reproducible crystal nucleation and growth.[16] Furthermore, such drops are convenient for macro- and microseeding.[17]

RNA crystals tend to decay rapidly when exposed to X rays. Thus, diffraction data are generally collected from RNA crystals at cryogenic temperatures. Compounds used in protein crystal stabilization are useful for RNA crystals as well.[18] These include sugars and nonvolatile alcohols such as glucose, glycerol, 2-methyl-2,4-pentanediol (MPD) and low molecular weight polyethylene glycols (PEGs). In addition, volatile alcohols such as ethanol and isopropanol have been useful in RNA crystal cryostabilization. Typically the cryoprotectant should constitute at least 15–20% (v/v) of the soaking solution for the crystals. Addition of the cryoprotectant should be in 1–2% increments by adding solution to the drop containing the crystal and resealing the coverslip to prevent evaporation. If the crystal appears unchanged after 2 min, another addition of the protectant can be made. Once the final concentration of cryoprotectant in the drop has been adjusted to high enough concentrations, the reservoir solution within the crystallization setup should be adjusted to the same composition, and the system resealed to equilibrate.

RNA crystals are highly sensitive to mono- and divalent salts and polyamines, and these additives must be adjusted in concert to find optimal stabilizing conditions. Often, as the concentration of divalent metal ions is decreased, the concentrations of polyamines such as spermine and spermidine need to be increased. Appropriate concentrations to use are highly empirical: the P4–P6 domain crystals cracked at concentrations of spermine higher than ~0.5 m*M*, whereas the HDV ribozyme crystals were optimal in 6–25 m*M* spermine. When searching for cryostabilizing conditions, one should consider the length of time required for later manipulations, i.e.,

[15] A. R. Ferre-d'Amare and J. A. Doudna, *Methods Mol. Biol.* **74,** 371 (1997).
[16] J. R. Luft and G. T. DeTitta, *Methods Enzymol.* **276,** 110 (1997).
[17] A. McPherson, *Methods Enzymol.* **114,** 112 (1985).
[18] D. W. Rodgers, *Methods Enzymol.* **276,** 183 (1997).

heavy atom soaks. The best conditions should not degrade crystal quality over time.

Note that RNA crystals are often highly susceptible to nonisomorphism, or crystal-to-crystal variation. In the case of the P4–P6 domain crystals, for example, one unit cell dimension fluctuated in length from 128 to 132 Å depending on the stabilizing conditions. This hinders all stages of phasing the structure factors because it changes the periodic sampling of the electron density in reciprocal space. In the case of P4–P6, the change in cell dimensions correlated with the percentage of isopropanol included in the cryostabilizer, up to approximately 5% 2-propanol. By increasing 2-propanol to 10%, we eliminated this variability.

Phasing by Heavy Atom Substitution

In protein crystallography, many of the methods for solving the phase problem are highly tuned. For proteins and protein–DNA complexes, brominated DNA and selenomethionyl-substituted proteins are used routinely for phasing the structure factors.[19] These methods apply to RNA as well: brominated RNAs were used to solve the structures of the hammerhead and loop E of 5S rRNA,[6,20,21] and selenomethionine was used to determine the structure of the HDV ribozyme–U1A complex.[11] We recommend using the technique of multiwavelength anomalous diffraction (MAD) for phase determination with RNA because it avoids the problems of nonisomorphism that are so often encountered (see later discussion). In this regard, iodinated RNAs are less useful because the absorption edge for iodine is at an inaccessible wavelength.[22] Larger RNAs may require a heftier signal than can be obtained from bromines or selenomethionine. For example, a molecule the size of P4–P6 (160 nucleotides) would require five or more bromine atoms to determine the structure by MAD.[22]

Two kinds of transition metal ions have been used with great success in RNA crystallography. First, lanthanide series cations contributed to solving the structures of tRNA, the hammerhead ribozyme, and the group I intron P4–P6 domain (see Ref. 23 for summary of conditions). Second, osmium(III) hexammine provided the high-resolution phasing for the P4–P6 domain structure.[24] Both types of ions are likely to bind to RNA at sites normally occupied by magnesium ions, though not exclusively to

[19] S. Doublie, *Methods Enzymol.* **276,** 523 (1997).

[20] W. G. Scott, J. T. Finch, R. Grenfell, J. Fogg, T. Smith, M. J. Gait, and A. Klug, *J. Mol. Biol.* **250,** 327 (1995).

[21] C. C. Correll, B. Freeborn, P. B. Moore, and T. A. Steitz, *J. Biomol. Struct. Dyn.* **15,** 165 (1997).

[22] W. A. Hendrickson, and C. M. Ogata, *Methods Enzymol.* **276,** 494 (1997).

[23] A. R. Ferre-D'Amare and J. A. Doudna, *Curr. Protocols Nucleic Acid Chem.,* in press (1999).

[24] J. H. Cate, and J. A. Doudna, *Structure* **4,** 1221 (1996).

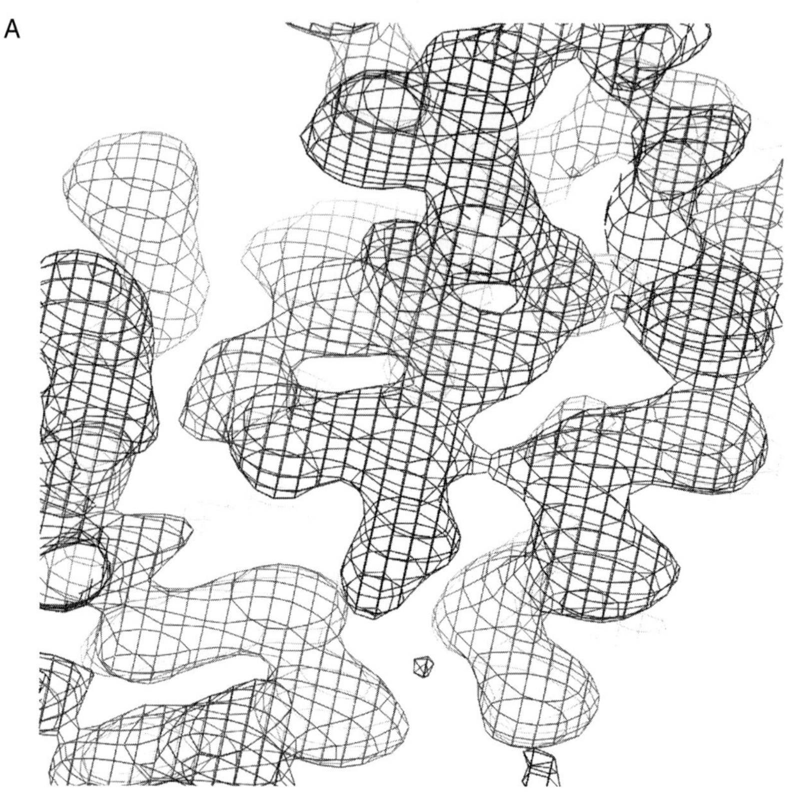

FIG. 3. Map quality of MAD versus MIR experiments. The A-rich bulge region of the P4–P6 domain electron density map is shown contoured at 1 standard deviation above the mean. Three data sets were used in the map calculations: Oshex λ1 and Oshex λ2, MAD data sets collected at the Os LIII absorption edge; Cohex 1, "native" data set.[4] (A) MAD treatment of the data. Heavy atom parameters were refined and phases were calculated with Oshex λ1 as the reference data set. (B) MIR treatment of the data. Heavy atom parameters were refined and phases were calculated with Cohex 1 as the reference data set.

these sites. One key difference between the lanthanides and osmium hexammine is the mode of binding to the RNA. Whereas the lanthanides can form inner-sphere coordination to the RNA, osmium hexammine can only form "outer-sphere" interactions, similar to fully hydrated magnesium ions. The best concentration ranges to use for these ions in the published structures varies from 100 μM to low millimolar. It is worthwhile screening all available lanthanides (all but Pm are commercially available) due to the change in ionic radius from Ce^{3+} to Lu^{3+}. In the case of the P4–P6 domain crystals, only Sm^{3+} provided a usable heavy atom derivative.[4]

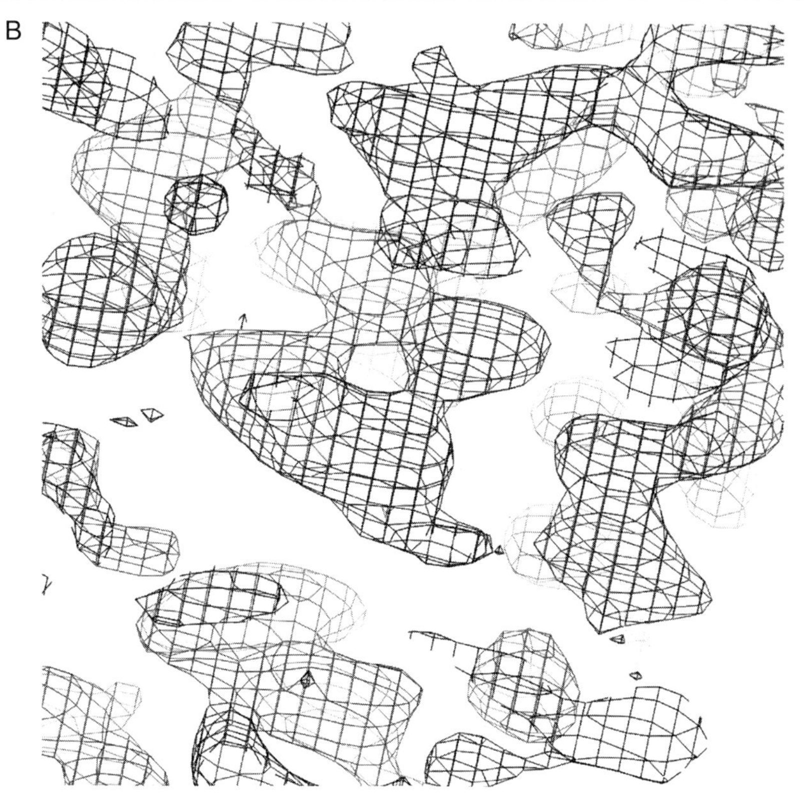

FIG. 3. (*continued*)

Osmium hexammine and the lanthanides, with a strong anomalous signal at the LIII absorption edge, have been exploited at synchrotron radiation sources to solve the structures of many proteins and the P4–P6 domain RNA by MAD phasing. Even slight nonisomorphism can interfere with multiple isomorphous replacement (MIR) phasing, while the same data sets treated as a MAD experiment yield a clean experimental electron density map. As an example, the same diffraction data sets used to produce the interpretable electron density map for the P4–P6 domain structure, when treated as MIR data, yielded a map with broken density (Fig. 3). Strategies for conducting a MAD experiment and determining phases from the heavy atom substitution are published elsewhere.[22,25]

[25] T. C. Terwilliger, *Methods Enzymol.* **276,** 530 (1997).

Map Interpretation and Model Building

RNA electron density maps have two strong features that can be used to place an initial model. First, at a resolution of approximately 3.5 Å or better, the phosphates appear as spheres in an electron density map contoured at a high signal-to-noise ratio. Second, purine bases form elongated disks at high signal-to-noise ratios. The hardest part of the map to fit is usually the ribose. As the weakest constraints are to the 5′ side of the ribose moiety, it is better to use the O-3′–P bond length to find the ribose pucker that best matches the density. Torsion angles can be used to position the ribose, with the base and surrounding phosphates held fixed, to try to match up the O-3′–P bond length. There is some flexibility around the C-5′–O-5′–P angle to position O5′.[26] The density around the phosphate may not always be centered on the phosphorous atom; it may "streak" along the P–O-5′–C-5′ direction toward the ribose.

After first positioning the phosphates and bases as purine–pyrimidine, a useful strategy is to use a ribose library to connect the phosphates with reasonable sugar puckers (Fig. 4).[24] After an initial round of manual placement or a refinement cycle, the library of ribose conformers is superimposed on C-1′, O-4′, and N-1 or N-9 of the attached base of each ribose in question to find the pucker that best fits the density. It may take some adjusting of the χ dihedral torsion angle (around N-1 or N-9 and C-1′ bond), with the base held fixed, to "position" the library by least-squares superposition in a graphical modeling program[27] such as O. At ~3.0 E resolution, C3′-endo and C2′-endo puckers are distinguishable for all but the most disordered residues. In the P4–P6 domain maps, the density for C2′-endo riboses was usually perpendicular to the plane of the base density (Fig. 4).

Refinement of the RNA model presents one major difference from that of protein structures. Because the furanose ring can take on either C3′-endo or C2′-endo puckering, one must use a more complicated restraint for the dihedral angles within and external to the ring. For example, CNS (or X-PLOR) provides a multiwell potential function that allows the ribose to go into either dominant pucker (X-PLOR manual). The depth of the wells around the C3′-endo and C2′-endo puckering may need some adjustment, depending on the resolution of the structure. In addition, we have found it necessary to choose the pucker of each ribose based on the experimental map prior to refinement, i.e., one may not be able to "flip" the pucker simply by least-squares refinement or simulated annealing procedures. If the pucker is chosen incorrectly, positive and negative density will appear

[26] W. Saenger, "Principles of Nucleic Acid Structure." Springer-Verlag, New York, 1984.
[27] T. A. Jones, J. Y. Zou, S. W. Cowan, and Kjeldgaard, *Acta Crystallogr. A,* **47,** 110 (1991).

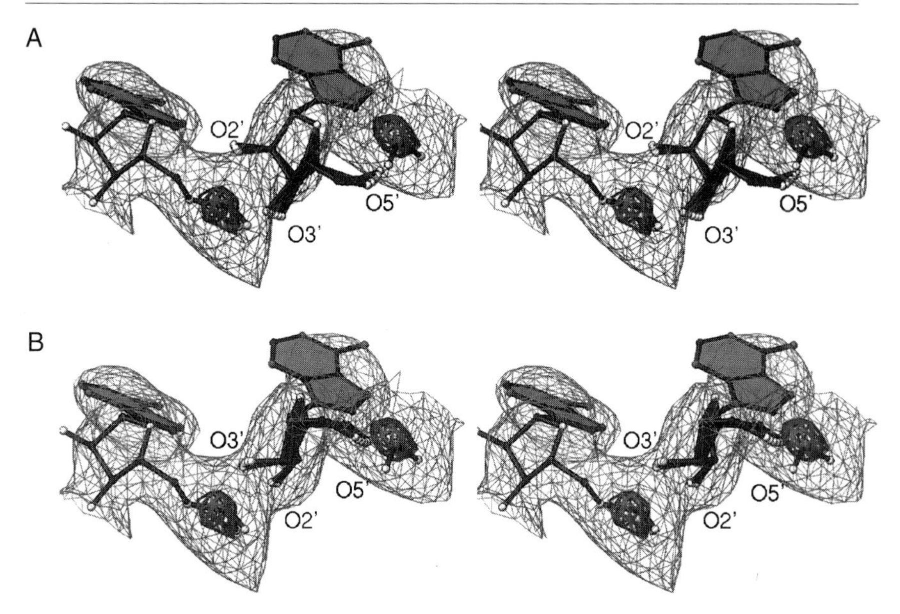

FIG. 4. Ribose rotamer library. Ribose moieties with ring puckers of 0–36 degrees (C3′-endo) or 145–181 degrees (C2′-endo), in 9-degree increments, are positioned to achieve a reasonable O3′–P bond distance. (A) C3′-Endo library positioned for nucleotide A206 in P4–P6 domain molecule A. Note that no torsional rotation of the ribose about the glycosidic bond could fit O2′, O3′, and C5′/O5′ in the density while maintaining the O3′–P distance. (b) C2′-endo library positioned to give a reasonable O3′-P bond length. All atoms of the ribose can be placed in the density.

above and below the ribose "plane" near the O4′ and C2′/O2′ positions in $(F_o - F_c)\varphi$ difference maps, where F_o and F_c are observed and calculated amplitudes, and φ are either experimental or calculated phases, depending on the stage of refinement. An additional clue comes from phosphate-to-phosphate distances. Phosphates are often approximately 5 Å apart when separated by a C3′-endo puckered ribose, whereas phosphates are often about 7 Å apart when separated by a C2′-endo puckered ribose.[26]

Interpretation of Solvent Electron Density Peaks

Metal ions play key roles in the folding and function of structured RNAs.[28] One of the best ways to identify these metal sites comes from RNA crystal structures. However, at the resolution of most of the known RNA structures (~3 Å), positively identifying solvent peaks in the electron

[28] A. M. Pyle, *Science,* **261,** 709 (1993).

density map proves difficult. Metal ions in the experimental electron density map of the P4–P6 domain crystal structure appeared at close to the same density level as the phosphates, initially complicating model building. To identify magnesium ions and other ions among the solvent electron density peaks may take a two-pronged approach, coupling crystallography and RNA biochemistry.

From the experimental electron density or refined model, indirect data can be obtained to identify metal ions. Although water molecules, magnesium ions, and sodium ions all have similar scattering properties, the coordination of the solvent peak to the RNA may provide clues.[29] Magnesium ions have octahedral coordination geometry and bond lengths of approximately 2.1 Å. Sodium ions have bond lengths of 2.4 Å, but may have variable geometry.[30,31] Water molecules can be identified by their involvement in hydrogen-bond length interactions with the RNA (2.7–3.0 Å). The differences in these geometries should be compared to the coordinate error for the RNA model, which may be on the order of 0.3–0.4 Å for a 3-Å resolution structure.[32] In practical terms, due to the close coordination of water molecules to magnesium and sodium, these hydrated metal ions may appear as larger blobs of electron density than that observed for single waters. Fully hydrated ions may appear as electron density peaks centered approximately 4 Å from the nearest RNA ligand.

Potassium ions, while having variable coordination geometry and bond lengths similar to those of water (2.7 Å),[30] scatter electrons more strongly than water or the other ions. This scattering difference may appear in the refined model as follows: if the solvent peak is originally modeled as water, its refined temperature factor may end up much lower than that of the surrounding RNA. In the case of the P4–P6 group I intron domain, putative waters in the major groove of AA platforms refined with temperature factors of 5–10 Å^2, whereas the temperature factors of the surrounding RNA hovered around 30–40 Å^2. However, at modest (~3-Å) resolution, unrestrained temperature factor refinement may be suspect and lead to model bias.

Crystallography provides a direct approach for identifying ions in a structure. Transition metal ions that can functionally substitute for lighter ions can be exploited for their anomalous scattering properties. In many cases, manganese ions partially or completely substitute for magnesium

[29] S. R. Holbrook, J. L. Sussman, R. W. Warrant, and S. H. Kim, *J. Mol. Biol.* **123,** 631 (1978).

[30] S. R. Holbrook, J. L. Sussman, R. W. Warrant, G. M. Church, and S. H. Kim, *Nucleic Acids Res.* **4,** 2811 (1977).

[31] J. L. Sussman, S. R. Holbrook, R. W. Warrant, G. M. Church, and S. H. Kim, *J. Mol. Biol.* **123,** 607 (1978).

[32] V. Luzzati, *Acta Crystallogr.* **5,** 802 (1952).

ions in RNA folding and activity.[33] Ion channel studies have shown that thallium ions (Tl^+) substitute for potassium ions.[34] Note that thallium is toxic and should be handled with gloves in a fume hood. Both Mn^{2+} and Tl^+ exhibit strong anomalous scattering that can be observed using a rotating anode X-ray generator and Cu-K_α radiation. To measure the anomalous data required to identify metal ion sites, the best approach is to use "inverse beam" geometry. From a single crystal, measure two complete data sets from starting rotations separated by 180°. In this way one is guaranteed of observing both Friedel mates for each unique reflection, while at the same time reducing systematic errors, i.e., in absorption. If crystal decay is significant, the data sets may have to be collected in 10° wedges. For example, the data might be collected from 0 to 10°, followed by 180 to 190°, and then back to 10 to 20°, and so on. We find it useful to exploit the Strategy option in the program MOSFLM to identify the rotations necessary to collect a complete data set, then repeat the suggested range ±180°.[35] See the *Methods in Enzymology* chapter by Hendrickson and Ogata for a more detailed discussion of anomalous data collection in the context of MAD experiments.[22]

Substitution of metal ions in RNA crystals requires some consideration of the competing effects of different salts. Manganese ions, for example, tend to bind soft ligands with higher affinity than Mg^{2+} ions do, i.e., to the N-7 position of guanosine. Thallium ions also bind soft ligands well. To overcome artifactual binding of these transition metals and to enhance their specificity, it may be necessary to drastically change crystal stabilization conditions. We found that by increasing the ratio of polyamine (spermine) to divalent cation, we could decrease the concentration of Mg^{2+} 10-fold while still maintaining reasonable diffraction from the P4–P6 domain crystals. With a much lower requirement for divalent cation, the crystals could then be transferred directly into Mn^{2+} with the complete removal of Mg^{2+}. Thallium tends to be insoluble as the chloride salt, thus requiring a change in stabilizer to one containing only acetate counterions, for example.

Conclusions

Although few RNA crystal structures have been determined, some general trends for solving RNA structures by crystallography have emerged. There are now proven methods for obtaining well-diffracting crystals of large RNAs. MAD phasing has become the most efficient approach for

[33] C. A. Grosshans and T. R. Cech, *Biochemistry* **28,** 6888 (1989).

[34] B. Hille, *J. Gen. Physiol.* **61,** 669 (1973).

[35] *Acta Crystallogr. D,* **50,** 760 (1994).

phasing structure factors and obtaining high-quality electron density maps. Crystallographic experiments coupled with biochemical tests have allowed the functional interpretation of RNA structures. As occurred in the protein field, the number of RNA crystal structures solved per year is likely to grow exponentially.

Acknowledgment

We thank Adrian Ferré-D'Amaré for helpful discussions about modeling ribose puckers in electron density maps. J.H.C. is a postdoctoral fellow of the Damon Runyon-Walter Winchell Cancer Research Fund; J.A.D. is an assistant investigator of the Howard Hughes Medical Institute, a Searle Scholar, a Beckman Young Investigator, and a Fellow of the David and Lucile Packard Foundation; this work was supported in part by a grant from the NIH.

[13] Conventional and Time-Resolved Ribozyme X-Ray Crystallography

By WILLIAM G. SCOTT and JAMES B. MURRAY

Introduction

Can X-ray crystallography help us to understand how ribozymes work? Although a number of biochemical techniques, perhaps most notably *in vitro* selection methods, have provided much needed insight into the nature and potentially broad spectrum of the catalytic capabilities of RNA, the problem of deducing the correlations between RNA three-dimensional structure and enzymatic activity remains a formidable one. Recent advances in crystallization methodologies and in RNA synthetic strategies have enabled us to begin to address this problem. In particular, the hammerhead ribozyme, hepatitis delta virus ribozyme, group I intron, hairpin ribozyme, RNase P, and other catalytic RNAs have been crystallized, and crystal structures of the first three have now been published. In addition, application of time-resolved crystallographic techniques to the hammerhead ribozyme allows us to begin to correlate the structure of a ribozyme directly with the chemistry of its catalysis. Here we discuss the methods used for both conventional and time-resolved crystallographic analyses of the hammerhead ribozyme. This ribozyme serves as a prototype for other analyses, because the techniques employed should be fairly generalizable for future work on other ribozyme systems.

0076-6879/00 $30.00

RNA Crystallizations

In general, the problem of crystallizing nucleic acids is more difficult than that of crystallizing proteins. The relative homogeneity of the phosphate backbone in nucleic acids and establishing helical end contacts between molecules make forming a crystal lattice difficult. Unlike protein crystallization, where a large number of structurally and chemically unique regions of surface are potentially available for the formation of crystal lattice contacts, RNA crystallization involves a much more limited repertoire of potential lattice-forming contacts. In crystals of simple helices, one often finds that two of the three or more contacts required to form a crystal lattice are composed of helical end contacts that establish formation of a quasi-continuous helix throughout the lattice. The base-stacking continues uninterrupted, but the phosphate backbone, which is not directly involved in such a contact, does not. (Minor variations on this theme, involving base triples at the lattice contacts, as well as 5′ to 5′ and 3′ to 3′ stacking arrangements, have also been observed, both in the context of DNA–protein complexes and in ribozyme crystallizations.)

The most definitive treatment of this problem is detailed in a crystallization paper of the CAP–DNA complex by Schultz, Shields, and Steitz.[1] Although the context is that of DNA–protein complex crystallizations, the principles are completely generalizable to RNA and RNA–protein complex crystallizations. The essential lesson from this treatment and from subsequent attempts at ribozyme crystallizations[2–5] is that one must often actively tamper with the nonessential sequences found at the helical termini until one finds, by trial and error, not only a sequence that is crystallizable, but one that produces crystals having suitable diffraction characteristics. This commonly involves synthesizing 10 or more trial sequence variants until one obtains the desired crystals. Rapid RNA chemical synthesis techniques make this approach much more tractable, at least in the case of relatively small RNAs that can be assembled from strands that are not longer than about 50 nucleotides.[3]

In addition to altering the sequences found at the ends of helices, it is generally helpful to experiment with nonessential regions within the molecule, especially to create potentially specific crystal packing contacts be-

[1] S. C. Schultz, G. C. Shields, and T. A. Steitz, *J. Mol. Biol.* **213,** 159 (1990).

[2] H. W. Pley, D. S. Lindes, C. DeLuca-Flaherty, and D. B. McKay, *J. Biol. Chem.* **268,** 19656 (1993).

[3] W. G. Scott, J. T. Finch, R. Grenfell, J. Fogg, T. Smith, M. J. Gait, and A. Klug, *J. Mol. Biol.* **250,** 327 (1995).

[4] B. L. Golden, E. R. Podell, A. R. Gooding, and T. R. Cech, *J. Mol. Biol.* **270,** 711 (1997).

[5] A. R. Ferre-D'Amare, K. Zhou, and J. A. Doudna, *J. Mol. Biol.* **279,** 621 (1998).

tween two elements of secondary structure, such as a GNRA tetraloop and an appropriate receptor sequence. Such contacts were found in both structures of the hammerhead ribozyme, although they were not planned. Interestingly, in some cases in which such contacts were deliberately designed into RNA molecules, crystals formed more readily but their diffraction properties were poor.[5] A subsequent variation on this theme, in which a loop that binds to an RNA-binding protein domain was grafted onto a nonessential region of the structure and the protein was employed in a cocrystallization, was much more successful at producing diffraction-quality crystals.[6] This success is probably due to the fact that potential protein–protein contacts are much more plentiful, and specific contacts did not have to be designed.

In essence, the difficulties in crystallizing nucleic acids arise from establishing good lattice contacts that allow the molecules to arrange themselves in a regular three-dimensional array. Molecules having only two separate helical ends still have a problem of establishing (at least) one more specific lattice contact, and the potential for forming such a third lattice contact may be enhanced by including a GNRA tetraloop along with known receptor sequences in the molecule, or alternately by including a bound protein that has the luxury of forming many different potential lattice contacts.

Several different "sparse matrix" crystallization screens have appeared recently that allow one to search for crystallization conditions efficiently.[3,4,7] Although these differ in detail, the differences are probably less important to successful crystallizations than the process of varying sequences to improve crystallizability.[3] Some general features for successful crystallizations of RNA, however, have emerged: whereas crystallizations of small duplexes and tRNA have relied on small organic molecules, particularly 2-propanol and MDP as precipitating agents, high concentrations of lithium and ammonium sulfate, and various molecular weights of PEGs and their derivatives in conjunction with 100 m*M* to 1 *M* monovalent salts, has proven to be particularly effective. The range of pH values employed for RNA crystallizations is generally more restricted than those used for protein crystallizations, because outside of the range of about pH 5–8.5, significant ionization of base functional groups may take place. Inclusion of Mg^{2+} or other divalent cations, such as Ca^{2+}, Co^{2+}, Mn^{2+} Cd^{2+}, and Zn^{2+}, at concentrations between 10 and 100 m*M,* is also worthwhile, especially if these cations are known to be required for biological activity. (Most of the softer metal ions form insoluble oxides at pH levels greater than 7.0 over the course of hours or

[6] A. R. Ferre-D'Amare, K. Zhou, and J. A. Doudna, *Nature* **395,** 6702 (1998).

[7] J. A. Doudna, C. Grosshans, A. Gooding, and C. E. Kundrot, *Proc. Natl. Acad. Sci. U.S.A.* **90,** 7829 (1993).

days, and therefore should be avoided at high pH). Inclusion of spermine or other such amines, while crucial for tRNA crystallization, does not in general appear to be a requirement for crystallization of most other RNA sequences (although its use as a nonspecific precipitation agent is sometimes incorporated in a crystallization condition). Within such parameters, it is generally most rewarding, and perhaps more rigorous in terms of reproducibility, to develop one's own sparse matrix screen tailored to the requirements of one's specific RNA, rather than to adhere slavishly to previously published conditions.

Structural Determination by Isomorphous Replacement

From the standpoint of macromolecular crystallography, solving an RNA structure by isomorphous replacement is essentially identical to solving a protein structure; the physical principles of data collection and solving the phase problem are identical in each case. These are discussed extensively in *Methods in Enzymology,* volumes 114, 115, 276, and 277. What differs in the case of RNA is the likelihood of success in various different approaches for forming heavy atom derivatives. Although finding conditions in which a heavy atom binds with high affinity to one or a few sites on an RNA molecule may be more difficult than with protein crystals, crystals of nucleic acids also present some advantages. If some or all of the RNA used in crystallizations has been obtained by chemical synthesis, 5-bromouridine and other brominated and iodinated phosphoroamidites, which are commercially available, can be used to incorporate covalently bound isomorphous heavy atom derivatives having complete specificity, to a predetermined number of sites on the RNA, and to full occupancy.[3] The ability to produce heavy atom derivatives having these characteristics is often crucial to the successful solution of a macromolecular crystal structure in a straightforward manner. The modified nucleotide in many cases need not be a ribonucleotide; often, the less expensive deoxy versions of these derivatives will do just as well.[3]

For small RNAs, covalently bound heavy atom derivatives can be incorporated without difficulty using standard RNA synthetic techniques. For larger RNAs that cannot be synthesized by standard chemical means, one proven approach is to make the large RNA in two parts, where the large and small parts can associate via conventional base-pairing interactions. The larger part, containing a terminal deletion of about 30 residues or less, can be transcribed from a template derived from the original full-length template, and the missing RNA sequence can then be synthesized chemically, allowing for incorporation of covalently attached heavy atoms. Although the two strands can then be ligated to form a single covalently linked RNA strand, this may not be required in practice for the formation of a stable complex, as was the case for a domain of the group I intron.[8]

In addition to brominated and iodinated phosphoroamidites, covalently bound heavy atoms may be incorporated into synthetic RNA using other commercially available phosphoroamidites, such as deoxyphosphorothioates that can be used to incorporate potential Hg binding sites in the RNA. (It is important to employ the deoxyamidite to prevent Hg-induced RNA strand cleavage involving the ribose adjacent to the modified phosphate.)

Data collection is typically carried out on area detector systems, as is the case with other macromolecular crystallography. However, nucleic acid X-ray diffraction data typically differ from protein diffraction data in that the lattice diffraction pattern is convoluted with the fiber-like diffraction that is the manifestation of the helical transform of RNA; i.e., the regular spacing of base pairs modulates the intensities of the Bragg reflections. As a consequence, the intensities typically do not tail off smoothly as a function of increasing resolution. Rather, the intensity modulation is periodic, and the reflections along the helical axes around 3.4 Å tend to be especially strong, while those perpendicular tend to be weaker. This gives rise both to anisotropy of the data as well as to very intense data, which contain much of the contribution to the electron density of the bases, that is quite difficult to process accurately. Conventional data processing in MOSFLM[9] or DENZO[10] involves defining an average spot profile; these reflections are often rejected due to poor profile fit unless the program is "tricked" into processing these very intense and seemingly highly mosaic reflections, either by relaxing the rejection criteria, or by processing the data a second time with the intent of only measuring these spots accurately, and then merging them with the conventionally processed data set after applying appropriate scaling correlations. These procedures are *ad hoc* and not particularly satisfactory, but have the merit of increasing the quality of the phases and therefore the electron density maps, at least in the case of the hammerhead RNA.

Cryoprotection of RNA crystals is often found to be essential, as most appear to be radiation sensitive. Cryoprotection of crystals for the purpose of multiwavelenght anomalous diffraction (MAD) data collection is especially recommended. (The apparent radiation sensitivity of RNA crystals may be correlated with the frequent use of cacodylate buffers in RNA

[8] B. L. Golden, A. R. Gooding, E. R. Podell, and T. R. Cech, *RNA* **2,** 1295 (1996).

[9] A. G. W. Leslie, "MOSFLM Users Guide." MRC Laboratory of Molecular Biology, Cambridge, UK, 1998.

[10] Z. Otwinowski and V. Minor, *Methods Enzymol.* **276,** 307 (1997).

crystallizations; merely changing the identity of a buffer has been shown to reduce dramatically the radiation sensitivity in the case of the hammerhead RNA.) The software used for isomorphous replacement phase calculation and refinement is identical to that commonly employed for protein crystallography, and the choices are generally a matter of personal preference. Maximum likelihood phasing algorithms such as MLPHARE[11] (distributed with the CCP4 crystallographic program suite, Ref. 12) or SHARP[13] have the advantage of not overestimating figures of merit and are thus more amenable to further improvement using solvent-flattening techniques. As with other macromolecular structures, noncrystallographic symmetry averaging can greatly improve the quality of initial electron density maps and structural refinement in cases where there are similar molecules in the asymmetric unit of the crystal. One must, however, be aware of the high degree of pseudosymmetry found in RNA molecules, because this can compound the problem of locating noncrystallographic symmetry axes. (It also makes the solution of RNA structures by molecular replacement quite difficult.)

Fitting a model of RNA to experimentally determined electron density can be rather more difficult than it is with protein structures, because one can generally distinguish side chains only as purines or pyrimidines unless the data are of high resolution. It is therefore extremely helpful to have both prior knowledge of the RNA secondary structure and convenient landmarks in the electron density. The latter can be obtained by using heavy atom derivatives that are covalently bound to known positions in the RNA sequence. An additional aid to the eye in terms of tracing the backbone of the RNA can be obtained by contouring the electron density map at about 2.0 to 3.0 times the rmsd (root mean square deviation) of the map; this procedure generally reveals the location of the phosphates unambiguously. These phosphate locations can often be verified using an anomalous differences Fourier map because the phosphorous atoms have detectable X-ray absorption even at the remote wavelengths of X rays typically employed for macromolecular structural determination. Elimination of (presumably spurious) anomalous differences greater than 3 times the rmsd tend to improve such maps dramatically.

Refinement of RNA structures again proceeds along the lines of the procedures typically used for protein crystallography. XPLOR[14] and its

[11] Z. Otwinowski, in "Isomorphous Replacement and Anomalous Scattering, Proceedings of Daresbury Study Weekend," pp. 80–85. SERC Daresbury Laboratory, Warrington, U. K., 1991.

[12] Collaborative Computational Project, N. 4, *Acta Crystallogr.* **D50,** 760 (CCP4, 1994).

[13] E. De La Fortelle and G. Bricongne, *Methods Enzymol.* **276,** 472 (1997).

[14] A. T. Brünger and L. M. Rice, *Methods Enzymol.* **277,** 243 (1997).

successor, CNS,[15] have proven to be successful in RNA structural refinement, especially when used with the new nucleic acid parameter library produced by Helen Berman and co-workers from the Nucleic Acids Structural Database.[16] It is generally recommended that the charges on the phosphates be reduced or set to zero during simulated annealing molecular dynamics refinement, because the counterion screening environment that neutralizes the RNA cannot be modeled readily. In practice, we have found that this does not seem to make an appreciable difference in the course of refinement, perhaps because the constraint of the electron density overcomes the potential electrostatic repulsion of the phosphate backbone in the model structure.

From Static RNA Structures to Reaction Dynamics

Although conventional X-ray crystallography is perhaps unparalled as a physical technique for elucidating RNA structure, it is limited in that it only provides a static representation of a molecule in a single state. In the case of the three currently elucidated ribozyme crystal structures, these states include the initial enzyme–substrate[17] or enzyme–inhibitor[18,19] complex, the free enzyme,[20] or the enzyme–product complex,[6] respectively. Although these structures provide starting points for inferring ribozyme mechanisms, their usefulness is inherently limited by the fact that they constitute single, and extremal, points in the reaction pathways of these ribozymes. However, reliable time-resolved crystallographic methods have now been developed in the context of protein enzymes that allow transient intermediates in the enzyme-catalyzed reaction that accumulate under steady-state conditions to high occupancy simultaneously throughout the crystal lattice to be observed rapidly (using Laue crystallography as a fast data collection technique) or trapped (either chemically or physically) and examined using monochromatic X-ray crystallography.[21–23] Because the

[15] A. T. Brünger, P. D. Adams, M. G. Clore, W. L. DeLano, P. Gross, R. W. Grosse-Kunstleve, R. W. Jiang, J.-S. Kuszewski, J. Nilges, N. S. Parru, R. J. Read, L. M. Rice, T. Simonson, and G. L. Warren, *Acta Crystallogr.* **D54,** 905 (1998).

[16] G. Parkinson, J. Vojtechovsky, L. Clowney, A. T. Brünger, and H. M. Berman, *Acta Crystallogr.* **D52,** 57 (1996).

[17] W. G. Scott, J. B. Murray, J. R. P. Arnold, B. L. Stoddard, and A. Klug, *Science* **274,** 2065 (1996).

[18] H. W. Pley, K. M., Flaherty, and D. B. McKay, *Nature* **372,** 68 (1994).

[19] W. G. Scott, J. T. Finch, and A. Klug, *Cell* **81,** 991 (1995).

[20] B. L. Golden, A. R. Gooding, E. R. Podell, and T. R. Cech, *Science* **282,** 259 (1998).

[21] D. W. J. Cruickshank, J. R. Helliwell, and L. N. Johnson, "Time-Resolved Macromolecular Crystallography." The Royal Society, Oxford Science Publications, 1992.

[22] K. Moffat and R. Henderson, *Curr. Opin. Struct. Biol.* **5,** 656 (1995).

[23] K. Moffat, *Methods Enzymol.* **277,** 433 (1997).

field of ribozyme crystallography is new, the techniques of time-resolved crystallography are only now beginning to be applied to the problem of ribozyme catalysis. For that reason we describe in some detail the methods that have proven successful in our laboratory in the context of the hammerhead ribozyme system with the hope that the lessons we have learned will be generalizable to other systems that have not yet been investigated.

Initiation of Ribozyme Catalysis in Crystals

A successful time-resolved crystallography experiment requires that the reaction catalyzed by a crystallized enzyme be initiated simultaneously throughout the crystal lattice so that the intermediate species that is to be observed can accumulate to high or full occupancy throughout the crystal simultaneously. This enables us to observe the intermediate by fast data collection techniques or to trap it physically. In the case of the hammerhead ribozyme sequence that we have crystallized, the catalytic turnover rate in the crystal is approximately 0.4 molecule/min when crystals grown at pH 5–6 are activated with the addition of a mother liquor solution buffered at pH 8.5 and containing 50–100 m*M* divalent metal ion. The time it takes for a fairly complex substrate (NADP) to diffuse into a crystal of isocitrate dehydrogenase measuring 0.5 mm in each dimension and to saturate the enzyme's active sites is approximately 10 sec when measured directly by video absorbance spectroscopy.[24] The corresponding time it takes a much smaller divalent metal ion to diffuse into and saturate the considerably smaller hammerhead ribozyme crystals ($0.3 \times 0.25 \times 0.25$ mm or smaller) is unlikely to be longer. Therefore, the diffusion time is sufficiently fast compared to the turnover rate to allow approximately synchronous initiation of the hammerhead ribozyme cleavage reaction throughout the crystal.

A hammerhead ribozyme cleavage reaction in the crystal therefore can be initiated by removing a crystal from the drop in which it has grown, using a small (approximately 0.4-mm in diameter) rayon loop mounted on a wire. The crystal is then immediately immersed into an artificial mother liquor solution containing divalent metal ions at a higher pH (typically 1.8 *M* Li_2SO_4, 50–100 mM divalent metal ion, and 50 m*M* Tris, pH 8.5, augmented with 20% glycerol as a cryoprotectant). The artificial mother liquor should be prepared immediately before use using freshly dissolved components. (This is especially critical when using the softer divalent metal ions because they slowly form insoluble metal oxides under even mildly basic conditions.) The crystal is placed into approximately 500 μl of the artificial mother liquor in a glass spot plate and observed under a polarizing microscope while being slowly wafted through the solution to aid in mixing at

[24] P. O'Hara, P. Goodwin, and B. L. Stoddard, *J. Appl. Crystallogr.* **28,** 829 (1995).

the solution–crystal interface. The crystal is subsequently removed from the solution in the same manner as before at the preordained time and is flash frozen in liquid nitrogen or liquid propane to trap any accumulated intermediate state. Monochromatic X-ray data can then be collected in the usual manner for cryoprotected crystals. The crystal may be removed from the X-ray source subsequent to data collection and stored frozen in liquid nitrogen until such time that assay of the extent of cleavage is convenient.

The relatively slow reaction rate of the hammerhead ribozyme[25] allows us to use this very simple method of freeze trapping. Depending on the turnover rate of other ribozymes whose reactions might be examined in the crystal [such as hepatitis delta virus (HDV), group I intron, or RNase P], it may be necessary to employ a flow cell to deliver a carefully timed pulse of substrate or divalent metal ion cofactor, or even to use a photochemically activatable "caged" substrate precursor in order to initiate the reaction rapidly throughout the crystal lattice.[26] Constructing a flow cell need not be a laborious process. A simple procedure that makes use of a conventional crystallographer's capillary connected to a syringe pump has been described previously.[27] Possible photoactivatable triggers include photolabile chelators for divalent catioins and RNA substrates that are caged with a photolabile *o*-nitrobenzyl moiety attached to the 2′-oxygen of the active site ribose. The latter has already been accomplished for the hammerhead ribozyme[28] and crystallization of the "caged" hammerhead ribozyme is currently under way in our laboratory.

Use of "Kinetic Bottleneck" Modifications to Trap Intermediates Chemically

To observe structural or chemical intermediates in crystallized enzymes, it is necessary for the intermediate to accumulate to high occupancy simultaneously throughout the crystal lattice. Otherwise, spatial and temporal averaging will make observation of the intermediate impossible, because the electron density will represent the entire ensemble of structures at a given time point. In practice, this means that there must be a rate-limiting step in the reaction scheme corresponding to the decomposition of the intermediate; i.e., a kinetic bottleneck in the reaction pathway must either

[25] T. K. Stage-Zimmermann and O. C. Uhlenbeck, *RNA* **4,** 875 (1998).
[26] I. Schlichting and R. S. Goody, *Methods Enzymol.* **277,** 467 (1997).
[27] G. A. Petsko, *Methods Enzymol.* **114,** 141 (1985).
[28] S. G. Chaulk and A. J. MacMillan, *Nucleic Acids Res.* **26,** 3173 (1998).

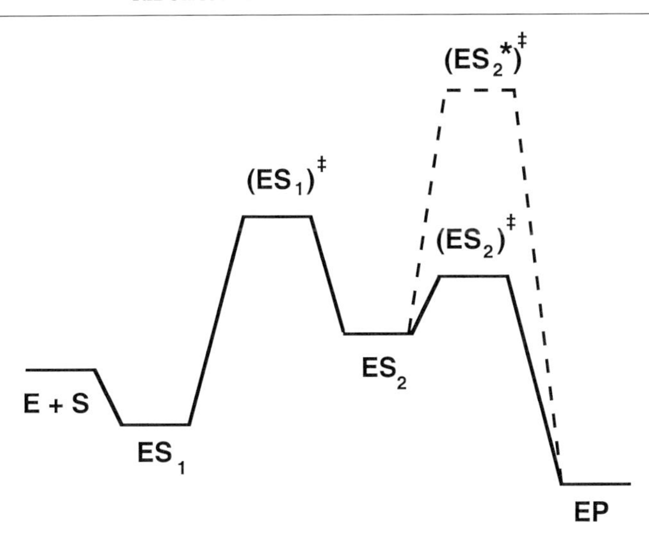

FIG. 1. An idealized potential energy diagram for an enzyme whose mechanism involves an intermediate species. The diagram depicts the relative energies of the free enzyme and substrate, the initial enzyme–substrate complex (ES_1), the enzyme–intermediate complex (ES_2), their corresponding transition states $(ES_1)^{\ddagger}$ and $(ES_2)^{\ddagger}$, and the enzyme–product complex (EP). A cross section of the potential energy surface along the reaction coordinate for the unmodified enzyme is shown as a solid line, and that for the idealized "kinetic bottleneck" mutant enzyme, where different, is shown as a dashed line. Note that for the unmodified enzyme, formation of the intermediate is the rate-limiting step, thus making observation of the intermediate species difficult. The ideal kinetic bottleneck modification designed for observing the intermediate will raise the energy of the transition state that follows formation of the intermediate $(ES_2{}^*)^{\ddagger}$, in such a way that decomposition of the intermediate now becomes rate limiting, but with minimal perturbation of the reaction equilibrium (i.e., the relative energy levels of ES_1, ES_2, and EP). In such a case the enzyme–intermediate complex (ES_2) will accumulate transiently during the course of a single-turnover reaction even though the equilibrium remains unchanged, and can then be observed using the techniques of time-resolved crystallography.

exist naturally, or it must be created through alteration of the reaction conditions or of the enzyme itself (Figs. 1 and 2).

Following the example of a time-resolved crystallography experiment with isocitrate dehydrogenase, in which observation of the chemical intermediate oxalosuccinate was made possible using a kinetic bottleneck that causes the decarboxylation step of the reaction to be rate limiting,[29,30] we

[29] B. L. Stoddard, B. E. Cohen, M. Brubaker, A. D. Mesecar, and D. E. Koshland, Jr., *Nature Struct. Biol.* **5,** 89 (1998).

[30] J. M. Bolduc, D. H. Dyer, W. G. Scott, P. Singer, R. M. Sweet, D. E. Koshland, Jr., and B. L. Stoddard, *Science* **268,** 1312 (1995).

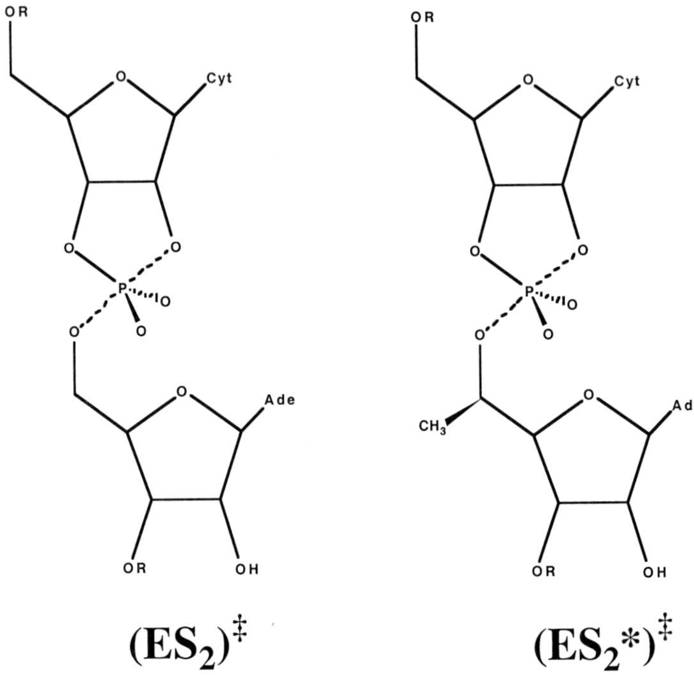

$(ES_2)^{\ddagger}$ $(ES_2{*})^{\ddagger}$

FIG. 2. Illustrations of normal (left) and kinetic bottleneck-modified (right) transition states of the hammerhead ribozyme cleavage reaction, corresponding to Fig. 1. Use of the methyl group adjacent to the leaving-group 5′-oxygen causes the bond-breaking step of the reaction to become rate limiting, enabling the conformationally changed structure, ES_2 of Fig. 1 to be trapped and observed crystallographically.

modified our hammerhead ribozyme sequence to make the bond cleavage step of the reaction rate limiting. On ionization of the 2′-hydroxyl at the cleavage site ribose of the natural hammerhead ribozyme, the charged oxygen initiates nucleophilic attack at the 3′-scissile phosphate. As the bond between the 2′-oxygen and the phosphorus forms, the bond between the phosphorus and the 5′-oxygen of the adjacent ribose breaks in a manner consistent with an $S_N2(P)$ mechanism. Though it is unclear whether this process is concerted or sequential, it is recognized that a conformational change must take place in the vicinity of the scissile phosphate (relative to the initial structure observed in hammerhead ribozyme crystals) in order for the chemical step or steps of the reaction to be initiated.[17–20] Therefore, the conformational intermediate state that is compatible with subsequent formation of an in-line transition state or chemical intermediate is of significant interest to those hoping to understand the stereochemical mechanism of hammerhead ribozyme catalysis. If the bond-breaking step of the reaction were to be made rate limiting, this would permit the transient accumulation of the conformational intermediate required for initiation of the reaction chemistry. By altering the properties of the 5′-leaving group, one could in principle engineer a ribozyme having a kinetic bottleneck at the bond-breaking step of the reaction. This has in fact been realized in the form of a hammerhead ribozyme that has an extra methyl group appended to the 5′-carbon adjacent to the leaving-group oxygen; such a ribozyme still exhibits the normal pH versus rate profile of the hammerhead ribozyme (suggesting that the reaction pathway has not been substantially altered), but cleavage is slowed by at least two orders of magnitude in the crystal. Use of this sort of modified RNA substrate at the active site of the ribozyme in single-turnover reactions enabled us to visualize the conformational change to a structure compatible with subsequent formation of an in-line transition state.[31] In this case (unlike isocitrate dehydrogenase) the lifetime of the intermediate was sufficiently long to enable conventional monochromatic X-ray data to be collected on a frozen crystal.

The particular form of the kinetic bottleneck modification used will be dependent on the reaction chemistry that is catalyzed by the ribozyme. However, all of the naturally occurring ribozymes catalyze phosphate chemistry in which either the 5′-oxygen or the 3′-oxygen is a leaving group, suggesting that modifications similar to the one employed with the hammerhead ribozyme should be useful in principle, though their usefulness in practice must of course be tested experimentally. Again, the main characteristics of a useful kinetic bottleneck modification are (1) that it creates a rate-limiting step corresponding to the breakdown of the intermediate that is to be observed; (2) that it alters the kinetic properties of the ribozyme without grossly perturbing the thermodynamics of the reaction (and therefore its equilibrium); and (3) that creation of the kinetic bottleneck does not cause the reaction to proceed via a different (aberrant) chemical mechanism.

Use of Cryocrystallographic Techniques to Trap Intermediates Physically

If the lifetime of a ribozyme intermediate species is on the order of seconds or more, physical trapping techniques can be used to further stabilize or immortalize the intermediate for the purpose of data collection.[22] The most straightforward approach is to simply flash freeze the crystal in

[31] J. B. Murray, D. P. Terwey, L. Maloney, A. Karpeisky, N. Usman, L. Beigelman, and W. G. Scott, *Cell* **92,** 665 (1998).

liquid propane or liquid nitrogen as one might normally do to collect data on a cryoprotected and frozen crystal.[32] Because one has normally determined the crystal structure of some initial state of an enzyme prior to embarking on a time-resolved crystallography experiment, it is likely that a set of conditions for cryoprotecting a crystal has already been found. In general, these conditions should be applicable to the case of freeze trapping without further alteration. In the case of the hammerhead ribozyme, simply augmenting the artificial mother liquor or the reaction mixture with 20% glycerol was sufficient for cryoprotection. Immersion of the crystal in the cryoprotectant solution for 15 sec or more was generally all that was required for subsequent flash freezing. Subsequent to flash freezing, the crystals were maintained at 100 K throughout data collection using an Oxford Cryostream system (Oxford Cryosystems, Oxford, UK). Further details on crystallographic freeze-trapping techniques and low-temperature data collection have been described elsewhere.[22,32]

Use of Polychromatic or Laue X-Ray Crystallography for Rapid Observation of Transient Intermediates

In the case of a ribozyme (such as RNase P, HDV ribozyme, or a group I intron) whose intermediate has a lifetime of less than 1 sec or so, it is unlikely that conventional physical trapping techniques such as flash freezing alone will allow crystallographic observation using conventional monochromatic data collection. Instead, fast data collection techniques such as Laue crystallography are required to observe more evanescent species as they accumulate transiently in the crystal subsequent to reaction initiation. In the case of Laue crystallography, polychromatic X rays generated from an intense synchrotron source may be used to collect a fairly complete data set in a single snapshot of picosecond to millisecond duration. Assuming a crystal can withstand repeated exposure to the intense X-ray beam, a series of time points may be obtained from a single crystal over the course of a reaction. If the lifetime of the intermediate that is to be observed is as long as, or is at least comparable to, the exposure duration of the X-ray snapshot, and if the homogenous synchronized accumulation of that intermediate can occur throughout the crystal at the time of the X-ray exposure, the transient intermediate should be observable, as has been shown to be the case for several protein enzyme crystals.

In addition to the requirements for synchronous accumulation of the intermediate to high occupancy throughout the crystal, the reaction must be initiated on a timescale that is fast compared to the lifetime of the

[32] D. W. Rogers, *Methods Enzymol.* **276,** 183 (1997).

intermediate, and the crystal must diffract sufficiently well throughout the course of the experiment in order to obtain useful data. Useful Laue data can be obtained only from crystals that are inherently well ordered and that are robust with respect to X-ray radiation damage. Lattice imperfections and even modest mosaicity can make such experiments impossible in practice. In the case of the hammerhead ribozyme, the crystals were not sufficiently perfect to allow processible Laue diffraction data to be collected. For that reason, we refer the interested reader directly to two comprehensive *Methods in Enzymology* discussions of Laue crystallography.[22,33] Hopefully crystals of other ribozymes, especially those having a much faster turnover rate, will be of sufficient quality to allow useful Laue data to be collected.

Analysis of Results of Time-Resolved Ribozyme Crystallography

Although detection of a conformational change or chemical intermediate using X-ray crystallography is fairly straightforward, it is important to eliminate possible sources of bias and to perform the proper experimental controls to prove that the crystal whose structure is being modeled does indeed contain the intermediate state one purports to observe. The first dictum is addressed here; the second is the subject of the next section, in which an assay for ribozyme catalysis in the crystal is described.

A chemical or conformational intermediate structure can be observed crystallographically using difference Fourier techniques. If, for example, a conformational change is to be observed, it is quite possible that simply re-refining the initial structure against the new data set will reveal the conformational change with no apparent ambiguity. Even if this is the case, it is extremely important that the parts of the structure that appear to be involved in the conformational change be omitted, and that the refinement then be repeated, beginning again with the initial-state structure. The phases generated from the omit-refinement should not contain any model bias based on the presumed conformational change. Omit difference Fouriers based on Sim-weighted or Sigma-A weighted coefficients will further reduce model bias,[34] and difference Fouriers having coefficients of the form $[F_{obs}^{(intermediate)} - F_{obs}^{(initial\ state)}]\ \exp\{i\Phi^{(initial\ state)}\}$ should be essentially free of bias with respect to the model of the intermediate state. The clarity of the difference electron density, however, may be severely compromised, depending on how much the true phases deviate from the calculated phases of the initial structure. Sigma-A weighted maps should improve this situa-

[33] I. J. Clifton, E. M. H. Duke, S. Wakatsuki, and Z. Ren, *Methods Enzymol.* **277,** 448 (1997).
[34] R. J. Read, *Acta Crystallogr.* **A42,** 140 (1986).

tion as well, and recently developed "holographic" methods for reconstructing electron density using real-space methods hold much promise for dramatically improving the quality of such electron density maps without introducing phase biases from either the initial structure or from models of the intermediate.[35] Finally, if some sort of landmark feature can be detected reliably in the new electron density, such as the location of a phosphate group, further confidence in a modeled conformational change can be obtained. In the case of modeling a chemical reaction intermediate, the above difference Fourier methods can be employed, and these are especially powerful if one wishes to locate functional groups that appear or disappear in the course of the reaction that forms the chemical intermediate, using maps of the form $\pm[F_{obs}^{(intermediate)} - F_{obs}^{(initial\ state)}] \exp\{i\Phi^{(initial\ state)}\}$, respectively.

Development of Assay for Crystalline Ribozyme Catalysis

It is also imperative that the reaction being monitored crystallographically be characterized using biochemically based enzymatic assays, such that the time-resolved crystallographic experiments can be calibrated correctly. In the case of the hammerhead ribozyme, cleavage in the crystal has been monitored using an assay developed specifically for such reactions. Conventionally, kinetics assays are performed on the hammerhead ribozyme using ^{32}P-labeled substrate-strand molecules in a cleavage reaction. The cleavage products can be separated unambiguously from reactants using polyacrylamide gel electrophoresis under denaturing conditions similar to those used for nucleic acid sequencing. The relative amounts of substrate and product can then be quantified using autoradiography. This procedure is somewhat cumbersome for assaying cleavage within crystals of RNA because it would require phosphorylation of the substrate either prior to crystallization or subsequent to dissolving the crystals; the former procedure introduces complications for purifying and crystallizing the RNA, and the latter introduces potentially large systematic errors because cleavage of the RNA may continue in solution under conditions that are amenable to phosphorylating the reaction mixture.

Alternatively, one can use gel electrophoresis combined with a sensitive RNA staining procedure such as silver staining or staining with 0.5% toluidine blue solution (followed by destaining with hot water) to assay cleavage in the crystal. This procedure has the advantage of avoiding the problems associated with radiolabeling the RNA in the crystal or after dissolution of the crystal, but accurate quantitation of the RNA stained on a gel is much

[35] A. Szoke, *Acta Crystallogr.* **A54,** 543 (1998).

more difficult than with autoradiography. In addition, dissolved crystal solutions contain high concentrations of EDTA to quench cleavage (see later section) as well as other salts. Samples of high ionic strength tend to cause severe compressions in nucleic acid sequencing gels, making separation of product from substrate strands more difficult.

For these reasons, as well as convenience, we developed a high-performance liquid chromatography (HPLC)-based assay for crystallized hammerhead ribozyme self-cleavage (Fig. 3). This procedure has the advantages of (1) speed and convenience, (2) being immune to the effects of having a high concentration of EDTA and other salts and buffers in the sample being assayed, and (3) providing output in a form that is readily quantifiable (i.e., the absorbance profiles of the eluates can be integrated to provide an accurate assay of RNA cleavage). The one disadvantage of HPLC analyses is that they require a higher concentration of RNA than do autoradiographic procedures. However, this is not a problem when dealing with crystallized

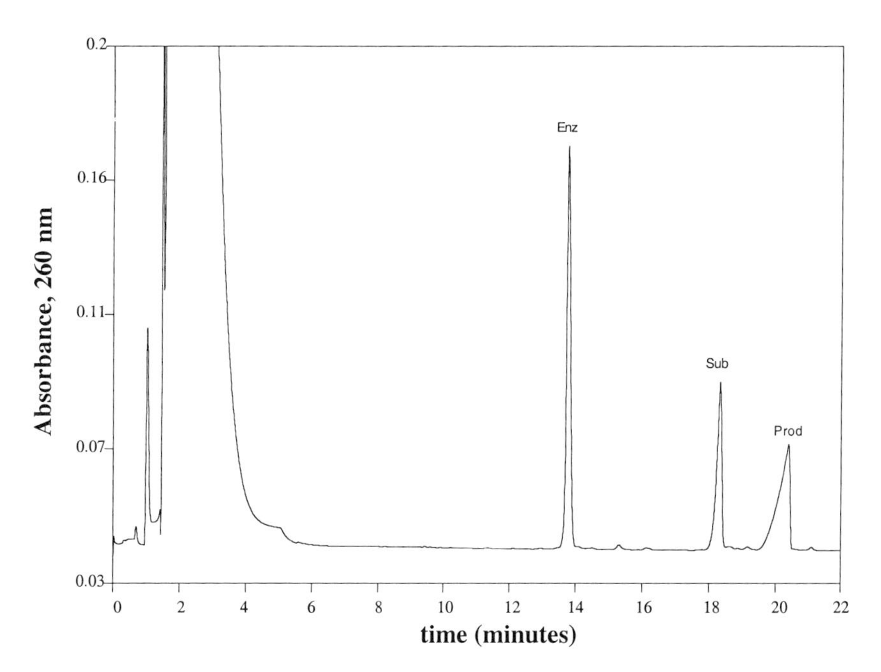

FIG. 3. A representative ion-exchange HPLC trace of a hammerhead ribozyme cleavage reaction monitored in the crystal, subsequent to quenching the reaction in 0.5 *M* EDTA as described in the text. The 16-mer enzyme strand elutes at about 14 min, the uncleaved (25-mer) at about 18 min, and the 20-mer cleavage product elutes later, at about 20 min, perhaps due to the greater charge density (see text). A large breakthrough peak is present that corresponds to the EDTA, buffers, and other salts in solution.

RNA; the RNA consumed in a typical crystallization experiment will drive most molecular biologists to tears. The use of HPLC to follow hammerhead ribozyme reactions in solution has previously been described.[36,37]

Termination of Ribozyme Catalysis in Crystals

The hammerhead cleavage reaction was initiated in the crystal as described earlier. The cleavage reaction can be arrested indefinitely by flash freezing, making observation of a trapped intermediate state possible. However, to assay the extent of cleavage in the crystal, the crystallized RNA must be dissolved. To obtain a reliable estimate of the extent of cleavage in the crystal, it is imperative that the RNA not be allowed to react further once it has been dissolved.

We established a procedure, based on trial and error, that terminates RNA cleavage in the crystal prior to dissolution and also prevents further reaction from taking place in solution. The most reliable procedure is to remove the crystal from the reaction solution (described earlier) using a rayon loop, and immediately transfer it to a 0.5-ml drop of 0.5 *M* EDTA at pH 6.5. In our experience, the crystal is relatively stable in such a solution for over half an hour, although some cracks in the crystal develop. The EDTA quickly enters the crystal and quenches the reaction. We have observed that concentrations of EDTA much less than 0.5 *M* are not reliable for quenching the reaction completely. Once the crystal has been allowed to soak for 30 min or more, it can be dissolved in the EDTA solution by manually disrupting the crystal with a micropipette. Once the crystal has been observed to have dissolved completely (making use of a polarizing microscope to detect small shards of undissolved crystal), the solution can then be analyzed directly by ion-exchange HPLC.

Analysis of Extent of Ribozyme Cleavage in Crystal by HPLC

The problems faced with in characterizing the amount of each RNA species in the dissolved crystals are large sample volumes (0.05–0.5 ml), high salt concentrations (0.5 *M* EDTA), and accurate quantification of the various RNA components. Ion-exchange HPLC coupled with UV absorbance detection is an ideal technique for dealing with these problems. Samples can be loaded in any size volume, which allows us to dilute the salt concentration to less than 100 m*M*. Quantification is straightforward, the area of each peak corresponds to the integrated absorbance intensity,

[36] L. Citti, L. Boldrini, S. Nevischi, L. Mariani, and G. Rainaldi, *Biotechniques* **23,** 898 (1997).
[37] R. Vinayak, A. Andrus, N. D. Sinha, and A. Hampel, *Anal. Biochem.* **232,** 204 (1995).

and therefore the concentration of RNA, assuming the molar extinction coefficient can be calculated accurately. The ion-exchange column of choice in many RNA synthesis laboratories is the Dionex DNA-PAC. For the hammerhead ribozyme experiments, we have used a resolving gradient of 0.35–0.65 *M* NH_4Cl in 22 min. This permits a sample to be analyzed every 40 min. The column should be maintained at 50°. This gives superior resolution relative to room temperature, allowing us to resolve the enzyme (16-mer), substrate (25-mer), and the 2′,3′-cyclic-phosphate product (20-mer) strands from one another. [The 5′-OH product (5-mer) can be observed by starting the gradient at lower NH_4Cl concentrations and by using larger loading volumes.] We found that the 20-mer product strand eluted at higher salt concentrations than did the 25-mer substrate strand. (The identities of the eluates were independently characterized subsequent to chromatographic separation by denaturing PAGE.) It is possible that the reversal of the expected order of elution is due to the fact that length to charge ratio of the product strand is less than that for the substrate strand, i.e., the 20-mer has 20 phosphates and the 25-mer has 24 phosphates, due to the presence of the 2′,3′-cyclic phosphate on the 3′ end of the 20-mer.

The proportion of substrate remaining was calculated from $A_{sub}/(A_{sub} + 1.25\,A_{prod})$. The 1.25 coefficient corrects for the differences in the extinction

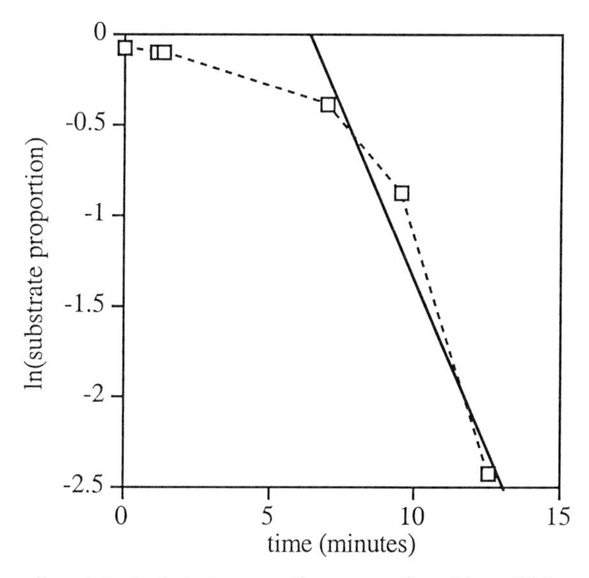

FIG. 4. A semilog plot of substrate proportion versus time. The solid line represents the best-fit linear slope following the characteristic lag time in the crystalline hammerhead ribozyme reaction, as measured by the HPLC assay.

coefficients between the substrate and product strands. The area of the substrate peaks serves as an internal control so that the $(A_{sub} + 1.25\, A_{prod})$ to A_{sub} ratio should remain constant. In our crystallized hammerhead ribozyme cleavage experiments, the ratio did in fact remain at 1.51 ± 0.03. A semilog plot of product production versus time reveals that a small proportion of the crystallized hammerhead RNA (about 7%) had cleaved during crystallization or subsequent storage and manipulation. Then after an initial lag phase of approximately 6 or 7 min, the remaining hammerhead RNA in the crystal undergoes cleavage at a rate of 0.4 min^{-1} (Fig. 4). The cause of the initial lag phase is unclear. Because the cleavage reaction is initiated by elevation of pH, and because the lag time increases with experiments done at lower pH values, it may be that the pH within the crystal is raised more slowly than would be expected based on the simple diffusion considerations described earlier, or it may reflect an activating conformational change that appears to take place in a concerted manner throughout the crystal lattice instead of in the stochastic manner one would expect from solution studies. These peculiarities, in addition to the fact that this particular RNA sequence actually cleaves several times *faster* in the crystalline lattice than it does in solution, illustrate the need to characterize the kinetics of the ribozyme cleavage reaction *in the crystal* when undertaking time-resolved crystallographic studies.

Acknowledgments

We thank the NIH for support of this research.

[14] Nuclear Magnetic Resonance Methods to Study RNA–Protein Complexes

By PETER BAYER, LUCA VARANI, and GABRIELE VARANI

Introduction

Structural studies of RNA-binding proteins and their RNA complexes are critical to our understanding of specificity in RNA–protein recognition and therefore how gene expression is regulated.[1,2] Considerable efforts have therefore been directed toward studying the structural principles underlying RNA–protein recognition, using both X-ray crystallography[3–5] and nuclear

[1] G. Varani, *Acc. Chem. Res.* **30,** 189 (1997).

[2] G. Varani and K. Nagai, *Ann. Rev. Biophys. Biomol. Struct.* **27,** 407 (1998).

[3] K. Nagai, *Curr. Op. Struct. Biol.* **6,** 53 (1996).

[4] C. Oubridge, N. Ito, P. R. Evans, C.-H. Teo, and K. Nagai, *Nature* **372,** 432 (1994).

[5] S. R. Price, P. R. Evans, and K. Nagai, *Nature* **394,** 645 (1998).

0076-6879/00 $30.00

magnetic resonance (NMR) methods. With specific regard to NMR, the relatively large molecular weight of such complexes has represented a serious limitation, despite some success.[6,7] One popular avenue has been the study of peptide models[8–11] that recapitulate important aspects of RNA recognition in some classes of RNA-binding proteins. However, these model peptides bind RNA tightly but not specifically. Furthermore, there is an ever more compelling need to extend structural studies beyond simple RNA–protein recognition, and toward studying the more complex macromolecular assemblies that constitute actual structural and functional units during gene expression. Progress in this area of structural biology requires extending existing NMR methods and developing new methodologies to determine structures of increased size and complexity. Here we review methods that have been used successfully in the study of protein–RNA complexes by NMR and introduce recent new approaches to study complex multimolecular assemblies.

Mapping Interaction Surfaces by NMR Chemical Shift Perturbation Analysis

A considerable strength of NMR is its ability to provide biochemically important information even before a complete structure determination is possible. If the structure of an RNA-binding protein domain and/or its RNA target is known, the intermolecular interface on the protein, RNA, or both can often be determined by comparing the NMR spectra of the free and bound forms of the two species (chemical shift perturbation analysis).[12,13] This approach can be extended to determine sites of protein–protein contacts that form when multiple proteins assemble cooperatively on specific mRNAs. This information can be of considerable importance in the design of biochemical and functional studies aimed at dissecting the biochemical basis of these interactions and their functional consequences.

[6] F.-H. T. Allain, C. C. Gubser, P. W. A. Howe, K. Nagai, D. Neuhaus, and G. Varani, *Nature* **380,** 646 (1996).

[7] P. W. A. Howe, F. H.-T. Allain, G. Varani, and D. Neuhaus, *J. Biomol. NMR* **11,** 59 (1998).

[8] P. Legault, J. Li, J. Mogridge, J. Greenblatt, and L. E. Kay, *Cell* **93,** 289 (1998).

[9] J. L. Battiste, H. Mao, N. S. Rao, R. Tan, D. R. Muhandiram, L. E. Kay, A. D. Frankel, and J. R. Williamson, *Science* **273,** 1547 (1996).

[10] J. D. Puglisi, L. Chen, S. Blanchard, and A. D. Frankel, *Science* **270,** 1200 (1995).

[11] R. N. De Guzman, Z. R. Wu, C. C. Stalling, L. Pappalardo, P. N. Borer, and M. F. Summers, *Science* **279,** 384 (1998).

[12] M. Görlach, M. Wittekind, R. A. Beckman, L. Mueller, and G. Dreyfuss, *EMBO J.* **11,** 3289 (1992).

[13] P. W. A. Howe, K. Nagai, D. Neuhaus, and G. Varani, *EMBO J.* **13,** 3873 (1994).

The task of defining the intermolecular interaction surface within an RNA-binding protein domain is generally straightforward once spectral assignments and the structure have been obtained for the RNA-free protein domain. If the RNA–protein complex is soluble, a set of correlated spectra (e.g., ^{1}H–^{15}N HSQC) at different protein : RNA ratios will identify any change in the immediate environment of NMR-active nuclei in the protein (e.g., most backbone and some side-chain NH groups; Fig. 1). This information can be obtained with small amounts of material (concentrations as low as 0.1–0.2 m*M* suffice, corresponding to ≈1 mg of protein and RNA). The protein backbone is particularly sensitive to any change that occurs on complex formation and ^{15}N-labeling is therefore sufficient for this task. Unless substantial conformational changes occur on complex formation, the information accessible from such samples is usually sufficient to define intermolecular interaction surfaces unambiguously. Additional information can be obtained by recording spectra on ^{13}C-labeled samples as well. The methyl region of these spectra contains relatively sharp and well-dispersed resonances, and is ideal for this analysis.

Data suitable for chemical shift perturbation analysis are best obtained by first recording a spectrum of the free protein, then progressively adding

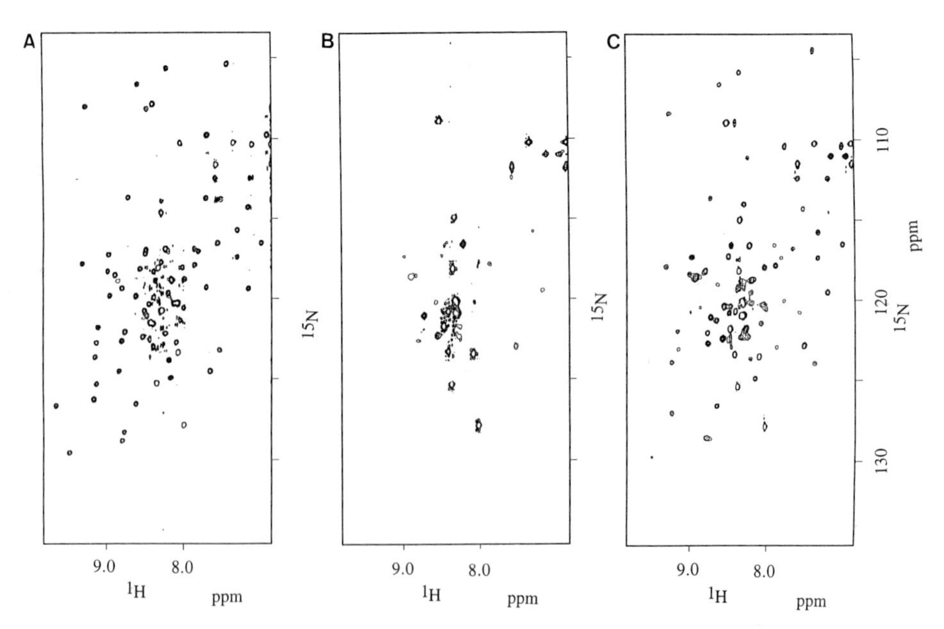

FIG. 1. Titration of *Drosophila staufen* dsRBD with increasing amounts of RNA. ^{1}H–^{15}N HSQC spectra of (A) free protein, (B) 2 : 1 protein : RNA complex; only resonances from the unfolded amino acids at both ends of the protein remain visible; and (C) 1 : 1 complex.

RNA until the protein : RNA ratio is 1 : 1. If the complex is only soluble when the RNA is in excess (as is sometimes the case), spectra would have to be recorded by increasing the protein concentration starting from an RNA-only sample. This approach provides information not only on structural properties, but on the kinetic properties of the interaction as well. In the spectrum shown in Fig. 1A, resonances for the free protein are relatively sharp, as expected for a folded small protein domain. When RNA is added, the protein resonances broaden considerably and many disappear (Fig. 1B). Most resonances sharpen up progressively as more RNA is added; the sharpest lines are found when the protein : RNA concentration is close to 1 : 1 (Fig. 1C). This textbook behavior corresponds to the so-called intermediate exchange regime, and is expected when the off rate for complex formation is in the millisecond timescale, i.e., for a weak ($K_d \approx \mu M$) complex. A more thorough analysis of the spectrum of the 1 : 1 complex reveals a more complicated behavior and a wide range of line shapes. Resonances corresponding to unfolded tails of the protein domain remain almost as sharp in the complex as in the free protein, whereas most resonances increase their line width roughly in proportion to the increase in molecular weight on complex formation. However, some resonances remain broad: the corresponding sites are likely to experience the effect of dynamic fluctuations between distinct conformational states in the complex. This observation reveals the presence of conformational flexibility at the RNA–protein interface, with motions occurring in the millisecond to microsecond (ms to μs) timescale.

The previous analysis can be applied to RNA-dependent protein–protein interfaces as well. Human U1A protein forms a tight ($K_d < 10^{-9}$ M) bimolecular complex when bound to a half-site polyadenylation inhibition element RNA.[14] Two U1A molecules then form a cooperative trimolecular complex on the complete regulatory element, with significant cooperativity provided by RNA dependent protein–protein interactions.[14,15] The RNA–protein interface is correctly mapped to the surface of the β sheet and two variable loops by the chemical shift perturbation analysis for the bimolecular complex (Fig. 2A).[6,13] The sites within U1A where chemical shift changes are observed between bimolecular and trimolecular complexes unambiguously map the protein–protein interactions to the C-terminal helix of the domain and to neighboring hydrophobic side chains (Fig. 2B).

[14] C. W. G. van Gelder, S. I. Gunderson, E. J. R. Jansen, W. C. Boelens, M. Polycarpou-Schwartz, I. W. Mattaj, and W. J. van Venrooij, *EMBO J.* **12,** 5191 (1993).

[15] S. I. Gunderson, K. Beyer, G. Martin, W. Keller, W. C. Boelens, and I. W. Mattaj, *Cell* **76,** 531 (1994).

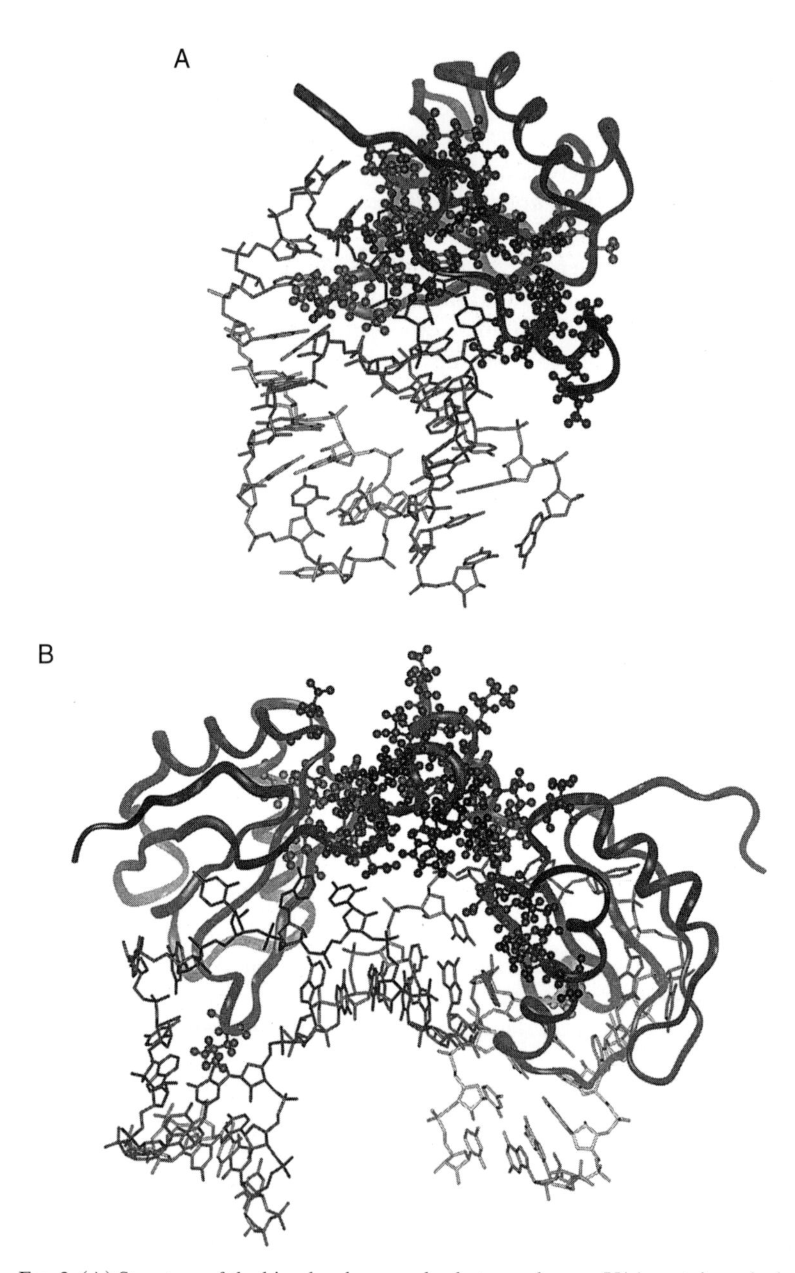

FIG. 2. (A) Structure of the bimolecular complex between human U1A protein and a half-site polyadenylation inhibition element RNA[6]; residues on the protein identified originally by chemical shift perturbation analysis[13] (side chains are explicitly shown only for these amino acids) map the RNA-binding surface of the protein. (B) In the cooperative trimolecular complex of two U1A molecules and the complete polyadenylation inhibition element, chemical shift perturbation analysis map protein–protein interactions to the C-terminal helix of the domain.

The methods illustrated in the previous paragraphs are far less useful and sometimes altogether misleading when applied to RNA. RNA chemical shift changes on complex formation are very often surprisingly modest, unless conformational rearrangement occurs. Even modest changes in intramolecular base stacking interactions are likely to lead to large chemical shift changes in comparison to those attributable to intermolecular contacts. As a consequence, even simple questions (e.g., whether an interaction occurs in the major or minor groove of an RNA double helix) cannot necessarily be addressed with confidence by this approach.

Structure Determination of Protein–RNA Complexes by Intermolecular NOE Distance Constraints

Structures of protein–nucleic acid and protein–protein complexes have been determined until now by extracting intermolecular distance constraints from Nuclear Overhauser Enhancement Spectroscopy (NOESY) experiments. This is accomplished by mixing isotopically labeled (^{13}C and/or ^{15}N) protein and unlabeled RNA and by recording appropriately filtered 2-D or 3-D experiments.[16–18] By using heteronuclear filters, it is possible to selectively observe NOE interactions between the isotopically labeled protein and unlabeled RNA. Labeling of the RNA provides a mirror image of this process and the opportunity to collect additional conformational constraints.[7] The best results are obtained by working with ^{13}C-labeled samples, since ^{15}N-attached RNA and protein resonances can generally be identified straightforwardly by characteristically distinct chemical shift and exchange behaviors.

This method is ideally suited to the study of multimolecular complexes as well, if samples with different labeled components can be prepared. For example, a ternary complex of two proteins and one RNA can be prepared by mixing labeled protein 1 with unlabeled protein 2 and RNA or with labeled protein 2 and unlabeled RNA and protein 1. If protein 1 is labeled, heteronuclear half-filtered experiments will provide protein 1–RNA interactions and protein 1–protein 2 contacts. If protein 2 is labeled, one would obtain protein 2–RNA interactions and the mirror image of the protein 1–protein 2 contacts. The study of symmetric dimers requires the preparation of samples containing an equal mixture of labeled and unlabeled proteins.[19] Filtered spectra will contain protein–RNA contacts as well as protein–protein interactions, but these latter will be present with reduced

[16] A. M. Petros and S. W. Fesik, *Methods Enzymol.* **239,** 717 (1994).

[17] M. Billeter, Y. Q. Qian, G. Otting, M. Müller, W. Gehring, and K. Wüthrich, *J. Mol. Biol.* **234,** 1084 (1993).

[18] G. Otting and K. Wüthrich, *Q. Rev. Biophys.* **23,** 39 (1990).

[19] P. J. M. Folkers, R. H. A. Folmer, R. N. H. Konings, and C. W. Hilbers, *J. Am. Chem. Soc.* **115,** 3798 (1993).

intensity since they will be generated by the fraction (statistically only 25% of the material) containing both labeled and unlabeled proteins (Fig. 3). If the protein–RNA off rate is slow and/or the complex strongly cooperative, it is essential to prepare these samples by premixing labeled and unlabeled proteins to avoid formation of complexes containing only either fully labeled or fully unlabeled proteins, but not the desired mixed samples.

The method described in the previous paragraphs is very powerful, but the sensitivity of carbon-filtered experiments decreases rapidly as the molecular weight increases, due to efficient relaxation (Fig. 3A). In addition, artifacts tend to appear in these spectra due to pulse imperfections and subtraction errors. These artifacts are more prominent for peaks close to the diagonal and for those involving sharp resonances, e.g., for methyl groups, and are very difficult to correct or eliminate completely (Fig. 3B). An alternative approach relies on the preparation of samples containing one protein component labeled with both ^{15}N and ^{2}H, leaving the other protein component and RNA unlabeled.[20] A ^{15}N-edited NOESY experiment recorded to selectively observe ^{15}N-attached resonances in one dimension will contain both intra- and intermolecular cross-peaks in the amide resonances region of the spectrum (typically downfield of 6.5 ppm). Even if a protein is prepared in D_2O to completely label all aliphatic side chains with deuterium, amide resonances will have exchanged back to ^{1}H during protein purification and sample preparation. The aliphatic region of the spectrum will instead contain only intermolecular NOE interactions, since all aliphatic protons have been substituted with deuterium by growing the protein in fully deuterated medium. This approach requires very high levels of deuteration (as close to 100% as possible), which are very difficult to obtain in practice. To achieve very high levels of deuteration, it is necessary to grow overexpressing bacteria in media containing deuterated glucose as nutrient. We observed many intramolecular cross-peaks in protein samples prepared in 100% D_2O but without deuterated glucose, but nearly all peaks disappeared when the protein was prepared in the presence of deuterated glucose. However, the possibility that leakage peaks due to intramolecular NOE interactions could obscure the sought-after intermolecular interactions must be investigated by recording spectra of the uncomplexed ^{15}N/^{2}H-labeled protein.[20] In this sample, amide aliphatic NOE cross-peaks can only correspond to intramolecular interactions and can be attributed to imperfect deuteration. If the two experiments are recorded under very similar conditions, any artifact due to incomplete ^{2}H-labeling can be corrected.

This approach has numerous advantages over the more traditional ex-

[20] K. J. Walters, H. Matsuo, and G. Wagner, *J. Am. Chem. Soc.* **119,** 5958 (1997).

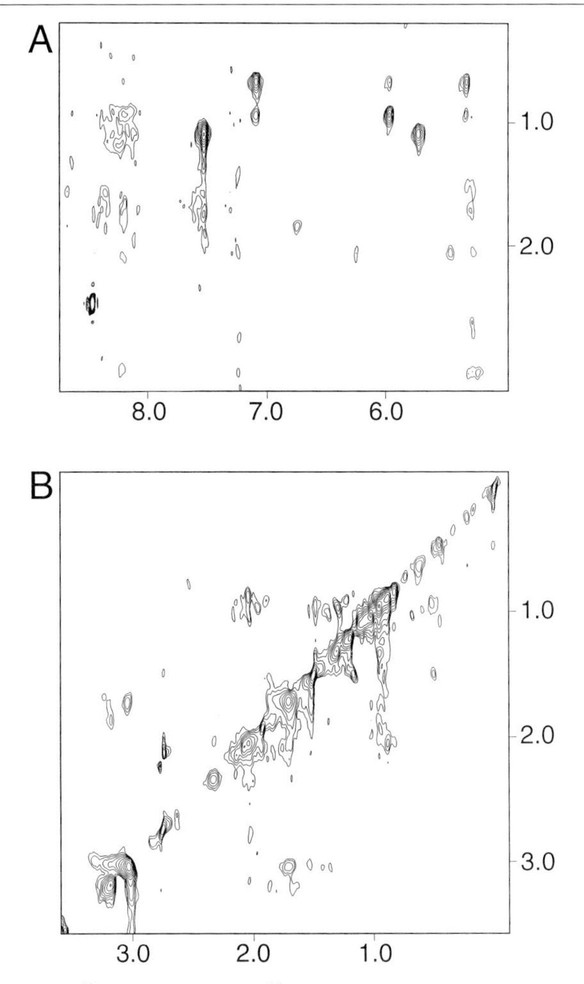

FIG. 3. One-half × ^{13}C-filtered spectrum[16] of the trimolecular complex of two U1A molecules and the complete polyadenylation inhibition element. (A) The region containing protein–RNA contacts; comparison of this spectrum with previously published data recorded on the smaller complex with the same protein[42] illustrates the reduction in signal as the molecular weight doubles. (B) The region containing protein–protein interactions. Many of the features seen in this region of the spectrum are artifacts due to imperfect cancellation of the sharp methyl resonances.

periments based on ^{13}C-labeling. First and foremost, the dynamic range of the experiment is very high: only intermolecular cross-peaks will be present if deuteration is complete, and any artifact due to incomplete deuteration can be accounted for as described in the previous paragraph. Second, ^{15}N-attached resonances relax more slowly than ^{13}C-attached resonances, and the high levels of deuteration further increase the relaxation times.[21] Therefore, this approach is less sensitive to relaxation at large molecular weights. On the other hand, the experiment relies on the observation of contacts between the backbone of one protein and the side chain of the other, rather than direct side chain–side chain interactions. These distances are very likely to be longer for hydrophobic interfaces defined by interactions mediated by residues carrying long aliphatic side chains, and therefore will often give rise to weak intermolecular NOE interactions.

Protein Deuteration in Study of Structure and Dynamics of Protein–RNA Complexes

The ability to prepare protein with different levels of deuteration allows numerous opportunities in the study of the structure and dynamics of protein–RNA complexes. In the preparation of deuterated protein samples, we use standard *Escherichia coli* BL21 (DE3) cells freshly transfected with the plasmid carrying the desired protein expression vector. A single colony from these plates is streaked on deuterated LA medium plates (DLA) and grown for about 15 hr. We tried to gradually adapt the cells to deuterium by growing them on plates with an increasing percentage of deuterium (50%, 75%, and finally 100%), but found no significant improvement in protein yields. Of course, this is likely to depend on the expression system used and the particular protein being overexpressed. A single colony from these plates is then inoculated into 5–10 ml of deuterated M9 media (DM9), with the desired percentage of deuterium and $^{15}NH_4Cl$ and ^{13}C glucose as sole sources of nitrogen and carbon, as desired. This starting culture is grown at 37° for about 15 hr, then added to 500 ml of prewarmed DM9 in a 2-liter flask, which is grown at 37° with vigorous shaking. The cells are induced with IPTG when they reach an absorbance of 0.6–0.8 at 600 nm and harvested 3.5 hr after induction. Cells grew approximately two to three times slower in DM9 compared to normal M9 media, but expression levels were comparable to those obtained with nondeuterated media. Protein yields were only reduced at the highest levels of deuteration. Deuterated media should not be autoclaved to minimize the loss of D_2O by evaporation. Thus, it is particularly important that good microbiological practice is followed at all times; the use of kanamycin-resistant expression vectors is also recommended because they provide more reliable expression levels and tighter regulation.

Having prepared samples with different levels of deuteration, we re-

[21] M. A. Markus, K. T. Dayie, P. Matsudaira, and G. Wagner, *J. Magn. Reson.* **B105,** 192 (1994).

70% deuterated sample

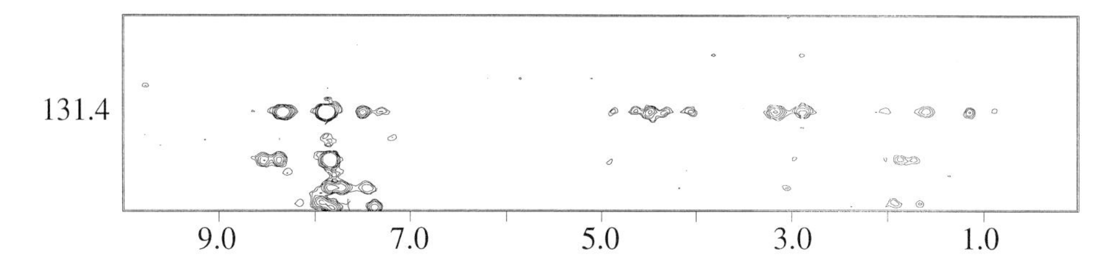

Fully protonated sample

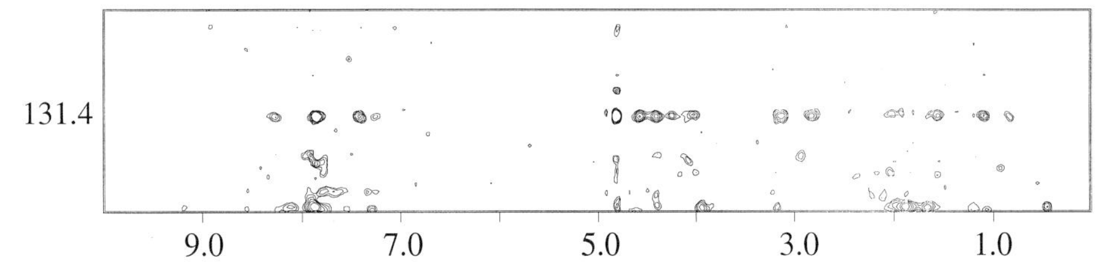

FIG. 4. Two-dimensional planes from a 3-D ^{15}N-edited NOESY spectrum at the chemical shift of residue Phe-101 of the trimolecular complex of two U1A molecules and the complete polyadenylation inhibition element. Both samples were fully ^{15}N–^{13}C labeled; the spectrum at the top was recorded on a 70% random fractional deuterated sample.

corded ^{15}N and ^{13}C-edited 3-D NOESY spectra with 50%, 70%, and 100% deuterated samples (these numbers are nominal and refer to the percentage of D_2O in the growing media rather than the actual level of deuteration, which is likely to be ≈10% lower in each case). The 70% deuterated samples provided the best results for the 40-kDa trimolecular complex of two human U1A proteins and the complete polyadenylation inhibition element. The aliphatic region is not worse in 70% deuterated samples by comparison to nondeuterated samples (Fig. 4, top), whereas the amide region is much better in the deuterated samples compared to protonated ones, as expected (Fig. 4, bottom). At this molecular weight, the loss of sensitivity in the aliphatic region of the spectrum (due to the lower number of protons) is compensated for by increased sharpness due to more favorable relaxation properties. The results of systematic simulations designed to estimate optimal levels of protein deuteration[22] are likely to be applicable to RNA as

[22] D. Nietlispach, R. T. Clowes, R. W. Broadhurst, Y. Ito, J. Keeler, M. Kelly, J. Ashurst, H. Oschkinat, P. J. Domaille, and E. D. Laue, *J. Am. Chem. Soc.* **118,** 407 (1996).

well, at least qualitatively. Isotope shifts degrade the quality of 2-D ^{13}C-edited HSQC spectra recorded on fractionally deuterated samples, but filter elements can be introduced in the NMR pulse sequence to distinguish methyl species containing 1 or 2 deuterons.[23] However, these effects are insignificant at the limited resolution of 3-D NOESY spectra and can be ignored in this context.

In addition to providing improved spectral quality, uniformly ^{13}C-labeled and fractionally deuterated protein–RNA samples can also be used to obtain information on the dynamics of protein side chains.[24] Deuteron relaxation is dominated by well-characterized quadrupolar interactions, making the interpretation of deuteron relaxation data more straightforward than for other side-chain groups. Experiments introduced to obtain this information select specifically for $^{13}CH_2D$ methyl groups using pulse sequence filter elements and make use of the high resolution provided by constant time ^{13}C–^{1}H correlation spectroscopy to measure the deuterium relaxation rate for methyl groups.[23] Fifty percent D_2O growing media gives the highest percentage of CH_2D groups among the various CH_3, CH_2D, CHD_2, and CD_3 combinations and are recommended for this application. In our experience, changes in protein dynamics on complex formation with RNA are far more informative when analyzing protein side chains by deuterium relaxation than analyzing the traditional amide backbone relaxation data.[25]

Obtaining Long-Range Intermolecular Distances Utilizing Paramagnetic Spin Labels

The methods described in the previous sections work well for systems characterized by tight binding ($K_d \approx 10^{-9}$ M or less) and highly specific recognition. However, many proteins of great biochemical interest bind RNA weakly ($K_d \approx 10^{-6}$ M) and with poor specificity. When these conditions occur, intermolecular NOE interactions can be quenched by dynamic processes at the interface (Fig. 1C), making high-resolution structure determination very difficult or altogether impossible. In this section, we describe a method to extract long-range intermolecular distance information in protein–RNA complexes based on electron–proton dipolar relaxation, which

[23] D. R. Muhandiram, T. Yamazaki, B. D. Sykes, and L. E. Kay, *J. Am. Chem. Soc.* **117,** 11536 (1995).

[24] L. E. Kay, D. R. Muhandiram, N. A. Farrow, Y. Aubin, and J. D. Forman-Kay, *Biochemistry* **35,** 361 (1996).

[25] T. Mittermaier, L. Varani, G. Varani, and L. Kay, *J. Mol. Biol.* (1999), in press.

is effective in overcoming some of the limitations of NOE-based experiments.[25a]

Paramagnetic spin labels have a long history in NMR spectroscopy, and have lately been rediscovered in studies of protein folding[26,27] and DNA–drug complexes.[28] Insertion of paramagnetic spin labels at specific sites on a protein or a ligand allows for the identification of resonances close to that site because the paramagnetic species increase the relaxation rate of NMR resonances in the vicinity of the unpaired electrons. Just as in NOE experiments, this effect is proportional to the inverse sixth power of the distance between the label and the reporter nucleus and can be quantitated by measuring the paramagnetic contribution to the relaxation properties of the nucleus of interest. In practice, intermolecular electron–proton relaxation can be detected by preparing complexes with spin-labeled RNA and isotopically labeled protein (or vice versa) and observing the intensity and line shape of protein (or RNA) resonances in heteronuclear correlated spectra.

This principle can be applied in the study of RNA–protein complexes either by labeling the protein or the RNA. If the RNA is to be labeled, techniques used to label proteins with paramagnetic groups has to be adapted to RNA chemistry.[25] Nitroxide spin labels can be attached to chemically synthesized RNAs containing single 4-thiouracyl bases at the desired position. After purification of the RNA by standard methods, 3-(2-iodoacetamido–proxyl) can be coupled to 4-thiouracyl by standard chemistry.[29] Progression of the coupling reaction can be followed by UV spectroscopy. The maximum in UV absorption for 4-thiouracyl is shifted to $\approx$320 nm, compared to $\approx$260 nm for uracyl; this peak reverts back to $\approx$260 nm when coupling occurs. The reaction is complete overnight at room temperature, without degradation of the RNA during the reaction or the subsequent purification of the spin-labeled RNA, but the sample must be kept in the dark at all times to avoid or at least minimize reduction of the photosensitive paramagnetic species. If the protein is to be labeled, spin labels can be attached to the side chain of an exposed cysteine.[26,27] This amino acid can either be present in the natural protein sequence or can be introduced at a desired location by site-directed mutagenesis. Iodoacetamide reacts efficiently with the free cysteine SH, but also with exposed Lys NH_2^+; phosphate is a good buffer for this reaction because of its poor

[25a] A. Ramos and G. Varani, *J. Am. Chem. Soc.* **120,** 10992 (1998).

[26] J. R. Gillespie and D. Shortle, *J. Mol. Biol.* **268,** 158 (1997).

[27] J. R. Gillespie and D. Shortle, *J. Mol. Biol.* **268,** 170 (1997).

[28] S. U. Dunham, C. J. Turner, and S. J. Lippard, *J. Am. Chem. Soc.* **120,** 5395 (1998).

[29] R. S. Coleman and E. A. Kesicki, *J. Am. Chem. Soc.* **116,** 11636 (1994).

nucleophilicity. At pH between 7 and 8 and room temperature, coupling to Cys side chains is more rapid, but if the reaction is allowed to progress for hours, coupling to Lys side chains will also occur. Working at lower pH can increase the selectivity for Cys, but the reaction becomes inefficient. It is essential to monitor progress of the reaction by mass spectrometry, which allows both to quantitate the extent of coupling and to ensure that a single side chain is modified. It is useful if the reaction is quantitative, but it is even more important that labeling occur at a single site: multiple labels would destroy the premise of this approach. If necessary, the level of labeling should be compromised in favor of specificity for the single engineered Cys side chain; methods to purify unreacted proteins containing free SH groups exist.[26,27]

The choice of site of labeling is crucial for the success of this approach. First of all, substitution with Cys or 4-thiouracyl and the subsequent labeling reaction must not cause misfolding of the protein or RNA. Secondly, the labeling site must lie outside the interaction surface to minimize (or prevent altogether) any effect on the complex, yet sufficiently close to allow the observation of intermolecular paramagnetic relaxation. In practice, it is essential to verify that attachment of the nitroxide spin labels does not significantly perturb the structure of the RNA or protein or of the protein–RNA complex. ^{1}H–^{15}N HSQC protein spectra or ^{1}H–^{13}C RNA HSQC should be recorded either with unmodified RNA samples or with spin-labeled RNAs; when compared, they must be very similar or identical.

An example of the results is shown in Fig. 5, which displays the heteronuclear ^{1}H–^{15}N HSQC spectra of the complex between a spin-labeled RNA substrate and a ^{15}N-labeled RNA-binding domain from *Drosophila staufen* protein. The two spectra were recorded under identical conditions before and after reduction of the paramagnetic label; the spectrum recorded after reduction of the paramagnetic species provides a crucial control. Certain resonances in the protein are broadened when the paramagnetic label is present (compare cross-peaks marked with residue labels in Figs. 5A and 5B; notice that some residues disappear from the spectrum altogether). Peaks that become sharper on reduction of the paramagnetic species can be unequivocally identified as belonging to amino acids located close to the spin label attached to the RNA. The intensity of the broadening effect is inversely proportional to the distance from the paramagnetic label, thereby providing a direct identification of all NH resonances from amino acids within ≈15 Å of the spin label.

The spectra of Fig. 5 were recorded at a concentration of only 0.05 m*M*. Increasing the concentration would allow the quantitative estimate of the paramagnetic contribution to the relaxation parameters T_1 and T_2,[27,28,30] and

[30] I. Bertini, A. Donaire, C. Luchinat, and A. Rosato, *Proteins* **29,** 348 (1997).

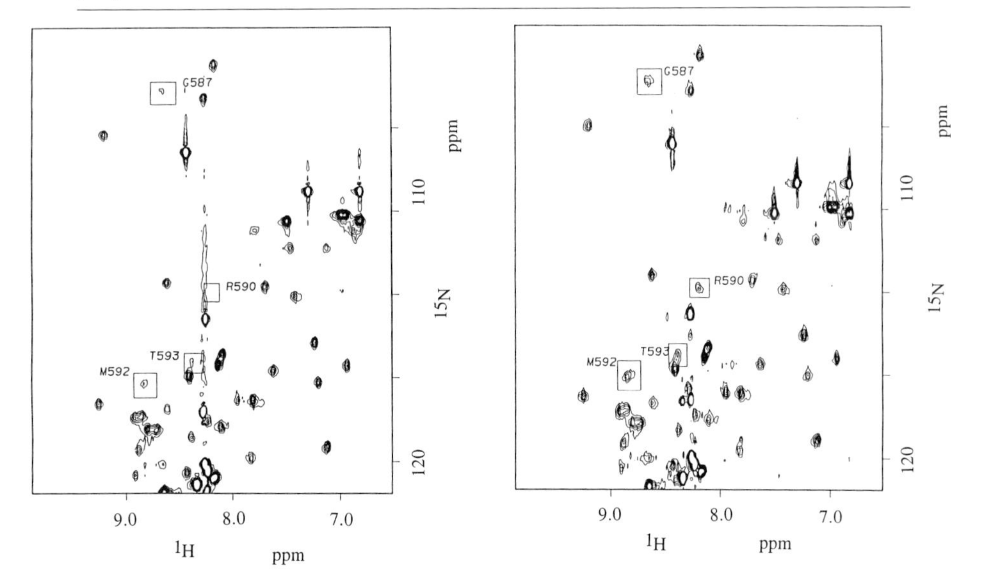

FIG. 5. ^{1}H–^{15}N HSQC spectra of the *D. staufen* dsRBD–RNA complex.[25] The spectra were recorded before (left) and after (right) reduction of the paramagnetic spin label on the RNA. Residues for which intermolecular paramagnetic protein relaxation occurs are explicitly identified in the figure.

this would in turn allow the extraction of quantitative long-range distance constraints (probably up to 20 Å). This approach remains effective at very high molecular weight, particularly using random fractional deuterated samples and line-narrowing techniques. In fact, it is ideally suited to be applied in conjunction with the TROSY principle[31] in the study of very large intermolecular complexes. Application of TROSY experiments to a multicomponent RNA–protein assembly, the 40-kDA U1A trimolecular complex, was done and compared with a ^{1}H–^{15}N TROSY spectrum recorded under identical condition. The sensitivity of the two sets of data was comparable, but the line width was much more favorable in the TROSY spectrum (data not shown). These spectra were recorded at 600 MHz, well below the optimum field for TROSY experiments (close to 1000 MHz), but already illustrate the power of this method.

The spin-labeling technique introduced here can provide long-range information for systems that can be studied using NOE-based methods. It can also be applied to complex assemblies with molecular weight well in excess of the current limits for NMR structure determination. This approach

[31] K. Pervushin, R. Riek, G. Wider, and K. Wüthrich, *Proc. Natl. Acad. Sci. U.S.A.* **94,** 12366 (1997).

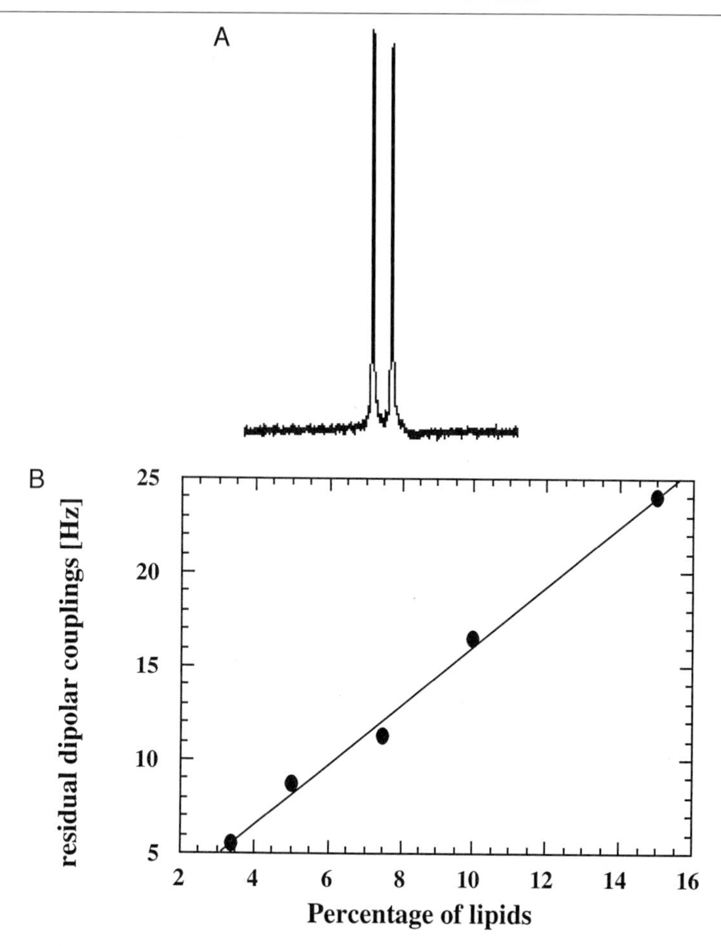

FIG. 6. (A) One-dimensional deuterium spectrum showing the splitting due to incompletely averaged quadrupolar interaction in a partially oriented sample. (B) The plot of residual deuterium quadrupolar splitting versus phospholipid concentration provides a calibration of the anisotropy in partially oriented samples.

is far more sensitive than the NOE-based approaches to detecting intermolecular interactions and is effective over much longer distances. The method is also robust enough to provide intermolecular distance constraints in cases where NOE interactions are unobservable due to dynamics at the intermolecular interface.[25] A strategy for the determination of structures where weak and/or poorly specific binding limits the NOE-based approach would use qualitative estimates of proton–electron distances to dock predetermined protein and RNA structures and obtain a first set of models for the complex. The starting structures could then be used to identify ambiguous NOE interactions and to provide quantitative estimates of electron–proton relaxation rates during the refinement of these initial structures, as demonstrated in studies of paramagnetic proteins.[30] This method should be most effective in the study of complex macromolecular assemblies. One could consider a strategy based on the determination of the structure of each individual component of a complex assembly. The structure of the entire complex could then be assembled by detecting intermolecular long-range distances by application of the spin-labeling methods in conjunction with TROSY-based experiments designed to measure quantitatively the paramagnetic contribution to T_1 and T_2 relaxation times.

Residual Dipolar Couplings in Analysis of Protein–RNA Complexes

Structure determination by NMR relies mainly on local information, but the quality of NMR structures can be significantly improved if observables that provide nonlocal information are measured as well. A class of such observables is represented by the orientation of NH, CH, or CC bond vectors relative to a reference axis. So-called dipolar interactions carry this information, which can be extracted by preparing partially oriented samples.[32] During the isotropic tumbling of a molecule in solution, dipolar interactions between pairs of spin average to zero over the time frame of the NMR experiment; in a solid instead, dipolar interactions are fully active, resulting in very broad spectra. When molecules are partially oriented in solution, dipolar interactions are not completely averaged to zero and their intensities bear this orientational information. Partial alignment can be obtained either by a strong magnetic field[33] or by introducing in solution larger molecules or macromolecular assemblies that can be ordered by an external magnetic field.[34] The incompletely averaged dipolar interactions induce splitting of resonance lines in the NMR spectrum, i.e., they appear as residual (dipolar) coupling constants. The values of the residual dipolar splittings depend on the degree of alignment and thus on the strength of the external field or on the concentration of molecules used to induce alignment in the sample, making it possible to scale the size of dipolar

[32] N. Tjandra, J. G. Omichinski, A. M. Gronenborn, G. M. Clore, and A. Bax, *Nature Struct. Biol.* **4,** 732 (1997).

[33] J. R. Tolman, J. M. Flanagan, M. A. Kennedy, and J. H. Prestegard, *Proc. Natl. Acad. Sci. U.S.A.* **92,** 9279 (1995).

[34] N. Tjandra and A. Bax, *Science* **278,** 1111 (1997).

couplings at will. If only a few percent of all molecules in solution are oriented, only part of the dipolar interaction remains active, allowing resonances to remain nearly as sharp as in the free-tumbling isotropic state. Individual couplings measured in these systems depend not only on the degree of alignment (anisotropy), but also on the polar coordinates with respect to the alignment axis and therefore provide direct geometric information on NH, CH, or CC bond vectors.

The best option to partially align biomolecules currently utilizes liquid crystalline phases composed of rod- or disk-shaped molecules capable of aligning under an external magnetic field (in practice, the field of the main magnet itself).[35] Nematic liquid crystals are characterized by a time-averaged spatial alignment of their components along one main axis and are therefore ideal to partially order biomolecules by sterically restricting their motion. The best media for such studies are provided by mixtures of phospholipids, for example 1,2-dihexanoyl-*sn*-glycero-3-phosphocholine (DHPC) and 1,2-dimyristoyl-*sn*-glycero-3-phosphocholine (DMPC). For protein–RNA complexes, the high temperatures required for establishing the liquid crystalline phase (>35°) over a long period of time can activate nucleases. Other liquid crystalline mixtures stable over a wider temperature range are becoming available; bacteriophages provide an ideal system to align negatively charged nucleic acid molecules (see [15] this volume).[35a] Sample preparation is critical to successfully study residual dipolar couplings in liquid crystalline fluids. A 1:3 (mol:mol) ratio of a 5.5–6.5% DHPC:DMPC mixture[35] provides good results for measuring the residual dipolar couplings in protein–RNA complexes of at least 40 kDa. The waxy DHPC has to be predried to avoid an underestimation of its weight caused by its hygroscopicity; underestimation of the dry weight of DHPC would lead to a final solution with the incorrect liquid crystalline properties. DHPC molecules are necessary to form the edges of the disk-shaped bicelles in the liquid crystals; if there is too little DHPC, bicelles or micelles with large diameters form, the fluid loses its transparency and appears white and milky. These milky phases are not stable over the long measurement times necessary for NMR, since gravitation induces phase separation. Very stable phases are obtained by determining the dry weight of DHPC in an Eppendorf and immediately adding the corresponding amount of DMPC to make a 1:3 mixture. The mixture is insoluble after the addition of water, but application of several vortexing/heating cycles (5 min at 4° followed by 10 min at 40°) leads to homogeneous fluids, clear at temperatures below 15°.

[35] N. Tjandra, D. S. Garrett, A. M. Gronenborn, A. Bax, and G. M. Clore, *Nature Struct. Biol.* **4,** 443 (1997).

[35a] M. R. Hansen, P. Hanson, and A. Pardi, *Methods Enzymol.* **317,** [15], 2000 (this volume).

The liquid crystalline phase forms by heating the sample at 38°. The solution is highly viscous with a blue taint originating from light scattering by the small disk-shaped bicelles. Concentrated stock solutions of DHPC : DMPC mixture (13–15%) can be stored in the cold for several months and diluted to the desired final concentration when needed. A solution containing the protein–RNA complex (10% D_2O, <50 m*M* NaCl, 10–20 m*M* phosphate buffer, pH 6–7) is added to the proper amount of the lipid mixture to form the partially oriented sample. However, vortexing/heating steps should be repeated before each measurement to improve the stability of the mixture.

Splitting of the deuteron resonance of D_2O (present in the sample as a lock reference) by incompletely averaged quadrupolar interactions can be measured in partially oriented phases. This splitting is proportional to the percentage of lipids in the phase and can be used to determine the active lipid concentration (Fig. 6).

Residual dipolar NH or CH couplings can be measured in standard $^{1}H-^{15}N$ or $^{1}H-^{13}C$ HSQC spectra. The difference in splitting of the NH or CH resonances measured at 20° (isotropic phase) and at 38° (liquid crystalline phase in the previously described lipid mixture) provides the residual dipolar coupling. Reference coupling constants can also be recorded at 38° in an aqueous buffered solutions. Couplings can be measured either in the direct or indirect dimension simply by not decoupling the heteronucleus of choice; improved methods to allow observation of one or the other of the components of a doublet have recently become available.[36,37] More accurate methods developed for small proteins[38] are ineffective at the molecular weight of protein–RNA complexes due to their fast relaxation properties.

The main use of residual dipolar couplings is in structure refinement (see below), but these observables can provide important structural information even when the structure is not known. Figure 7A shows residual NH dipolar couplings for the RNA complex of the third RNA-binding domain of *D. staufen* protein, together with the secondary structure of the protein in the complex. Negative residual couplings are found for the α-helical elements and positive values for the β strands, demonstrating that the α-helical NHs (roughly perpendicular to the α helix axis) align parallel to the β sheet. This information can be combined with residual coupling

[36] P. Andersson, K. Nordstrand, M. Sunnerhagen, E. Liepinsh, I. Turovskis, and G. Otting, *J. Biomol. NMR* **11,** 445 (1998).

[37] M. Ottiger, F. Delaglio, and A. Bax, *J. Magn. Reson.* **131,** (1998).

[38] A. Bax, G. W. Vuister, S. Grzesiek, F. Delaglio, A. C. Wang, R. Tschudin, and G. Zhu, *Methods Enzymol.* **239,** 79 (1993).

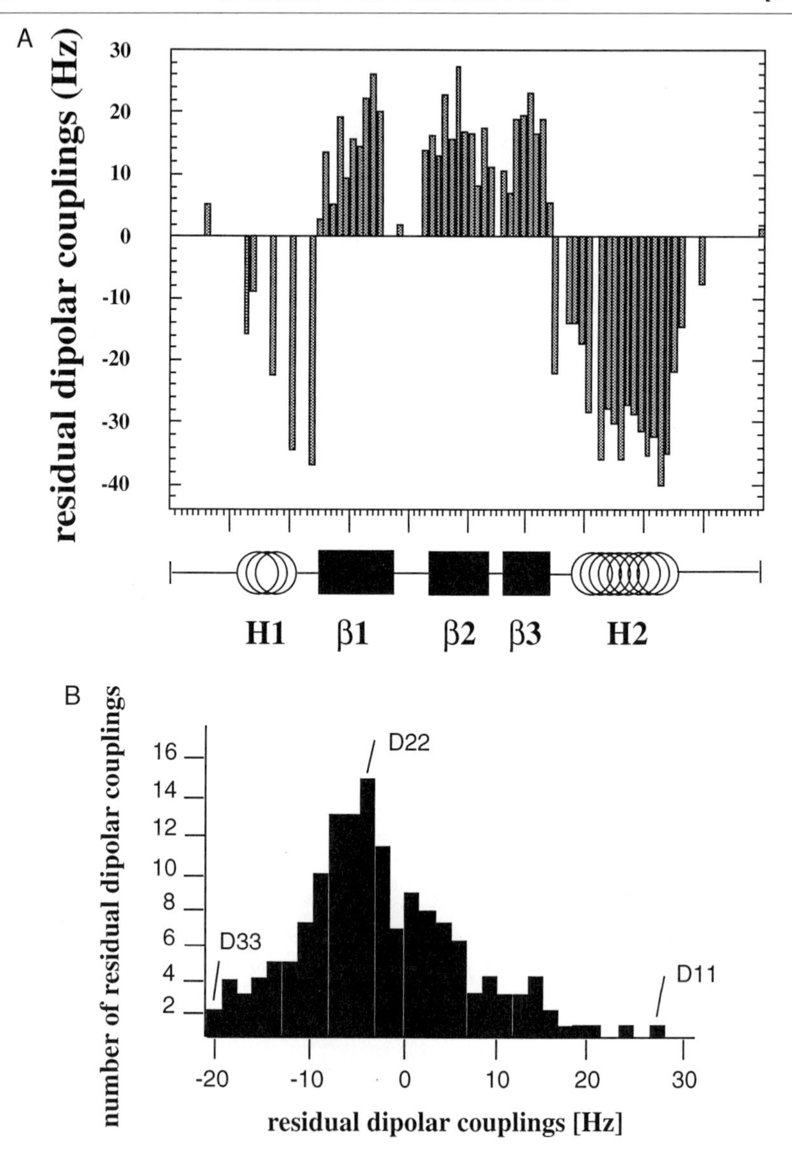

FIG. 7. (A) Residual dipolar couplings for backbone NH resonances for the *D. staufen* dsRBD–RNA complex in an ≈5.5% phospholipid solution. The secondary structure of the protein is shown at the bottom. (B) "Powder"-like histogram of the residual dipolar couplings; this plot was generating by reporting the number of times a given residual coupling is observed versus the residual coupling itself.

constants obtained on the labeled RNA in the complex. Residual couplings for imino NH and base CH bonds are all positive. In an A-form double-stranded RNA, the bases are roughly perpendicular to the helix axis. Positive values of CH and NH couplings within the RNA, compared to the negative couplings for the protein α-helical NHs, show that the protein is bound to the RNA with its helices roughly parallel to the RNA double helical axis.

Residual dipolar couplings can be used quantitatively in structure refinement, and existing protocols have been adapted to make use of this information. The residual dipolar coupling of two nuclei X and H is given by the following equation[32,35,39]:

$$D_{XH} = D_a^{XH}\{(3\cos^2\theta - 1) + 1.5D_r^{XH}(\sin^2\theta\cos 2\phi)\}$$

where D_a^{XH} represents the axial component of a tensor $\mathbf{D}$; $R = 1.5\ (D_1^{XH}/D_a^{XH})$ is called asymmetry and D_r^{XH} is called the rhombic part of the tensor $\mathbf{D}$. The symbol θ describes the angle between the XH bond vector and the z axis of the tensor, ϕ the angle between the projection of the same vector onto the xy plane and the x axis. Symmetric tensors are fully defined by six independent variables; three of them describe the direction of the tensor and three others represent asymmetry, anisotropy, and trace. These latter quantitites can be defined in terms of the diagonal components D_{11}, D_{22}, and D_{33} of the tensor:

$$D_{11}^{XH} = 2D_a^{XH}$$
$$D_{22}^{XH} = \{D_a^{XH}(1 - 1.5R)\}$$
$$D_{33}^{XH} = \{D_a^{XH}(1 + 1.5R)\}$$

Residual dipolar couplings depend on the angles θ and ϕ as well as on D_a^{XH} and R; therefore, geometrical information over the XH bond vectors can only be obtained when D_a^{XH} and R (or D_{11}^{XH}, D_{22}^{XH}, and D_{33}^{XH}) are known. This can easily be done if a sufficient number of XH bond vectors is measured, and if these bonds are distributed in such a way that every possible direction in space is occupied.[40] When this occurs, the histogram of Fig. 7B, where residual dipolar couplings are plotted against their number of appearance, allows extraction of D_{11}^{XH}, D_{22}^{XH}, and D_{33}^{XH}. D_{11}^{XH} corresponds to the highest absolute value in the powder pattern, the other extreme is D_{33}^{XH}; D_{22}^{XH} is the highest probable value for the histogram (Fig. 7B). For small proteins or nucleic acids, the measurable

[39] G. M. Clore, A. M. Gronenbron, A. Szabo, and N. Tjandra, *J. Am. Chem. Soc.* **120,** 4889 (1998).

[40] G. M. Clore, A. M. Gronenborn, and N. Tjandra, *J. Magn. Reson.* **131,** 159 (1998).

number of residual dipolar couplings is too small and their distribution does not contain all possible values. In this case, the value of D_a can only be estimated, whereas R can be found by systematically varying it during structure calculations and simultaneously monitoring the value of the target function (total energy, energy of residual dipolar couplings, NOE violations, and so on) that represents the quality of the structure.[40] R can be obtained from the value that provides the best agreement with the target function.

The improvement in the quality of the structure of an RNA–protein complex provided by the use of residual dipolar couplings is illustrated in Fig. 8. No significant improvement was obtained when only residual dipolar couplings measured for the protein component of the complex were measured; the U1A protein globular structure is already well determined by a large set of NOE-based distance constraints.[6,7] However, a significant improvement in the RNA structure in the complex is clear in the comparison of the superposition of structures calculated before or after inclusion of the RNA residual dipolar couplings. The regions of the RNA in contact with the protein are already well determined by numerous intra- and intermolecular NOE distances,[6,7] but the regions of the RNA double helix away from the protein interface are much better defined when residual dipolar couplings are included. Although the local structure of the RNA double helical regions is not affected by the new constraints, the orientation of the two helices in the RNA with respect to the protein and to each other is significantly better defined on inclusion of the additional constraints. Thus, this method allows for the improvement of the long-range quality of RNA structure and alleviates the main weakness of NMR-based RNA structures.[41,42]

Conclusions and Perspectives

In this article, we have reviewed the state-of-the art methods to study the structure and dynamics of protein–RNA complexes using NMR spectroscopy. We have emphasized relatively recent approaches, including methods of great promise that have yet to be used in the determination of biologically important structures. The future of structural molecular biology increasingly lies in the study of molecular recognition and macromolecular assemblies. Studies of RNA–protein recognition during gene expression are progressively moving away from the study of isolated protein domains or RNA motives, and even from bimolecular protein–RNA complexes.

[41] F. H.-T. Allain and G. Varani, *J. Mol. Biol.* **267,** 338 (1997).

[42] C. C. Gubser and G. Varani, *Biochemistry* **35,** 2253 (1996).

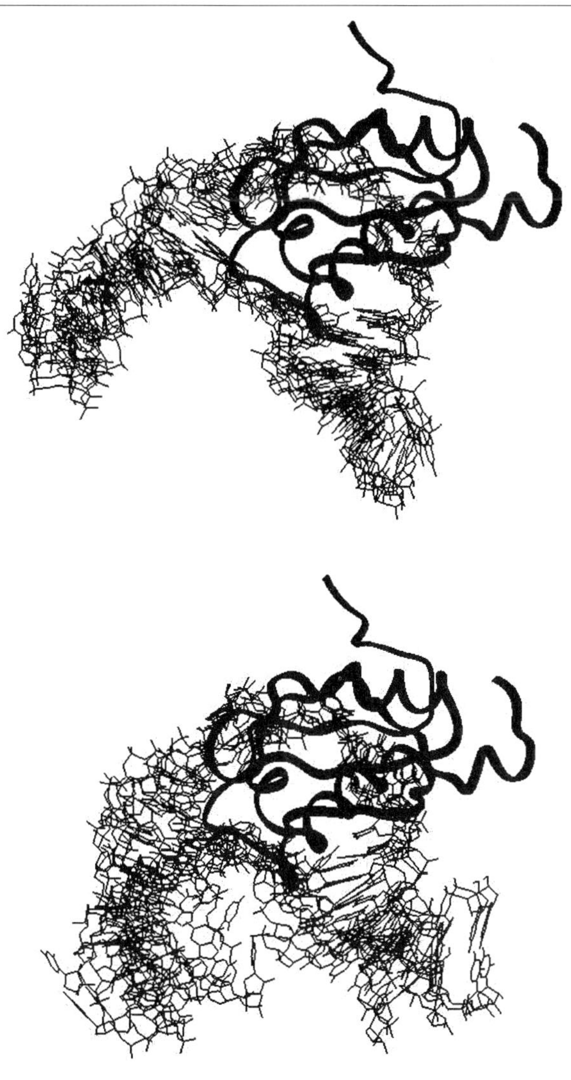

FIG. 8. Superposition of five low-energy structures of the U1A bimolecular protein–RNA complex. Inclusion of residual dipolar couplings (top) improves the definition of the regions of the RNA structure distant from the protein–RNA interface[6] (bottom).

This area of structural biology will increasingly shift its emphases toward the characterization of the complex multimolecular assemblies that constitute structural and functional units during gene expression and its regulation. We have highlighted here the recent technical progress that is allowing

NMR studies of the structure and dynamics of ever more complex RNA–protein complexes and their assemblies.

Acknowledgments

It is a pleasure to thank Dr. Andres Ramos for the results illustrated in Figs. 1 and 4 and Dr. David Neuhaus for many very helpful discussions. P.B. and L.V. acknowledge the support of European Union training fellowships.

[15] Filamentous Bacteriophage for Aligning RNA, DNA, and Proteins for Measurement of Nuclear Magnetic Resonance Dipolar Coupling Interactions

By MARK R. HANSEN, PAUL HANSON, and ARTHUR PARDI

Introduction

Solution nuclear magnetic resonance (NMR) spectroscopy has emerged as an important method for obtaining structural information on RNA and protein–RNA systems.[1,2] A present limitation in solution structure determinations is that although the local structure of a molecule can usually be determined with high precision, the global conformation is often poorly determined. This is a result of the short-range nature of the ^1H–^1H nuclear Overhauser effect (NOE) distance constraints used for determining structures of macromolecules in solution. Thus if there are no NOEs between residues distant in the primary structure, errors in the local structure propagate, making it impossible to determine accurately long-range structural features such as bending or interhelical angles. For example, in an RNA molecule that has an upper and a lower helix connected by a bulged base or an internal loop, it is extremely difficult to determine the relative orientations of the helices by standard NMR techniques. This situation has recently changed where the use of dipolar coupling interactions allows measurement of long-range structural information[3–5] and therefore promises to greatly improve the solution structure determinations of RNAs. One of the require-

[1] A. Pardi, *Methods Enzymol.* **261,** 350 (1995).

[2] G. Varani, F. Aboul-ela, and F. H. T. Allain, *Prog. Nucl. Magn. Reson. Spectrosc.* **29,** 51 (1996).

[3] N. Tjandra and A. Bax, *Science* **278,** 1111 (1997).

[4] A. Bax and N. Tjandra, *J. Biomol. NMR* **10,** 289 (1997).

[5] J. H. Prestegard, *Nature Struct. Biol.* **5 Suppl.,** 517 (1998).

0076-6879/00 $30.00

ments for obtaining dipolar coupling information is preparing a system where the macromolecule is partially ordered in solution. In this work we describe the application of filamentous bacteriophage as a versatile tool for generating partially ordered solutions of nucleic acids and proteins. The Pf1 phage system described here is an ideal method for obtaining a tunable degree of alignment of RNA and DNA oligomers and some proteins in solution,[6] which then allows measurement of homonuclear and heteronuclear dipolar coupling interactions.

Dipolar couplings are a valuable source of structural information that contain both an angle and a distance component.[3–5] Until recently, however, dipolar couplings were only observed in solid-state NMR. This is because for most molecules in solution, molecular tumbling averages the dipolar couplings to zero, therefore they are difficult or impossible to observe by solution NMR techniques. Some molecules have a natural anisotropic magnetic susceptibility that leads to slight alignment in a magnetic field and under these circumstances the dipolar couplings do not completely average to zero.[7] The helical structure of nucleic acids leads to an anisotropic magnetic susceptibility, but the resulting residual dipolar couplings are only a few hertz, and therefore difficult to measure.[8] The advent of higher field spectrometers helps these cases because the degree of alignment increases as the square of the magnetic field strength[7]; but even at the highest fields available these residual dipolar couplings are normally too small to be useful in solution structure determinations.

A number of approaches can be used to induce a higher degree of alignment in macromolecules and therefore increase the size of the residual dipolar couplings. Paramagnetic ions have large magnetic moments and thus for proteins or nucleic acids that have anisotropic metal binding sites, paramagnetic ions can lead to partial alignment and large dipolar coupling constants.[9–11] This paramagnetic-induced alignment is limited to molecules with a specific metal binding site(s) and thus cannot be generally applied. Bax and co-workers developed a general technique that uses liquid crystalline media to align macromolecular cosolutes.[3,4] At the appropriate ratios

[6] M. R. Hansen, L. Mueller, and A. Pardi, *Nature Struct. Biol.* **5,** 1065 (1998).

[7] A. A. Bothner-By, *in* "Encyclopedia of Nuclear Magnetic Resonance" (D. M. Grant and R. K. Harris, eds.), pp. 2932. John Wiley & Sons, Chichester, U.K., 1995.

[8] H. C. Kung, K. Y. Wang, I. Goljer, and P. H. Bolton, *J. Magn. Reson. B* **109,** 323 (1995).

[9] J. R. Tolman, J. M. Flanagan, M. A. Kennedy, and J. H. Prestegard, *Proc. Natl. Acad. Sci. U.S.A.* **92,** 9279 (1995).

[10] J. R. Tolman, J. M. Flanagan, M. A. Kennedy, and J. H. Prestegard, *Nature Struct. Biol.* **4,** 292 (1997).

[11] R. D. Beger, V. M. Marathias, B. F. Volkman, and P. H. Bolton, *J. Magn. Reson.* **135,** 256 (1998).

and concentrations, mixtures of certain lipids go through a temperature-dependent phase transition to form large disk-shaped bilayer-type aggregates known as bicelles.[12] These bicelles have a large magnetic anisotropy and align in magnetic fields with their normal perpendicular to the magnetic field. Interactions with the aligned bicelles can impart a small degree of alignment on macromolecules in the solution, leading to readily measured dipolar couplings. The bicelle system has been successfully applied to both proteins and DNA oligomers but the liquid crystalline phase for the bicelles is restricted to certain concentrations of the lipids and a limited range of temperatures (~30°–40°).[3,4,12] Thus the liquid crystal phase transition dictates the conditions under which solution structural data can be obtained.

Another general method for generating large residual dipolar couplings of macromolecules in solution employs magnetically aligned filamentous bacteriophage as the ordering agent.[6,13,14] We have recently shown that low concentrations of Pf1 phage can be used to induce a tunable degree of alignment of protein or nucleic acid cosolutes.[6,13] The phage form a liquid crystalline-type medium in solution where the long axis of the phage align parallel to the magnetic field.[15,16] However, unlike the bicelles, the phage fully align at any concentration and temperature compatible with biological systems. Specifically the phage readily align macromolecules over a wide range of temperatures (5° to >45°) and phage concentrations (<1–50 mg/ml). Thus the phage represent an extremely versatile approach for inducing alignment of macromolecules in solution.

The Pf1 filamentous phage are particularly well suited for studies of dipolar coupling interactions of RNA and DNA oligomers. Like nucleic acids, these phage have a negatively charged surface at physiologic pH ($pI \sim 4.0$)[17]; therefore, the negatively charged nucleic acids do not bind to the phage particles. Under these conditions, alignment results from collisions that occur between biomolecules and the surface of the magnetically aligned phage particles. It might seem surprising that such low concentrations of phage (<5% by mass) can impart sufficient alignment on macromolecular cosolutes to yield measurable dipolar couplings. However, it is important to realize that solution NMR studies require relatively small residual dipolar couplings (<40 Hz). Because the largest dipolar couplings in a fully oriented system would be >10 kHz, the optimal situation for

[12] C. R. Sanders and J. P. Schwonek, *Biochemistry* **31,** 8898 (1992).
[13] M. R. Hansen, M. Rance, and A. Pardi, *J. Am. Chem. Soc.* **120,** 11210 (1998).
[14] G. M. Clore, M. R. Starich, and A. M. Gronenborn, *J. Am. Chem. Soc.* **120,** 10571 (1998).
[15] J. Torbet and G. Maret, *J. Mol. Biol.* **134,** 843 (1979).
[16] J. Torbet and G. Maret, *Biopolymers* **20,** 2657 (1981).
[17] K. Zimmermann, H. Hagedorn, C. C. Heuck, M. Hinrichsen, and H. Ludwig, *J. Biol. Chem.* **261,** 1653 (1986).

solution NMR studies is a very small degree of alignment, approximately 0.1%, which yields only a small fraction of the total dipolar coupling. If the residual dipolar couplings are much larger, multiple ^1H–^1H and ^1H–^{31}P couplings would lead to such extensive line broadening and such a complex network of dipolar interactions that the spectra would be impossible to interpret. In addition the phage particles have an extremely large surface area. The Pf1 phage particles are 20,000 Å long with a diameter of ~60 Å, and a molecular weight of ~40 MDa.[18,19] Thus 30 mg of Pf1 phage (~4.5 × 10^{14} particles) expose ~17 m^2 of surface area oriented parallel to the magnetic field. This very high degree of oriented surface area, combined with the very low degree of alignment required for solution NMR studies, makes it possible for relatively low concentrations of phage to induce sufficient alignment of macromolecules in solution to observe useful dipolar couplings.

Filamentous bacteriophage are broken down into two classes based on their structure: class I consists of the *Escherichia coli* infecting strains M13, fd, fl, IKe, and Ifl; class 2 consists of Xf, which infects *Xanthomonas oryzea,* and Pf1 and Pf3, which infect separate strains of *Pseudomonas aeruginosa.*[20,21] Although all work presented here describes Pf1 filamentous phage with which we have had the greatest success, we also tested M13, which can be prepared and purified by identical methods to those used for Pf1.[6] Solid-state NMR studies on Pf1 and fd phage show that the Pf1 phage particles undergo rapid reorientation (rates $>10^4$ sec^{-1}) about their long axis at pH 8.0, whereas the fd phage (which differs from M13 by a single amino acid in the phage coat protein) is comparatively immobile on this same timescale, suggesting that it is more susceptible to aggregation.[22] We have studied both M13 and Pf1 phage and under our conditions, M13 showed significantly more problems associated with aggregation than Pf1, which is consistent with this solid-state NMR study. Clore and co-workers have shown that fd phage and tobacco mosaic virus can align proteins for measurement of residual dipolar couplings in solution,[14] indicating that a variety of phage and viral particles may be suitable for inducing alignment of macromolecules.

Pf1 filamentous bacteriophage (shown in Fig. 1) was isolated more than 40 years ago and is known to only infect a single host, *P. aeruginosa* strain

[18] D. F. Hill, N. J. Short, R. N. Perham, and G. B. Petersen, *J. Mol. Biol.* **218,** 349 (1991).
[19] R. Nambudripad, W. Stark, and L. Makowski, *J. Mol. Biol.* **220,** 359 (1991).
[20] P. Model and M. Russel, *in,* "The Bacteriophages" (R. Calendar, ed.), p. 375. Plenum Press, New York, 1988.
[21] D. A. Marvin, *Curr. Opin. Struct. Biol.* **8,** 150 (1998).
[22] P. Tsang and S. J. Opella, *Biopolymers* **25,** 1859 (1986).

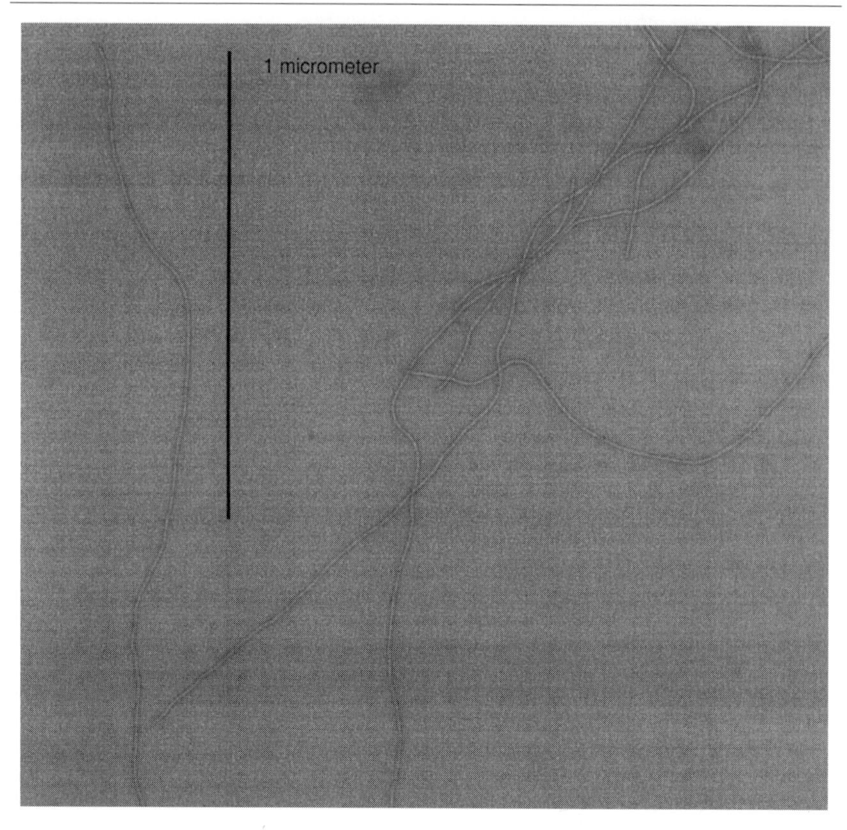

FIG. 1. An electron micrograph of Pf1 filamentous phage. Pf1 was stained with 1% uranyl acetate and examined at a magnification of 26,000×. Bar: 1μM (10,000 Å).

K, that displays polar sex pila that act as receptors for phage particles.[20,23,24] Strains that do not display these polar sex pila are resistant to infection. Pf1 is the longest characterized filamentous bacteriophage measuring 20,000 Å, consisting of a 7.4-kb circular single-strand DNA genome with approximately one coat protein per nucleotide.[18] The phage particle is a closely packed network of interlocking helical coat proteins that lie approximately 15° relative to the phage axis.[19] The result of this repeating coat protein motif is a large net magnetic susceptibility, which causes the phage to align readily in magnetic fields. Tunable macromolecular alignment is achieved by varying the Pf1 concentrations between 1 and 50 mg/ml to observe residual dipolar couplings up to ±50 Hz for ^{1}H–^{15}N, ^{1}H–^{13}C, and ^{1}H–^{1}H

[23] K. Takeya and K. Amako, *Virology* **28,** 163 (1966).

[24] D. E. Bradley, *Can. J. Microbiol.* **19,** 623 (1973).

in DNA, RNA, and protein macromolecules.[6,13] Pf1 phage have been used to align molecules over a wide range of temperatures and buffer conditions and their application to RNA and DNA systems is discussed here.

Preparation of Pf1 Phage

Pf1 filamentous phage are prepared using methods similar to those described previously.[25,26] *P. aeruginosa* host and Pf1 phage were obtained from American Type Cell Collection (Rockville, MD, ATCC 25102 and ATCC 25102-B1, respectively). The host cells are resuspended according to ATCC protocols and then divided into 100-μl aliquots, which are frozen for future use ($-70°$). *P. aeruginosa* hosts are grown at 37° in 50 ml LB media (10 g tryptone, 5 g yeast extract, and 10 g NaCl) enriched with 20 g/liter glucose in a 250-ml shaker flask to an OD_{600} of ~0.7. Standard sterile techniques are carefully followed because the host has no antibiotic markers. Special care is taken to store and grow *P. aeruginosa* cultures in a Pf1-free environment to prevent the development of resistance to infection by Pf1. These host cells are then infected with Pf1 phage by adding 50–100 μl of either resuspended Pf1 from ATCC or previously purified Pf1 (50 μg/ml). Pf1 infected *P. aeruginosa* are grown at 37° for a further 15 min and then used to inoculated multiple 2-liter baffle flasks containing 750 ml of media each. These are then grown 14–16 hr at 37° to stationary phase. Infected host cells extrude Pf1 phage particles without undergoing lysis and the cell growth is slowed to roughly half their normal rate.[27]

Phage are harvested by centrifuging the host cells out of solution at 11,300*g* for 45 min at 5° (Beckman JA10 rotor at 8,000 rpm). Cell pellets are discarded and phage are precipitated from the supernatant by the addition of 60 g/liter of NaCl and 20 g/liter of polyethylene glycol (8000 average molecular weight) and stirred for ~3 hr at 5° until the polyethylene glycol and NaCl are completely dissolved. Addition of polyethylene glycol and NaCl causes the solution to become extremely cloudy with precipitating phage. The phage are pelleted out of solution by a second centrifugation at 11,300*g* for 30 min at 5° (Beckman JA10 rotor at 8000 rpm) and the supernatant is discarded. Each pellet is resuspended in 5 ml of distilled water and then transferred into a 50-ml Oakridge centrifuge tube. After the phage have completely dissolved, any remaining cell debris is removed by centrifuging at 23,500*g* for 45 min at room temperature (15,000 rpm in a Beckman JA20 rotor).

[25] T. A. Cross and S. J. Opella, *Biochemistry* **20,** 290 (1981).
[26] T. A. Cross, P. Tsang, and S. J. Opella, *Biochemistry* **22,** 721 (1983).
[27] K. Horiuchi, *J. Mol. Biol.* **169,** 389 (1983).

Purification of phage is achieved using a KBr density gradient. Gradients are poured as block layers of 50, 42, 36, 28, and 20% KBr with the phage dissolved in the top layer (20% KBr). Gradients are poured from lowest density (20%) to highest density (50% KBr), each successive heavier layer carefully injected with a syringe below the previous layer.[28] Each centrifuge tube (Beckman, Palo Alto, CA, 1- × 3.5-inch polyallomer centrifuge tube) is loaded with 8–10 ml of the 20 and 50% layers and 5–8 ml of each of the intermediate layers. A set of six centrifuge tubes containing these density gradients is used to purify up to 200 mg of Pf1 by ultracentrifugation at 82,700*g* for 18–24 hr at 25° (25,000 rpm in a Beckman SW28 rotor). Phage form white aggregates in the density gradient at low temperature, so it is important to perform this step at 25° to achieve proper separation. The phage appear as a ~1-cm blue-gray band in the density gradient that is most easily observed when viewed against a black background. The uppermost band, which should contain the most material and correspond to homogeneous length Pf1, is carefully extracted with a syringe while avoiding contamination with any bands or debris located lower in the gradient. Purified phage are then extensively dialyzed into 1.0 m*M* EDTA, 10 m*M* Tris, pH 8.0. The phage concentrations are determined by UV absorbance at 270 nm using an extinction coefficient for Pf1 of 2.25 cm mg^{-1} ml^{-1}.[29] It can be difficult to determine precisely the concentration of the phage due to the high viscosity of concentrated phage solutions and because the phage have a tendency to stick to surfaces of the tubes and pipettes. Thus we estimate our errors in phage concentrations are ~10–20%. Typical yields of Pf1 phage range from 50 to 100 mg/liter of initial growth media. To prevent microbial growth, phage solutions are brought to 0.05% NaN_3 and the phage are pelleted out of solution by centrifuging at 475,000*g* for 1 hr at 5° in a Beckman tabletop ultracentrifuge (95,000 rpm in a Beckman TLA-100.3 rotor). Alternatively, if a tabletop ultracentrifuge is not available, the phage can be pelleted by ultracentrifugation for 3–6 hr in a VTi50 rotor at 40,000 rpm as previously described.[25] The resulting phage pellet will be glassy in appearance and highly viscous. Phage pellets can be stored in this state indefinitely at 5°, but should not be frozen.

Under stress, Pf1 phage can mutate to produce mini- and macrophage particles with the macrophage being up to twice the size of the wild-type Pf1.[27,30] We observed this phenomenon and although these phage prepara-

[28] J. Sambrook, E. Fritsch, and T. Maniatis, "Molecular Cloning, A Laboratory Manual," 2nd Ed., pp. 2.73. Cold Spring Harbor Laboratory Press, Plainview, New York, 1989.

[29] L. G. Kostrikis, D. J. Liu, and L. A. Day, *Biochemistry* **33,** 1694 (1994).

[30] L. Specthrie, E. Bullitt, K. Horiuchi, P. Model, M. Russel, and L. Makowski, *J. Mol. Biol.* **228,** 720 (1992).

tions were not characterized in detail, they clearly had suboptimal properties for aligning NMR samples. Therefore, it is important to determine the homogeneity of each phage preparation. A small amount of phage (<1 mg) are extracted with phenol/chloroform to remove the coat proteins, and the DNA is ethanol precipitated.[28] The extracted DNA is suspended in buffer (1.0 m*M* EDTA, 10 m*M* Tris, pH 8.0), 1–3 μg of Pf1 DNA is diluted with loading dye, placed into an 0.8% agarose gel, and run for 45 min at 110 V. The gel is stained with ethidium bromide (1 μg/ml) for 4 hr and visualized with a UV source. The size of the phage is directly proportional to the length of its genome. After staining, a single dominant band (>99%) should appear on the gel corresponding to wild-type length phage. Figure 2A shows a 0.8% agarose gel contain wild-type Pf1 phage and Fig. 2B shows a phage preparation that had mutated in multiple rounds of serial growths, where the phage particles have gotten longer that the wild-type phage. The size of the phage DNA can be compared to the DNA markers or to previous preparations of wild-type phage. The results of these gels clearly stress the importance of beginning each preparation with pure, wild-type phage, as opposed to propagating heterogeneous length phage.

Wild-type Pf1 phage dissolved in 10 m*M* Tris at pH 8.0 should align instantaneously in the magnetic field even at concentrations as high as 50 mg/ml. This is determined by acquiring a 1-D ^{2}H spectrum of the phage solutions containing D_2O (see Fig. 3). As discussed later, the aligned phage induce a net alignment on the surrounding water that results in residual quadrupole coupling of the ^{2}H nuclei in water molecules. This coupling is diagnostic of the alignment properties of the phage solutions, and no changes should be observed in the line shape, linewidth, or splitting of the deuterium water resonance over time. However, for phage preparations that contain heterogeneous length phage (such as those illustrated in Fig. 2B), we have observed multiple problems including changes in the residual quadrupole splitting of the ^{2}H spectrum over time (many hours), large linewidth of the deuterium water resonance, or multiple doublets of the deuterium water resonance. If the phage solutions show any of these properties, the phage DNA should be carefully checked by gel electrophoresis.

Prior to use in NMR studies, the phage are changed into the desired buffer by resuspending the pellet in NMR buffer (10 or 99% D_2O) and then repelleting the phage by ultracentrifugation (as done earlier). This procedure is repeated twice with the resulting supernatant drawn off and discarded. Some salt (>20 m*M*) must be present in the new buffer because we observed that the phage will pellet poorly from distilled water. As an alternative to pelleting, the phage may be dialyzed into the desired buffer conditions. The relative ability of the phage to orient in the magnetic field appears similar between the two methods (see results). We would normally

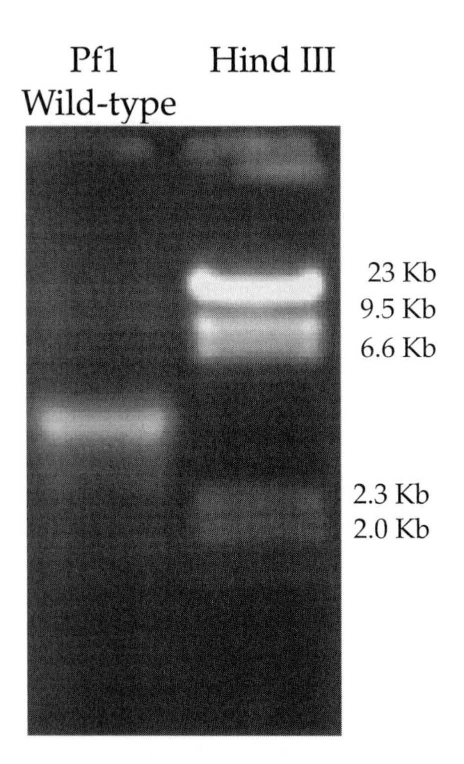

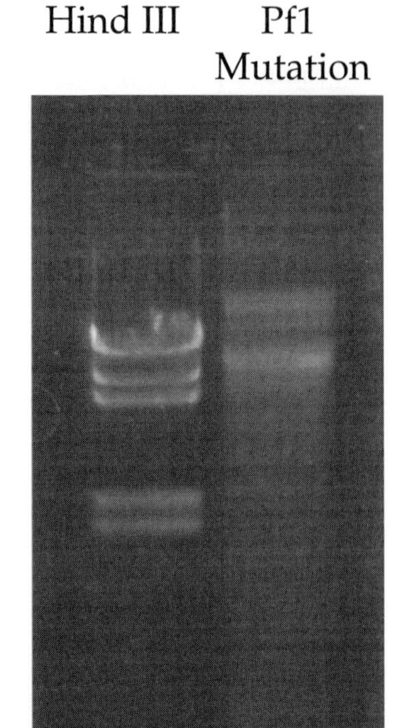

FIG. 2. Ethidium-stained 0.8% agarose gels of DNA isolated from (A) wild-type Pf1 phage and (B) a Pf1 preparation containing macrophage that has undergone mutations. Both gels also contain *Hin*dIII fragments of λ-DNA that were used as markers; however, it is not possible to make accurate estimates of the phage DNA length from these markers because the Pf1 is a single-stranded circular DNA, and the λ-DNA restriction fragments are double stranded. The wild-type Pf1 DNA runs between the 2.3- and 4.3-kb markers under these conditions, whereas the Pf1 that has undergone mutation has bands above the 9.5- and 23-kb markers.

exchange buffers into NMR samples by ultrafiltration with a Centricon (Amicon, Danvers, MA) spin column; however, we found that this procedure does not work with Pf1 phage, which appear to clog the ultrafiltration membrane. Thus the phage must by exchanged into the appropriate buffer by pelleting in a ultracentrifuge or by dialysis.

When mixed at high concentrations, some macromolecules cause the phage and/or macromolecules to precipitate out of solution. To determine

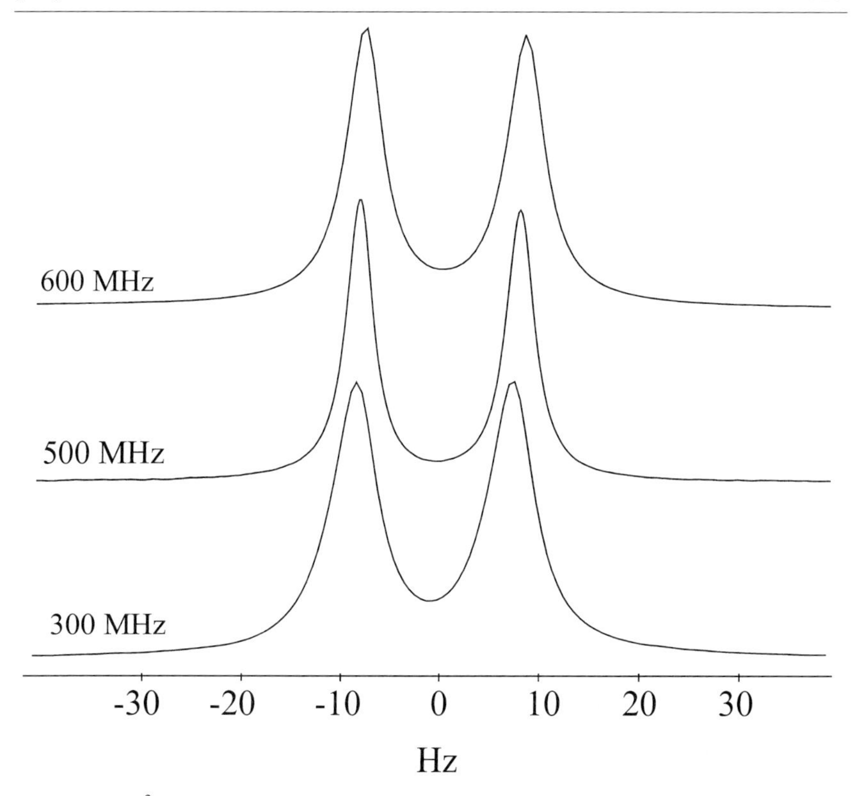

FIG. 3. 1-D ^2H spectra of D_2O as a function of magnetic field in a solution containing ~17 mg/ml Pf1 in 100 m*M* NaCl, 10 m*M* Tris (pH 8.0), 0.1 m*M* EDTA. Spectra were collected at 25° using four scans and a sweep width of 20,000 Hz. The 1-D ^2H spectra of this sample are shown at 300, 500, and 600 MHz ^1H NMR field strengths. The vertical scales of the individual spectra were adjusted so that all peaks have the same height.

the appropriate concentrations for the phage and the macromolecule, 5 μl of each is mixed in a drop on a glass slide and the solution should be clear with no precipitate. The macromolecule, phage, and buffer concentrations can then be adjusted to achieve appropriate mixing conditions for the whole sample. For DNA and RNA oligomers, the optimal conditions are usually <50 mg/ml Pf1 and <3.5 m*M* oligonucleotide.

P. aeruginosa bacteria are an opportunistic human pathogen that do not represent a risk to healthy individuals, but are a potential risk to individuals with cystic fibrosis or compromised immune systems. The ATCC classifies the *P. aeruginosa* strain employed here at only a biosafety level 1, but appropriate precautions should be used in handling and disposing of the *P. aeruginosa* bacteria.

Pf1 Phage as Tool for Generating Partially Aligned Solutions of Macromolecules

The binding of water molecules to magnetically aligned phage leads to a slight overall alignment of the water, which can be used to assess indirectly the alignment properties of the phage. The bound and free water molecules are in fast exchange on the NMR timescale, which leads to a net alignment of the water and therefore incomplete averaging of the quadrupole coupling of the deuterium water resonance. As shown in Fig. 3, the ^{2}H quadrupole coupling for the water leads to splitting of its ^{2}H signal. The size of the residual quadrupole coupling is proportional to the concentration of aligned phage in solution (see Fig. 4A) and therefore is a simple indicator of the degree of alignment of the phage. We have observed a general correlation between the size of the residual dipolar couplings measured in the macromolecule and the size of the residual quadrupole coupling of the ^{2}H water resonance in the phage solutions.

Figure 3 shows the residual quadrupole coupling of the ^{2}H water resonance for a 17 mg/ml Pf1 phage solution at magnetic fields corresponding to 300, 500, and 600 ^{1}H frequencies. The coupling is the same (within error) at all three fields, demonstrating that the phage are fully aligned even at the lowest magnetic field. This means that ultrahigh magnetic fields are not required for aligning phage solutions and in fact no additional alignment is obtained at the higher fields.

The temperature dependence of the residual quadrupole coupling for the ^{2}H water resonance is shown in Fig. 4B illustrating that there is only modest change in ^{2}H splitting of the water resonance from 10° to 45°. The ^{2}H splitting is essentially unchanged above 15° but there is a small decrease in the residual quadrupole coupling below 15°. This decrease could result from reduced alignment of the phage at lower temperature or could arise from changes in the properties of the water bound to the phage at lower temperature. As discussed later, the sizes of the ^{1}H–^{15}N dipolar couplings in phage-aligned IRE-I RNA show little temperature dependence from 5° to 35°, again indicating that the degree of alignment of the phage is not significantly affected by temperature. These results illustrate an important advantage of the phage alignment system, that NMR studies can be performed over a wide range of temperatures (5° to >45°).

Alignment of macromolecules with Pf1 phage has been achieved under a variety of buffer conditions. Although 10 m*M* Tris, pH 8, and 10 m*M* NaH_2PO_4, pH 8, are the easiest to work with, many other conditions work well. Table I gives a summary of various buffer conditions that have been employed with Pf1 phage. Also given is a list of the normalized residual ^{2}H quadrupole coupling of deuterium water resonance (per 10 mg/ml of

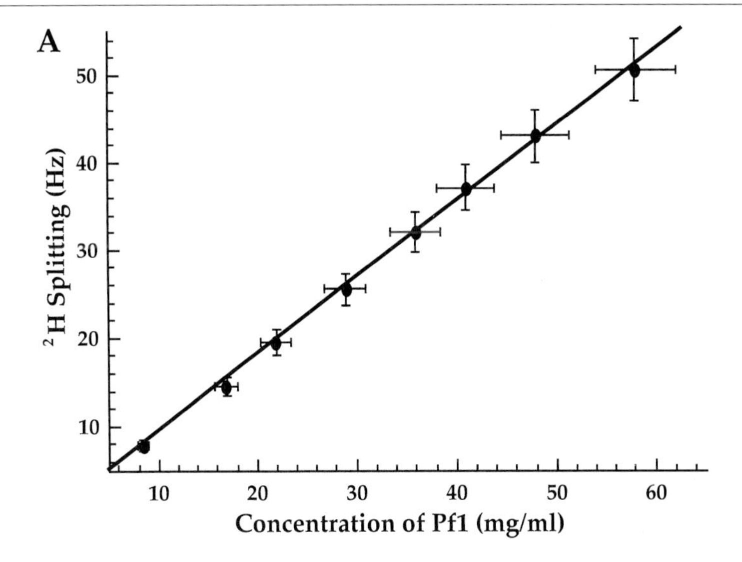

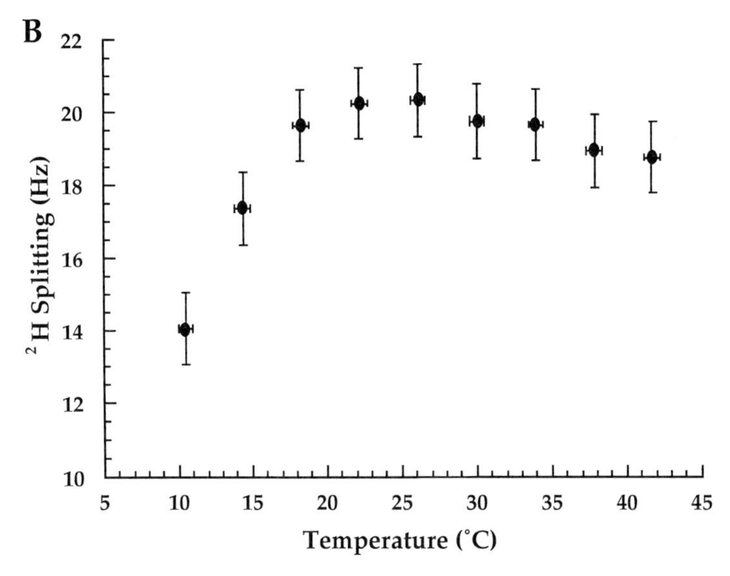

FIG. 4. (A) Plot of the residual ^{2}H quadrupole coupling of a 90% H_2O/10% D_2O sample as a function of Pf1 concentration. Spectra were collected at 25° on solutions of Pf1 in 10 m*M* Tris, pH 8.0, using four scans and a sweep width of 20,000 Hz. (B) Plot of the residual ^{2}H quadrupole coupling of the ~28 mg/ml Pf1 sample as a function of temperature. Spectra were collected in 5° increments from 10° to 45° allowing the sample to equilibrate at each temperature for 10 min.

TABLE I
BUFFER CONDITION USED FOR ALIGNMENT OF MACROMOLECULES WITH Pf1 FILAMENTOUS PHAGE

Buffer conditions	Normalized ^{2}H residual coupling (Hz)[a]
10 m*M* NaH_2PO_4, 1.5 m*M* EDTA, pH 8, ~17 mg/ml Pf1, 0.3 m*M* IRE RNA	8.2
40 m*M* NaH_2PO_4, 0.5 m*M* EDTA, pH 8, ~13 mg/ml Pf1, 1.8 m*M* DNA 16-mer	6.0
10 m*M* NaH_2PO_4, 100 m*M* NaCl, pH 6.8, ~6 mg/ml Pf1, 1 m*M* DNA 10-mer	7.7
10 m*M* Tris(d_{11}), 0.5 m*M* EDTA, pH 8, ~22 mg/ml Pf1, 0.9 m*M* IRE RNA	6.1
10 m*M* imidazole(d_4), 5 m*M* KCl, 0.5 m*M* EDTA, pH 6.5, ~22 mg/ml Pf1, 1.9 m*M* apocalmodulin	7.2
10 m*M* succinate, 0.1 m*M* EDTA, pH 5.5, ~19 mg/ml Pf1	8.2 (7.3)
10 m*M* succinate, 100 m*M* NaCl, 0.1 m*M* EDTA, pH 5.5, ~19 mg/ml Pf1	7.8 (5.0)
10 m*M* succinate, 100 m*M* NaCl, 2.0 m*M* $MgCl_2$, 0.1 m*M* EDTA, pH 5.5, ~20 mg/ml Pf1	6.9 (4.6)
10 m*M* Tris(d_{11}), 0.1 m*M* EDTA, pH 8, ~15 mg/ml Pf1	8.8
10 m*M* Tris(d_{11}), 0.1 m*M* EDTA, pH 7, ~15 mg/ml Pf1	5.5
10 m*M* Tris(d_{11}), 0.1 m*M* EDTA, pH 6, ~15 mg/ml Pf1	4.4

[a] The residual quadrupole coupling constants for the D_2O signal were measured by 1-D ^{2}H spectra. Because all buffer conditions did not have the same Pf1 concentration, the values given are normalized to a Pf1 phage concentration of 10 mg/ml. Values in parentheses indicate that the sample was prepared by dialysis, whereas the other samples were prepared by pelleting the phage in the given buffer.

phage), which provides a qualitative indication of the degree of alignment of the phage under these conditions. Note that at pH < 4.5 or at high salt and low temperature, Pf1 visibly aggregate as white precipitate. We have also observed reduced residual quadrupole coupling of the ^{2}H water resonance below pH 5.5, depending on the salt and buffer concentrations. Thus it is likely that there is some aggregation of the phage at these low pH values. Phage samples contaminated with macrophage (described earlier) are much more sensitive to pH and salt concentrations. They appear to aggregate more readily, leading to aberrant ^{2}H spectra at pH < 7.0 and salt concentrations above 20 m*M*.

In our experience the Pf1–macromolecule samples are stable indefinitely and the alignment does not degrade over time. The naturally occurring phage are extremely stable to temperature and solution conditions and therefore represent an optimal medium for performing NMR studies of macromolecules. In addition, it is very easy to quantitatively recover

RNA, DNA, and protein samples from the phage solutions by ultracentrifugation. The phage from a pellet can be reused in future experiments and the macromolecule remains in the supernatant (see earlier discussion). Thus, we have routinely recycled both the phage and nucleic acids/protein molecules, which is a critical issue for expensive $^{13}C/^{15}N$-labeled samples.

Use of Dipolar Coupling for Generating Distance and Angle Information in Macromolecules

For macromolecules aligned by phage the dipolar coupling between two nuclei i and j has the form[3];

$$D_{ij}(\theta, \phi) = -\frac{\mu_0}{4\pi}\gamma_i\gamma_j hS[A_a(3\cos^2\theta - 1) + \frac{3}{2}A_r\sin^2\theta\cos 2\phi]/4\pi^2 r_{ij}^3 \quad (1)$$

where A_a and A_r are the axially symmetric and rhombic components of the molecular alignment tensor of the macromolecule, respectively; S is the generalized order parameter for internal motions of the ij vector; θ and ϕ are the spherical coordinates describing the orientation of the ij vector in the principal axis system of the alignment tensor; γ_i and γ_j are the gyromagnetic ratios of nuclei i and j, respectively; r_{ij} is the distance between the interacting nuclei; and μ_0 and h are the magnetic permeability of vacuum and Planck's constant, respectively. The size of the residual dipolar coupling depends on the degree of alignment of the macromolecule, the gyromagnetic ratio of the nuclei involved, the distance between these two nuclei, and the angle between the internuclear vector for these nuclei and the molecular alignment axes of the molecule. For many DNA and RNA molecules, the alignment is axially symmetric and the rhombic term in Eq. (1), the A_r term, is zero, which simplifies the analysis. For nuclei that are at a fixed distance, such as $^1H-^{15}N$ and $^1H-^{13}C$ bonded partners, the angular constraints in macromolecules can be obtained by measurement of their residual dipolar couplings. This type of information has been used to help improve solution structures of proteins[31] and is also being applied to RNA and DNA oligomers.[6] The angular data represent one of the more valuable pieces of structural information that is obtained from measurement of dipolar coupling interactions in nucleic acids.

Pf1 phage were used to align a uniformly ^{15}N-labeled IRE-I RNA (sequence shown in Fig. 5A) in order to measure residual one bond $^1H-^{15}N$ dipolar couplings. A stock solution of 50 mg/ml of Pf1 phage was titrated

[31] N. Tjandra, J. G. Omichinski, A. M. Gronenborn, G. M. Clore, and A. Bax, *Nature Struct. Biol.* **4,** 732 (1997).

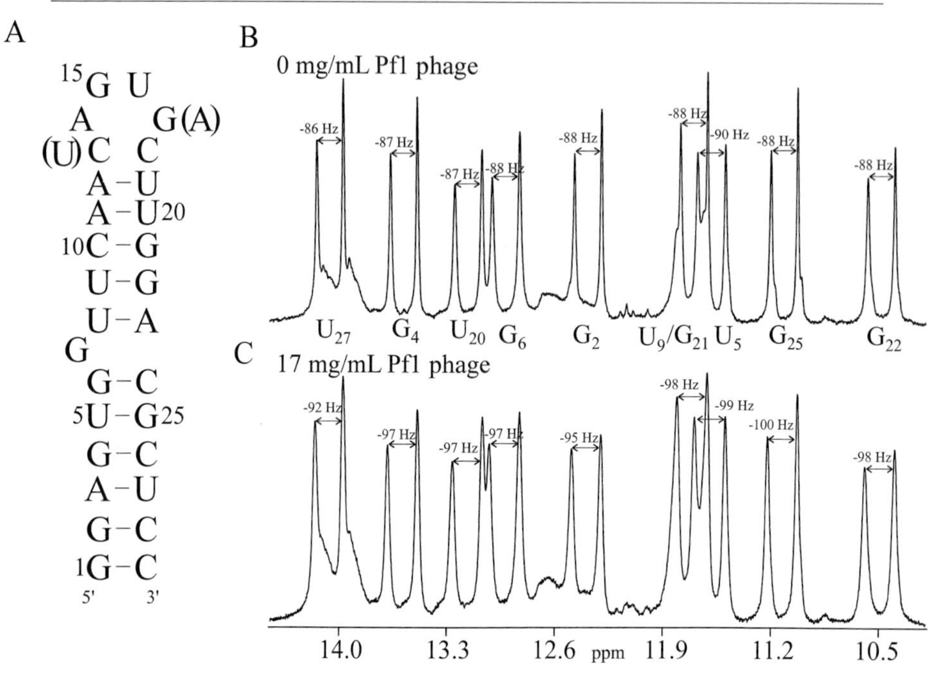

FIG. 5. (A) The sequence and secondary structure the IRE RNA hairpins used here.[32] The IRE-I hairpin sequence is shown here and the IRE-II sequence has two changes in the hairpin loop C_{13} and G_{17}, to U_{13} and A_{17} (shown in parentheses). (B) 1-D imino proton spectra of the ^{15}N-labeled IRE-I RNA (0.3 m*M*) in 10 m*M* NaH_2PO_4, pH 8.0, with no phage and (C) with 17 mg/ml Pf1 phage. These spectra were collected on a Varian Inova 500 spectrometer with *z*-axis pulsed-field gradients at 25° with a flip-back WATERGATE sequence[34,35] to suppress the water resonance and with no ^{15}N decoupling. The measured splittings for the imino protons are shown and represent the ^{1}H–^{15}N scalar coupling in the spectrum with no phage and the sum of the ^{1}H–^{15}N scalar and dipolar coupling in the spectrum with phage. The assignments for the imino proton resonance are as shown.[32] (Adapted from Hansen *et al.*[6])

into the sample and 1-D imino proton spectra without ^{15}N decoupling were collected for a range of phage concentrations. Figs. 5B and 5C show the imino proton spectra at 0 and 17 mg/ml Pf1 phage, respectively. The magnitude of the dipolar coupling is the difference in the observed splitting between the isotropic (no phage) and partially oriented (with phage) spectra. The ^{1}H–^{15}N dipolar couplings for all the resolved G and U imino proton resonances increase proportionally with phage concentration, yielding a value of -9.0 ± 3.0 Hz at 17 mg/ml Pf1 phage. No change in these residual ^{1}H–^{15}N dipolar couplings was observed with temperature from 5° to 35° (not shown). Inspection of these 1-D spectra reveals an apparent increase in the resonance linewidths in the spectrum collected with Pf1 phage. Such line broadening can arise from an increase in the rotational correlation time of the molecule that would be consistent with binding of the RNA to the phage.[6] To test for changes in the rotational correlation time of the RNA on addition of phage, the ^{15}N $T_{1\rho}$ and T_2 relaxation times for the imino nitrogens in the IRE were measured on the samples with and without phage. These relaxation times are a sensitive function of the rotational correlation time of the molecule where longer correlation times lead to shorter $T_{1\rho}$ and T_2 relaxation times. The measured ^{15}N $T_{1\rho}$ and T_2 relaxation times for the imino nitrogens in the IRE are unchanged (within experimental error) with and without phage.[6] These data demonstrate that the phage do not effect the rotational correlation time of the RNA. This is an important result because it means that standard high-resolution NMR experiments can be performed on macromolecules aligned by Pf1 phage.

The addition of phage leads to a significant increase in the viscosity of the solution, which might at first seem detrimental to acquisition of high-resolution NMR spectra. However, as previously discussed,[6] these long rod-shaped phage lead to an increase in the *macroscopic* viscosity of the solution but have no measurable effect on the *microscopic* environment of RNA molecules in solution. This is because the phage particles are >300 Å apart in a 30 mg/ml solution of Pf1 phage and therefore the time between collisions of an RNA molecule with phage particles is orders of magnitude greater than the rotational correlation time of the macromolecule. Thus the higher viscosity has no effect on the rotational correlation time of a RNA or DNA in phage solutions.

The source of the increased linewidth in the IRE spectrum on addition of phage (see Fig. 5) is the ^{1}H–^{1}H dipolar couplings between the imino proton and the surrounding protons in the RNA. Specifically, a G imino proton located in the Pf1-aligned ^{15}N-IRE RNA sample lies 1.04 Å from the imino nitrogen and 2.17 and 2.48 Å from the G and C amino protons in a Watson–Crick G–C base pair. From Eq. (1), if the ^{1}H–^{15}N dipolar coupling for the imino proton is 10 Hz, then the ^{1}H–^{1}H couplings between the imino proton and the G and C amino protons would be 11 and 7 Hz, respectively. These ^{1}H–^{1}H dipolar couplings are not resolved in this spectrum (Fig. 5C), thus residual ^{1}H–^{1}H dipolar couplings are manifested as broader lines. While greater degrees of alignment yield larger and more easily measured ^{1}H–^{15}N or ^{1}H–^{13}C dipolar couplings, increased linewidths caused by additional ^{1}H–^{1}H couplings reduce the signal-to-noise ratio of the NMR spectra. As a consequence, these two considerations must be carefully balanced when preparing samples for studies of dipolar interactions in RNA, DNA and proteins.

As seen in Fig. 5 the sizes of the residual ^1H–^{15}N dipolar couplings for the base-paired G and U imino groups in the RNA are all the same within experimental error (~−9 Hz). This indicates that alignment of the IRE is axially symmetric and that the imino ^1H–^{15}N vectors are aligned at approximately the same angle with the helical axis of the IRE. Additional angular constraints can be obtained from measurement of the residual ^1H–^{13}C dipolar couplings for base and sugar resonances.[6] These dipolar couplings can be determined from comparison of 2-D (^{13}C, ^1H) HSQC experiments with and without phage acquired with no ^1H decoupling in t_1, as illustrated in Fig. 6. The base H8/C8 or H6/C6 and sugar H1′/C1′ residual dipolar couplings were measured for the uniformly $^{15}N/^{13}C$-labeled IRE-

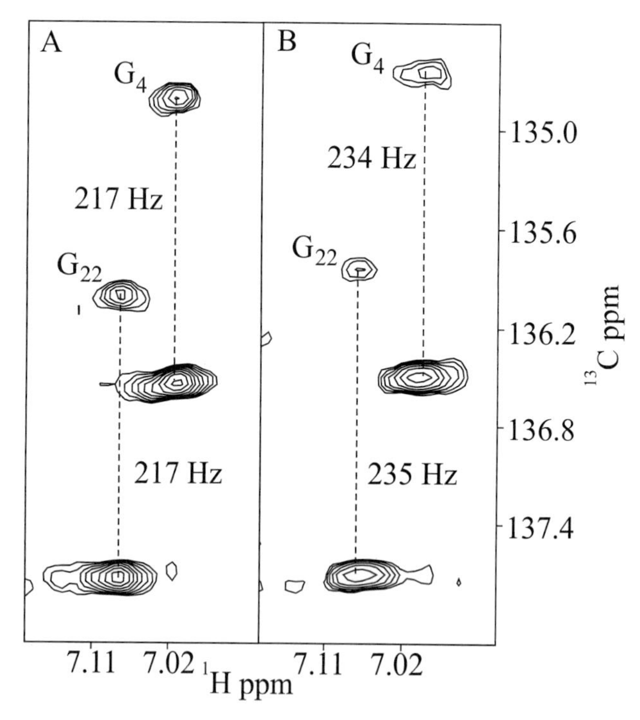

FIG. 6. (A) A portion of the aromatic region of a 2-D (^{13}C, ^1H) constant time HSQC experiment[36,37] collected with no Pf1 and (B) with ~22 mg/ml Pf1. The H8/C8 resonances for G_4 and G_{22} are shown. The spectra were collected on samples containing ~0.5 m*M* $^{13}C/^{15}N$ labeled IRE-II RNA hairpin (sequence given in Fig. 5A) with no ^1H decoupling in the t_1 evolution period using a constant time delay of 30 ms. Ninety complex points were collected in t_1 with a sweep width of 3000 Hz using the States-TPPI method for quadrature detection[38] and a sweep width of 4406 Hz in t_2. The spectra were collected on a Varian Inova 500 spectrometer with *z*-axis pulsed-field gradients at 25° with buffer conditions of 10 m*M* Tris(d_{11}), pH 8.0.

II RNA and showed large residue specific couplings ranging from −14 to +22 Hz.[6] As previously discussed,[6] for some residues there is good agreement between the measured dipolar couplings and the dipolar couplings predicted from the NMR structure of this IRE RNA[32]; however, for other residues the agreement is not very good. Thus these measured dipolar coupling constants provide additional structural constraints that can be used to further refine the solution structure of this RNA.

Due to the large gyromagnetic ratio for ^1H nuclei, proton pairs can interact strongly in a partially aligned system, which can lead to observable dipolar coupling interactions. These ^1H–^1H dipolar coupling interactions yield both *distance* and *angle* information and represent a complementary method to the NOE for measurement of through-space ^1H–^1H interactions in solution.[3,6,13,33] We have recently shown that proton–proton interactions at distances >7.0 Å can be observed by dipolar coupling interactions (DPI) using a dipolar coupling spectroscopy (DCOSY) experiment.[13] This is a much longer distance than can be seen from standard NOESY experiments and reflects the r^{-3} distance dependence of the dipolar coupling [see Eq. (1)] compared to the r^{-6} distance dependence of the NOE. These ^1H–^1H through-space DPIs in macromolecules can also be observed from COSY-type or HNHA spectra on partially aligned macromolecules.[3,6,33] In isotropic solution, a 2-D COSY spectrum only shows cross-peaks between *J*-coupled protons. However in an anisotropic solution cross-peaks are also observed between protons that have nonzero ^1H–^1H dipolar couplings. Thus for phage-aligned solutions, COSY spectra yield information on through-space interactions between protons.[6,33] This is illustrated on a DNA duplex aligned with 20 mg/ml Pf1 phage, where Fig. 7A shows the aromatic proton to H1′ region of a DQF-COSY experiment. In a sample of this DNA without phage (not shown) the only cross-peaks in this region arise from the pyrimidine H5–H6 *J*-coupling interactions. However, in the aligned DNA duplex there are multiple cross-peaks corresponding to through-space interactions between the H8/H6 base protons and the H1′ ribose protons.

The signal-to-noise ratio (SNR) suffers in the DQF-COSY spectrum due to the cancellation of the antiphase multiplets in the cross-peaks. To help overcome this cancellation, we developed the DCOSY experiment,

[32] K. J. Addess, J. P. Basilion, R. D. Klausner, T. A. Rouault, and A. Pardi, *J. Mol. Biol.* **274,** 72 (1997).
[33] P. J. Bolon and J. H. Prestegard, *J. Am. Chem. Soc.* **120,** 9366 (1998).
[34] M. Piotto, V. Saudek, and V. Sklenár, *J. Biomol. NMR* **2,** 661 (1992).
[35] S. Grzesiek and A. Bax, *J. Am. Chem. Soc.* **115,** 12593 (1993).
[36] J. Santoro and G. C. King, *J. Magn. Reson.* **97,** 202 (1992).
[37] G. W. Vuister and A. Bax, *J. Magn. Reson.* **98,** 428 (1992).
[38] D. Marion, M. Ikura, R. Tschudin, and A. Bax, *J. Magn. Reson.* **85,** 393 (1989).

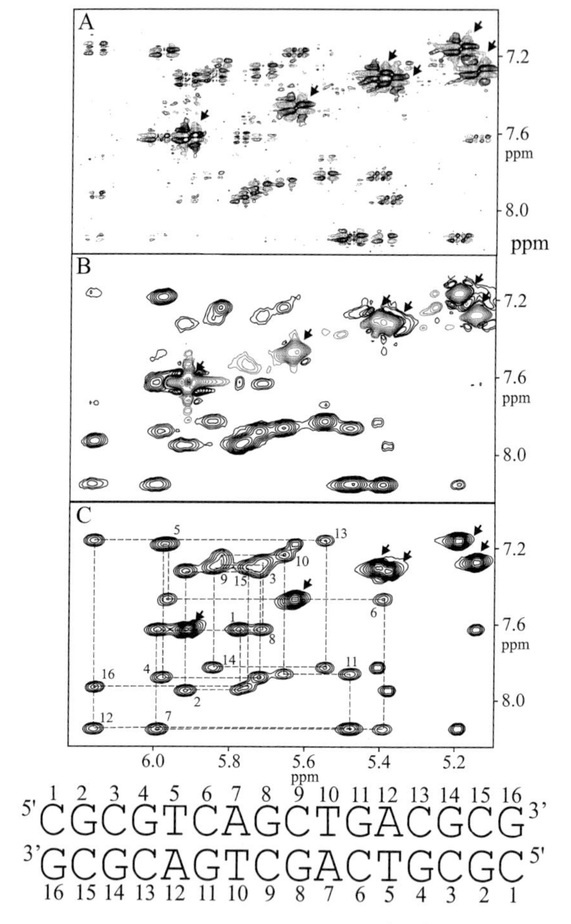

FIG. 7. (A) The H1′ to aromatic proton region of a ^{1}H–^{1}H DQF-COSY spectrum of a 16-mer DNA (1.9 m*M* duplex) in 40 m*M* NaH_2PO_4, pH 8, with ~20 mg/ml Pf1 phage. This region of the spectrum shows ^{1}H–^{1}H dipolar coupling interactions between the aromatic protons (pyrimidine H6 and purine H8) and the ribose H1′ protons. Cross-peaks associated with the *J*-coupled pyrimidine H5–H6 protons are marked with arrows and are the only cross-peaks observed in this region of the DQF-COSY spectrum of this DNA duplex with no phage (not shown). (B) The same region of a DIPSI-2 DCOSY spectrum[13] collected on the 16-mer DNA with 20 mg/ml Pf1 phage. The pyrimidine H5–H6 cross-peaks that arise from *J*-coupling are again marked with arrows and are opposite sign to the DCOSY cross-peaks. This spectrum is phased with the diagonal negative. (C) The same region of a 200-ms NOESY spectrum collected on the 16-mer DNA with no Pf1 phage. The spectra were collected on a Varian Inova 500 spectrometer with *z*-axis pulsed-field gradients at 25°. The sequence of the 16-mer DNA is shown. (Adapted from Hansen *et al.*[6,13])

which employs a TOCSY-type pulse sequence and leads to in-phase cross-peaks for the dipolar coupling interactions and therefore generally has higher SNR than the DQF-COSY spectrum.[13] The DCOSY and DQF-COSY spectra both probe through-space ^1H–^1H interactions as illustrated by comparison with the NOESY spectrum collected without phage (Fig. 7C). It is interesting to note the differences in intensity between the NOE and the dipolar coupling interactions cross-peaks. While many of the same peaks are observed, the DCOSY and the DQF-COSY clearly show the angle dependence of the dipolar interaction. For an axially symmetric system such as a DNA duplex the dipolar coupling depends on θ, the angle formed between an internuclear vector and the molecular alignment axis [see Eq. (1)]. Thus these ^1H–^1H dipolar coupling interactions have both an r^{-3} distance dependence and a $(3 \cos^2 \theta - 1)$ angle dependence, whereas the NOE interactions have a r^{-6} distance dependence but no angle dependence. Therefore, at favorable angles (θ near 0 or 90°), the DCOSY can show intense cross-peaks but at unfavorable angles (θ near the magic angle of 54.7°) the dipolar coupling goes to zero and no cross-peaks are observed. Thus even a rather qualitative analysis of the DCOSY spectra can yield valuable distance and angle constraints, which can be used to improve the NMR solution structure determination of a macromolecule.

Conclusion

The Pf1 filamentous phage system is ideally suited for obtaining a tunable degree of alignment in RNA and DNA molecules, and acidic proteins, for observation of dipolar couplings.[6] The phage particles are fully aligned at magnetic fields as low as 300 MHz and are stable indefinitely at a range of temperatures and buffer conditions. These dipolar couplings provide long-range structural information that is presently unavailable by standard solution NMR techniques; therefore, in the future dipolar coupling data may prove to be as indispensable as NOEs in solution structure determinations of macromolecules. This should be especially true for structure determinations of RNA and DNA oligomers where the low density of protons compared to proteins leads to fewer useful NOE distance constraints and therefore generally less well-defined structures. For example, the orientation of a base within an RNA molecule can be determined with only a few dipolar coupling constraints even if the other residues around it are not well defined. This is a quite different situation from NOE data where all of these distance constraints are strictly related to the local structure of the molecule. In addition, the current solution NMR techniques are generally unable to determine global structural features, such as bending in nucleic acids, due to the local nature of the NOE distance constraints. However, the dipolar coupling constraints should make it possible to structurally relate distant regions of the molecule. In this way the addition of dipolar coupling data should lead to improvements in both the local and global conformations of proteins and nucleic acids determined by solution NMR techniques.

Acknowledgments

This work was supported by grants from the NIH and the Colorado RNA Center (A.P.) and a Leukemia Society of America fellowship (M.R.H.). We would like to thank E. Mollova, L. Mueller, M. Rance, A. J. Wand, J. Wank, and D. S. Wuttke for valuable discussions and technical assistance

[16] Biochemical and Nuclear Magnetic Resonance Studies of Aminoglycoside–RNA Complexes

By STEPHEN R. LYNCH, MICHAEL I. RECHT, and JOSEPH D. PUGLISI

Introduction

Nuclear Magnetic Resonance (NMR) spectroscopy has become a powerful tool for studying RNA–ligand interactions. Our laboratory has studied complexes formed between aminoglycoside antibiotics and an RNA oligonucleotide that mimics the ribosomal RNA target of these drugs. Size limitations of approximately <30,000 molecular weight (100 nucleotides) prevent the study of the whole biological system, the ribosome, by NMR. Aminoglycoside antibiotics bind both with high affinity and specificity to short oligonucleotide sequences, which makes the system ideal for NMR. Structural studies have revealed features of the drug and RNA required for binding.

The aminoglycosides are effective antibiotics because they bind to prokaryotic ribosomes more tightly than to eukaryotic ribosomes. Aminoglycoside antibiotics bind to the aminoacyl-tRNA site (A site) of 16S rRNA on the 30S subunit[1] and interfere with protein synthesis by inducing codon misreading and inhibiting translocation.[2] Structural studies of an aminoglycoside–RNA complex provide insights into miscoding and reveal the origins of specific aminoglycoside binding to prokaryotic ribosomal RNA, thus leading to the possibility of using the three-dimensional (3-D) structure of

[1] D. Moazed and H. F. Noller, *Nature,* **327,** 389 (1987).

[2] E. F. Gale, E. Cundliffe, P. E. Reynolds, M. H. Richmond, and M. J. Waring, "The Molecular Basis of Antibiotic Action." John Wiley & Sons, London, 1981.

0076-6879/00 $30.00

the complex for structure-based drug design. This article focuses first on how to use biochemical methods to identify a minimal RNA sequence that mimics the aminoglycoside binding site on the ribosome, then on how to apply NMR spectroscopy to identify the site of interaction of the antibiotic on the oligonucleotide and subsequently determine the 3-D structure of the antibiotic–RNA complex.

Choice and Design of Antibiotic–RNA Oligonucleotide Complex

Several factors made the aminoglycoside–rRNA complex amenable to NMR study. First, the interaction of aminoglycosides with the ribosome had been previously characterized biochemically.[1] Second, the site of interaction of the drug is localized to a small region of the 16S rRNA. Third, this region of rRNA does not appear to interact with ribosomal proteins.[3] These factors suggested that the local structure of this region of rRNA could be maintained in an oligonucleotide in the absence of any of the ribosomal proteins.

Before NMR studies were started, it was necessary to determine if the model system was an accurate representation of the aminoglycoside–rRNA interaction. Figure 1A (see color insert) shows the secondary structure of 16S rRNA where the aminoglycoside antibiotic paromomycin binds and where mRNA decoding occurs. Figure 1B presents the oligonucleotide designed to mimic binding of the antibiotic. Because large quantities of sample are required for NMR, 2 G–C base pairs were added to the 5′ end to enable high yield of RNA by *in vitro* transcription with T7 RNA polymerase. An ultrastable [5′ UUCG 3′] tetraloop was also added[4] to help ensure that the oligonucleotide forms the desired secondary structure with two 4-bp stems on either side of the asymmetric internal loop. Chemical probing experiments identical to the type used to characterize the drug–ribosome interaction were used to assay the interaction of the antibiotic with the oligonucleotide.[5] Chemical probing and primer extension analysis of oligonucleotides requires that several changes be made to the standard ribosome modification experiments.[6] These methods are outlined in the following section.

[3] T. Powers and H. F. Noller, *RNA* **1,** 194 (1995).
[4] C. Cheong, G. Varani, and I. Tinoco Jr., *Nature* **346,** 680 (1990).
[5] M. I. Recht, D. Fourmy, S. C. Blanchard, K. D. Dahlquist, and J. D. Puglisi, *J. Mol. Biol.* **262,** 421 (1996).
[6] S. Stern, D. Moazed, and H. F. Noller, *Methods Enzymol.* **164,** 481 (1988).

Methods

Chemical Modification

Chemical modification reactions are performed on the oligonucleotide in a buffer similar to that used for characterizing the interaction of the ribosome with aminoglycoside antibiotics. The duration of modification and/or concentration of modifying reagent must be adjusted for each oligonucleotide construct. It is important to ensure that, on average, there is no more than one modification on each oligonucleotide molecule. This can be estimated following primer extension as follows: if the band corresponding to full-length oligonucleotide is more intense than the sum of all reverse transcriptase stops below it, then the RNA has not been overmodified.

In the case of the A-site oligonucleotide, modification reactions (20 μl) are performed in 80 m*M* potassium cacodylate, pH 7.0, with 75 n*M* RNA oligonucleotide. To ensure proper folding, the RNA is heated at 90° for 1 min then placed on ice for 5 min. Antibiotics (Sigma, St. Louis, MO) are added and the modification performed by addition of 1 μl dimethyl sulfate (1:10 dilution in ethanol, final dilution 1:200) followed by incubation at room temperature for 5 min. Reactions are stopped by addition of 1/4 volume 1 *M* Tris–acetate, pH 7.5, 1.5 *M* sodium acetate, 1 *M* 2-mercaptoethanol followed by ethanol precipitation.[5] To allow detection of modifications at the N-7 of guanine, sodium borohydride reduction and aniline-induced strand scission are performed as described[7] except that lyophilization steps are replaced by phenol extraction and ethanol precipitation.

Modified RNA is resuspended in 10 μl 1 *M* Tris-HCl, pH 8.2. On addition of 10 μl of freshly prepared 0.2 *M* sodium borohydride, the samples are incubated on ice in the dark for 30 min. The reaction is quenched by addition of 300 μl 0.3 *M* sodium acetate followed by ethanol precipitation. Pellets are dissolved in 20 μl 1.0 *M* aniline/acetate, pH 4.5, followed by incubation in the dark for 20 min at 60°. The reaction is terminated by addition of 110 μl 0.3 *M* sodium acetate, pH 5.5, and 110 μl phenol/chloroform/isoamyl alcohol (25:24:1) followed by vigorous mixing and centrifugation. The RNA is concentrated by ethanol precipitation of the aqueous phase and pellets washed with 100 μl 70% ethanol.

The RNA is resuspended in water (~1 pmol/μl) and primer extension is performed as described[6] with the following modifications: (1) 5′-labeled primer is annealed at a 2-fold excess to RNA, (2) 4 μl of 1 m*M* dNTPs is added to each tube prior to addition of reverse transcriptase, and (3) five units (0.2 μl) of avian myeloblastosis virus (AMV) reverse transcriptase

[7] D. A. Peattie, *Proc. Natl. Acad. Sci. U.S.A.* **76,** 1760 (1979).

A

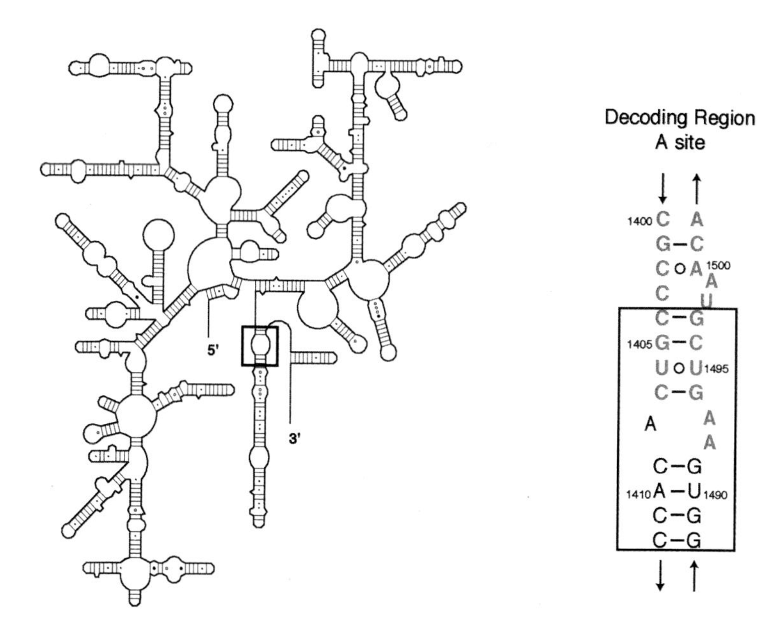

B

5' 3'
G−C
G−C
C−G
1405 G−C
U o U 1495
C−G
A A
A
C−G
A−U
1410 C−G 1490
C−G
U G
U C

FIG. 1. (A) Secondary structure of *Escherichia coli* 16S rRNA. The A-site region is boxed. Nucleotides that are conserved in all 16S-like rRNAs are indicated in green. (B) Sequence of the A-site oligonucleotide used in NMR studies. Nucleotides critical for high-affinity binding of the aminoglycoside paromomycin to the oligonucleotide are indicated in red.

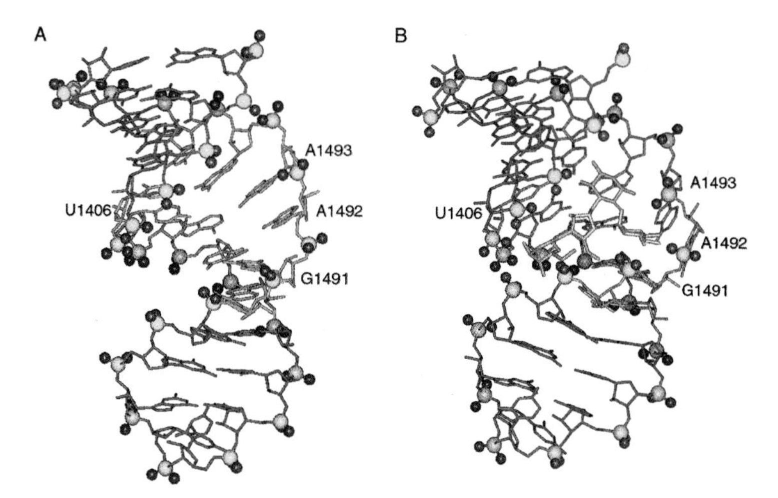

FIG. 4. Representative structures of (A) the free form and (B) the paromomycin-bound form of the A-site oligonucleotide. Nucleotides of the aminoglycoside binding site are shown in blue. All other nucleotides are green. Paromomycin is tan.

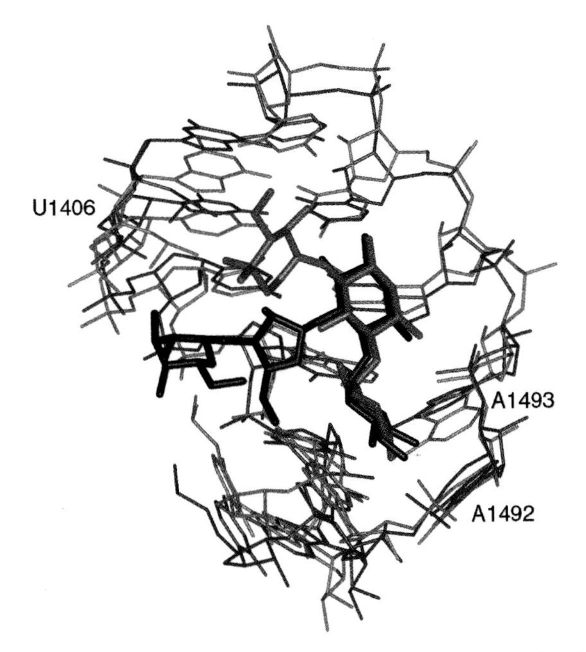

FIG. 5. Superposition of the paromomycin (black)– and gentamicin (red)– A-site RNA structures. Rings I and II of the antibiotics bind in a similar manner and induce the same conformational change in the RNA. In paromomycin, ring III is linked to the 5 position of ring II. Gentamicin has ring III linked to position 6 of ring II.

(Seikagaku, Ijamsville, MD) per tube was used in the extension. These changes decrease the background of spontaneous reverse transcriptase stops. A primer annealing site must be present at the 3′ end of the oligonucleotide. In the case of the A-site oligonucleotide, a 17-nucleotide sequence was added to the original construct. Priming sites as short as 10 nucleotides may be used. Additional nucleotides are not required if one does not wish to monitor nucleotides near the 3′ end of the molecule. Extension is performed for 30 min to 1 hr, and the reaction is stopped by addition of 20 μl 0.1 *M* sodium hydroxide followed by incubation for 15 min at 90° to digest remaining RNA. For RNA molecules up to 50 nucleotides, failure to treat with base prior to electrophoresis results in a smearing of bands. The DNA is precipitated and pellets washed with 70% (v/v) ethanol. DNA is suspended in 10 μl 7 *M* urea loading dyes, heated to 90° for 1–2 min, and 2.5 μl is loaded on a 20% (w/v) polyacrylamide (19:1 acrylamide:bis acrylamide) 7 *M* urea gel.

Comparison of Model System with Ribosome

Because the same chemical probing experiments were used to study the interaction of paromomycin with the A-site oligonucleotide and the ribosome, the results can be directly compared. Paromomycin causes protections of the same set of nucleotides at the same drug concentration on both the A-site oligonucleotide and the ribosome. Data from chemical probing experiments in which the antibiotic is titrated over the relevant concentration range can be used to determine an approximate K_D for the interaction of the drug with both the oligonucleotide and the ribosome. Band intensities are converted to fraction of RNA bound to antibiotic. These experiments indicated a K_D for the interaction of paromomycin with the A-site RNA of $\approx 0.2\ \mu M$,[5] which is similar to that determined for the paromomycin–ribosome interaction by equilibrium dialysis.[8]

Mutant Oligonucleotides

The specificity of the aminoglycoside antibiotic–A-site RNA interaction is studied by making mutations in the RNA sequence. For a short oligonucleotide, such as the A-site RNA, this is simply a matter of synthesizing appropriate DNA templates to be used for *in vitro* transcription. The advantage of performing mutational analysis on the oligonucleotide is that it is a strictly *in vitro* assay for the rapid screening of large numbers of mutants. Chemical footprinting experiments as a function of antibiotic concentration can be performed on the mutant oligonucleotides to determine an approxi-

[8] D. Lando, M. A. Cousin, T. Ojasoo, and J. P. Raymond, *Eur. J. Biochem.* **66,** 597 (1976).

mate binding constant. Nucleotides critical for the antibiotic–A-site RNA interaction were identified using this approach (Fig. 1B). Once interesting mutants in the oligonucleotide have been identified, these can be introduced *in vivo*.

It is possible to introduce mutations into plasmid-encoded copies of the 16S rRNA. We observed a direct correlation between nucleotides critical for the binding of paromomycin to the A-site oligonucleotide and to the ribosome,[9] demonstrating the validity of the model system.

Purification of RNA Sample for NMR Studies

Large quantities of both antibiotic and RNA oligonucleotide are required for structural studies. Many antibiotics can be purchased at sufficient purity for low cost from chemical companies, but the RNA oligonucleotide must be synthesized and purified. The methods for synthesis and purification of the RNA NMR sample have been described elsewhere.[10] After synthesis and purification of the RNA oligonucleotide, the NMR sample is dialyzed into buffer. A typical NMR buffer is 10 m*M* sodium phosphate at pH 6.0–7.0. Phosphate is commonly used because it lacks nonexchangeable protons that would interfere with the NMR spectrum of the RNA oligonucleotide and is a good buffer near neutral pH. Ideally, the concentration of the RNA in the NMR tube should be ~3 m*M*. The duration of an experiment can be greatly reduced with highly concentrated samples; to achieve an equivalent signal-to-noise ratio (SNR) with a sample half as concentrated as another, the NMR experiment needs to be run four times as long. However, some RNA oligonucleotides aggregate or dimerize at high concentrations, so spectra for these samples need to be acquired at lower concentrations or different sample conditions.

$^{13}C/^{15}N$-Labeled RNA Samples

A sample uniformly labeled with ^{13}C and ^{15}N is critical for high-resolution structure determination of RNA–antibiotic complexes. Complete assignment of the NMR spectrum of the RNA is only possible with heteronuclear experiments involving ^{13}C or ^{15}N or both. Labeling is accomplished at only moderate expense by initially isolating ^{13}C- and ^{15}N-labeled ribonucleotides from bacteria grown in media with $^{13}CH_3OH$ as the sole carbon source and $^{15}NH_4SO_4$ as the only source of nitrogen. The labeled nucleotides are purified using published protocols.[11] Overall, approximately 50 mg of

[9] M. I. Recht, S. Douthwaite, K. D. Dahlquist, and J. D. Puglisi, *J. Mol. Biol.* **286,** 2247 (1999).
[10] J. D. Puglisi and J. R. Wyatt, *Methods Enzymol.* **261,** 323 (1995).
[11] R. T. Batey, J. L. Battiste, and J. R. Williamson, *Methods Enzymol.* **261,** 300 (1995).

labeled NTPs can be isolated per liter of cell culture. Because the media contains 0.8 g of $^{13}CH_3OH$ per liter, the NTP yield is approximately 62 mg NTPs/1 g $^{13}CH_3OH$.

Labeled nucleotides can be used in place of unlabeled nucleotides in the *in vitro* transcription reaction. Yield of labeled RNA from the transcription reaction is often lower than the yield of unlabeled RNA, because the molar ratio of the four labeled NTPs is not equivalent. However, the transcription reaction can be optimized by titrating all components of the reaction to make the labeled NTPs the limiting reagent. An NMR sample can thus be purified from as little as 100 mg of NTPs for a small RNA molecule (<20 nucleotides) that synthesizes well. The cost of sample preparation may increase by as much as 10- to 15-fold for larger molecules and sequences that synthesize poorly. Heteronuclear NMR experiments that take advantage of the additional spin 1/2 nuclei are used to assign the proton NMR spectrum of the labeled RNA oligonucleotide. In addition, assignment of nuclear Overhauser effects (NOEs) that are overlapped in the ^1H NOESY spectra is easier in a 3-D NOESY experiment with separation by carbon or nitrogen chemical shift in the third dimension.

NMR Assignments of RNA

The first step after sample purification is to assign proton resonances of both the RNA and the antibiotic in their respective free forms. Characterization of the free RNA and antibiotic by NMR facilitates analysis of the RNA–drug complex, and possible conformational changes that occur on complex formation. Initial resonance assignment of the RNA in the absence of antibiotic is accomplished through application of well-established homonuclear and heteronuclear NMR experiments.[12–15] Assignment strategies for nonexchangeable protons, which include the protons on the ribose sugar except for the 2′-OH, the H-5 and H-6 protons on pyrimidine bases, the H-8 of purine bases, and the H-2 proton on the adenosine base, have been discussed elsewhere.[13–15]

Exchangeable Proton Assignment

Recent NMR developments have greatly aided through-bond assignment of base exchangeable protons. Correlation experiments that apply

[12] K. Wüthrich, "NMR of Proteins and Nucleic Acids." John Wiley & Sons, New York, 1986.
[13] A. Pardi, *Methods Enzymol.* **261,** 350 (1995).
[14] J. G. Pelton and D. E. Wemmer, *Ann. Rev. Phys. Chem.* **46,** 139 (1995).
[15] G. Varani, F. Aboul-ela, and F. H.-T. Allain, *Prog. Nuclear Magn. Reson. Spectrosc.* **29,** 51 (1996).

heteroTOCSY pulse schemes in water on uniformly $^{13}C/^{15}N$-labeled samples enable resonance assignment of all base protons that do not exchange rapidly with water. HeteroTOCSY experiments correlate exchangeable and nonexchangeable base protons: guanosine NH (imino) to H-8,[16–18] uridine NH (imino) to H-6,[17,19] cytidine NH_2 (amino) to H-6,[19] and adenosine NH_2 (amino) to the H-2 and H-8.[18,20] Each of these experiments is valuable for assigning the water exchangeable resonances on the bases.

An example of guanosine imino proton (H-1) to H-8 proton correlation is shown in Fig. 2. This experiment was performed on a uniformly $^{13}C/^{15}N$ labeled samples of a single base mutant of the oligonucleotide presented in Fig. 1 (nucleotide A1408G) in water at 5°. A single correlation is observed for each guanosine imino proton to its own H-8; previously one depended on through-space correlation in NOESY experiments to assign these protons.

Imino and amino protons often provide long-range NOEs. In RNA stems, NOEs from imino, amino, and adenosine H-2 protons are observed across the helix, unlike NOEs involving all other nonexchangeable protons.[21] Furthermore, imino and amino protons in loops often make interesting contacts across loops; in the UUCG tetraloop, NOEs from the cytidine amino proton to sugar protons of the two tetraloop uridines indicate an amino hydrogen bond to a phosphate oxygen.[4,22] Amino groups are often hydrogen bond donors in RNA-ligand recognition.[23–25] The two amino protons can be hydrogen bond donors to the antibiotic and an RNA base simultaneously. Importantly, the cytidine and adenosine amino protons face the major groove, whereas those of guanosine face the minor groove (see later section).

Resonance Assignments of Antibiotic

Given its small size, resonance assignment of aminoglycosides would seem trivial compared to that of the RNA. However, there is significant

[16] J.-P. Simorre, G. R. Zimmermann, L. Mueller, and A. Pardi, *J. Biomol. NMR* **7,** 153 (1996).
[17] V. Sklenar, T. Dieckmann, S. E. Butcher, and J. Feigon, *J. Biomol. NMR* **7,** 83 (1996).
[18] R. Fiala, F. Jiang, and D. J. Patel, *J. Am. Chem. Soc.* **118,** 689 (1996).
[19] J.-P. Simorre, G. R. Zimmermann, A. Pardi, B. T. I. Farmer, and L. Mueller, *J. Biomol. NMR* **6,** 427 (1995).
[20] J.-P. Simorre, G. R. Zimmermann, L. Mueller, and A. Pardi, *J. Am. Chem. Soc.* **118,** 5316 (1996).
[21] G. Varani and I. Tinoco, Jr., *Q. Rev. Biophys.* **24,** 479 (1991).
[22] F. H.-T. Allain and G. Varani, *J. Mol. Biol.* **250,** 333 (1995).
[23] L. Jiang and D. J. Patel, *Nature Struct. Biol.* **5,** 769 (1998).
[24] P. Fan, A. K. Suri, R. Fiala, D. Live, and D. J. Patel, *J. Mol. Biol.* **258,** 480 (1996).
[25] J. L. Battiste, H. Mao, N. S. Rao, R. Tan, D. R. Muhandiram, L. E. Kay, A. D. Frankel, and J. R. Williamson, *Science* **273,** 1547 (1996).

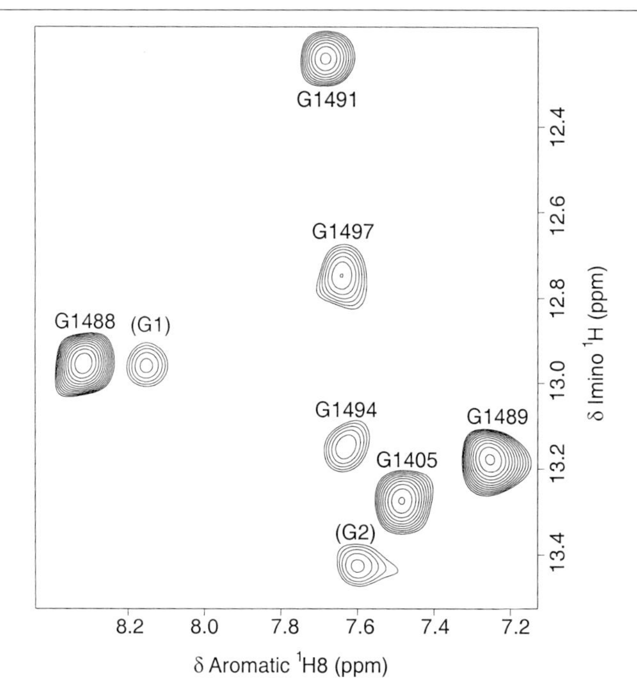

FIG. 2. The NMR spectrum of an H(NC)-TOCSY-(C)H experiment on the A1408G mutant of the A-site oligonucleotide. This spectrum was acquired by the method of Simorre *et al.*[16] on a 2.0 m*M* uniformly $^{13}C/^{15}N$ labeled NMR sample in 5 m*M* sodium phosphate buffer, pH 6.2, at 5° on a Varian Inova 500-MHz spectrometer. The experiment accomplishes through-bond correlation of the guanosine imino H-8 protons. A peak is observed in the spectrum for each guanosine residue except G1408, which is not observed due to line broadening; the peak for the guanosine in the tetraloop (G15) is shifted outside this region of the spectrum. The total experimental time was 24 hr for 1024 complex t_2 points, 80 complex t_1 points, and 512 scans per FID.

overlap of proton resonances. Isotopic labeling, which is invaluable in assignment of the RNA molecule, is difficult to achieve for an antibiotic. Two possible methods of isotopic labeling exist: organic synthesis from a labeled small molecule precursor or isolation of the antibiotic from a producing organism after growth in uniformly labeled media. To date, neither method has been feasible in our laboratory. First, the antibiotics are too complicated to synthesize readily from a molecule such as glucose using chemical methods. Second, in our hands, the paromomycin producer, *Streptomyces rimosus paromomycinus,* will produce antibiotic in rich, but not minimal, medium.

Without the aid of the heteronuclear correlation experiments applied on isotopically labeled samples, assignments of NMR resonances for the antibiotic must be accomplished with homonuclear experiments and natural abundance $^{1}H/^{13}C$ correlation experiments. The primary experiments to apply are DQF-COSY, TOCSY, and ^{13}C–^{1}H HMQC. In the DQF-COSY, protons three bonds apart can be correlated to each other, allowing for resonance assignment by correlating protons attached to adjacent carbon atoms. This experiment uses the $^{3}J_{HH}$, which is dependent on the torsion angle between the two protons. The $^{3}J_{HH}$ varies between 12 and 1 Hz.[21] Cross-peaks are only observed when the J_{HH} is greater than approximately 1 Hz. The TOCSY experiment with a long mixing time (approximately 80 ms) complements the DQF-COSY with total correlation of all protons within a spin system. Dynamic averaging of the ring conformation leads to measurable coupling constants for all protons on adjacent carbons. An alternative to the DQF-COSY is a TOCSY experiment with a short mixing time (approximately 20 ms). The short mixing time allows for only three-bond correlation as in the DQF-COSY, while a longer mixing time allows magnetization transfer through the whole ring.

Assignment of the antibiotic is greatly aided by natural abundance ^{13}C–^{1}H HMQC. With sample concentrations of approximately 3 m*M*, single-bond proton–carbon correlations are obtained with a sufficient SNR in 8–12 hr on a modern NMR spectrometer. The HMQC experiment disperses proton resonances by carbon chemical shift, which is indicative of the type of proton. Aromatic carbon atoms such as those on the purine and pyrimidine bases of nucleic acids have downfield chemical shifts (100–160 ppm).[12,21,26] The anomeric carbon of ribose and aminoglycoside sugars has the most downfield carbon chemical shift typically observed in the range of 85–100 ppm.[21,27–29] Methyl carbon atoms are typically the most upfield shifted carbon atom, resonating in the 20–40 ppm range.[26–29] The other carbon atoms in aminoglycoside antibiotics resonate between 40 and 90 ppm, similar to the C-2′, C-3′, C-4′, and C-5′ of RNA.[21] Separation of the proton resonances by type of proton in an HMQC experiment eliminates ambiguities in assignment of the different types of protons that overlap in ^{1}H chemical shift. The exchangeable proton resonances are often not observed due to rapid exchange with solvent.

[26] A. J. Gordon and R. A. Ford, "The Chemist's Companion: A Handbook of Practical Data, Techniques, and References." Wiley-Interscience, New York, 1972.

[27] D. G. Reid and K. Gajjar, *J. Biol. Chem.* **262,** 7967 (1987).

[28] L. Szilagyi and Z. S. Pusztahelyi, *Magn. Reson. Chem.* **30,** 107 (1992).

[29] G. I. Eneva and S. L. Spassov, *Spectrosc. Lett.* **28,** 69 (1995).

Forming RNA–Antibiotic Complex

NMR Titration of Antibiotic into RNA

Following completion of the assignment of the NMR resonances of each molecule free in solution, work can begin on the RNA–antibiotic complex. To form a complex, antibiotic is titrated into the RNA sample in water. The titration of the antibiotic into the RNA sample accomplishes three separate purposes. The first is to establish a drug–RNA complex suitable for further NMR experiments. The second purpose is to establish stoichiometry of the interaction, and the third purpose is to identify resonances that change on antibiotic binding.

The most downfield protons in the NMR spectrum, the NH (imino) protons of guanosine and uridine are easily followed by 1-D NMR, which makes them the most convenient marker for binding of the antibiotic to the RNA. Titration of the antibiotic paromomycin into an NMR sample of the A-site oligonucleotide (Fig. 3) yields an additional set of resonances,

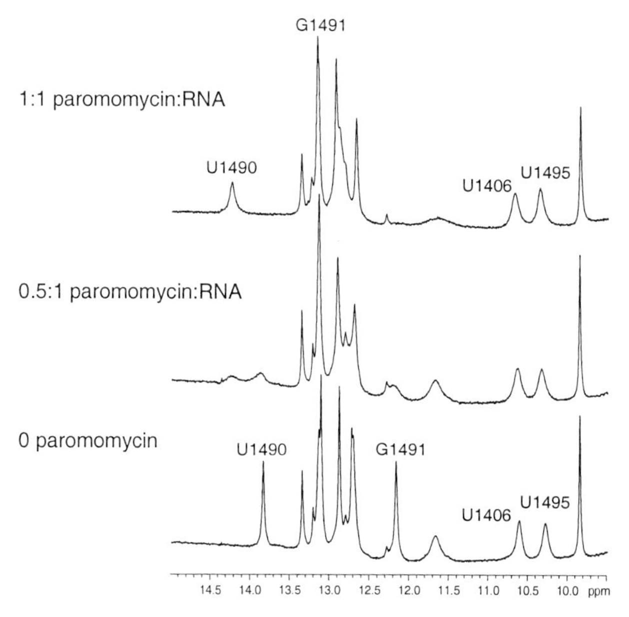

FIG. 3. Series of imino proton NMR spectra of the A-site RNA for free RNA, 0.5:1 paromomycin:RNA, and 1:1 paromomycin:RNA. Spectra were recorded at 25° in 10 m*M* sodium phosphate, pH 6.4, on a Varian Unity + 500-MHz spectrometer. Imino proton resonances that shift on addition of paromomycin are indicated, as are those of U1406 and U1495, which form a base pair.

which results from imino protons on the RNA bound to the antibiotic. As antibiotic is added, the resonances from RNA in the free form disappear, and resonances in the bound form appear. At a stoichiometry of 1 : 1 RNA to antibiotic, the only resonances observed are those from RNA imino protons from RNA bound to antibiotic.

In the case presented in Fig. 3, the stoichiometry of the interaction was 1 : 1. However, multiple drugs could bind to one RNA molecule. For example, the titration of distamycin into a DNA oligonucleotide demonstrated that two drug molecules bind one DNA helix.[30]

The third purpose of a drug–RNA titration is to identify protons that change chemical shift on binding. In the paromomycin-oligonucleotide titration, two residues exhibit large chemical shift changes on antibiotic binding. This information enables one to map where the drug is binding on the RNA without any additional NMR experiments. Despite the low resolution of these data, they suggest further experiments and answer important questions such as whether different drugs induce equivalent or different effects on the RNA. Unfortunately, binding site mapping might be the only information that can be determined with NMR if the resonances in the complex are broadened. Broad resonances can make proton assignment and the identification of critical intermolecular NOEs much more difficult. The broadening of resonances primarily results from two phenomena: intermediate exchange of the complex on the NMR timescale and slower tumbling due to the larger size of the complex relative to the oligonucleotide alone.

Affinity

The affinity of the RNA-antibiotic interaction is the critical factor affecting what can be determined by NMR. A high-affinity interaction will likely be amenable to high-resolution structure determination of the complex, whereas a lower affinity interaction may be amenable to only low-resolution characterization. The effect of affinity on NMR data can be understood by analyzing the relationship between binding constants and chemical exchange.

The chemical reaction for the antibiotic binding to the oligonucleotide is

$$\text{RNA} + \text{drug} \rightleftharpoons \text{RNA} \cdot \text{drug}$$

For this reaction the binding constant of association (K_A) is defined as:

$$K_A = \frac{[\text{RNA} \cdot \text{drug}]}{[\text{RNA}][\text{drug}]}$$

[30] J. G. Pelton and D. E. Wemmer, *Proc. Natl. Acad. Sci. U.S.A.* **86,** 5723 (1989).

where the [drug] is the concentration of free drug in solution and [RNA] is the concentration of RNA free in solution. The dissociation of the complex would be the inverse reaction, and thus the dissociation constant would be $1/K$. Thus, the expression for the dissociation constant, K_D would be

$$K_D = \frac{[\text{RNA}][\text{drug}]}{[\text{RNA} \cdot \text{drug}]}$$

It follows that the K_D would be equal to [drug] when 1/2 the sites on the oligonucleotide are occupied. A high-affinity interaction of RNA and drug would have a K_D on the order of 1 μM. In the NMR tube with an initial RNA concentration of approximately 2 mM, the free RNA and free drug concentrations would be almost zero for a high-affinity complex. However, complexes that have a K_D on the order of 2 mM would obviously be harder to study as half of the drug molecules would be free in solution at equilibrium.

The K_D can be related to the rate of a simple one-step reaction. For the reaction:

$$\text{RNA} + \text{drug} \rightleftharpoons \text{RNA} \cdot \text{drug}$$

$$K_D = \frac{k_{-1}}{k_1}$$

where k_1 is the rate of the forward reaction and k_{-1} is the rate of the reverse reaction.

In a diffusion-limited reaction, k_1 is approximately $10^7\ M^{-1}\text{s}^{-1}$.[31] Solving the equation for k_{-1}, using a value of 1 μM for K_D, results in a rate constant for the reverse reaction of approximately 10 per second or a time constant on 0.1 sec. A weaker affinity complex of 100 μM would therefore have a k_{-1} of approximately 1000 per second or a time constant of 1 ms. These results can be related to chemical exchange observed by NMR and line-widths for NMR resonances of the complex.

Chemical Exchange

Three possibilities exist for the rate at which the complex dissociates as observed by NMR: slow exchange, intermediate exchange, and fast exchange. In slow exchange, the dissociation rate of the complex is much slower than the frequency difference between the free and bound resonances; in other words, the time constant or lifetime is much larger for the dissociation of the complex. For example, a ^{1}H chemical shift change of

[31] A. Fersht, "Enzyme Structure and Mechanism." W. H. Freeman and Company, New York, 1985.

0.5 ppm for an RNA proton on binding to antibiotic on a 500-MHz (11.7 tesla) NMR spectrometer has a frequency difference of 250 Hz, or a time constant of 4 ms. A smaller chemical shift change of only 0.1 ppm would result in a time constant of 10 ms. As discussed earlier, the time constant for dissociation of a high-affinity (~1 μM K_D) complex is approximately 100 ms. Thus, high-affinity binding results in slow exchange. Slow exchange results in the observation of two resonances at a substoichiometric ratio of antibiotic to RNA,[32] one resulting from the free form and one from the bound form as observed in the titration shown in Fig. 3.

Intermediate exchange occurs when the dissociation rate is on the same timescale as the frequency difference of the two resonances. A time constant of 5–10 ms is approximately the timescale for the frequency difference between two resonances as determined earlier; 5 ms would translate into a K_D of approximately 50–100 μM. This type of exchange results in broadening of the resonances at the binding site.

The third type of exchange is fast exchange where the dissociation rate of the antibiotic–RNA complex is fast on the NMR timescale, less than a millisecond, which results from weak binding (>100 μM). This type of exchange results in the observation of one resonance at a chemical shift between the resonance in the free form and in the bound form. The chemical shift of the resonance reflects the ratio of RNA bound to antibiotic to that of free RNA. The resonances affected by antibiotic binding continue to shift until all RNA molecules in the solution are saturated with antibiotic, which usually occurs at greater than 1:1 antibiotic to RNA stoichiometry.

Slow exchange is the optimal situation for NMR studies of an RNA–antibiotic complex. Typically, high-affinity RNA–aminoglycoside complexes on the ribosome have dissociation constants on the order of 1–5 μM, whereas those of aminoglycoside–RNA aptamers are often 2 orders of magnitude smaller (~10 nM).[33] NMR studies on a tobramycin–RNA aptamer complex (K_D = 9 nM) show the aminoglycoside binding tightly in the major groove of the RNA.[23,34] High-affinity RNA aptamer–aminoglycoside complexes are in slow exchange making them amenable to NMR studies. Although it is possible to determine high-resolution structures of complexes not in slow exchange, it is much more difficult, so mapping the binding site may be the only realistic goal.

In cases of intermediate exchange, it is imperative that conditions of the NMR sample be changed to attempt to achieve slow exchange or fast exchange for the complex. NMR conditions that can be changed to escape intermediate exchange include salt concentration, pH, temperature, and

[32] H. Gunther, "NMR Spectroscopy." John Wiley & Sons, Chichester, UK, 1995.
[33] Y. Wang and R. R. Rando, *Chem. Biol.* **2,** 281 (1995).
[34] L. Jiang, A. K. Suri, R. Fiala, and D. J. Patel, *Chem. Biol.* **4,** 35 (1997).

TABLE I
PROTONS IN MAJOR AND MINOR GROOVES OF A-FORM RNA

Major groove	Minor groove
H-3′	H-1′
Pyrimidine H-5	H-4′
Pyrimidine H-6	H-5′
Purine H-8	H-5″
Cytidine H-4 (amino)	2′-OH
Adenosine H-6 (amino)	Adenosine H-2
	Guanosine H-2 (amino)

even sample concentration. The broad resonances resulting from intermediate exchange are difficult to assign, and few NOEs may be observed, leading to few distance constraints for structure calculations. Nonetheless, broadening due to intermediate exchange can reveal resonances at the antibiotic binding site.

Mapping of Binding Site of Antibiotic on RNA

Chemical Shift

The 1-D NMR titration data suggest RNA residues that are in proximity to the antibiotic in the complex. Chemical shift is sensitive to local electronic environment. A positively charged drug binding to the RNA induces a significant change in the local environment of protons on the RNA and drug, which is reflected in different chemical shifts in the complex. Extensive analysis of chemical shift effects requires complete resonance assignment of both the RNA and antibiotic through application of the same NMR techniques discussed earlier.

With complete assignments, difference of chemical shifts in the free and bound form can be analyzed. Specific resonances that change on complex formation are grouped by nucleotide and by type of proton. The nucleotides whose resonances are most affected often constitute the binding site on the RNA. The type of proton that is affected indicates where the drug is binding on each nucleotide. If the drug binds in the major groove, changes in chemical shift for the easily identified pyrimidine H-5 and H-6 protons and purine H-8 protons are expected. If the drug binds in the minor groove, changes in chemical shift for adenosine H-2 protons and ribose H-1′ protons are expected. The groups that face the minor and major grooves in A-form RNA are detailed in Table I.[35]

[35] W. Saenger, "Principles of Nucleic Acid Structure." Springer-Verlag, Berlin, 1984.

Whereas chemical shift is a sensitive method for analyzing the binding site, two problems exist for its exclusive use. First, conformational changes in the RNA structure on binding can also lead to changes in chemical shift for residues that are far from the binding site. A conformational change in the RNA on binding is an interesting result, but it is hard to distinguish by chemical shift alone between its effect and an effect of drug proximal to the binding site. Second, chemical shift is only low-resolution data. While theoreticians are trying to understand how to use chemical shift to determine tertiary structure, at this moment, chemical shift has not yet been successfully applied as a structural constraint for nucleic acids. Thus, more detailed analysis of the interaction must be applied to obtain a high-resolution structure.

Intermolecular NOEs

NOEs provide the most critical restraints for structure determination. Homonuclear and heteronuclear NOESY experiments applied to the RNA in the free form can also be applied to the complex. The determination of the structure of the RNA–antibiotic complex depends especially on intermolecular NOEs. Intramolecular NOEs can help define the structure of either of the individual components, but intermolecular NOEs define the conformation of the complex. Furthermore, the intermolecular restraints can act as long-range restraints for the RNA, actually improving the structure of the RNA oligonucleotide with the antibiotic acting as a rigid axis that fits into the oligonucleotide.

Intermolecular NOEs are distinguished from intramolecular NOEs by two methods. The first simply uses ^{1}H chemical shift to demonstrate whether an NOE is intramolecular or intermolecular. RNA has defined regions where protons normally resonate (4–5 ppm for ribose protons, 5.1–6.2 ppm for anomeric protons and pyrimidine H-5 protons, 7–8.5 ppm for aromatic protons, 6–9 ppm for amino protons, and 10–14.5 for imino protons). Similarly, antibiotics have defined ranges where protons resonate; in the case of aminoglycosides, that range is primarily 2–5 ppm for all protons except the anomeric protons, which resonate between 5 and 6 ppm. Thus, any NOEs observed from a proton with a chemical shift upfield of 4 ppm to a proton downfield of 7 ppm would be expected to be an intermolecular NOE. In addition RNA has many more protons in the 5–6 ppm range, so an NOE observed from a proton in that region to one in the 2–4 ppm range could also be easily identified as an intermolecular NOE. These intermolecular NOEs are identified in a simple 2-D ^{1}H–^{1}H NOESY. Thus, one can quickly identify the potential quality of a complex structure by scanning the NOESY spectrum of the complex in search of intermolecular NOES.

A second method for identification of intermolecular NOEs is through heteronuclear-filtered NOESY experiments. These experiments work by filtering out resonances resulting from either ^{13}C or ^{12}C (similarly ^{15}N or ^{14}N).[36–43] Although the sensitivity of X-filtered experiments is less than that of ^{1}H-only NOESY experiments, they can be invaluable in distinguishing between intermolecular and intramolecular NOEs. The X-filtered NOESYs work with a complex formed with uniformly $^{13}C/^{15}N$-labeled RNA bound to unlabeled drug; alternatively, the labeled drug bound to unlabeled RNA would also work, but labeled RNA is much easier to purify as discussed earlier. The NMR pulse sequence then filters out peaks resulting from the unwanted nucleus, so it is possible to have a spectrum with only NOEs from ^{13}C–^{13}C (RNA–RNA), ^{13}C–^{12}C (RNA–drug), or ^{12}C–^{12}C (drug–drug). The intermolecular (^{13}C–^{12}C) NOEs are the critical constraints necessary for positioning the drug onto the RNA structure during structure calculations.

Structure Calculations

After intermolecular and intramolecular NOEs are assigned and torsion angles are determined, the structure of the complex is calculated using a simulated annealing protocol followed by restrained molecular dynamics and energy minimization. The calculation is performed in the same way as RNA in the free form. Many different computer programs have been applied to structure refinement based on NMR data through application of a simulated annealing protocol followed by restrained molecular dynamics and energy minimization. These include X-PLOR,[44] CNS,[45] and NMRArchitect (Molecular Simulations Incorporated, San Diego, CA). In this laboratory, NMRArchitect and X-PLOR[44] have been applied.

[36] C. Zwahlen, P. Legault, S. J. F. Vincent, J. Greenblatt, R. Konrat, and L. E. Kay, *J. Am. Chem. Soc.* **119,** 6711 (1997).

[37] G. Otting and K. Wüthrich, *Q. Rev. Biophys.* **23,** 39 (1990).

[38] G. Otting and K. Wüthrich, *J. Magn. Reson.* **85,** 586 (1989).

[39] H. Kogler, O. W. Sorensen, G. Bodenhausen and R. R. Ernst, *J. Magn. Reson.* **55,** 157 (1983).

[40] M. Ikura and A. Bax, *J. Am. Chem. Soc.* **114,** 2433 (1992).

[41] P. J. M. Folkers, R. H. A. Rolmer, R. N. H. Konings, and C. W. Hibers, *J. Am. Chem. Soc.,* 3798 (1993).

[42] G. Genmecker, E. T. Olejniczak, and S. W. Fesik, *J. Magn. Reson.* **96,** 199 (1992).

[43] K. Ogura, H. Terasawa, and F. Inagaki, *J. Biomol. NMR* **8,** 492 (1996).

[44] A. T. Brünger, "X-PLOR Version 3.1. A System for X-ray Crystallography and NMR." Yale University, New Haven, Connecticut, 1993.

[45] A. T. Brünger, P. D. Adams, G. M. Clore, W. L. Delano, P. Gros, R. W. Grosse-Kunstleve, J.-S. Jiang, J. Kuszewski, M. Nilges, N. S. Pannu, R. J. Read, L. M. Rice, T. Simonson, and G. L. Warren, *Acta Crystallogr.* **D54,** 905 (1998).

The critical difference between structure calculations of the biomolecular complex and unimolecular systems is the ability to treat the two molecules separately and simultaneously. In that way, the structures of each individual components are not determined separately and then docked after refinement. Direct calculation of the complex structure is a more robust method for dealing with molecules with many degrees of freedom like the aminoglycoside antibiotics and RNA. NMRArchitect and X-PLOR both have that capability.

In X-PLOR, the two molecules need to be separately named with a file for each molecule defining the atoms connected by covalent bonds, and a file specifying the bond lengths and bond angles. These files are then converted into a structure file with all atoms having proper covalent geometry. The calculations then start from 100 structures with random torsion angles for both the RNA oligonucleotide and aminoglycoside. The random structures are subjected to a global fold protocol by first heating to 1000 K, while hydrogen bonding, distance, and dihedral constraints are gradually increased to full value over 40 ps of molecular dynamics.[46] The RNA and drug are subsequently cooled to 300 K for 10 ps and energy minimized in the presence of Lennard–Jones potentials, but in the absence of electrostatics. Electrostatics are not used in the calculation because the pK_a on each amino group is not known, and electrostatics may bias the calculated structure away from the observed NMR data. The total energies of the structures are analyzed, and the structures that converge to low energy are subjected to a structure refinement protocol. During refinement, the RNA–drug complex is heated to 1000 K for 30 ps of molecular dynamics with all hydrogen bonding, distance, and torsion angle constraints, and cooled to 300 K over 10 ps, again in the presence of Lennard–Jones potentials and without electrostatics. A final minimization step is performed in the same way. The resulting structures are finally superimposed to analyze the similarity of the structures.

Building on Structure

Comparison with Chemical Modification Data

Once the structure has been solved by NMR spectroscopy it can be compared to biochemical data. Chemical probing experiments have identified specific atoms that are modified in the absence of aminoglycoside and are protected from chemical modification in the presence of the antibiotic. Protected atoms should be buried in the structure, most likely at the RNA–

[46] D. Fourmy, S. Yoshizawa, and J. D. Puglisi, *J. Mol. Biol.* **277,** 333 (1998).

antibiotic interface. Atoms that are modified in footprinting analysis of the complex would be exposed to solvent in the structure. Atoms could also be enhanced in modification, although no atoms in the antibiotic–oligonucleotide complex studied in this laboratory showed enhancement. These atoms should have greater solvent accessibility in the complex than in the free form.

Chemical probing experiments demonstrated that the model system mimics the aminoglycoside–rRNA interaction. If the structure does not agree with probing data, then the 3-D structure may be different under biochemical and NMR experimental conditions. Conditions ideally would be the same in both types of experiments. However, NMR requires sample concentrations on the order of 1 m*M*, whereas biochemical experiments require orders of magnitude lower concentrations of RNA (less than 1 μM). In addition, normal buffer conditions are different between the two sets of experiments, because biochemical experiments are often done at higher pH than those of NMR. Higher pH conditions in NMR experiments lead to faster solvent exchange of water-exchangeable protons and faster sample degradation over long time periods. Experimental conditions that mimic the biochemical experiments as closely as possible should then be tested by 1-D NMR and 2-D NOESY to determine whether the structure is truly different at different conditions. If experimental conditions have not led to different structures, then the data should be reanalyzed for possible discrepancies in interpretation.

Comparison with Mutational Data

The structure must be compared to results from mutational and modification-interference experiments. Important contacts defined in the structure should involve nucleotides that, when mutated or chemically modified, interfere with the binding of the antibiotic to the oligonucleotide.[5,47] The antibiotic makes specific contacts to the conserved nucleotides of the internal loop. Nonspecific contacts are made to the phosphate backbone of nonconserved nucleotides in the lower stem.

Biochemical data indicated that specific nucleotides in the aminoglycoside binding site could be replaced by only one or two of the other three nucleotides with no effect on affinity. Ideally, the structure can explain these results. The location of a specific functional group on the base may be critical for binding of the antibiotic to the RNA. For example, the position 1495 of the A-site oligonucleotide (Fig. 1) can be G or U and still bind paromomycin with high affinity, since both sequences present a

[47] S. C. Blanchard, D. Fourmy, R. G. Eason, J. D. Puglisi, *Biochemistry* **37,** 7716 (1998).

carbonyl at this position.[5] The structure of this complex shows that a carbonyl at this position forms a hydrogen bond to an amino group on the antibiotic, in agreement with the mutational result.[48]

Testing Biological Predictions by NMR

If the biochemical and NMR data agree, then interesting features of the RNA–antibiotic complex are analyzed. In Fig. 4 (see color insert), two structures determined in this laboratory of the RNA oligonucleotide presented in Fig. 1B in its free form and complexed to the aminoglycoside antibiotic paromomycin are presented.[48] In the 7-nucleotide asymmetric internal loop, three base pairs were identified: a U1406–U1495 pair, a Watson–Crick C1407–G1494 pair, and an A1408–A1493 pair. In addition, the N-1 atoms of A1492 and A1493 are pushed toward the minor groove, exposed to solvent, which is consistent with a model of hydrogen bonding of 2′-OH protons on the mRNA to A1492 and A1493.[48] The antibiotic binds in the major groove, which is deep but narrow in RNA helices.[35] This observation suggests a testable model of minor groove recognition of the mRNA by the rRNA since aminoglycosides cause misreading but do not competitively inhibit tRNA binding. While solution structures of RNA–antibiotic complexes can yield many interesting results, a structure of the complex is not the end of the project, because the structure will suggest new biochemical and structural experiments.

Use of Mutant Oligonucleotides

The 3-D structure of an RNA–antibiotic complex indicates which nucleotides on the oligonucleotide interact with specific positions on the drug. Mutations that should have no effect on binding of the antibiotic can be incorporated in the oligonucleotide. Chemical shift changes in the mutant RNA can be analyzed by 1-D titration and intermolecular NOEs can be analyzed by 2D NOESY and compared to the wild-type sequence. Mutations that affect binding are analyzed similarly by NMR. The intermolecular NOEs reveal whether the interaction with the antibiotic is disrupted only at the site of the mutation or throughout the RNA molecule.

Point mutants are often readily studied by NMR. A single base change is likely to affect only the chemical shift of resonances in the immediate vicinity of that nucleotide. In an RNA helix, nucleotides that are more than two base pairs away would not change in chemical shift unless the global structure of the mutant is different. Thus, the spectrum of the mutant

[48] D. Fourmy, M. I. Recht, S. C. Blanchard, and J. D. Puglisi, *Science* **274,** 1367 (1996).

oligonucleotide is mostly unchanged from that of the wild-type oligonucleotide. Nonetheless, analysis of mutant RNA–drug complexes is often difficult. The K_D for the aminoglycoside–RNA mutant complex is often higher, leading to intermediate chemical exchange rather than slow exchange as observed for the wild-type interaction. The NMR resonances become broader and harder to assign. Additionally, fewer intermolecular NOEs are observed for the weaker interaction, which makes high-resolution structure determination too difficult.

One biologically relevant mutation of the decoding site is A1408G. Position 1408 is not a universally conserved nucleotide. It is a guanosine in all eukaryotic ribosomes and an adenosine in all prokaryotic ribosomes. In the structure discussed earlier, A1408 forms a base pair with A1493. G1408 cannot replace A1408 in the same orientation and still form a base pair with A1493. Biochemical data demonstrated that this mutation weakens aminoglycoside binding affinity, as ring I sits in a binding pocket formed by the A1408 · A1493 base pair. Structure analysis of this mutant will further show how a guanosine at position 1408 can produce a functional ribosome and confer resistance to aminoglycosides.

Studies with Different Antibiotics

Although the various aminoglycosides contain some conserved chemical groups on rings I and II, variations both in the substituents on ring I and the total number of rings in the drug do exist. Studying the interaction of a variety of aminoglycosides with the oligonucleotide can reveal which of these groups are critical for the high-affinity binding of the drug to the RNA and if a common mode of interaction for the various aminoglycosides exists. Complexes of the RNA oligonucleotide with of a series of antibiotics similar to paromomycin have been studied.[49] Neamine, which consists of only rings I and II of neomycin, binds specifically to the oligonucleotide and has antibacterial activity.[50] It binds with lower affinity to the RNA, presumably because it lacks the additional contacts observed in the structure afforded by the presence of rings III and IV in neomycin and paromomycin. The lower affinity interaction of neamine with the oligonucleotide causes a more dynamic complex in fast exchange, making this complex harder to study by NMR. However, many of the chemical shift changes and intermolecular NOEs observed for the paromomycin complex are also observed for the neamine complex. Thus, neamine is still capable of recognizing

[49] D. Fourmy, M. I. Recht, and J. D. Puglisi, *J. Mol. Biol.* **277,** 347 (1998).

[50] J. Woodcock, D. Moazed, M. Cannon, J. Davies, and H. F. Noller, *EMBO* **10,** 3099 (1991).

the critical nucleotides in the RNA sequence and inducing the important conformational changes.

This laboratory has determined 3-D structures of the oligonucleotide bound to either paromomycin or gentamicin as shown in Fig. 5 (see color insert).[45,51] The primary difference between these two antibiotics is the linkage between rings II and III; ring III of gentamicin is linked to position 6 of ring II but this linkage is at position 5 in paromomycin.

The two structures determined by NMR show that rings I and II of the antibiotic interact with the RNA in a similar manner. A network of intermolecular hydrogen bonds lines the pocket in the major groove where the antibiotic recognizes the RNA oligonucleotide. The features on the RNA critical for recognition of antibiotic, including the U1406–U1495 base pair and the A1408–A1493 base pair, are observed in both structures. Ring III in gentamicin makes additional base specific contacts to G1405, which are not observed in the paromomycin–RNA complex. One of the most striking features is the similar conformational change observed between the free RNA and RNA bound with either paromomycin or gentamicin. On binding of either antibiotic, the N-1 of both A1492 and A1493 are displaced toward the minor groove. The structures of the two different classes of aminoglycosides have revealed how common chemical groups on the drug contribute to specific binding to the ribosomal target, and how the binding of either drug induces a similar conformational change in the RNA.

Drug Design

An obvious application of the RNA–antibiotic structures is structure-based drug design. Both the elements of the antibiotic that are critical for binding and the shape of the binding site on the RNA are known. Molecular modeling can be used to design novel drugs that contain the required chemical groups but which are attached to an alternate scaffold. Likewise, drugs could be designed to include the aminoglycoside scaffold but contain additional functional groups to increase specificity. These drugs can be made via organic synthesis and tested both for binding to the ribosomal target and antibiotic activity. Drugs that possess higher affinity and specificity for their ribosomal target can be refined and further tested. Although structure-based drug design does not always work, the potential for identifying new drugs in this manner is possible with the increasing number of structures of RNA–drug complexes and with the improving resolution of these structures.

[51] S. Yoshizawa, D. Fourmy, and J. D. Puglisi, *EMBO J.* **17,** 6437 (1998).

Conclusion

Structure determination of RNA–antibiotic complexes by NMR provides a way to understand how an RNA sequence defines its structure, how a specific RNA structure relates to its biological function, and how a drug can be designed to fit into that structure. The methodology for RNA structure determination by NMR is now well established and can be applied to RNA–antibiotic complexes. Although an RNA free in solution can be quite dynamic, and sometimes difficult to study by NMR, high-affinity complexes of RNA oligonucleotide and antibiotic are less dynamic. With the vast improvement in NMR technology in the past few years, structures of RNA free in solution and in complexes with small ligands including aminoglycoside antibiotics will greatly increase in both number and quality.

[17] Experimental Prerequisites for Determination of tRNA Binding to Ribosomes from *Escherichia coli*

By FRANCISCO J. TRIANA-ALONSO, CHRISTIAN M. T. SPAHN, NILS BURKHARDT, BEATRIX RÖHRDANZ, and KNUD H. NIERHAUS

Introduction

The tRNA binding features of 70S ribosomes from *Escherichia coli* have been extensively studied by several research groups. It has been established that these ribosomes contain three tRNA binding sites. The A site, which accepts the aminoacyl-tRNA during the first step of the elongation phase in protein biosynthesis, can also accept peptidyl-tRNA (or peptidyl-tRNA analogs); the P site, where the three possible forms of a tRNA, peptidyl-, aminoacyl-, or deacyl-tRNA can bind; and the E site, which shows a strong specificity for deacylated tRNA.[1–5] The functional links between the three sites have been also established for the elongation phase of protein biosynthesis. The growing peptide chain is prolonged by one amino acid via three basic reactions. (1) the first reaction is the occupation of the A site by an aminoacyl-tRNA, which separates into a selection

[1] H.-J. Rheinberger and K. H. Nierhaus, *Biochem. Intl.* **1,** 297 (1980).
[2] H.-J. Rheinberger, H. Sternbach, and K. H. Nierhaus, *Proc. Natl. Acad. Sci. U.S.A.* **78,** 5310 (1981).
[3] R. A. Grajevskaja, Y. V. Ivanov, and E. S. Saminksy, *Eur. J. Biochem.* **128,** 47 (1982).
[4] S. V. Kirillov, E. M. Makarov, and Y. P. Semenkov, *FEBS Lett.* **157,** 91 (1983).
[5] R. Lill, J. M. Robertson, and W. Wintermeyer, *Biochemistry* **23,** 6710 (1984).

0076-6879/00 $30.00

step and a tight-binding step. The selection process analyzes mainly the correctness of codon–anticodon interaction.[6] This step occurs at the decoding center that is located on the small ribosomal subunit and is part of the A site. The binding substrate is a ternary complex of aminoacyl-tRNA · EF-Tu · GTP. During the occupation of the A site the GTPase center of the elongation factor Tu cleaves the GTP and the binary complex EF-Tu · GDP leaves the ribosome. (2) After occupation of the A site, peptide-bond formation occurs: the peptidyl moiety attached to the P-site-bound tRNA is transferred to the free amino group of the aminoacyl-tRNA at the A site via a peptide bond. This reaction is catalyzed by the peptidyltransferase center located on the large subunit of the ribosome. (3) The third reaction is the EF-G · GTP–dependent translocation of the new peptidyl-tRNA from the A to the P site. Simultaneously the deacylated tRNA bound to the P site is cotranslocated to the E site.

This functional description clearly implies that during protein biosynthesis only one growing peptidyl-tRNA can reside on the ribosome, oscillating between the A and the P sites. Accordingly, tRNA saturation experiments performed *in vitro* indicated that only one peptidyl-tRNA analog, the *N*-acetyl-Phe-tRNA^{Phe} could be bound to poly(U) programmed 70S ribosomes from *E. coli.* In contrast, the binding of aminoacyl-tRNA and deacylated tRNA saturate at two and three molecules per ribosome, respectively, in the presence of poly(U).[2,7] Essentially the same features were found for archebacterial[8] and eukaryotic ribosomes.[9,10] This remarkable discrimination power of the ribosome has been formulated as an "exclusion principle" for the binding of peptidyl-tRNA or peptidyl-tRNA analogs to programmed 70S. It is thought that the exclusion principle reflects the existence of one channel or tunnel for the nascent peptide chain. If the tunnel is occupied by one peptidyl-tRNA, a second peptidyl-tRNA cannot bind anymore to the ribosome, although two binding sites for the tRNA moiety exist, namely, the A and the P site.[2,11]

Reports from other research groups dealing also with the tRNA binding capabilities of *E. coli* ribosomes, coincide with our results with respect to both saturation with deacylated tRNA and the maximum binding of aminoacyl-tRNA. However, saturation values of AcPhe-tRNA^{Phe} in the

[6] A. P. Potapov, F. J. Triana-Alonso, and K. H. Nierhaus, *J. Biol. Chem.* **270,** 17680 (1995).

[7] S. Schilling-Bartetzko, F. Franceschi, H. Sternbach, and K. H. Nierhaus, *J. Biol. Chem.* **267,** 4693 (1992).

[8] H. Saruyama and K. H. Nierhaus, *Mol. Gen. Genet.* **204,** 221 (1986).

[9] F. Triana, K. H. Nierhaus, and K. Chakraburtty, *Biochem. Mol. Biol. Intl.* **33,** 909 (1994).

[10] F. J. Triana-Alonso, M. Dabrowski, J. Wadzack, and K. H. Nierhaus, *J. Biol. Chem.* **270,** 6298 (1995).

[11] U. Geigenmüller, T. P. Hausner, and K. H. Nierhaus, *Eur. J. Biochem.* **161,** 715 (1986).

presence of poly(U) programmed ribosomes from *E. coli* were reported to approach two molecules per ribosome.[5,12] The reasons for this discrepancy are analyzed in the present work and protocols are reported that allow an unequivocal determination of the maximal binding number of AcPhe-tRNAs, a simple analog of a peptidyl-tRNA. Three experimental parameters might explain the conflicting reports: (1) quality of the tRNA preparations, (2) specific activity of the radioactive label used to follow the tRNA binding, and (3) mode of preparation of the 70S ribosomes.

Prerequisites: Preparation of S100 Enzymes, Acylated tRNAs, and tRNA Binding to Ribosomes

The isolation of tightly coupled ribosomes and crude S100 enzymes still containing RNAs such as tRNAs followed Bommer *et al.*[13] and the isolation of reassociated ribosomes from ribosomal subunits are described by Blaha *et al.*[14]

S100 Enzymes Freed of tRNAs

The postribosomal fraction containing the aminoacyl-tRNA synthetase activities (crude S100 enzymes)[13] is concentrated and made free of tRNA by batch treatment with DEAE-cellulose of the supernatant. This treatment also removes most of the RNases. The following procedure is described for 150 ml crude S100 enzymes in 70S buffer $H_{20}M_6N_{30}SH_4$.

Fifteen grams of DE-52 cellulose (Whatman, Clifton, NJ) is resuspended in prewarmed $H_{20}M_{10}K_{500}$ (20 m*M* HEPES–KOH, pH 7.6, at 0°, 10 m*M* magnesium acetate and 500 m*M* KCl; 90°) and incubated at 90° for 30 min. The supernatant is decanted from the sedimented cellulose matrix. This procedure is repeated twice with $H_{20}M_{10}K_{500}$ at 90° and thrice with $H_{20}M_{10}K_{150}$ at room temperature. After the last extraction the supernatant should have a pH between 7.0 and 8.0 and should be removed as completely as possible. The cellulose matrix is left for ≥2 hr (e.g., overnight) at 4° and then mixed with 150 ml of crude S100 enzymes. The suspension is left at 0° for 2 hr under occasional swinging and then centrifuged at 10,000*g* at 4° for 30 min (e.g., Dupont HB-4 rotor). The supernatant is fraction I, the pelleted matrix is treated sequentially with 40 ml of $H_{20}M_{10}K_{150}$,

[12] S. V. Kirillov and Y. P. Semenkov, *FEBS Lett.* **148,** 235 (1982).

[13] U. Bommer, N. Burkhardt, R. Jünemann, C. M. T. Spahn, F. J. Triana-Alonso, and K. H. Nierhaus, *in,* "Subcellular Fractionation. A Practical Approach," (J. Graham and D. Rickwoods, eds.), p. 271. IRL Press, Washington, DC, 1996.

[14] G. Blaha, U. Stelzl, C. M. T. Spahn, R. K. Agrawal, J. Frank, and K. H. Nierhaus, *Methods Enzymol.* **317,** [19], 2000, (this volume).

$H_{20}M_{10}K_{200}$, and $H_{20}M_{10}K_{500}$, an equilibration of 2 hr is followed by a centrifugation step in each case. The resulting fractions II, III, and IV are dialyzed four times against 2 liters of ribosome buffer $H_{20}M_6N_{30}SH_4$ for 45 min each at 4° and centrifuged for 30 min at 10,000*g* and at 4° to remove residual matrix. The buffer can contain 5–10% of glycerol to preserve activity during storage. The fractions are tested analytically for aminoacylation activity (see next section). Normally fractions II and III contain the highest activities; they are aliquotized (100-μl samples) and stored at −80°.

Acylation of tRNA^{Phe} with Phe and Acetylation of Phe-tRNA Yielding AcPhe-tRNA

Materials. tRNA^{Phe} (*E. coli* MRE 600) is purchased from Boehringer Mannheim. [^{14}C]Phe (465 mCi/mmol), and [^{3}H]Phe (4.6 Ci/mmol) are from Amersham-Buchler, Braunschweig. Adenosine triphosphate (ATP) and most of the chemicals are from Sigma (Deisenhofen, Germany).

Method. The acylation procedure is the same whether [^{3}H]Phe or [^{14}C]Phe is used. For example, [^{3}H]Phe-tRNA^{Phe} is prepared by incubation of 10–50 A_{260} units of tRNA^{Phe} with five-fold molar excess of [^{3}H]phenylalanine in 5–10 ml of 20 m*M* HEPES, pH 7.6, at 0°, 100 m*M* KCl, 10 m*M* magnesium acetate, 3.5 m*M* ATP, 5 m*M* 2-mercaptoethanol, and an optimal amount of an S100 enzyme preparation freed of tRNA (usually about 400 μg of protein per 10 A_{260} units of tRNA). After 15 min of incubation at 37°, the reaction mixture is extracted with one vol of 70% phenol and the [^{3}H]Phe-tRNA recovered by ethanol precipitation before the final purification step (next paragraph). Ac[^{3}H]Phe-tRNA^{Phe} is prepared by acetic anhydride treatment of nonpurified [^{3}H]Phe-tRNA^{Phe}[15]: Acetic anhydride is added (1/30th of the sample volume) and the mixture incubated at 0° for 15 min. This procedure is repeated three times. The AcPhe-tRNA is precipitated with two vol of ethanol and resuspended in H_2O. When indicated, the AcPhe-tRNA is made free of Phe-tRNA contamination via selective enzymatic deacylation (see later section) before the purification described in the next paragraph.

AcPhe-tRNA^{Phe} and Phe-tRNA^{Phe} are made free of nucleotides and other low molecular weight materials by a gel-filtration step performed on NAP-25 columns (Pharmacia, Piscataway, NJ) before purification by reversed-phase HPLC on a Nucleosil 300-5 C_4 column using a programmed binary gradient of buffers A (20 m*M* ammonium acetate, pH 5.0, 10 m*M* magnesium acetate, 400 m*M* NaCl) and B (60% v/v methanol in buffer A).

Analytical aminoacylation assays are performed with 0.01–0.03 A_{260}

[15] A. Haenni and F. Chapeville, *Biochim. Biophys. Acta* **114,** 135 (1966).

units of $tRNA^{Phe}$ in the presence of 15-fold mol/mol [^{14}C]Phe in 30 μl of reaction mix under the conditions, stated earlier. After a 15-min incubation at 37°, the amount of Phe incorporated is determined by precipitation with ice-cold trichloroacetic acid (10%), filtration through glass-fiber filters, and counting in a scintillation counter.

Site-Specific tRNA Binding to 70S Ribosomes

Materials. Poly(U) was from Pharmacia.

Method. The tRNA binding to 70S ribosomes has been described in detail[13,14] and is therefore only briefly summarized specifically addressing the binding assays described here.

For P-site saturation experiments with AcPhe-$tRNA^{Phe}$, 20 pmol of tightly coupled 70S ribosomes is incubated in 50-μl aliquots of binding buffer $H_{20}M_6N_{150}Spd_2Spm_{0.05}SH_4$ (20 m*M* HEPES, pH 7.6, at 0°, 6 m*M* magnesium acetate, 150 m*M* NH_4Cl, 2 m*M* spermidine, 0.05 m*M* spermine, 4 m*M* 2-mercaptoethanol) with 200 μg of poly(U) for 10 min at 37°. The volume is adjusted to 100 μl by addition of the indicated amounts of Ac[^{14}C]Phe-$tRNA^{Phe}$ (or a mix of Ac[^{14}C]Phe-$tRNA^{Phe}$, 1030 dpm/pmol, and [^{3}H]Phe-$tRNA^{Phe}$, 1900 dpm/pmol) in binding buffer $H_{20}M_6N_{150}Spd_2Spm_{0.05}SH_4$ and incubated for 20 min at 37°. The binding is determined by nitrocellulose filtration. Control assays without ribosomes yielded the filter backgrounds that are subtracted from the binding data. The A-site saturation experiments are performed in a similar way including 25–30 pmol of deacylated $tRNA^{Phe}$ (1500 pmol/A_{260} unit) in the first incubation containing poly(U) and ribosomes. The radioactivity measurements and the corresponding isotope discrimination in double label experiments are done using a LKB Rack Beta 1219 liquid scintillation counter (Pharmacia).

The Problem: Two Different Artifacts Can Distort Saturation Curves of Binding AcPhe-tRNA to Poly(U) Programmed Ribosomes

We have identified two different kinds of deviations of AcPhe-tRNA binding to tight-coupled 70S ribosomes from *E. coli* in the presence of poly(U). The first case is related to the quality of the AcPhe-tRNA preparation, the second deals with the specific activity of the radioactive label used to follow the binding. The results presented in Fig. 1 are representative of these cases. Figure 1A shows the results of an AcPhe-tRNA binding experiment where the saturation curve deviates from the shape, observed in previous reports,[2,7,11] in the following aspects:

1. A maximum of binding followed by a moderate decrease is often observed between 2 and 3 AcPhe-tRNA added per ribosome.

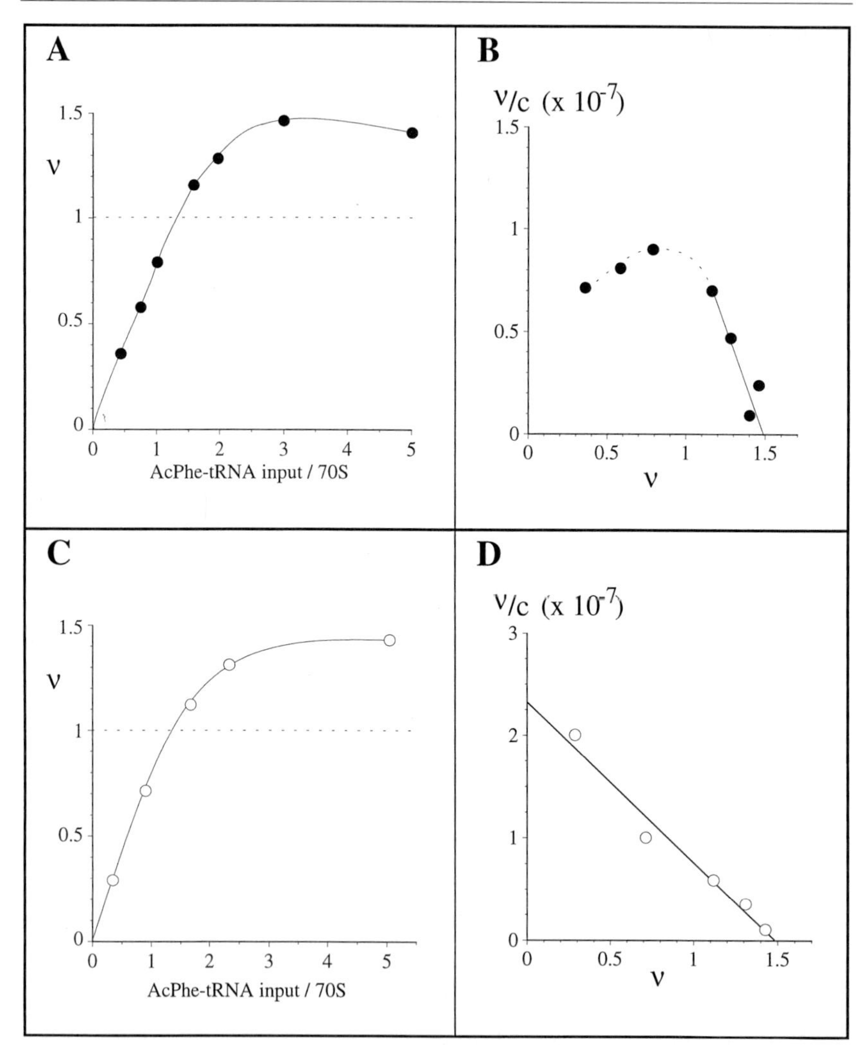

FIG. 1. Ac[^{14}C]Phe-tRNA^{Phe} saturation of poly(U) programmed 70S ribosomes from *E. coli.* (A) Saturation curve with an AcPhe-tRNA containing a Phe-tRNA contamination. The indicated amounts of Ac[^{14}C]Phe-tRNA^{Phe} (1030 dpm/pmol) were added to the binding assay containing 10 pmol of ribosomes in a final volume of 50 μl. The binding was determined by nitrocellulose filtration. (B) Scatchard plot of the binding data presented in (A) (C) and (D). Corresponding experiments with an Ac[^{14}C]Phe-tRNA^{Phe} preparation with an incorrect specific activity.

2. The apparent saturation is variable but always significantly higher than the upper limit of 1.0 ± 0.1 predicted by the exclusion principle (the apparent ν_{max}, the number of tRNAs bound per ribosome, ranges from 1.2 to 1.6).
3. The saturation curve often shows an inflection point at low AcPhe-tRNA input (in the range of 0.2 to 1.0 AcPhe-tRNA added per 70S; not clearly seen in Fig. 1A).

Also in contrast with the data presented in previous reports,[2,7] the analysis of the binding data according to Scatchard[16] showed a maximum indicating an apparent positive cooperativity in the binding process (Fig. 1B). We demonstrate in the next section that this type of artifact is caused by the presence of residual amounts of Phe-tRNA in the AcPhe-tRNA preparation.

Another type of artifact is demonstrated in Fig. 1C that was observed with an AcPhe-tRNA preparation that had been freed of residual amounts of Phe-tRNA with a procedure described in the next section. This Ac[^{14}C]-Phe-tRNA^{Phe} batch saturated poly(U) programmed ribosomes at 1.4 molecules per ribosome. Interestingly, the saturation curve showed none of the abnormal features described earlier, and the Scatchard plot analysis does not show evidence of positive cooperativity (Fig. 1D). The explanation for this deviation was found when the specific activity of the [^{14}C]phenylalanine label used to follow the binding was checked systematically using a test where its biochemical properties are closely involved in the measurements.

Phe-tRNA Contamination Simulates Anomalous Binding of AcPhe-tRNA to Poly(U) Programmed Ribosomes

To demonstrate whether a Phe-tRNA contamination in the AcPhe-tRNA can be responsible for abnormalities in the observed binding, a double-label experiment using a mixture of pure Ac[^{14}C]Phe-tRNA^{Phe} and [^{3}H]Phe-tRNA^{Phe} is performed. Figures 2A and 2B show the results obtained in a parallel ribosome saturation experiment with a pure Ac[^{14}C]Phe-tRNA^{Phe} preparation and one "contaminated" with [^{3}H]Phe-tRNA^{Phe} (the Phe-tRNA addition is adjusted to 16% of the AcPhe-tRNA in the mix). When the amounts of Phe-tRNA and AcPhe-tRNA bound to the ribosome are added (filled circles), the saturation curve shows the previously detected anomalies (Fig. 1A), i.e., the saturation level is significantly higher than 1 tRNA per ribosome (about 1.3 in this case), and a tendency to an inflection at low tRNA input is visible, indicating a complex binding process. More-

[16] G. Scatchard, *Ann. N.Y. Acad. Sci.* **51,** 660 (1949).

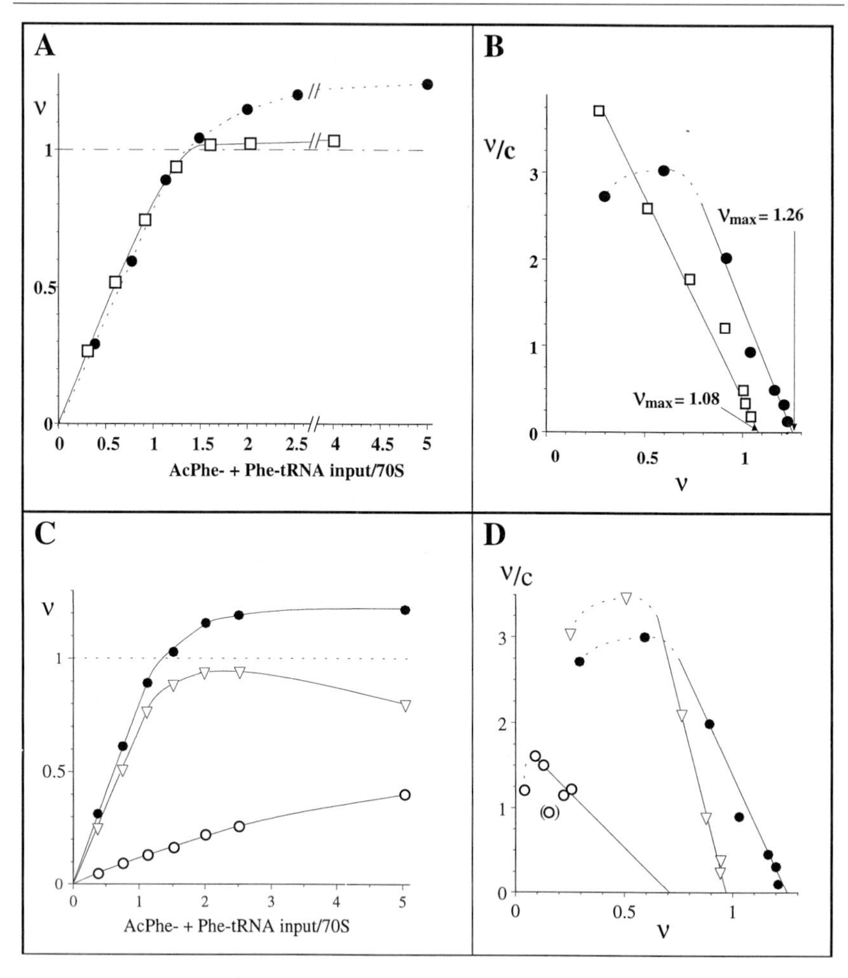

FIG. 2. Binding of AcPhe-tRNA^{Phe} to poly(U) programmed 70S ribosomes in the presence and absence of a Phe-tRNA^{Phe} contamination. The binding assays were performed as described before using the indicated amounts of purified Ac[^{14}C]Phe-tRNA^{Phe} (1050 dpm/pmol) freed of a Phe-tRNA contamination or a mix of this preparation with high-performance liquid chromatography (HPLC) purified [^{3}H]Phe-tRNA^{Phe} (1900 dpm/pmol; 16% of the AcPhe-tRNA content). (A) Binding curve of the purified Ac[^{14}C]Phe-tRNA^{Phe} (□) and the mixture of Ac[^{14}C]Phe-tRNA^{Phe} and [^{3}H]Phe-tRNA^{Phe} (●). (B) Scatchard plots of the data presented in (A) using the same symbols. (C) and (D), Presentation of the binding curves (C) and the Scatchard plot analysis (D) of the individual radioactive compounds of the double-label experiment with the mixture of Ac[^{14}C]Phe-tRNA and [^{3}H]Phe-tRNA. The binding curves and the Scatchard plots are presented for the combined AcPhe- and Phe-tRNA (●) as well as for the AcPhe-tRNA (▽) and Phe-tRNA (○) binding separately, respectively.

over, the Scatchard plot shows clear evidence of a positive cooperativity in the binding process. In contrast, the control experiment using only pure AcPhe-tRNA shows again a saturation close to 1 AcPhe-tRNA per ribosome and the Scatchard plot is consistent with a single binding site.

The individual binding curves for Ac[^{14}C]Phe-tRNA and [^{3}H]Phe-tRNA derived from the double-label experiment of Fig. 2A (filled circles, AcPhe-tRNA plus Phe-tRNA) show several interesting features (Figs. 2C and 2D). The ^{14}C-binding data show a maximum at an input of 2–2.5 acylated tRNAs per 70S (open triangles in Figs. 2C and 2D; saturation at 0.94 pmol of AcPhe-tRNA/pmol of 70S). After the maximum, the binding started to decrease smoothly. In contrast, the ^{3}H-binding data show a constant tendency to increase (open circles in Figs. 2C and 2D). This binding tendency strongly suggests the formation of peptides during the time of incubation. The fact that dipeptidyl- and tripeptidyl-tRNA formed on the ribosome are prone to dissociate from the ribosome in contrast to longer peptides[17] can explain the slight decrease of the ^{14}C-binding. The Scatchard plot for [^{3}H]Phe-tRNA (open circles) cannot be evaluated since the points are clustered. The plots for both AcPhe (triangles) and (AcPhe + Phe; filled circles) indicate positive cooperativity. The former intersects the *x* axis near 1 demonstrating one binding site for AcPhe, the latter intersects at a significant higher value of 1.26.

Removal of Phe-tRNA Contamination from AcPhe-tRNA Preparation: Enzymatic Deacylation of Aminoacyl-tRNA

Selective deacylation of [^{14}C]Phe-tRNA is accomplished by reversion of the aminoacylation reaction catalyzed by synthetases present in S100 enzymes in the presence of AMP and pyrophosphate.[18–20]

The analytical assays are performed in 50 μl of deacylation mix containing 50–100 pmol of aminoacyl-tRNA, 5–10 μg of an S100 preparation freed of tRNA in 30 m*M* HEPES, pH 7.6, 10 m*M* magnesium acetate, 100 m*M* KCl, 5 m*M* 2-mercaptoethanol, 6 m*M* adenosine monophosphate (AMP), and 6 m*M* pyrophosphate (PP_i, added from a freshly prepared 100 m*M* Na_4PP_i stock solution). The PP_i is always added shortly before the S100 enzymes in order to reduce the risk of depletion by precipitation with Mg^{2+}. After incubation for 5 min (or for the indicated times) at 30°, the amount of remaining aminoacyl-tRNA is determined via cold TCA precipi-

[17] H. J. Rheinberger and K. H. Nierhaus, *Eur. J. Biochem.* **193,** 643 (1990).
[18] P. Berg, F. H. Bergmann, E. J. Ofengand, and M. Dieckmann, *J. Biol. Chem.* **236,** 1726 (1961).
[19] T. Ishida, J. L. Arceneaux, and N. Sueoka, *Methods Enzymol.* **20,** 98 (1971).
[20] W. M. Stanley, *Methods Enzymol.* **29,** 530 (1974).

tation of 10-μl samples of the incubation mixture. Figure 3A demonstrates deacylation kinetics of [^{3}H]Phe-tRNA. After 2 min, more than 90% of the Phe-tRNA is cleaved.

Preparative deacylation assays are performed under the same conditions increasing the tRNA concentration to 4 pmol per microliter of deacylation reaction mixture. In this case it is important to eliminate the ATP remaining from the aminoacylation step by a previous gel filtration on a NAP-25 column.

Comment. The Ac[^{14}C]Phe-tRNA preparation that yielded more than 1.4 tRNAs bound per ribosome as shown in Figs. 1A and 1B was subjected to the enzymatic deacylation procedure (Fig. 3B). About 40% of the input tRNA could be deacylated; the process was completely dependent on the presence of the S100 enzymes and grossly dependent on the presence of pyrophosphate (PP_i). Obviously, AcPhe-tRNA is resistant to the deacylation reaction as demonstrated in Fig. 3C. An analytical deacylation of the AcPhe-tRNA before and after the preparative deacylation demonstrates that after the preparative step the remaining AcPhe-tRNA is resistant against the enzymatic deacylation step (less than 2% compared to about 40% before deacylation). We routinely tested various AcPhe-tRNA preparations and found that 10–40% of the input tRNA could be deacylated.

When poly(U) programmed ribosomes were saturated with the Ac[^{14}C]-Phe-tRNA preparation deacylated as shown in Fig. 3B, both the binding curve and the Scatchard plot demonstrated a maximal binding of about 1 AcPhe-tRNA per ribosome, an affinity constant of $K_a = 3.9 \times 10^7\ M^{-1}$ (Fig. 4, open circles), and no indications of a positive cooperativity. The binding of this AcPhe-tRNA to the A site yielded a straight Scatchard plot as well, with a maximal binding of about 0.8 and an association constant of $K_a = 6.2 \times 10^6\ M^{-1}$. These results are in good agreement with corresponding data collected at 10 m*M* Mg^{2+} and 160 m*M* NH_4Cl in the absence of polyamines.[7]

The Specific Activity of Commercially Available [^{14}C]Phe Can Deviate Up to 40% from the Manufacturer's Specifications

We sometimes observed significantly different saturating values with the same ribosome preparation using different AcPhe-tRNA preparations freed of Phe-tRNA contaminants. We suspected that the specific activities of the various batches of [^{14}C]Phe were not identical in contrast to the information provided by the producer. The specific activity of commercial [^{14}C]phenylalanine was determined via aminoacylation of $tRNA^{Phe}$ under saturation conditions (15-fold molar excess of phenylalanine over tRNA) with isotopic dilution. The saturation point was determined in a previous series of analytical aminoacylation assays with a fixed amount of $tRNA^{Phe}$

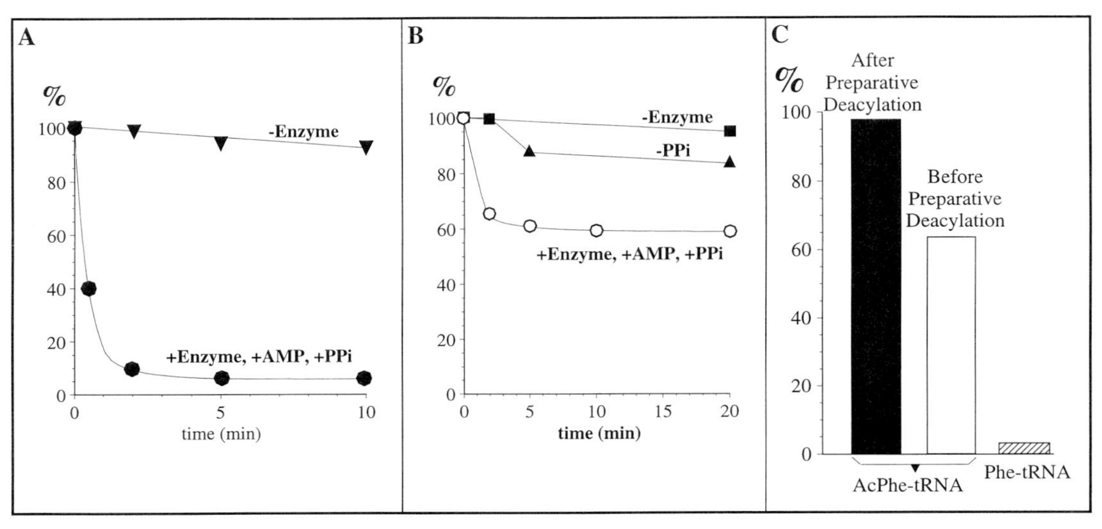

FIG. 3. Enzymatic deacylation of [^{3}H]Phe-tRNA^{Phe} and different preparations of Ac[^{14}C]Phe-tRNA^{Phe} from *E. coli.* (A) 40 pmol of [^{3}H]Phe-tRNA^{Phe} (1900 dpm/pmol) were incubated at 30° in 40 μl of 30 m*M* HEPES–KOH, pH 7.6, at 0°, 10 m*M* magnesium acetate, 100 m*M* KCl, and 4 m*M* 2-mercaptoethanol, in the presence of 6 m*M* PP_i, 6 m*M* AMP, S100 enzymes freed of tRNAs (5 μg per 40 μl of deacylation mixture) or $H_{20}M_{10}$ (−Enzyme, assays without S100 enzymes). 8-μl samples were withdrawn at the indicated times and subjected to cold TCA precipitation. The 100% control corresponds to the amount of precipitable radioactivity before the deacylation reaction. (B) Deacylation as that shown in (A) but using an Ac[^{14}C]Phe-tRNA^{Phe} (1050 dpm/pmol) that gave a ν_{max} = 1.42 of tRNAs bound per ribosome as shown in Fig. 1A. −Enzyme and −PP_i, controls. (C) Analytical deacylation of an AcPhe-tRNA preparation before and after a preparative deacylation reaction. Phe-tRNA was used as control.

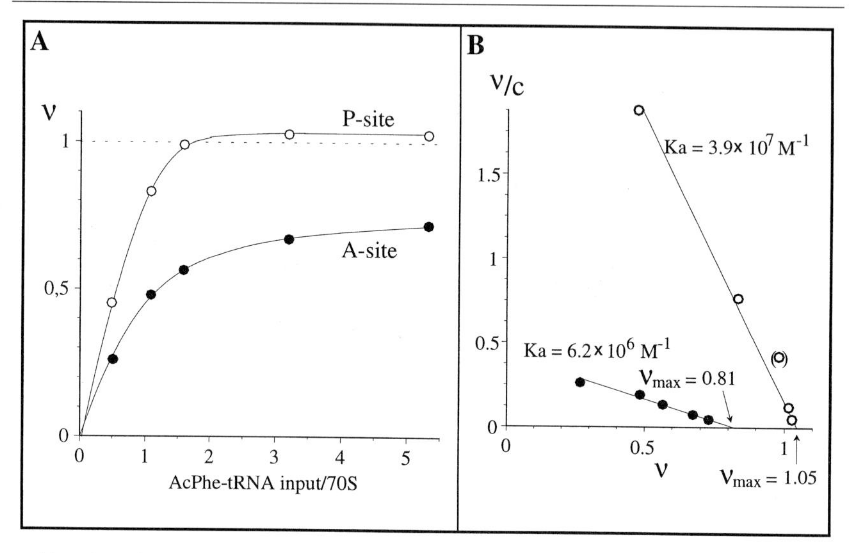

FIG. 4. A-site and P-site saturation of poly(U) programmed 70S ribosomes with the an Ac[^{14}C]Phe-tRNA^{Phe} preparation freed of Phe tRNA contamination. (A) Saturation curves. For A-site binding deacylated tRNA^{Phe} (1650 pmol/A_{260} unit) was added in an 1.5 molar excess over 70S. The P-site and A-site binding results are plotted as tRNA bound per 70S ribosome (ν) versus the Ac[^{14}C]Phe-tRNA^{Phe} input per 70S. (B) Scatchard plots of the data in (A) using the same symbols.

and variable input of the amino acid. Maximal aminoacylation took place at 7-fold excess of Phe over tRNA, and this level is unchanged up to a 30-fold excess of amino acid (data not shown). Therefore, a 15-fold excess was chosen as a safe value for the isotopic dilution experiments (taking into account possible deviations of the [^{14}C]Phe concentration given in the commercial product).

Method. The commercial radioactive amino acid shows high chemical purity when analyzed with thin-layer chromatography (TLC) techniques (data not shown). Analytical assays of tRNA^{Phe} aminoacylation at saturation conditions (15-fold excess of amino acid over tRNA) and at different dilutions of the commercial [^{14}C]Phe with a standard solution of nonlabeled phenylalanine are performed. The radioactivity incorporated into the TCA precipitable material at different isotopic dilutions can be used to test and, if necessary, to correct the specific activity of the commercial [^{14}C]phenylalanine using the following expression:

$$\frac{TCA_u}{TCA_d} - 1 = Sa\left(\frac{\text{Phe}_{\text{cold}}}{\text{dpm}_{\text{T}}}\right)$$

where TCA_u is the TCA-precipitable radioactivity (dpm) after aminoacylation with undiluted [^{14}C]Phe; TCA_d, the TCA-precipitable radioactivity (dpm) after aminoacylation with diluted [^{14}C]Phe; Phe_{cold}, pmols of cold phenylalanine added to the assays with isotopic dilution; dpm_T, total dpm of [^{14}C]Phe added to the assay; and *Sa,* correct specific activity of the commercial [^{14}C]Phe (in dpm/pmol).

This specific activity can be easily derived from data collected from the isotopic dilution, taking into account that the amount of synthesized Phe-tRNA is the same with or without dilution (for this reason the assay is performed under saturation conditions). This expression implies that the correct specific activity *(Sa)* of a radioactive amino acid can be experimentally determined as the slope of a straight line passing through the origin when the values ($[TCA_u/TCA_d] - 1$) for several isotopic dilutions of the [^{14}C]Phe are represented versus the corresponding [Phe_{cold}/dpm_T] values (see Fig. 5). In cases where a commercial preparation contains a fraction of radioactivity not belonging to the amino acid molecule (it can be easily shown via TLC or HPLC techniques), the formula can still be applied by

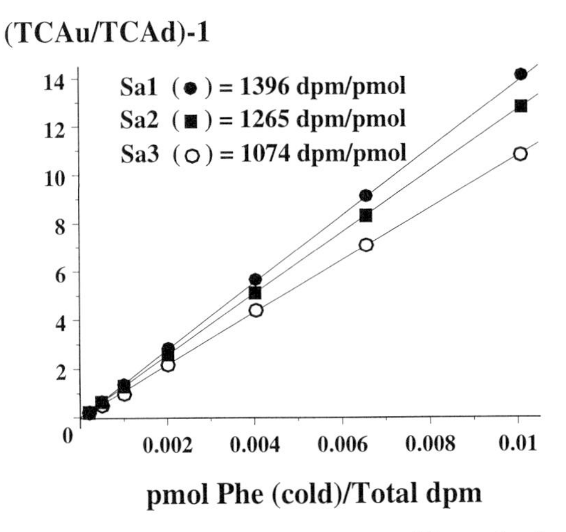

FIG. 5. Determination of the specific activity of three different batches of [^{14}C]Phe by analytical aminoacylation with isotopic dilution. Three different batches of [^{14}C]phenylalanine should be of the same specific activity according to the manufacturer. The isotopic dilutions were prepared by adding 0.25, 0.5, 1.0, 2.0, 4.0, 7.5, and 10.0 mol of nonlabeled phenylalanine per mole of [^{14}C]phenylalanine. The solution of nonlabeled amino acid was made with highly pure phenylalanine and carefully confirmed by spectrophotometric measurements. The value Sa_1 (1396 dpm/pmol) corresponds to the AcPhe-tRNA preparation used in Fig. 1C without correcting the specific activity that yielded an apparent saturation of about 1.4 AcPhe-tRNA per 70S label.

changing the dpm_T term to the corresponding fraction of radioactivity associated with the free amino acid.

Figure 5 shows the specific activities determined for three different batches of [^{14}C]phenylalanine purchased from the same producer claiming that all batches should have a specific activity of around 1000 dpm per pmol. The [^{14}C]Phe from batch 1 (Sa_1 = 1396 dpm/pmol) was used for the AcPhe-tRNA preparation taken for the saturation experiment shown in Fig. 1C. Without correction of the specific activity, the binding of this AcPhe-tRNA apparently saturated at about 1.5 molecules per ribosome (Figs. 1C and 1D). Taking into account the correct specific activity obtained from Fig. 5, the saturation point is 1.07 AcPhe-tRNA molecules per 70S, in close agreement with the previous results. Figure 5 also shows two other [^{14}C]Phe batches. The specific activity determined for batch 3 (1074 dpm/pmol) is very close to the one indicated in the manufacturer's specifications, whereas batch 2 has an intermediate value between batch 1 and batch 3. It follows that it is advisable for quantitative determinations to check carefully the specific activity of the radioactive material. The same problem is observed with samples of other amino acids from different suppliers.

Associated Ribosomes Can Bind Significantly More AcPhe-tRNA Than Tightly Coupled 70S Ribosomes

A third factor affecting the apparent amount of tRNA able to bind to 70S ribosomes is the preparation procedure. Ribosomes prepared via zonal centrifugation under conditions allowing selection for stable 70S ribosomes at 6 m*M* magnesium, i.e., tightly coupled ribosomes,[21,22] were compared in their ability to bind AcPhe-tRNA to 70S particles obtained after reassociation of 50S and 30S subunits. To this end, binding assays were performed in the presence of poly(U) and AcPhe-tRNA at saturating concentration (threefold molar excess over ribosomes). As can be seen in Table I, the reassociated 70S particles were able to bind about 30% more AcPhe-tRNA than the tightly coupled 70S, clearly violating the exclusion principle (1.30 versus 0.98 Ac-Phe-tRNA/70S for reassociated and tightly coupled 70S, respectively). We note that the level of deviation in the saturation binding varies from one preparation to another (data not shown). This tendency to higher tRNA binding values of the reassociated ribosomes has also been observed (although with variable strength) in saturation experiments using model heteropolymeric mRNAs (F. J. Triana-Alonso and K. H. Nierhaus,

[21] B. Hapke and H. Noll, *J. Mol. Biol.* **105,** 97 (1976).

[22] H.-J. Rheinberger, U. Geigenmüller, M. Wedde, and K. H. Nierhaus, *Methods Enzymol.* **164,** 658 (1988).

TABLE I
Ac[^{14}C]Phe-tRNA^{Phe} BINDING TO TIGHT COUPLES AND REASSOCIATED 70S RIBOSOMES UNDER SATURATION CONDITIONS[a]

70S ribosomes	AcPhe-tRNA binding under saturation (pmol AcPhe-tRNA/pmol 70S)
Tightly coupled	0.98 ± 0.05
Reassociated	1.30 ± 0.06

[a] Ac[^{14}C]Phe-tRNA^{Phe} binding to poly(U) programmed ribosomes was determined at three-fold AcPhe-tRNA input over ribosomes (saturation condition). The binding assays were performed as described in the Materials and Methods sections using 10 pmol of ribosomes and 30 pmol of Ac[^{14}C]Phe-tRNA^{Phe} (1074 dpm/pmol) per determination. The results are the mean values of triplicate samples.

unpublished) suggesting that the natural discrimination power of the ribosome for tRNA binding is indeed affected by the preparation procedure. It is likely that the deviation from the exclusion principle is caused by a considerable amount of loose coupled ribosomes in the 70S preparation, since the deviation is negligible when we isolate the heavy shoulder of reassociated 70S from the sucrose gradient, where the tightly coupled 70S ribosomes accumulate (not shown).

Conclusions

We have identified three parameters that, if not carefully controlled, can violate the exclusion principle for peptidyl-tRNA analogs such as AcPhe-tRNA. The most common artifact is due to residual amounts of Phe-tRNA in an AcPhe-tRNA preparation. As shown here, erroneous saturation curves can be avoided by an enzymatic deacylation step that specifically deacylates the aminoacyl-tRNA contaminant in the AcPhe-tRNA preparation. We recommend that a deacylation step be routinely added before an AcPhe-tRNA preparation is purified via HPLC.

For critical cases of saturation experiments a vetting of the specific activity of the radioactive component might be necessary. For acylation assays of tRNA, a simple method is given in this paper.

Reassociated 70S ribosomes are comparable with tightly coupled ribosomes concerning the saturation level of binding of AcPhe-tRNA, if the heavy shoulder of reassociated 70S ribosomes are used. The latter preparation has the additional advantage over standard tightly coupled ribosomes in that they do not contain any residual amounts of tRNAs or mRNA fragments.

If the precautions listed here are obeyed, reproducible and unequivocal results are obtained demonstrating the validity of the "exclusion principle" for the binding of peptidyl-tRNAs and its analogs.

Acknowledgments

We thank Sean Connell and Uli Stelzl for help and discussions.

[18] Three-Dimensional Cryoelectron Microscopy of Ribosomes

By JOACHIM FRANK, PAWEL PENCZEK, RAJENDRA K. AGRAWAL, ROBERT A. GRASSUCCI, and AMY B. HEAGLE

Introduction

Cryoelectron microscopy (cryo-EM) of single ribosomes embedded in ice has been very successful in yielding information on the ribosomal structure[1–6] and functional binding behavior.[7–14] For ribosomes of eukaryotes,

[1] J. Frank, P. Penczek, R. A. Grassucci, and S. Srivastava, *J. Cell Biol.* **115,** 597 (1991).
[2] P. Penczek, R. A. Grassucci, and J. Frank, *Ultramicroscopy* **53,** 251 (1994).
[3] J. Frank, J. Zhu, P. Penczek, Y. Li, S. Srivastava, A. Verschoor, M. Radermacher, R. A. Grassucci, R. K. Lata, and R. K. Agrawal, *Nature* **376,** 441 (1995).
[4] H. Stark, F. Müller, E. A. Orlova, M. Schatz, P. Dube, T. Erdemir, F. Zemlin, R. Brimacombe, and M. van Heel, *Structure* **3,** 815 (1995).
[5] R. K. Lata, R. K. Agrawal, P. Penczek, R. A. Grassucci, J. Zhu, and J. Frank, *J. Mol. Biol.* **153,** 979 (1996).
[6] I. S. Gabashvili, R. K. Agrawal, R. A. Grassucci, and J. Frank, *J. Mol. Biol.* **5,** 1285 (1999).
[7] R. K. Agrawal, P. Penczek, R. A. Grassucci, Y. Li, A. Leith, K. H. Nierhaus, and J. Frank, *Science* **271,** 1000 (1996).
[8] H. Stark, E. A. Orlova, J. Rinke-Appel, N. Jünke, F. Müller, M. Rodnina, W. Wintermeyer, R. Brimacombe, and M. van Heel, *Cell* **88,** 19 (1997a).
[9] H. Stark, M. Rodnina, J. Rinke-Appel, R. Brimacombe, W. Wintermeyer, and M. van Heel, *Nature* **389,** 403 (1997b).
[10] R. K. Agrawal, P. Penczek, R. A. Grassucci, and J. Frank, *Proc. Natl. Acad. Sci. U.S.A.* **95,** 6134 (1998).
[11] A. Malhotra, P. Penczek, R. K. Agrawal, I. S. Gabashvili, R. A. Grassucci, R. Jünemann, N. Burkhardt, K. H. Nierhaus, and J. Frank, *J. Mol. Biol.* **280,** 103 (1998).
[12] J. Frank, *Am. Sci.* **86,** 428 (1998a).
[13] J. Frank, *J. Struct. Biol.* **124,** 142 (1998b).

0076-6879/00 $30.00

there has been similar progress, even though with more modest resolution.[15–18]

The three-dimensional (3-D) density maps resulting from such studies can be used to determine the binding positions of ligands such as tRNA, EF-G, EF-Tu whose X-ray structure is known, or they can be used for the low-resolution phasing of X-ray data for an entire subunit.[19]

Even though the development of methods is far from complete, and the resolution has not progressed beyond 10 Å, it is possible to outline the different steps of specimen preparation, electron microscopy, and image processing for a large class of important projects aimed at characterizing the binding of large ligands to the ribosome.

The premise of the single-particle reconstruction approach is that the macromolecule investigated, embedded in a thin layer of ice, exists in many identical copies in a wide range of orientations. Thus, a snapshot of a specimen field contains the same information as would be otherwise obtained by tilting a single copy of the macromolecule into many orientations. The advantage of collecting the information through a snapshot of multiple molecule images is that the electron dose can be kept to a minimum as it is spread out among the different molecules. The disadvantage is that the relative orientations are not readily known, but have to be obtained through a variety of mathematical procedures.

Isolation and Purification of Ribosomes

The quality of biochemical specimens is one of the prerequisites for obtaining a reliable 3-D cryo-EM map. Because the final 3-D map is obtained by merging several thousand projections from different molecules (see later section), interpretation of results from a heterogeneous specimen would be misleading. Therefore, ensuring the homogeneity of the specimen is of crucial importance. In most cases, prokaryotic 70S ribosomes are isolated from the RNase I^- *Escherichia coli* strain MRE 600. Ribosomes

[14] R. K. Agrawal, P. Penczek, R. A. Grassucci, N. Burkhardt, K. H. Nierhaus, and J. Frank, *J. Biol. Chem.* **274,** 8723 (1999).

[15] A. Verschoor, S. Srivastava, R. Grassucci, and J. Frank, *J. Cell Biol.* **133,** 495 (1996).

[16] A. Verschoor, J. R. Warner, S. Srivastava, R. A. Grassucci, and J. Frank, *Nucleic Acids Res.* **26,** 655 (1998).

[17] R. Beckmann, D. Bubeck, R. A. Grassucci, P. Penczek, A. Verschoor, G. Blobel, and J. Frank, *Science* **278,** 2123 (1997).

[18] P. Dube, M. Wieske, H. Stark, M. Schatz, J. Stahl, F. Zemlin, G. Lutsch, and M. van Heel, *Structure* **6,** 389 (1998).

[19] N. Ban, B. Freeborn, P. Nisson, P. Penczek, R. A. Grassucci, R. Sweet, J. Frank, P. B. Moore, and T. A. Steitz, *Cell* **93,** 1105 (1998).

are purified in two ways. In the most popular, traditional method (e.g., Agrawal and Burma[20]), after the first ultracentrifugation at ~100,000*g*, the ribosomes are washed with high salt concentration (varies from laboratory to laboratory, but usually is in the range of 0.5 *M* ammonium chloride) to remove the loosely bound ribosome-associated factors. It is believed that under those conditions most of the ribosomal proteins are intact; however, the possibility that in a fraction of ribosomes some of the loosely bound ribosomal proteins are washed out during the salt wash cannot be ruled out. In the second method, high salt wash is not used based on the view that salt wash is not needed to obtain factor-free ribosomes (e.g., Bommer *et al.*[21]; also see article by Blaha *et al.*,[22] in this volume). However, in that case, it is difficult to say that ribosomes are completely free from some of the associated factors.

In both methods of ribosome preparation, homogeneity is ascertained by 5–30% sucrose gradient centrifugation and the major 70S peak is pooled as the intact ribosome. Isolation of ribosomal subunits is always done following standard protocols, by dialyzing the 70S ribosomes at low concentration of Mg^{2+} ($\leq$1.0 m*M*) and subsequently subjecting them to 5–30% sucrose gradient centrifugation. The peaks corresponding to 30S and 50S subunits are pooled, pelleted, and then dissolved and dialyzed in the requisite buffer conditions. Final preparations of ribosomes are stored in small aliquots in −80°. Different labs use somewhat different buffer conditions during the isolation procedure as well as for the storage. The most commonly used buffer is referred to as *conventional buffer* (CB), whereas the more complex buffer, which is expected to represent conditions close to *in vivo* conditions, is referred to as *polyamine buffer* (PB). (For more details on buffer compositions, see Agrawal *et al.*[14]; Blaha *et al.*,[22] this volume.)

We have obtained 3-D maps of ribosomes isolated by using both methods[20,21] and found the results obtained very similar. At least in the 15- to 20-Å resolution range, no obvious difference is noticed. For example, the 3-D map presented by Malhotra and co-workers[11] was from a ribosomal preparation obtained by the method described by Bommer and co-workers[21] under the PB conditions, whereas the map presented by Agrawal and co-workers[10] was from a ribosomal preparation obtained by the method described by Agrawal and Burma[20] under the CB conditions. Both maps look very similar. The difference in resolution between the two maps (15

[20] R. K. Agrawal and D. P. Burma, *J. Biol. Chem.* **271,** 21285 (1996).

[21] U. Bommer, N. Burkhardt, R. Jünemann, C. M. T. Spahn, F. J. Triana-Alonso, and K. H. Nierhaus, *in,* "Subcellular Fractionation—A Practical Approach," (J. Graham and D. Richwood, eds.), pp. 271–301. IRL Press, Washington, DC, 1997.

[22] G. Blaha, U. Stelzl, C. M. T. Spahn, R. K. Agrawal, J. Frank, and K. H. Nierhaus, *Methods Enzymol.* **317,** [19], 2000 (this volume).

versus 20 Å) could be the result of the large difference in the size of the data set (30,000 versus 7000 single particles, respectively) used to obtain the 3-D maps. However, depending on buffer conditions used, we do see a significant difference in the positions of tRNAs bound to the ribosome, or in the partitioning of occupancies among possible binding sites (see Agrawal *et al.*[14]; Blaha *et al.*,[22] this volume). This difference appears to be due to buffer-dependent changes in microenvironment in the L1-protein region (Agrawal and Frank[23] in preparation).

In addition to the overall structure of the ribosome, a major interest of our group is to study the 3-D binding positions of various ligands on the ribosome. Significant progress in this direction has been made in the past 4 years by our group and others. In the next section of this article, we present the procedures used to obtained various ribosome–ligand complexes and the rationale behind adopting these procedures in obtaining the best results from 3-D cryo-EM.

Preparation of Ribosome–Ligand Complexes for Cryo-EM

As mentioned in the previous section, the homogeneity of the specimens is of crucial importance in obtaining a reliable 3-D cryo-EM map that can be used to locate the interaction sites of ligands on the ribosome. One of the practical problems is that one never achieves 100% occupancy, i.e., it is not possible to obtain a specimen in which every single ribosome is bound with a given ligand. However, by using certain tactics, one can maximize the occupancy (see later discussion). Usually, a fraction (15–20%) of the isolated ribosomal preparation is not very active in terms of ligand binding and another, greater fraction (20–40%) is not very active in protein synthesis. This means that any standard preparation method yields a heterogeneous population, either as the result of one or more proteins missing in a fraction of the preparation, or due to the existence of different conformational states of the ribosome in any given preparation, e.g., tight and loose couples.[20]

To prepare a ribosome–ligand complex for cryo-EM, binding of ligands is checked by using either a radiolabeled ligand or a radioactive cofactor that binds with the ligands (e.g., radioactive tRNA or radioactive ATP, GTP, or their nonhydrolyzable analogs). This can be done either by a filter-binding assay, gel filtration assay, or by sucrose gradient analysis. Once the condition to obtain the highest binding level is known, it can be used for the preparation of the ligand–ribosome complex. However, the results of these binding experiments can easily differ in the occupancy of the ligand

[23] R. K. Agrawal and J. Frank, *Curr. Opin. Struct. Biol.* **9,** 215 (1999).

in the complex actually used for the cryo-EM purpose. The reason for this is that most of the ligands interact with the ribosome in the micromolar to nanomolar concentration range. For example, tRNA binds to the ribosome at micromolar concentration range, whereas elongation factors bind in a close to nanomolar concentration range. The ribosome concentration needed for optimum spread on the EM grid is between 30 and 35 *nM*. That means ribosomes are substantially diluted during the application to the EM grid, which could easily result in the dissociation of ligands due to a shift of equilibrium. We should point out that loosely bound ligands can also dissociate from the ribosome at the time of the grid preparation because of rapid change in the surface tension. Therefore, to obtain a ligand–ribosome complex with high ligand occupancy, different strategies are used depending on the nature of the ligand.

One way is to incubate the ribosome in an appropriate buffer for the requisite amount of time with the requisite amount of ligand and apply the complex directly on the EM grid. In such a case, one would need to add a large excess of ligand over the ribosome to shift the equilibrium toward the formation of the complex. Attempts to purify such complexes on a sucrose gradient result in lower occupancy of the ligand in the complex (e.g., Agrawal *et al.*[7]). However, a quick operation using gel-fitration spun columns (containing Sepharose 6B or Sephacryl 300) can be used in some cases, where the affinity of ligand is comparatively higher, without making any significant change in the occupancy (e.g., Malhotra *et al.*[11]). In the experiment carried out by Malhotra and co-workers,[11] tRNA was bound to the ribosomal P site, where the binding affinity of tRNA is approximately 10-fold higher than those for A and E sites. For the P site, occupancy of tRNA is generally between 81%[11] and 90%.[14] The same spun-column procedure does not work quite well for binding tRNAs to the A and E sites, where highest tRNA occupancy is expected to be lower than that in the P site, in the range of 60%. Therefore, to obtain the maximum occupancy at A and E sites of the ribosome for EM visualization, an unpurified tRNA–ribosome complex is more useful. We should point out that the high occupancy of tRNA at the ribosomal P site brings most of the ribosomes into a unique homogeneous conformation, which does not seem to be much affected by the subsequent binding of the tRNA at the A site. Thus, even though the complex has lower occupancy at the A site, the ribosome appears to be in a very homogeneous conformational state.

Another way could be to use an antibiotic, which inhibits the dissociation of a particular ligand and thus shifts the equilibrium toward the formation of the complex. For example, kirromycin binds to the aminoacyl-tRNA · EF-Tu · GDP · 70S ribosome complex and prevents dissociation of EF-Tu from the complex. Similarly, fusidic acid binds to the EF-G · GDP · 70S ribosome complex and prevents dissociation of EF-G from the ribosome. Both of these strategies have been successfully used to locate the detailed binding positions of both elongation factors on the ribosome.[9,10]

Electron Microscopy

Preparation of Specimen Grid

The grids are prepared using a modified protocol (described in Wagenknecht *et al.*[24]). Grids of 300-mesh copper (or molybdenum) are used with thick (approximately 800-Å) holey carbon on top of which is floated a fresh layer of thin (approximately 200-Å) carbon. The grids are then glow discharged with air for 30 sec to ensure that the surface is uniformly hydrophilic.

Plunge-Freezing

The grid is mounted on a freeze-plunging apparatus (Fig. 1). A droplet of the ribosome dispersion is put on the grid, followed by blotting of excess buffer from both sides as suggested by Cyrklaff and co-workers.[25] Following the release of the plunger in the guillotine-like apparatus, the grid is rapidly immersed in liquid ethane or propane, then transferred into the cryoholder, which is inserted into the specimen chamber through an airlock. During the transfer and the subsequent experiment, the specimen grid is kept at a temperature (approx. $-180°$) close to that of liquid nitrogen ($-193°$). Suitable fields of the specimen grid are recorded on cut film (e.g., Kodak S0163, Rochester, NY) at around ×50,000 magnification, a value that ensures that the optical density obtained for low-dose imaging (see later discussion) is above the fog level.[26]

Low-Dose Electron Microscopy

Data are collected on a transmission electron microscope equipped with a low-dose kit. The image is focused in an area of the specimen adjacent to where the micrograph is taken. The dose of each exposure is about 10 $e^-/Å^2$. An exposure is taken at a defocus between 1.0 and 2.0 μm. The exact defocus is later determined on the optical diffractometer and in the computer. If necessary, more data are collected to boost any underrepresented spatial frequencies.

[24] T. Wagenknecht, R. A. Grassucci, and J. Frank, *J. Mol. Biol.* **199,** 137 (1988).
[25] J. Cyrklaff, M. Adrian, and J. Dubochet, *J. Electron Microsc. Tech.* **16,** 351 (1990).
[26] P. N. T. Unwin and R. Henderson, *J. Mol. Biol.* **94,** 425 (1975).

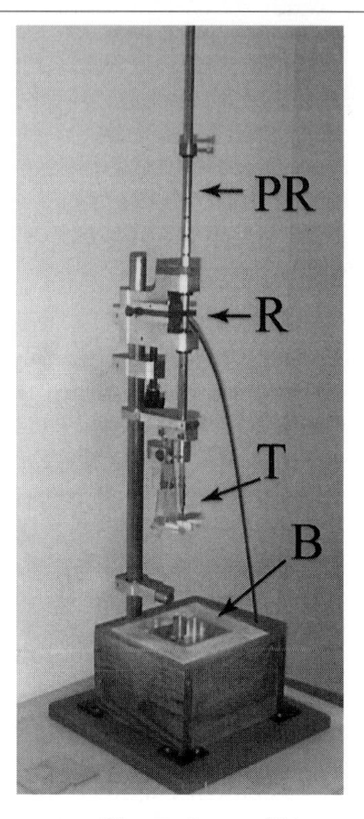

FIG. 1. Plunge-freezing apparatus. The specimen grid is mounted between tweezers (T), which are in turn mounted on the plunging rod (PR). After the sample is put on the grid and blotted, the foot pedal-operated release (R) is pushed and is plunged into a bath (B) filled with the cryogen.

Optical Diffraction

Micrographs are inspected to ascertain that the number of particles is sufficient—it should be in the hundreds per micrograph for best efficiency of data collection—and then checked in the optical diffractometer for quality. A good micrograph taken by microscopes equipped with conventional cathodes (hairpin or Lab6) will show circular Thon rings out to 12-Å resolution, and those taken with a field emission gun (FEG) will show rings out to as far as 5 Å (Fig. 2). Ellipticity of the rings is a sign for uncorrected astigmatism; striping or unidirectional limitation of the optical diffraction pattern indicates specimen drift. Micrographs showing such anomalies in the optical diffractometer are rejected.

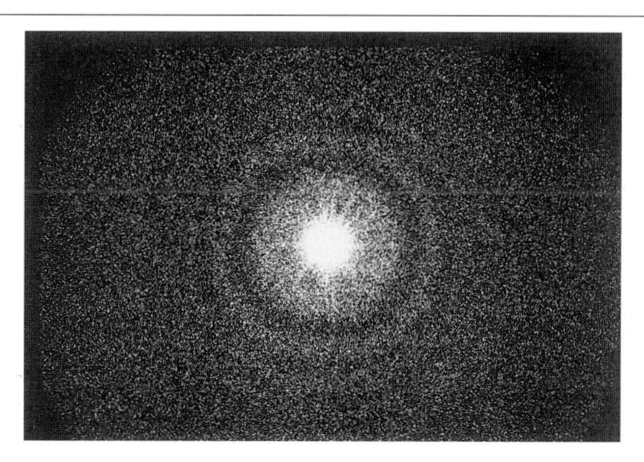

FIG. 2. Optical diffraction pattern of a cryoelectron micrograph, showing the presence of Thon rings that represent a signature of the contrast transfer function. Perfect roundness of the rings indicates absence of astigmatism, and the position of the rings reflects the defocus setting.

Image Processing

Reconstruction Strategies

The goal of single-particle reconstruction is to obtain a faithful 3-D distribution of the Coulomb potential of a macromolecule, given its 2-D projections in random, unknown orientations (the Coulomb potential distribution is closely related to, but not identical with, the electron density distribution obtained by X-ray crystallography). The main problem is the determination of the orientations of views in 3-D space. There are two possible situations to consider: either the orientations have to be derived from the data set itself, or a reference structure is already known that can be used to obtain initial estimates of the orientations. In the first case, there are a number of strategies available. *Random conical tilt* reconstruction is the most robust method and historically the first one used[27] (for a detailed account, see Radermacher[28]). It gives approximate orientations without any handedness ambiguity. For the same grid area, pairs of micrographs are collected, one with the specimen tilted, the other untilted. The particles from the untilted-specimen micrographs are first aligned and then classified. The alignment provides the azimuthal Eulerian angle of the particle, while classification divides the data set into homogenous classes, where each class combines all particles that have the same orientation on the grid, and the different classes reflect the different orientations that coexist on the

[27] M. Radermacher, T. Wagenknecht, A. Verschoor, and J. Frank, *J. Microsc.* **146,** 113 (1987).
[28] M. Radermacher, *J. Electron Microsc. Tech.* **9,** 359 (1988).

specimen grid. The two remaining Eulerian angles are established based on the geometric relations between tilted and untilted micrographs, and the angles are assigned to the corresponding tilted-particle images. For the actual 3-D reconstruction, only particles from the tilted-specimen micrographs are used. The 3-D reconstruction is calculated using either a real space-based algebraic iterative technique[29,30] or a Fourier space-based weighting technique.[31–33] In our experience, the former method provides superior results, even if the time of calculations is significantly longer than for Fourier methods.

More recently, strategies that do *not* require collection of tilt data have been used. They employ a direct *common lines-based* determination of orientation parameters.[34–37] These techniques cannot resolve an ambiguity of handedness, so some additional information must be available. As with the random-conical method, an additional refinement step is required after the initial structure has been obtained.[37–39]

In the second case, when a reference structure is already available from earlier studies—which is the case we are exclusively dealing with when we study the ribosome (see Frank *et al.*[1,3])—there exists a relatively simple technique of finding the orientations, and this technique is identical to the technique used in the refinement step (see later discussion): by comparing the projections with a set of reference projections of the existing structure.

Practical Steps

In the following step-by-step description of the practical approach, we assume the second-case scenario: (1) selection and scanning of micrographs, (2) estimation of the contrast transfer function (CTF) parameters, (3) particle picking and density normalization, (4) reconstruction of the initial 3-D density map, (5) refinement of the 3-D density map, and (6) visualization

[29] H. Gilbert, *J. Theor. Biol.* **36,** 105 (1972).
[30] J. Zhu, P. Penczek, R. Schröder, and J. Frank, *J. Struct. Biol.* **118,** 197 (1997).
[31] M. Radermacher, T. Wagenknecht, A. Verschoor, and J. Frank, *J. Microsc.* **141,** RP1 (1986).
[32] G. Harauz and M. van Heel, *Optik* **73,** 146 (1986).
[33] M. Radermacher, *in,* "Electron Tomography," (J. Frank, ed.). Plenum, New York (1992).
[34] M. van Heel, *Ultramicroscopy* **21,** 111 (1987).
[35] N. Farrow and F. P. Ottensmeyer, *Ultramicroscopy* **52,** 141 (1993).
[36] M. Radermacher, *Ultramicroscopy* **53,** 121 (1994).
[37] P. Penczek, J. Zhu, and J. Frank, *Ultramicroscopy* **63,** 205 (1996).
[38] T. S. Baker and R. H. Cheng, *J. Struct. Biol.* **116,** 120 (1996).
[39] M. van Heel, G. Harauz, E. V. Orlova, R. Schmidt, and M. Schatz, *J. Struct. Biol.* **116,** 17 (1996).

and interpretation of the 3-D map. Occasional references are made to the SPIDER image processing system developed here.[40,41]

Selection and Scanning of Micrographs. First, electron micrographs are inspected for quality. Visual inspection answers the following questions: Is the contrast in the field sufficiently high? Do the particles look crisp? Are they clustered together? Is the number of particles sufficiently high to make scanning worthwhile? Is there any noticeable drift in the field? Moreover, the micrographs are analyzed in the optical diffractometer for the presence of Thon rings. Those found acceptable are scanned using a step size ranging from 2.5 to 5.0 Å on the object scale, depending on the type of project and the resolution expected. The Shannon sampling theorem, according to which the step size should be at least half the size of the smallest detail present in the image, sets the theoretical limit of the resolution. In practice, due to the interpolations used in many image manipulation steps, the resolution achievable is always worse than the theoretical. An additional limiting factor is the contrast transfer function of the electron microscope, more specifically the envelope function part of it.[42] High spatial frequencies are strongly suppressed and the highest resolution achieved so far for ribosomes does not exceed 11.5 Å (Gabashvili, Spahn, Agrawal, Grassucci, Frank, Penczek, unpublished results).

Estimation of Contrast Transfer Function. The CTF parameters[42] are calculated based on the power spectrum calculated/estimated from the micrograph. The calculation is done using a dedicated, semiautomatic procedure designed by Zhu and co-workers.[30] To improve the statistical properties of the power spectrum estimate, the method of averaged overlapping periodograms is employed.[43] The micrograph field (usually approximately 2500×3500 pixels in size) is divided into a number of small windows (512×512 pixels) with an overlap between adjacent windows of 50%. The size of the small windows and the amount of overlap have to be adjusted for the particular experimental conditions.[44] The periodogram is calculated for each small window, and all periodograms thus obtained are averaged to produce a 2-D estimate of the power spectrum. This 2-D image is rotationally averaged and the resulting 1-D curve is finally used to estimate such parameters as the fall-off of the envelope function and the defocus.

[40] J. Frank, A. Verschoor, and M. Boublik, *Science* **214,** 1353 (1981).

[41] J. Frank, M. Radermacher, P. Penczek, J. Zhu, Y. Li, M. Ladjadj, and A. Leith, *J. Struct. Biol.* **116,** 190 (1996).

[42] J. Frank, ed., "Three-Dimensional Electron Microscopy of Macromolecular Assemblies." Academic Press, San Diego, 1996.

[43] P. D. Welsh, *IEEE Trans. Audio Electroacoust.* **AU-15,** 70 (1967).

[44] J. J. Fernandez, J. R. Sanjurjo, and J. M. Carazo, *Ultramicroscopy* **68,** 267 (1997).

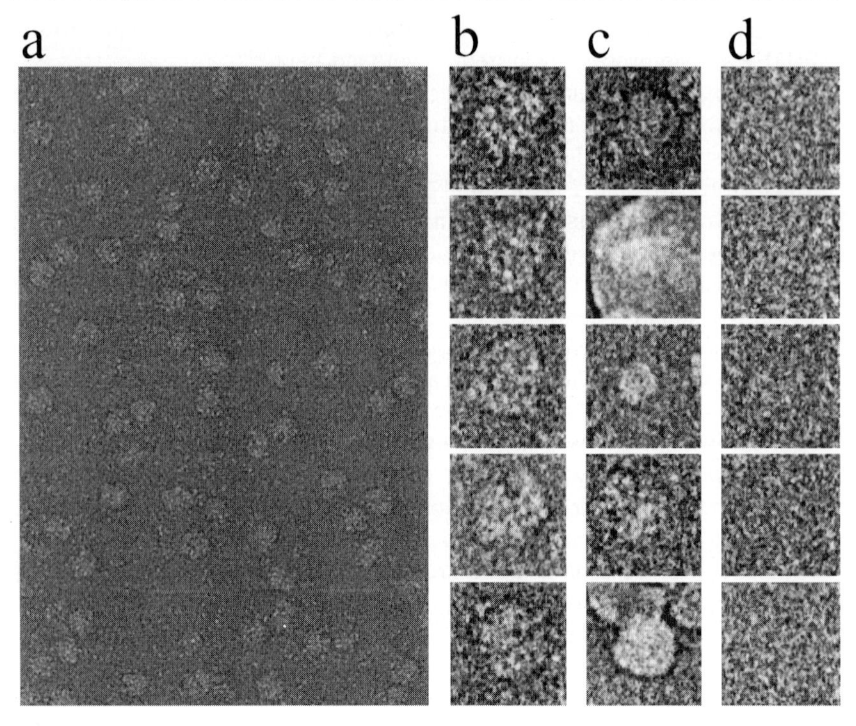

FIG. 3. Particle picking from electron micrographs. (a) A portion of an electron micrograph containing frozen, hydrated 70S *E. coli* ribosomes to which EF-G is bound. (b–d) Examples of micrograph images selected by the automatic particle picking program. The selected particle candidates are interactively classified as (b) particles, (c) junk, and (d) noise. Based on this information, the computer continues to make selections of particles.

Once the CTF parameters are known, the micrographs are sorted into defocus groups for further processing.

Particle Picking and Optical Density Normalization. The particles are picked from digitized images using an automated particle picking procedure developed by Lata and co-workers,[45] that proved to be sufficiently efficient to become a standard tool (Fig. 3). This procedure evaluates textural parameters for data windows containing particle candidates. The efficiency of the method is about 80%; thus additional screening of selected particles is required to eliminate false positives using an interactive "selection of particles" option within the WEB graphics user interface of the SPIDER system. After acceptable particle candidates are selected, a density normalization is applied to each data window to compensate for varying electron microscopy conditions, such as electron dose, uneven illumination by the electron beam,

[45] R. K. Lata, P. Penczek, and J. Frank, *Ultramicroscopy* **58,** 381 (1995).

and ice thickness. The procedure comprises two steps. The first step compensates for uneven illumination; a "ramp" function is subtracted from the particle window. Second, a linear transformation of pixel densities is applied. This transformation is based on the assumption that, although the variance of the particle densities can be different for different views of the same object, the statistics of the surrounding noise should remain the same. Thus, a reference histogram of pixel values is estimated from a reference window containing only noise. Next, for every particle window, parameters of the transformation are adjusted in such a way that a histogram of pixel values in the window corners (presumably containing noise) matches the reference histogram in the least squares sense.[46]

Reconstruction of Initial 3-D Density Map. The availability of a reference structure makes it possible to generate a small set of reference projections. Based on quasi-even distribution of angular directions (for details see Penczek *et al.*[2]), we decided that an angular step of 15 degrees provides a reasonable trade-off between speed and accuracy for the initial reconstruction. This step size results in 83 unique projection directions. In terms of Eulerian angles, these directions cover, for nonsymmetric objects, the range of $\phi \in \langle 0.0, 360.0 \rangle$ and $\theta \in \langle 0.0, 90.0 \rangle$. Projections with angular directions of $\theta > 90.0$ are mirrored versions of projections with $\theta \leq 90.0$, and since the projection matching operations in SPIDER can verify both positions very efficiently given the unique set only, there is no need to generate the mirrored set. Thus, all the experimental projections are aligned with all the reference projections using a dedicated procedure that estimates rotation angle and translation parameters simultaneously.

In addition, a number of most similar reference projection is determined. Because the Eulerian angles used to generate the reference projections are known, one can easily assign them to the experimental projections and calculate first a low-resolution version of the 3-D structure. The resolution is obviously limited by the small number of projection directions used. Nevertheless, it is already possible at this stage to check the structure for anisotropic artifacts and generate a histogram showing the distribution of angular directions. Data with lowest quality are eliminated, as determined by the correlation coefficient.

This is done for every defocus group separately, and for each group a separate resolution curve (FSC; see later discussion) is calculated. The initial estimation of defocus is verified and defocus curves are checked for complementarity, i.e., it is estimated whether the CTF correction and merging of all the individual volumes could possibly result in a close to uniform resolution curve. Major gaps in the resolution curve indicate the

[46] N. Boisset, P. A. Penczek, J. Taveau, V. You, F. de Haas, and J. Lamy, *Ultramicroscopy* **74,** 201 (1998).

need for additional data sets with modified defocus settings. If all the tests give favorable results, the final step, i.e., the 3-D projection alignment, is performed.

Refinement of 3-D Density Map. In the refinement, a much finer angular step is used in order to improve the resolution. The size of the step is estimated using simple geometrical considerations. The sampling theorem dictates that the coverage of space need not to be finer than one pixel. It follows that for a structure with radius of R pixels, the angular step is small enough when $\Delta\theta \leq \arcsin(1/R)$. In our work on the 70S *Escherichia coli* ribosome, we used a lax step size of $\Delta\theta = 2.0$ degrees resulting in a total of 5088 reference projections.[2] The reference projections are calculated using the 3-D structure modified by the CTF estimated for a given defocus group. High efficiency of the procedure is achieved by separating the search for the Eulerian angles from the search for the translational parameters. First, all the projection data are compared with all the reference projections using the rotational cross-correlation function in polar coordinates. Two Eulerian angles are determined by the angular direction of the most similar reference projection. The third, in-plane, rotation angle is calculated from the position of the maximum of the rotational cross-correlation function.

Second, the translational cross-correlation function between each experimental projection and its already determined reference counterpart is calculated and the translation parameters are determined and applied to the projection data. After all the parameters have been determined, the new, refined 3-D structure is calculated. Both alignment and 3-D reconstruction are repeated for each defocus group, each time using the reference structure modified by the respective CTF. Each cycle of the refinement procedure is concluded by the CTF correction step. In this step, the 3-D volumes from different defocus groups (Fig. 4) are merged using a Wiener filtration approach.[47] The merged structure is used as a new reference structure in the subsequent refinement cycle. During the refinement procedure, changes in angular assignments of projection data are being monitored, and the process is terminated when the reassignments decrease below a specified threshold. Finally, the projection data set is randomly split into two groups, separate volumes are calculated and compared in Fourier space using the Fourier shell correlation (FSC).[48,49] The spatial frequency at which

[47] J. Frank and P. Penczek, *Optik* **98,** 125 (1995).

[48] M. van Heel, W. Keegstra, W. Schutter, and E. F. L. van Bruggen, *in* "Structure and Function of Invertebrate Respiratory Proteins" (E. J. Wood, ed.), pp. 69–73. Harwood Academic, Reading, UK (1982).

[49] M. van Heel, *Optik* **73,** 83 (1986).

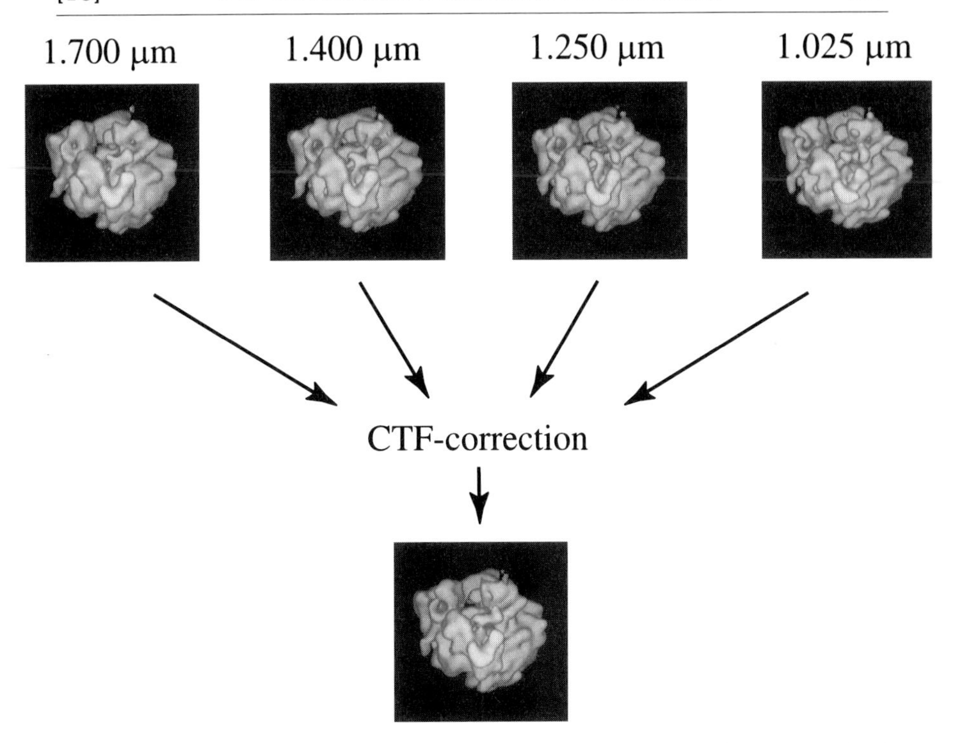

FIG. 4. Surface representations of 3-D reconstructions of a 70S *E. coli* ribosome · EF-G complex. Four separate reconstructions were obtained for defocus groups 1.7, 1.4, 1.25, and 1.025 μm, respectively. The reconstructions are merged using Wiener filtration and a CTF-corrected structure is thereby obtained.

FSC decreases below the 0.5 threshold[11,50] is reported as the resolution achieved (Fig. 5).

Visualization and Interpretation of 3-D Map. The FSC curve obtained for the merged, CTF-corrected volume is the basis for the choice of filtration for the 3-D map. The purpose of filtration is to decrease (or entirely eliminate) the influence of spatial frequency regions found unreliable according to the FSC analysis. The spatial frequency region within which FSC has values close to one should be left intact, while information within spatial frequency regions with FSC values close to zero should be suppressed. The filtration is performed using the Butterworth filter,[51] with the pass-band

[50] B. Böttcher, S. A. Wynne, and R. A. Crowther, *Nature* **386,** 88 (1997).

[51] S. D. Stearns and R. A. David, "Signal Processing Algorithms." Prentice-Hall, Englewood Cliffs, New Jersey, 1988.

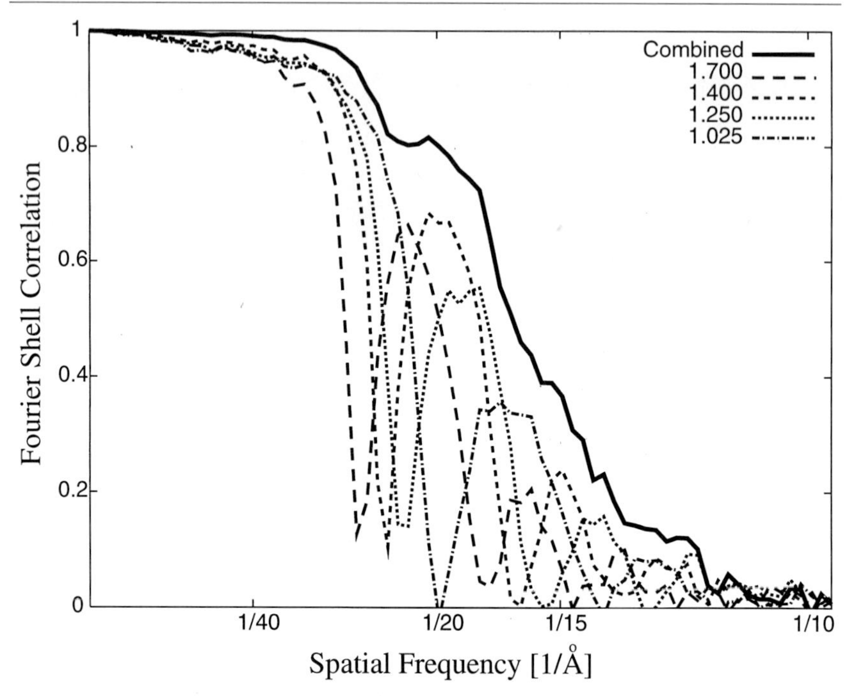

FIG. 5. Estimation of the effective resolution of the 70S *E. coli* ribosome reconstruction based on Fourier shell correlation curves. The curve obtained for the CTF-corrected reconstruction is shown in bold. The FSC curves for individual reconstructions decrease to zero at spatial frequencies corresponding to zeros of the respective CTFs. The spatial frequency for which FSC drops below 0.5, in this case (unpublished data corresponding to Fig. 4) 1/17 Å, is designated the resolution of the merged reconstruction.

and stop-band frequencies set according to FSC-determined values. Next, the resulting map is normalized, so the comparisons with structures obtained in other experiments become possible.

The visualization of the 3-D map is done using either the surface option of the WEB graphical user interface[44] or graphics software such as EXPLORER (Numerical Algorithms Group, Inc., Downers Grove, IL) or AVS (Advanced Visual Systems, Inc., Waltham, MA). The appropriate density threshold for the representation is decided based on molecular mass data and the densities of protein and RNA.[30] In addition, preservation of the spatial continuity of the structure is taken into account. Detailed interpretation of 3-D maps and, if available, docking of X-ray crystal structures is done using the program O.[52] When installed on an SGI (Silicon Graphics, Inc., Mountainview, CA) workstation equipped with LCD stereo

[52] T. A. Jones, J. Y. Zhou, S. W. Cowan, and M. Kjeldgaar, *Acta Crystallogr.* **47,** 110 (1991).

viewing eyewear, this program is able to display 3-D density maps and structures imported from the Protein Data Bank in stereo. This feature facilitates the precise determination of conformational changes and locations of bound ligands.

Conclusion

Even though the method of 3-D cryo-EM is still under development, this article covers the way to obtain density maps in a resolution range (10–20 Å) that has proven to be extremely useful in identifying binding positions. What are the present hurdles to realizing higher resolution that can in principle be obtained?[53] We can depict the situation as a succession of three hurdles: (1) conformational heterogeneity, (2) problems in EM imaging, and (3) the difficulties connected with the collection and processing of very large data sets required for statistical reasons.

Conformational homogeneity is an absolute prerequisite to obtaining meaningful, unblurred reconstructions—in other words, all ribosomes should be in the same binding state. To what extent this can be achieved in a particular case is difficult to assess, but our experience has shown that high occupancy of a functionally meaningful state makes success more likely.[11]

Provided the specimen is conformationally homogeneous, the next hurdle is the resolution limitation inherent to the imaging with the electron microscope. Here the new generation of electron microscopes with exceptional coherence and stability will likely provide a breakthrough, as already demonstrated in recent ground-breaking results with viruses.[50]

Finally, if such a microscope is available, and high-resolution images can be obtained, it is still necessary to collect images in sufficient quantities to realize that resolution in a reconstruction. Recent experience[11] (Gabashvili, Spahn, Agrawal, Grassucci, Frank, Penczek, unpublished results) indicates that the number of images to be collected may well have to exceed the 100,000 mark to realize resolutions better than 10 Å. This is well in excess of the theoretical numbers estimated by Henderson,[53] with the discrepancy being due to many factors including residual instabilities and drift of the electron microscope and the inefficiency of practical data collection. For this reason, the automation of data collection in the electron microscope and of image processing will prove pivotal for realizing high resolution in the reconstruction of single-particle specimens.

Acknowledgments

This work was supported, in part, by NIH grant R37 GM29169 to J.F.

[53] R. Henderson, *Q. Rev. Biophys.* **28,** 171 (1995).

[19] Preparation of Functional Ribosomal Complexes and Effect of Buffer Conditions on tRNA Positions Observed by Cryoelectron Microscopy

By GREGOR BLAHA, ULRICH STELZL, CHRISTIAN M. T. SPAHN, RAJENDRA K. AGRAWAL, JOACHIM FRANK, and KNUD H. NIERHAUS

Introduction

Electron microscopy and immunoelectron microscopy have played an important role in establishing the gross shape of ribosomes and in localizing of proteins within the ribosome during the 1970s and 1980s.[1–3] Due to the development of three-dimensional cryoelectron microscopy (3-D cryo-EM) of single particles[4–6] the ribosome is now seen with much higher quality: reconstructions of the highly asymmetric particles reached a resolution of 20–25 Å.[7,8] In the light of this new resolution it was possible to study functional complexes, i.e., ribosomes carrying tRNAs,[9,10] and clongation factors such as EF-Tu[11] and EF-G.[12] A further improvement in resolution was achieved by using programmed ribosomes with high occupancy at the P site (80%) with fMet-tRNA: for the first time, a large portion of the tRNA was directly visible, thus allowing an unambiguous localization and orientation of the tRNA within the intersubunit space. A large part of

[1] J. A. Lake, *J. Mol. Biol.* **105,** 131 (1976).
[2] V. D. Vasiliev, O. M. Selivanova, V. I. Baranov, and A. S. Spirin, *FEBS Lett.* **155,** 167 (1983).
[3] M. Stöffler-Meilicke and G. Stöffler, *Methods Enzymol.* **164,** 503 (1988).
[4] J. Frank, P. Penczek, R. A. Grassucci, and S. Srivastava, *J. Cell Biol.* **115,** 597 (1991).
[5] P. Penczek, M. Radermacher, and J. Frank, *Ultramicroscopy* **40,** 33 (1992).
[6] P. Penczek, R. A. Grassucci, and J. Frank, *Ultramicroscopy* **53,** 251 (1994).
[7] J. Frank, J. Zhu, P. Penczek, Y. H. Li, S. Srivastava, A. Verschoor, M. Radermacher, R. Grassucci, R. K. Lata, and R. K. Agrawal, *Nature* **376,** 441 (1995).
[8] H. Stark, F. Mueller, E. V. Orlova, M. Schatz, P. Dube, T. Erdemir, F. Zemlin, R. Brimacombe, and M. van Heel, *Structure* **3** (1995).
[9] R. K. Agrawal, P. Penczek, R. A. Grassucci, Y. Li, A. Leith, K. H. Nierhaus, and J. Frank, *Science* **271,** 1000 (1996).
[10] H. Stark, E. V. Orlova, J. Rinke-Appel, N. Junke, F. Mueller, M. Rodnina, W. Wintermeyer, R. Brimacombe, and M. van Heel, *Cell* **88,** 19 (1997).
[11] H. Stark, M. V. Rodnina, J. Rinke-Appel, R. Brimacombe, W. Wintermeyer, and M. van Heel, *Nature* **389,** 403 (1997).
[12] R. Agrawal, P. Penczek, R. Grassucci, and J. Frank, *Proc. Natl. Acad. Sci. U.S.A.* **95,** 6134 (1998).

0076-6879/00 $30.00

this improvement in resolution has been attributed to the high degree of conformational stability of the ribosome–tRNA complex.[13]

This article covers the isolation of the ribosomes and the preparation of functional complexes followed by a short overview of the possibilities for analyzing ribosomal complexes. Further, we summarize and discuss the results of recent cryoelectron microscopy studies that reflect the effect of buffer conditions. Although previous studies have established that the ribosome has three tRNA binding sites, 3-D cryo-EM has surprisingly revealed five different tRNA positions on the ribosome, classified as A, P, P/E, E, and E2.[14–16] The occupancy of some of these positions strongly depends on the buffer conditions used and the charge state of the tRNA. We will show that in the presence of the polyamine buffer, mimicking the *in vivo* conditions, only occupancy of A, P, and E sites is observed in complexes of the initiating and elongating ribosomes.

Preparation of 70S Ribosomes

The procedure described here for the small-scale isolation of tightly coupled ribosomes yields highly active and intact ribosomes, an important prerequisite for the preparation of functional complexes. For large-scale isolation using zonal centrifugation see Bommer *et al.*[17] In addition we describe the isolation of ribosomal subunits that can be used to prepare reassociated ribosomes. Reassociated ribosomes show a more efficient tRNA binding, as compared to tightly coupled ribosomes, because saturation of tRNA binding is reached at molar ratios slightly above stoichiometric ones. This can be attributed to at least two factors, (1) a selective pressure for active particles in the reassociation step and (2) the loss of residual amounts of tRNAs (about 0.8 per 70S ribosomes)[18] and of mRNA fragments.

[13] A. Malhotra, P. Penczek, R. K. Agrawal, I. S. Gabashvili, R. A. Grassucci, R. Jünemann, N. Burkhardt, K. H. Nierhaus, and J. Frank, *J. Mol. Biol.* **280,** 103 (1998).

[14] R. Agrawal, P. Penczek, A. Malhotra, R. Grassucci, I. Gabashvili, A. Heagle, S. Srivastava, N. Burkhardt, R. Jünemann, K. Nierhaus, and J. Frank, *Proc. 14th Intl. Congr. Electron Microsc.* **1,** 717 (1998).

[15] R. K. Agrawal, P. Penczek, R. A. Grassucci, N. Burkhardt, K. H. Nierhaus, and J. Frank, *J. Biol. Chem.* **274,** 8723 (1999).

[16] C. M. T. Spahn and K. H. Nierhaus, *Biol. Chem.* **379,** 753 (1998).

[17] U. Bommer, N. Burkhardt, R. Jünemann, C. M. T. Spahn, F. J. Triana-Alonso, and K. H. Nierhaus, *in* "Subcellular Fractionation. A Practical Approach" (J. Graham and D. Rickwoods, eds.), p. 271. IRL Press, Washington, DC, 1996.

[18] J. Remme, T. Margus, R. Villems, and K. H. Nierhaus, *Eur. J. Biochem.* **183,** 281 (1989).

Materials

To minimize damage by RNases, we use the strain CAN/20-E12 derived from *Escherichia coli* K12, which is deficient in RNases BN, II, D, I.[19]

Bacto-tryptone and yeast extract, DIFCO Laboratories (Detroit, MI); Alcoa A-305, Serva (Heidelberg, Germany); sucrose (ultrapure), GibcoBRL (UK); all other chemicals mentioned are from Merck (Darmstadt, Germany) and Roche Diagnostics (Mannheim, Germany).

Buffers

Ribosome buffer $H_{20}M_6N_{30}SH_4$: 20 m*M* HEPES–KOH, pH 7.5, at 4°, 6 m*M* magnesium acetate, 30 m*M* NH_4Cl, and 4 m*M* 2-mercaptoethanol. (Tightly coupled ribosomes withstand these conditions, whereas loosely couple ribosomes dissociate into subunits. Tightly coupled ribosomes are functionally competent in contrast to the loosely coupled 70S.)[20]

Dissociation buffer $H_{20}M_1N_{200}SH_4$: 20 m*M* HEPES–KOH, pH 7.5, at 4°, 1 m*M* magnesium acetate 200 m*M* NH_4Cl, and 4 m*M* 2-mercaptoethanol (lowering Mg^{2+} and raising monovalent ions cause dissociation of ribosomes).

Reassociation buffer $H_{20}M_{20}K_{30}SH_4$: 20 m*M* HEPES–KOH, pH 7.5, at 4°, 20 m*M* magnesium acetate 30 m*M* KCl, and 4 m*M* 2-mercaptoethanol.

Growth of Bacteria

1 liter of LB media [10 g Bacto-tryptone, 5 g yeast extract, 5 g NaCl, and 25 ml of a 20% (w/v) glucose solution per 1 liter] is inoculated with 10 ml of an overnight culture of *E. coli* CAN/20-E12. Fermentation is performed under continuous agitation at 37°. Cell growth is stopped at 0.5 A_{560} units per milliliter, ensuring that mid-log phase had not been passed. The activity of ribosomes depends considerably on the time at which the cells are harvested. Early mid-log phase ribosomes have proved to be optimal in tRNA binding and elongation. After centrifugation for 15 min at 10,000 rpm (about 30,000*g* at 4°) in a Sorvall GSA rotor, wet cells (usually 1–1.2 g per 1 liter) can be stored at −80°.

[19] M. P. Deutscher, C. W. Marlor, and R. Zaniewski, *Proc. Natl. Acad. Sci. U.S.A.* **81,** 4290 (1984).

[20] B. Hapke and H. Noll, *J. Mol. Biol.* **105,** 97 (1976).

Isolation of Crude 70S

The following procedure is performed on ice or at 4°, and sterilized glassware and tubes are used. The cells are resuspended in 30 ml of ribosome buffer $H_{20}M_6N_{30}SH_4$ and centrifuged for 15 min at 10,000 rpm (about 30,000*g*) in a Sorvall GSA rotor, mixed with Alcoa A-305 (twofold cell weight) and ground in a prechilled mortar and pestle for 2 min. The cell paste is resuspended in ribosome buffer $H_{20}M_6N_{30}SH_4$ (2 ml per 1g cells). The paste is subjected to two low-speed centrifugation steps (20 min at 12,000 rpm in a Sorvall SS34 rotor to remove the Alcoa A-305 followed by 60 min at 16,000 rpm in a Sorvall SS34 rotor to remove the cell debris; 15,000 and 30,000*g*, respectively). The resulting supernatant is centrifuged for 17 hr at 40,000 rpm (about 110,000*g*) in a Beckman 70.1 Ti rotor. The crude ribosomal pellet is rinsed with ribosome buffer $H_{20}M_6N_{30}SH_4$ buffer when tightly coupled 70S are to be isolated, or with dissociation buffer $H_{20}M_1N_{200}SH_4$ when ribosomal subunits are to be isolated. The pellet is then resuspended in the same buffer with continuous shaking for about 1 hr. The resuspended crude ribosomes are clarified (5 min at 10,000 rpm in a Eppendorf centrifuge 5415) and their concentration is determined. Usually, the yield is 300–400 A_{260} units per gram of cells and the concentration is between 500 and 1,000 A_{260} units per milliliter.

Isolation of Tight-Coupled 70S

The crude ribosomes in ribosome buffer $H_{20}M_6N_{30}SH_4$ are loaded on a sucrose gradient (10–30% sucrose; about 100 A_{260} units per bucket) in ribosome buffer $H_{20}M_6N_{30}SH_4$ and centrifuged for 16 hr at 19,000 rpm (about 48,000*g* at 4°) in a Beckman SW 28 rotor. After centrifugation the tightly coupled 70S ribosomes are separated from the subunits. The gradient is fractionated and the fractions of the 70S peak are pooled and pelleted (24 hr at 24,000 rpm, about 47,000*g*, in a 55.2 Ti rotor at 4°).

Ribosomal pellets are resuspended in ribosome buffer $H_{20}M_6N_{30}SH_4$ (100 μl per tube) by gently shaking for about 60 min at 4°. The resuspended particles are clarified by a low-speed centrifugation, and their concentrations are determined from the A_{260}.

Isolation of 30S and 50S Subunits

The crude ribosomes are loaded on a sucrose gradient (10–30% sucrose; about 100 A_{260} units per bucket) in dissociation buffer $H_{20}M_1N_{200}SH_4$ and centrifuged for 17 hr at 19,000 rpm (about 48,000*g* at 4°) in a Beckman SW 28 rotor. The centrifugation separates the subunits from tRNA (4 to 5S) and mRNA fragments. The gradient is fractionated and the 50S and 30S

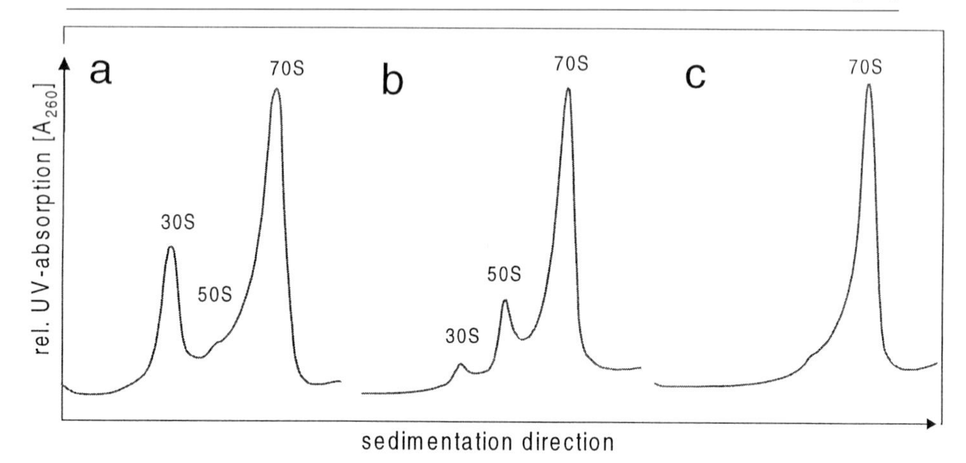

FIG. 1. Analytical sucrose gradient runs in reassociation buffer $H_{20}M_{20}K_{30}SH_4$. (a) A_{260} profile after incubating the 30S plus 50S mixture, (b) after isolating the 70S peak from the sucrose gradient, and (c) after a final incubation at 40° for 20 min.

peaks are individually pooled and pelleted (20 hr at 48,000 rpm, about 190,000*g*, in a 55.2 Ti rotor at 4°).

Ribosomal pellets are resuspended in reassociation buffer $H_{20}M_{20}K_{30}SH_4$ (100 μl per tube) by gently shaking for about 60 min at 4°. The resuspended particles are cleared by a low-speed centrifugation and their concentrations are determined from A_{260}.

Preparation of Reassociated 70S Ribosomes

The 50S and 30S subunits are mixed in a 1 : 1 ratio of A_{260} units, diluted to a final concentration of 40–140 A_{260}/ml in reassociation buffer $H_{20}M_{20}K_{30}SH_4$, and incubated for 40–60 min at 40° (Fig. 1a). By using an excess of 30S the amount of free 50S is minimized, thus improving the separation of the reassociated 70S ribosomes from the 50S subunits in the following gradient centrifugation. After incubation (10 min at 4°) the particles are subjected to a gradient centrifugation (10–30% sucrose; about 100 A_{260} units per bucket) in reassociation buffer $H_{20}M_{20}K_{30}SH_4$ and centrifuged for 17 hr at 18,000 rpm (about 35,000*g* at 4°) in a Beckman SW 28 rotor. The gradient is fractionated and the 70S peak is pooled and pelleted by centrifugation in a 50.2 Ti rotor for 27 hr at 21,000 rpm, about 40,000*g* at 4°. The use of higher centrifugation speeds is not recommended because it may lead to pressure-induced dissociation of the ribosomes.[21] The ribosomal

[21] M. Gross, K. Lehle, R. Jaenicke, and K. H. Nierhaus, *Eur. J. Biochem.* **218,** 463 (1993).

pellets are resuspended in reassociation buffer $H_{20}M_{20}K_{30}SH_4$ (Fig. 1b) and incubated 20 min at 40° (Fig. 1c). After the solution has been cleared by low-speed centrifugation, the concentration is determined from the A_{260}. Small aliquots (50 μl) are shock frozen in liquid nitrogen and stored at −80°. A homogenous population of reassociated ribosomes is obtained with a yield of around 75% (100% is the input of 50S in A_{260} units multiplied with 1.5). When using reassociated ribosomes in the poly(U)-dependent poly(Phe) synthesis or in the preparation of functional complexes, the ionic conditions have to be adapted to the ribosome buffer $H_{20}M_6N_{30}SH_4$.

Routinely, the quality of the preparation is checked using three assays: (1) A SW 40 run is performed (gradient of 10–30% sucrose in reassociation buffer $H_{20}M_{20}K_{30}SH_4$, 16 hr at 22,000 rpm, about 34,000*g*, and 4°) in order to test the homogeneity of the reassociated 70S (see Fig. 1c). (2) RNA gels are run in order to test the intactness of the ribosomal RNA. 16S and 23S RNA have to be essentially free from breaks. (1) and (2) establish the structural integrity of the particle. (3) The activity in the poly(U)-dependent poly(Phe) synthesis system is an important criterion for estimating the activity of the reassociated 70S preparation. Bommer *et al.*[17] describe this assay system in detail: it uses the postribosomal supernatant (S-100) as the source of enzymes, i.e., tRNA synthetases, the elongation factors EF-Tu, EF-G, and EF-Ts, etc. The ionic conditions are identical to those of the binding buffer $H_{20}M_6N_{150}SH_4Sp_{0,05}Spd_2$ that are also used for the preparation of functional complexes.

Preparation of Defined Functional Complexes

The two main states of the ribosome in the ribosomal elongation cycle, namely, the state before translocation or PRE state and the state after translocation or POST state,[22] are prepared and assayed in an experimental approach based on a procedure described by Watanabe in 1972.[23] The ability to produce homogeneous PRE and POST states is due to the high energy barrier between the two states. During elongation EF-Tu and EF-G catalyze the transition from POST to PRE and from PRE to POST, respectively.[22] Some antibiotics can fix the elongation factors on the ribosome. In particular, kirromycin stalls EF-Tu on the ribosome[11] and fusidic acid prevents dissociation of EF-G after translocation.[12]

The reaction scheme in Fig. 2 allows a controlled stepwise execution of one complete elongation cycle, thus providing a tool for the precise determination of the occupation of the three tRNA binding sites on the

[22] K. H. Nierhaus, *Nature (Lond.)* **379,** 491 (1996).

[23] S. Watanabe, *J. Mol. Biol.* **67,** 443 (1972).

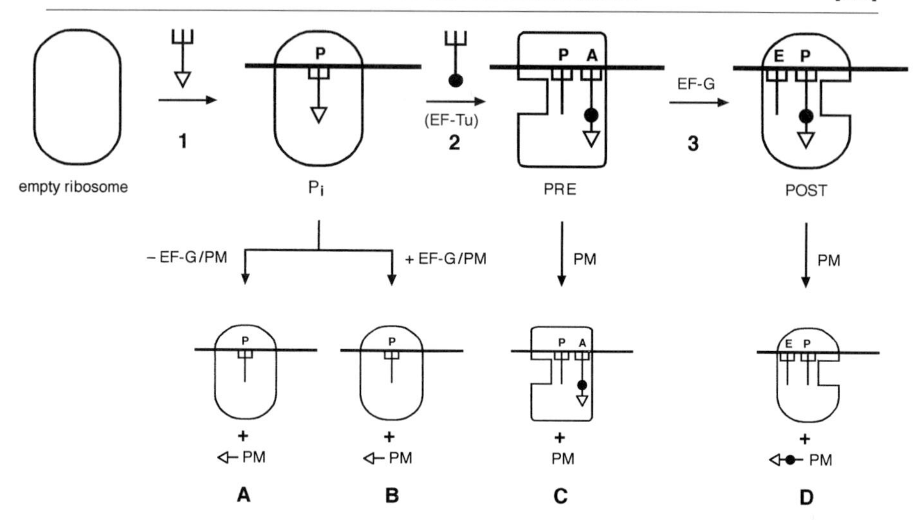

FIG. 2. Reaction scheme for the preparation of ribosomal functional complexes carrying tRNAs in defined positions. The P_i complex carries only one tRNA, namely, a peptidyl-tRNA in the P site, the PRE complex has a deacyl-tRNA in the P site and a peptidyl-tRNA in the A site, which move to the E and P sites on translocation (3), respectively. (A–D) Products of the puromycin reaction.

various complexes. In the first step, a 70S · mRNA · tRNA complex is formed, in which the tRNA is located in the ribosomal P site (note that the first tRNA to be bound to ribosomes usually occupies the P site). Either AcPhe-tRNA^{Phe} or fMet-tRNA^{fmet} are used, the α-amino group of which are blocked by an acetyl and formyl group, respectively, and therefore behave like a peptidyl-tRNA. A programmed ribosome that carries only one tRNA occupying the P site mimics an initiation complex and is called P_i complex (**i** for **i**nitiation). The P_i complex is assayed with and without EF-G in the puromycin reaction. No difference in reactivity with puromycin (Fig. 2, products A and B) indicates 100% P-site occupation. A similar highly defined P_i complex carrying an fMet-tRNA^{fMet} led to a cryoelectron microscopy reconstruction of the ribosome with the highest resolution obtained so far with a functional complex (better than 15 Å).[13]

In the second step the A site is filled with the corresponding cognate aminoacyl-tRNA. Peptidyl transfer is a fast step in the elongation cycle and occurs immediately. The PRE complex is translocated in the third step by the addition of EF-G. In the POST complex the tRNAs are now located in the P and E sites, respectively. The efficiency of the translocation reaction and the binding state of the tRNAs can be determined by means of the puromycin reaction: the P site bound peptidyl-tRNA will react (Fig. 2, product D), while the peptidyl-tRNA in the A site will not (Fig. 2, product C). PRE and POST states show typically about 80% homogeneity.[24] Binding values (tRNAs per ribosome) for each complex are assayed via nitrocellulose filter binding.

Single Reactions of the Elongation Cycle Produce Defined Functional States

We use a fully homologous system with respect to the source of ribosomes, tRNAs, factors, and enzymes in order to avoid ambiguities concerning tRNA binding features. The use of heteropolymeric mRNAs containing unique codons is to ensure unequivocal assignments of the bound tRNA to the various ribosomal sites,[25] although homopolymeric mRNAs, e.g., poly(U), in general yield higher binding values for tRNAs. Design and synthesis of heteropolymeric mRNAs are described by Bommer *et al.*[17] Near-physiologic conditions (polyamine buffer) with respect to the concentrations of the important ions are essential (see later discussion). The different mixes and buffers listed next are used to maintain the conditions of the binding buffer $H_{20}M_6N_{150}SH_4Sp_{0,05}Spd_2$ during the assay.

Buffers

Mix I: $H_{60}M_{18}N_{690}SH_{12}Sp_{0.25}Spd_{10}$: 60 m*M* HEPES–KOH (pH 7.5 at 0°), 18 m*M* $MgCl_2$, 690 m*M* NH_4Cl, 12 m*M* 2-mercaptoethanol, 0.25 m*M* spermine, 10 m*M* spermidine

Mix II: $H_{100}M_{30}N_{750}SH_{20}Sp_{0.25}Spd_{10}$: 100 m*M* HEPES–KOH (pH 7.5 at 0°), 30 m*M* $MgCl_2$, 750 m*M* NH_4Cl, 20 m*M* 2-mercaptoethanol, 0.25 m*M* spermine, 10 m*M* spermidine

Mix IIE (**E** for **e**nzymatic A-site occupation): $H_{40}M_{12}N_{300}SH_8Sp_{0.125}Spd_5$: 40 m*M* HEPES–KOH (pH 7.5 at 0°), 12 m*M* $MgCl_2$ 300 m*M* NH_4Cl, 8 m*M* 2-mercaptoethanol, 0.125 m*M* spermine, 5 m*M* spermidine

Mix III: $H_{66.7}M_{20}N_{500}SH_{13.3}Sp_{0.25}Spd_{10}$: 66.7 m*M* HEPES–KOH (pH 7.5 at 0°), 20 m*M* $MgCl_2$, 500 m*M* NH_4Cl, 13.3 m*M* 2-mercaptoethanol, 0.25 m*M* spermine, 10 m*M* spermidine

HMK buffer $H_{20}M_6K_{150}SH_4$: 20 m*M* HEPES–KOH (pH 7.5 at 0°), 6 m*M* $MgCl_2$, 150 m*M* KCl, 4 m*M* 2-mercaptoethanol

Ribosome buffer $H_{20}M_6N_{30}SH_4$: 20 m*M* HEPES–KOH (pH 7.5 at 0°), 6 m*M* $MgCl_2$, 30 m*M* NH_4Cl, 4 m*M* 2-mercaptoethanol

[24] J. Wadzack, N. Burkhardt, R. Jünemann, G. Diedrich, K. H. Nierhaus, J. Frank, P. Penczek, W. Meerwinck, M. Schmitt, R. Willumeit, and H. B. Stuhrmann, *J. Mol. Biol.* **266,** 343 (1997).

[25] A. Gnirke, U. Geigenmüller, H.-J. Rheinberger, and K. H. Nierhaus, *J. Biol. Chem.* **264,** 7291 (1989).

Binding buffer $H_{20}M_6N_{150}SH_4Sp_{0,05}Spd_2$: 20 m*M* HEPES–KOH (pH 7.5 at 0°), 6 m*M* $MgCl_2$, 150 m*M* NH_4Cl, 4 m*M* 2-mercaptoethanol, 0.05 m*M* spermine, 2 m*M* spermidine

P Site Binding: Construction of the Pi Complex

For construction of a P_i complex, incubate 110 pmol of 70S ribosomes (10 pmol per single determination) in 275 μl of binding buffer containing a 5- to 10-fold molar excess of heteropolymeric mRNA over ribosomes and fMet-tRNA in a 1.2- to 1.5-fold molar ratio to ribosomes for 30 min at 37°. The optimal amount of heteropolymeric mRNA and tRNA are determined according to their ability to saturate specific tRNA binding in filter binding assays. MF-mRNA is used as a heteropolymeric mRNA in many cases. This mRNA is 46 nucleotides long (a sequence of this length is covered by the ribosome) and contains in the middle the codons AUG (coding for Met)–UUC (Phe). We describe the preparation of 11 aliquots although only 10 will be used. The extra aliquot allows for minor pipetting imprecisions.

The 275 μl is pipetted in the following sequence:

27.5 μl of mix I $H_{60}M_{18}N_{690}SH_{12}Sp_{0.25}Spd_{10}$

27.5 μl of mix II $H_{100}M_{30}N_{750}SH_{20}Sp_{0.25}$ Spd_{10} (the separation into two mixes I and II is required for A-site binding as described in the next section; if only P-site binding is planned, these two mixes can be easily combined)

27.5 μl H_2O containing MF-mRNA

137.5 μl H_2O containing f[^{3}H]Met-tRNA^{fMet}

55 μl of ribosomes in ribosome buffer $H_{20}M_6N_{30}SH_4$

Incubate 30 min at 37° and then add 55 μl of a mix composed of mix III, H_2O, and a guanosine triphosphate (GTP) solution (2.5 m*M*, pH 6 with KOH) in a volume ratio of 3:4:3. Divide into 10 aliquots, 30 μl each. Continue as described in the Translocation section.

A-Site Binding: Construction of PRE Complex

When constructing a PRE complex the first site to be occupied with a tRNA is the P site. After this has been accomplished an acyl-tRNA is added in a 0.8- to 2-fold molar ratio to ribosomes in case of nonenzymatic A-site binding. For enzymatic A-site binding a ternary complex (aminoacyl-tRNA · EF-Tu · GTP) is formed immediately before its addition to the binding assay: aminoacyl-tRNA (1–2 pmol per picomole of 70S ribosomes), 0.5 m*M* GTP, and EF-Tu (1.2 pmol per picomole of aminoacyl-tRNA) is preincubated for 2 min at 37° under the ionic conditions of the binding buffer and is added to the reaction mixture.

First the P site is occupied by mixing

27.5 μl of mix I $H_{60}M_{18}N_{690}SH_{12}Sp_{0.25}Spd_{10}$

27.5 μl of H_2O containing the MF-mRNA

27.5 μl of H_2O with the fMet[^{3}H]tRNA^{fMet}

55 μl of ribosome buffer $H_{20}M_6N_{30}SH_4$ containing 110 pmol of ribosomes and incubating 10 min at 37°.

For nonenzymatic A-site occupation prepare a mix containing 27.5 μl of mix II $H_{100}M_{30}N_{750}SH_{20}Sp_{0.25}Spd_{10}$ and 110 μl of H_2O with [^{14}C]Phe-tRNA^{Phe}, and add to the ribosome mixture. For enzymatic A-site occupation prepare a mix containing 55 μl of mix IIE $H_{40}M_{12}N_{300}SH_8Sp_{0.125}Spd_5$, 33 μl of H_2O with [^{14}C]Phe-tRNA^{Phe}, 27.5 μl of HMK buffer containing EF-Tu, and 22 μl of a GTP solution (2.5 mM, pH 6). Incubate at 37° for 2 min and add to the ribosome mixture.

Incubate either of these mixes for 20 min at 37° and add 55 μl composed of 16.5 μl of mix III $H_{66.7}M_{20}N_{500}SH_{13.3}Sp_{0.25}Spd_{10}$, 22 μl H_2O, and 16.5 μl of a GTP solution (2.5 m*M*, pH 6 with KOH). Divide into 10 aliquots of 30 μl each.

The stability of the ternary complex can be checked by cold TCA precipitation. Samples of the preparation of the ternary complex are precipitated at various times (up to 20 min) and compared to samples of aminoacyl-tRNA without the factor under the same conditions. More than 90% of the initial radioactivity is precipitated after 10 min when EF-Tu is present, whereas a significant reduction (≥40%) is detected when the factor is not added. The precipitable material indicates intact aminoacyl-tRNA.

Translocation: Construction of POST Complex

POST complexes are constructed *via* an EF-G dependent translocation of the PRE complex. Add 2.5 μl of HMK buffer $H_{20}M_6K_{150}SH_4$ containing EF-G (0.3 pmol/pmol 70S) to five aliquots and incubate for 10 min at 37°. The other five samples are incubated with identical amounts of HMK buffer lacking EF-G.

Filter Binding Assay

Take two samples (30 μl each = 9.23 pmol ribosomes) with and without EF-G to measure the binding of tRNAs by nitrocellulose filtration. Mix the complexes with 2 ml ice-cold binding buffer $H_{20}M_6N_{150}SH_4Sp_{0,05}Spd_2$ and filtrate immediately through nitrocellulose filters that have been preincubated with binding buffer for 30 min. Wash twice with ice-cold binding buffer. The radioactivity retained on the filter is determined by liquid scintillation counting.

Include binding assays without ribosomes as standard controls in order to determine the filter background. This background is normally low (below 10% of the binding signal) and directly proportional to the concentration of the radioactive component in the assay.

Puromycin Reaction

The six remaining samples are used in the puromycin reaction: Add 2.5 μl of binding buffer to two control samples (*i.e.,* without puromycin and $\pm$ EF-G) as background for the puromycin reaction, and 2.5 μl of puromycin stock solution (10 m*M* in binding buffer $H_{20}M_6N_{150}SH_4Sp_{0,05}Spd_2$, final concentration 0.7 m*M*) to four samples (i.e., two with and two without EF-G) in order to determine the amount of A-site occupation. Incubate at 0° for about 12 hr and stop the reaction by adding 32.5 μl of 0.3 *M* sodium acetate, pH 5.5, saturated with $MgSO_4$. Extract with 1 ml of ethyl acetate: 1 min mixing, incubate 10 min at 0°, centrifuge briefly, and measure the radioactivity contained in 700 μl of the ethyl acetate phase by liquid scintillation counting (corresponds to 7-pmol ribosomes).

The radioactivity extracted in the controls (minus puromycin) is subtracted from the sample containing puromycin in order to calculate the amount of fMet-puromycin formed.

A successful puromycin reaction depends critically on the way in which the puromycin stock solution is prepared and handled. Two basic rules for the preparation of a puromycin stock solution must be observed to achieve maximal activity: (1) The pH of the solution must be neutral. Since the puromycin is obtained commercially as a hydrochloride, the pH of the solution has to be neutralized by adding 1 *M* KOH (2% of the volume). (2) The puromycin stock solution must be kept at room temperature (otherwise it precipitates, lowering the effective concentration). Under these conditions the stock solution retains its maximum activity for about 1 hr.

For cryoelectron microscopy analysis the complexes are diluted with binding buffer $H_{20}M_6N_{150}SH_4Sp_{0,05}Spd_2$ to 1.4 A_{260}/ml (about 35 pmol/ml), the concentration that is used to produce sample grids. This is done just before the preparation of the grids. Due to the strong dilution of the sample, the ribosomes are well separated from unbound ligands, and therefore purification of the ribosome complexes is usually not necessary. In cases that require further purification suitable methods are described in the next section.

Purification of Ribosomal Complexes and Determination of Stoichiometry of Ligands

Ribosomal complexes can be separated from free ligands by several methods such as centrifugation of the complex through a sucrose cushion, gel filtration by gravity flow, or spun columns. The determination of the stoichiometry of tRNA binding is particularly easy, since it can be achieved by nitrocellulose filtration if the tRNA is radioactively labeled, as in the examples given in the preceding sections. (Charging of tRNAs with radioactive amino acids is described by Rheinberger *et al.*[26] and Triana-Alonso *et al.*,[26a] this volume.) For cases such as factor binding to ribosomes, separation and evaluation of the binding stoichiometry is described later.

Isolation of Ribosome Complexes via Centrifugation through Sucrose Cushion

A 100-μl sample containing 5–10 A_{260} of ribosomes or ribosomal subunits in binding buffer $H_{20}M_6N_{150}SH_4Sp_{0.05}Spd_2$ is loaded on a 200-μl 15% sucrose cushion in binding buffer (sucrose ultrapure, GibcoBRL, UK). After centrifugation for 3 hr at 4° and 40,000 rpm (about 57,000*g*) in a Beckman TLA 120.1 rotor, the pellet is resuspended in 20 μl of binding buffer (yield 30–40% of input).

Isolation of Ribosomal Complexes via Gel Filtration

The choice of the gel matrix depends on the size of the ligand. Here we describe a procedure using Sephadex G-100 DNA grade (Pharmacia, Uppsala, Sweden) that effectively separates ligands of up to 100 kDa from the ribosome.

The sample (100 μl, 5–6 A_{260} of complexes with ribosomes or ribosomal subunits) is applied to a small column (Pasteur pipette closed with glass wool) containing about 2 ml of gel matrix. The column matrix was preequilibrated with binding buffer $H_{20}M_{6,}N_{150}SH_4Sp_{0.05}Spd_2$. Fractions (100 μl) are collected under gravity flow (ribosomal complexes elute in fractions 4–8).

Isolation of Ribosome Complexes via Spun Column

A 100-μl sample containing 2–6 A_{260} of ribosomal complex is loaded onto a cDNA spun column S300 (Pharmacia, Uppsala, Sweden) that has been preequilibrated with binding buffer $H_{20}M_6N_{150}SH_4Sp_{0.05}Spd_2$. The column is centrifuged for 2 min at 1500 rpm at 4° in a Sorvall HB4 rotor. The flow-through (fraction 1) is collected in a microcentrifuge tube. The

[26] H.-J. Rheinberger, U. Geigenmüller, M. Wedde, and K. H. Nierhaus, *Methods Enzymol.* **164,** 658 (1988).

[26a] F. J. Triana-Alonso, C. M. T. Spahn, N. Burkhardt, B. Rhosdamy, and K. H. Nierhaus, *Methods Enzymol.* **317,** [17], 2000.

next fractions are collected by loading 100 μl of binding buffer $H_{20}M_6N_{150}SH_4Sp_{0.05}Spd_2$ and repeating the centrifugation procedure (ribosomal complexes are eluted in fractions 1–3).

Determination of Stoichiometry of Ribosomal Complexes

The amount of ribosomes in a sample is usually assessed via A_{260} measurement. The following equivalence rules might be helpful: 1 A_{260} 70S ribosomes = 24 pmol 70S; 1 A_{260} 50S subunits = 36 pmol 50S; 1 A_{260} 30S subunit = 72 pmol 30S.

Radioactively Labeled Ligand. If the specific activity of the ligand is known, the molar ratio of ligand to ribosome can be determined easily. The A_{260} reflects the amount of ribosomes while the radioactivity indicates the amount of the ligand in the same volume element. If the ligand is a protein, the main challenge is to isolate the protein in a radioactively labeled form; this can be done *in vivo* and *in vitro.*

In vivo incorporation of radioactive labeled amino acids can be achieved via supplying a labeled amino acid to the medium of a strain that is auxotrophe with respect to the same amino acid or, alternatively, the biosynthesis of the corresponding amino acid can be inhibited.[27] An easy and established method for *in vitro* labeling is the reductive methylation or alkylation of ε-NH_2 groups on the lysine side chains of a protein.[28,29]

If a ligand interacts with a secondary ligand (e.g., GTP) the ratio (ligand : ribosome) can be determined via this secondary ligand if it is labeled. In the case of Tet(O), a protein conferring resistance to tetracycline, [^{35}S]GTPγS was used as a secondary ligand.[30]

Identification of Ligand Binding via Staining Methods. Additionally, determination of the protein–ribosome stoichiometry is possible by using SDS–PAGE following by Coomassie staining or Western blotting and digitalization of the stained bands including the ligand–protein and some ribosomal proteins such as L2. This method requires that standards with different and known ratios of protein : ribosome be applied to the same SDS gel. A well-tried procedure for the standard calibration curve is done as follows: After digitalization of the bands the ratio (density of the band of the protein ligand : density of the L2 band) is plotted against the respective ratio (input of the protein ligand : input of ribosomes). The ratio of the density of the band from the bound ligand to that of L2 of the binding ribosomes reliably gives the stochiometry. L2 is the largest ribosomal protein of the 50S subunit

[27] S. Doublié, *Methods Enzymol.* **276,** 523 (1997).

[28] G. M. Wystup and K. H. Nierhaus, *Methods Enzymol.* **59,** 776 (1979).

[29] N. Jentoft and D. G. Dearborn, *Methods Enzymol.* **91,** 570 (1983).

[30] C. A. Trieber, N. Burkhardt, K. H. Nierhaus, and D. E. Taylor, *Biol. Chem.* **379,** 847 (1998).

and the second largest of 70S ribosomes; L2 is clearly separated from the other ribosomal proteins. It is not advisable to use the large ribosomal protein S1 of the small ribosomal subunit for this method, since S1 can be easily sheared from the ribosome. S4 is a good candidate for small ribosomal subunits to be used as a reference protein.

The stoichiometry of ligands bound to ribosomes can also be estimated by cryoelectron microscopy: The average density of the ligand is compared with a distinct part of the ribosome, for example, the L1 protuberance.[12]

In the rest of this article, we present and discuss results of 3-D cryo-EM reconstructions that illustrate the importance of buffer conditions in determining the positions of tRNAs on the ribosomes.

Location of tRNAs on Ribosome as Seen by 3-D Cryoelectron Microscopy

Three tRNA binding sites are known on the ribosome. Based on the puromycin reaction, two tRNA binding sites are operationally defined; acyl-tRNAs at the A site (A for acceptor) do not react with puromycin. The A site harbors the decoding center that is located on the small ribosomal subunit. Acyl-tRNAs at the P site (P for peptidyl) do react with puromycin. The third tRNA binding site, the E site (E for exit), is specific for deacyl-tRNA. A detailed description and a discussion of the current models of the ribosomal elongation cycle can be found in reviews.[16,31]

In early cryoelectron microscopy studies, three primary tRNA positions were visualized on the ribosome.[9,10] The A site was identified close to the L7/L12 stalk of the ribosome, the P-site tRNA spans the intersubunit space from the neck of the small subunit to the 50S subunit, and the E-site tRNA was observed close to the mushroom-shaped L1 protuberance (Fig. 3a; see color insert). Both studies placed the anticodon of the E-site tRNA far from the anticodon of the P-site tRNA; however, the overall positions of the E-site tRNA derived from the two studies were remarkably different. Subsequent studies[14,32] assigned the E-site tRNA to a position drastically different from those two, namely, close to the P-site tRNA of a POST complex, placing the E-site and P-site tRNAs in a similar mutual arrangement as the P- and A-site tRNAs in a PRE complex (Fig. 3b). Because the latter E site, which was derived by EF-G-dependent translocation of a PRE complex, was considered to be the authentic E site (see later section), the former one derived from the first visualization study[9] was termed E2

[31] K. S. Wilson and H. F. Noller, *Cell* **92,** 337 (1998).

[32] R. K. Agrawal, A. B. Heagle, P. Penczek, R. A. Grassucci, and J. Frank, *Nature Struct. Biol.,* **6,** 643 (1999).

TABLE I
ION CONCENTRATIONS OF POLYAMINE BUFFER AND CONVENTIONAL BUFFER SYSTEMS WITH CORRESPONDING CONCENTRATIONS *In Vivo*[a]

			Polyamines		
System	Mg^{2+}	K^+, NH_4^+	Spermidine	Spermine	Putrescine
Conventional	7–20	100	None	None	None
Polyamine	3–6	150	2	0.05	None
In vivo	~4	~150	1–4	~0.03	20

[a] Concentrations (m*M*) of ions and polyamines are important for ribosomal functions *in vitro* in various buffer systems and *in vivo*. The comparison shows that the polyamine buffer[b] matches *in vivo* conditions[c–f] closely. Conventional buffers do not contain polyamines but higher Mg^{2+} concentrations.[34] K^+ and NH_4^+ are more or less equivalent in *in vitro* systems of protein synthesis due to their similar ionic radii. Putrescine has no effects on protein synthesis *in vitro* (B. Lewicki and K. H. Nierhaus, unpublished).

[b] A. Bartetzko and K. H. Nierhaus, *Methods Enzymol.* **164,** 650 (1988).

[c] J. E. Lusk, R. J. P. Williams, and E. P. Kennedy, *J. Biol. Chem.* **243,** 2618 (1968).

[d] M. Kamekura, K. Hamana, and S. Matsuzaki, *FEMS Microbiol. Lett.* **43,** 301 (1987).

[e] B. Richey, D. S. Cayley, M. C. Mossing, C. Kolka, C. F. Anderson, T. C. Farrar, and M. T. J. Record, *J. Biol. Chem.* **262,** 7157 (1987).

[f] C. W. Tabor, and H. Tabor, *Microbiol. Rev.* **49,** 81 (1985).

site.[16] Furthermore, a deacylated tRNA that should be present at the P site according to general wisdom was found at two different positions when the ribosomes carried only one tRNA.[33] That study showed that an fMet-tRNA binds to the P site in the same position under different buffer conditions (Figs. 4a and 4b; see color insert), whereas the position of a deacylated tRNA varies with the buffer condition; it binds mostly in a "hybrid" P/E position when the conventional buffer (Table I) is used (Fig. 4c), and at the P site, in a position identical with that for fMet-tRNA (Fig. 4d), when polyamine buffer is used.

Importance of Charging State of tRNA and the Buffer Conditions for Position of tRNAs on Ribosome

The cryo-EM studies, which revealed five distinct tRNA locations, were done under a variety of conditions differing mainly in the charging state of the tRNA and the ionic conditions. Apparently, both parameters can drastically influence the position of tRNAs.

[33] R. K. Agrawal, P. Penczek, C. M. T. Spahn, R. A. Grassucci, K. H. Nierhaus, and J. Frank, in preparation (1999).

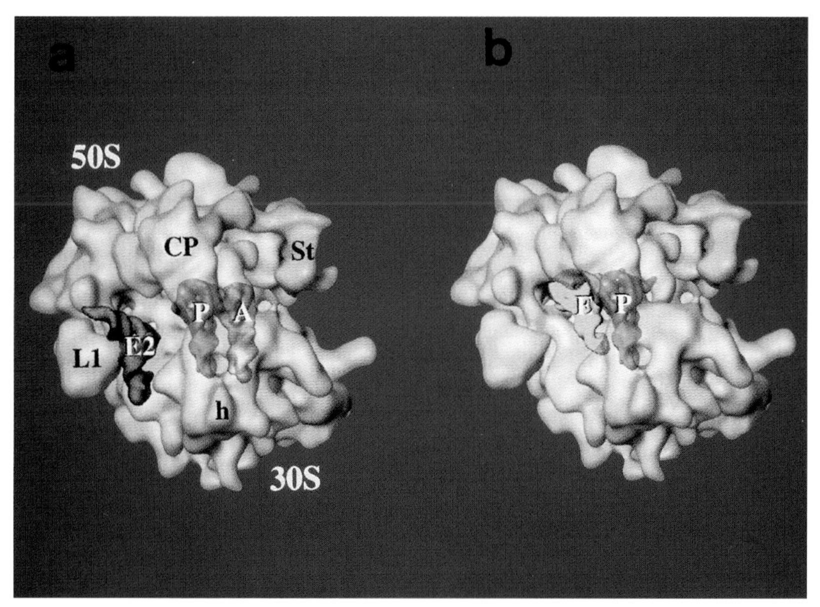

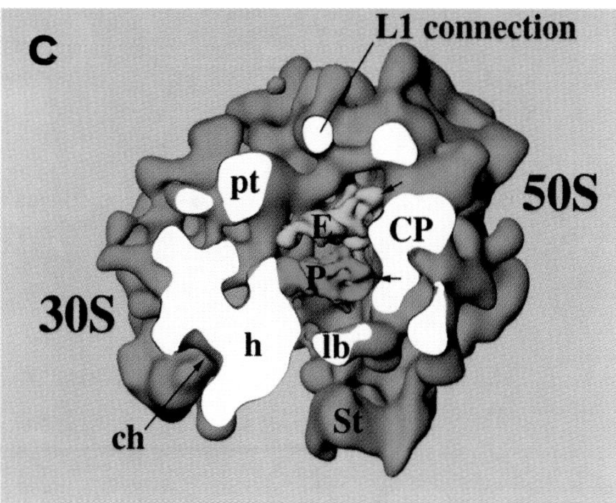

FIG. 3. tRNA positions identified by cryoelectron microscopy when more than one tRNA was bound to programmed ribosomes. (a) The complex was formed under conventional buffer conditions. Occupation of the E site is not observed but rather a tRNA bound to L1 in a presumably labile fashion, in a position where no codon–anticodon interaction is possible. This position is termed "E2" (brown) in order to avoid confusion with the E site. (b) The POST complex was formed under the conditions of the polyamine buffer. Both tRNAs in P (green) and E (yellow) sites can undergo codon–anticodon interactions. We note that acyl-tRNAs are found in the same A (pink) and P (green) positions under both buffer conditions. The view is from the top of the transparent 70S ribosome, with the 30S ribosomal subunit in front (h, head) and the 50S subunit at the back (L1, L1 protuberance; CP, central protuberance; St, L7/L12 stalk). Without transparent display of the ribosome, the 30S head and the central protuberance of the 50S subunit would cover the tRNAs at the A and P sites completely. (Reproduced with permission from Agrawal *et al.*[14]). (c) Cut-open view of a ribosome in the POST state with tRNAs at the P (green) and E (yellow) sites. Note that the tRNA at the E site is surrounded by ribosomal matrix.

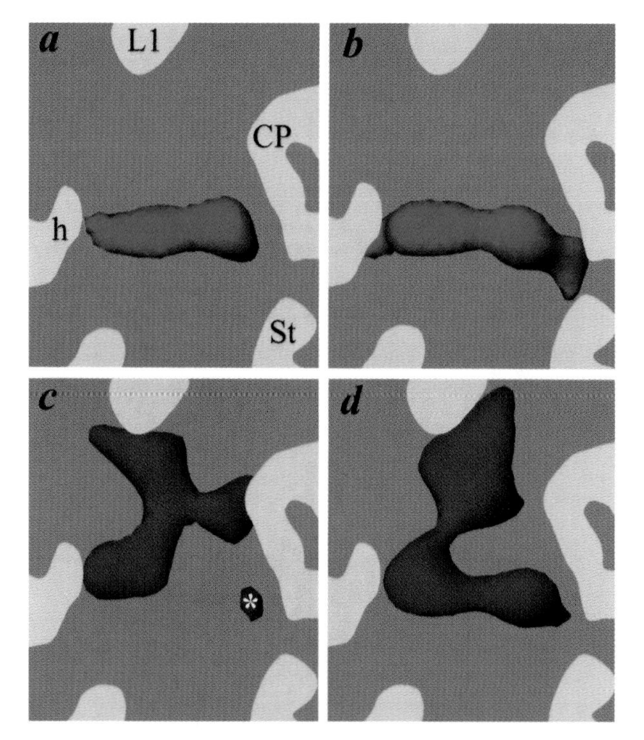

FIG. 4. Difference maps of programmed ribosomes. fMet-tRNA was bound to the P site under conditions of (a) a conventional buffer system and (b) the polyamine buffer. Deacylated tRNA was bound to the so-called P site in the presence of (c) conventional buffer and (d) polyamine buffer.

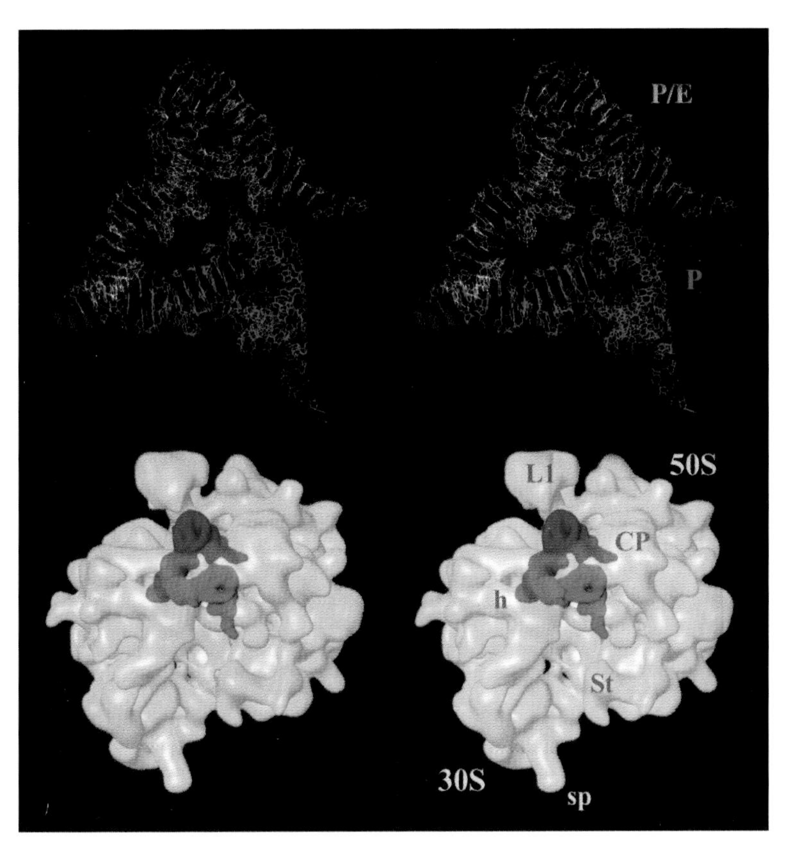

FIG. 5. A comparison of tRNAs observed at the P site and the hybrid site P/E. Note that the anticodon regions (red) of both tRNAs overlap. (Reproduced with permission from Agrawal *et al.*[14])

E and E2 Position. It is known from biochemical data that the features of the E site are strongly dependent on the buffer conditions.[34–36] Deacyl-tRNA is stably bound in the polyamine buffer and it interacts with the mRNA via codon–anticodon interaction after translocation whereas in the conventional buffer system these features are not seen. The different behaviors of a deacyl-tRNA bound to the E site are reflected by the two different positions seen in cryoelectron microscopy reconstructions:

1. In the conventional buffer system the tRNA is found in the E2 site apparently mainly interacting with the L1 stalk.[9] A number of arguments suggest that this tRNA binding site is of secondary importance: (a) L1 is not essential for cell viability[37]; (b) deacyl-tRNA does not form a stable complex with *isolated* ribosomal protein L1 (U. Stelzl, G. Blaha and K. H. Nierhaus, unpublished observations); (c) labile binding to the E site, which was claimed to be characteristic for this site,[34] is not observed in native polysomes where most of the ribosomes carry a deacylated tRNA at the E site in a stable fashion.[18,36]
2. As shown in Fig. 3b, a ribosomal POST state in the polyamine buffer shows the E-site tRNA with the anticodon tips as well as the CCA ends of both P- and E-site tRNAs in close mutual proximity.[14,32] Neutron scattering analysis of a PRE and POST complex reveals a similar mutual arrangement of these tRNAs on the ribosome,[38] in good agreement with the E position. A number of arguments support the view that this site is the authentic E site: (a) A deacylated tRNA at the E site is found to bind into a pocket-like structure in the 50S subunit making intimate contacts with the ribosome (Fig. 3c), in contrast to a tRNA at the E2 site where it has only contacts with the L1 protuberance mainly along the inner bent of its elbow region. In excellent agreement, the iodine-induced cleavage pattern of a deacylated tRNA bound to the P site remains the same during translocation to the E site.[39] These intensive contacts explain the tight binding of tRNA in the E site in native polysomes. (b) The E site, and not the E2 site, is seen in POST complexes when the polyamine buffer is used, although the mass of density representing the E-site tRNA is always weak.

[34] Y. P. Semenkov, M. V. Rodnina and W. Wintermeyer, *Proc. Natl. Acad. Sci. U.S.A.* **93,** 12183 (1996).

[35] H. J. Rheinberger and K. H. Nierhaus, *J. Biomol. Struct. Dyn.* **5,** 435 (1987).

[36] K. H. Nierhaus, R. Jünemann, and C. M. T. Spahn, *Proc. Natl. Acad. Sci. U.S.A.* **94,** 10499 (1997).

[37] A. R. Subramanian and E. R. Dabbs, *Eur. J. Biochem.* **112,** 425 (1980).

[38] K. H. Nierhaus, J. Wadzack, N. Burkhardt, R. Jünemann, W. Meerwinck, R. Willumeit, and H. B. Stuhrmann, *Proc. Natl. Acad. Sci. U.S.A.* **95,** 945 (1998).

[39] M. Dabrowski, C. M. T. Spahn, M. A. Schäfer, S. Patzke, and K. H. Nierhaus, *J. Biol. Chem.* **273,** 32793 (1998).

P and P/E Positions. The different P-site positions of deacylated tRNA have been systematically examined.[15] In a conventional buffer, a deacyl-tRNA is seen in a position that may be interpreted as the P/E hybrid site postulated by Moazed *et al.*[40] The anticodon mass shows up in the anticodon region of the P-site tRNA, the elbow and CCA end point to the E site (Fig. 5; see color insert). In the polyamine buffer, both the deacyl-tRNA and the charged fMet-tRNA are positioned in the P site,[15] even though a small fraction of deacylated tRNA is always present in the P/E position when decylated tRNA was bound to the so-called P site.

Conclusions

It has been suggested that the buffer conditions (e.g., Mg^{2+} concentration) have a strong influence on the tRNA binding behavior of the ribosome.[35,41] Cryo-EM has now shown that different buffers lead to markedly different observed positions of the ribosome-bound tRNA. In addition to positions that can be attributed to the A, P, and E sites, two new positions, termed P/E and E2, are observed almost exclusively in the conventional buffer system, with E-site occupancy markedly reduced.

These results are interesting in the context of the long-standing controversy regarding the role of the E site in protein synthesis. According to one view,[42] the E site plays a minor role as it is seen to merely facilitate the dissociation of the tRNA from the P site where it is bound too tightly to allow an energetically easy release from the ribosome. According to the other view,[43,44] the role of the E site is essential in protein synthesis, because codon–anticodon interaction at this site is seen to have two important functions: that it helps to maintain the reading frame during the translocation reaction, and that it signals the ribosome to adopt the POST state. The POST state so defined, in turn, is a prerequisite for the accurate selection of the cognate ternary complex during the decoding process[43,45]

The cryo-EM results have a bearing on the controversy in the following way: Whether or not tRNA in a given assay forms codon–anticodon contact at the third tRNA binding site (which is possible according to the position of the anticodon of the observed E-site tRNA) is evidently dependent on which site is predominantly occupied, which in turn (again according to cryo-EM, as we have seen) depends on the buffer conditions. In the conven-

[40] D. Moazed and H. F. Noller, *Nature* **342,** 142 (1989).
[41] R. Lill and W. Wintermeyer, *J. Mol. Biol.* **196,** 137 (1987).
[42] J. M. Robertson and W. Wintermeyer, *J. Mol. Biol.* **196,** 525 (1987).
[43] K. H. Nierhaus, *Biochemistry* **29,** 4997 (1990).
[44] K. H. Nierhaus, *Mol. Microbiol.* **9,** 661 (1993).
[45] U. Geigenmüller and K. H. Nierhaus, *EMBO J.* **9,** 4527 (1990).

tional buffer system, the weak occupancy of the E site yet strong occupancy of other sites termed P/E and E2 would lead to the conclusion that codon–anticodon interaction at the third tRNA binding site plays no essential role. However, a significant influence of codon–anticodon interaction on the occupation of the third tRNA binding site has been observed under assay conditions, where binding to the E2 site is favored,[41] a site where the anticodon is too far away to contact the condon. It is therefore possible that the buffer conditions do not influence the tRNA position per se, but that E and E2 site are in equilibrium with each other, such that, dependent on the buffer conditions, the equilibrium is shifted toward the E site (polyamine buffer) or the E2 site (conventional buffer). Thus, depending on the buffer conditions used, one would come to opposite conclusions regarding the properties of the third tRNA binding site.

As to the significance of the observed P/E and E2 positions, it is quite possible that both represent transitional states that are favored under the conventional buffer conditions, while only transiently occupied under the conditions of the polyamine buffer and probably also in vivo. Thc P/E position might in fact represent a "frozen" hybrid state,[40] while the E2 position, with the tRNA attached to the L1 protein, might reflect a deacyl-tRNA released from the E site and on its way to leave the ribosome.

Acknowledgments

We thank Sean Connell for help and discussions. This work was supported in part, by the Volkswagen-Stiftung (1/72 341) and the Deutsche Forschungsgemeinschaft (Ni 174/8-2) to K.H.N. and by NIH grants 1R37 GM29169 and 1RO1 GM55440 (to J. F.).

Section III

Techniques for Monitoring RNA Conformation and Dynamics

A. Solution Methods
Articles 20 through 22

B. Electrophoretic and Spectroscopic Methods
Articles 23 through 26

[20] Probing RNA Structure and Function by Circular Permutation

By TAO PAN

Introduction

A circularly permuted (CP) RNA is generated by connecting its normal 5′ and 3′ ends and cleaving the ribose phosphate backbone at another site (Fig. 1A). An RNA can therefore have the same number of CP isomers as its total number of nucleotide residues. Circular permutation can affect the structure and function of RNA in several ways: (1) A trivial case involves an RNA where the normal 5′ or 3′ ends participate directly in the function. The 3′ end of a transfer RNA is strictly required for aminoacylation, so that circular permutation cannot be used for the analysis of tRNA function. However, the tertiary folding of a tRNA does not involve the normal 5′ or 3′ ends, so that circular permutation can be applied to analyze tRNA folding.[1] (2) Breaking the RNA backbone at another site changes the chemical composition of the phosphodiester bond. The 5′-phosphate, 3′-phosphate, and 2′-OH group are converted to other functional groups (Fig. 1B). Circular permutation is therefore expected to affect interactions involving the natural phosphodiester bond or the 2′-OH group at the CP position. (3) Circular permutation can affect the thermodynamic and kinetic properties of the wild-type RNA. Interestingly, breaking the RNA backbone does not necessarily destabilize the RNA structure. For example, a break in the middle of an RNA helix increases the stability of the helix.[2] A CP RNA can have a different folding rate or even altered folding pathway as demonstrated for a bacterial RNase P RNA in which the folding kinetics for the wild-type RNA is limited by disruption of a kinetic trap.[3] Circular permutation at a particular position destabilizes the trapped species so that folding becomes much faster and goes through a different pathway.[4]

Although circular permutation can be considered as a form of backbone modification, it has two unique aspects compared to other modification methods: (1) An experimental procedure, termed circular permutation analysis (CPA), allows structure–function analysis of all CP RNAs in a single

[1] T. Pan, R. R. Gutell, and O. C. Uhlenbeck, *Science* **254,** 1361 (1991).
[2] A. E. Walter, D. H. Turner, J. Kim, M. H. Lyttle, P. Muller, D. H. Mathews, and M. Zucker, *Proc. Natl. Acad. Sci. U.S.A.* **91,** 9218 (1994).
[3] T. Pan and T. R. Sosnick, *Nature Struct. Biol.* **4,** 931 (1997).
[4] T. Pan, X.-W. Fang, and T. R. Sosnick, *J. Mol. Biol.* **286,** 721 (1999).

0076-6879/00 $30.00

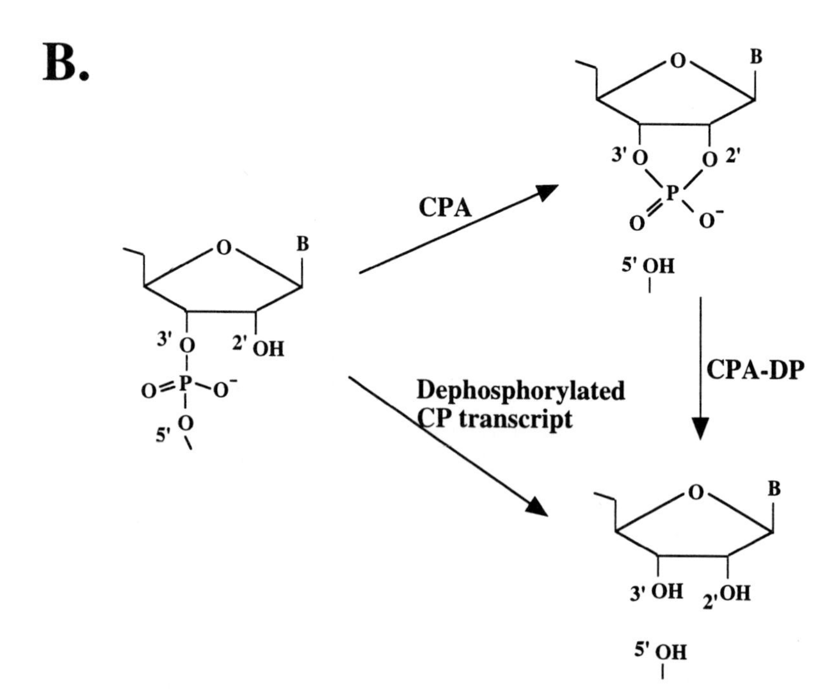

FIG. 1. (A) The idea of circular permutation as shown for hypothetical RNA. (B) The chemical composition of the RNA backbone after circular permutation.

experiment.[5] This procedure vastly improved our ability to quickly examine the property of all ribose phosphate positions. A modified procedure, termed CPA coupled with dephosphorylation (CPA-DP), can determine the function of CP isomers through increased activity, allowing the results to be interpreted with greater confidence.[6] In contrast, most chemical modification procedures assess the structure–function of ribose phosphate positions through decreased activity.[7] (2) Similar to base mutations, a CP RNA can be easily made in large quantities, so that the property of such a backbone modification can be tested at will. Synthesis of large amounts of an RNA containing a different kind of modified backbone is still difficult to accomplish.

This article describes how to apply circular permutation to probe the structure and function of RNA. Five examples are provided that describe the application of CPA in the analysis of RNA folding, protein–RNA interaction, substrate binding, and catalysis by a bacterial RNase P RNA. These examples illustrate the feasibility of using circular permutation to analyze the structure and function of biologically relevant RNAs.

Circular Permutation Analysis

Circular permutation implies covalent linkage of the normal 5′ and 3′ ends of an RNA. The first consideration in applying circular permutation is to make the circular RNA and to assay for its structure and function. This task is easier for an RNA whose secondary structure around the normal 5′ and 3′ ends is known. Ideally, the 5′ and 3′ ends are located at one site of an RNA helix so that only a new hairpin loop is generated in the circular RNA. If the termini are not known to be located this favorably, an artificial linker region may be added. The size and the sequence of the linker will depend on the particular RNA under study.

Circular RNA can be synthesized by linking the normal 5′ and 3′ ends of an RNA with either T4 RNA ligase[8,9] or T4 DNA ligase and an DNA oligonucleotide splint[10] or through the action of a ribozyme.[11,12] The effi-

[5] T. Pan and O. C. Uhlenbeck, *Gene* **125,** 111 (1993).

[6] T. Pan, A. Loria, and K. Zhong, *Proc. Natl. Acad. Sci. U.S.A.* **92,** 12510 (1995).

[7] C. Ehresmann, F. Baudin, M. Mougel, P. Romby, J.-P. Bel, and B. Ehresmann, *Nucleic Acids Res.* **15,** 9109 (1987).

[8] T. E. England, A. G. Bruce, and O. C. Uhlenbeck, *Methods Enzymol.* **65,** 65 (1980).

[9] T. Pan and C. U. Uhlenbeck, *Biochemistry* **31,** 3887 (1992).

[10] M. J. Moore and P. A. Sharp, *Science* **256,** 992 (1992).

[11] P. R. Bohjanen, R. A. Colvin, M. Puttaraju, M. D. Been, and M. A. Garcia-Blanco, *Nucleic Acids Res.* **24,** 3733 (1996).

[12] M. Puttaraju and M. D. Been, *J. Biol. Chem.* **271,** 26081 (1996).

ciency and the choice of ligase depend on the structure and the sequence of the 5′ and 3′ ends. T4 RNA ligase is very efficient for RNAs with one unstructured 5′ nucleotide and three unstructured 3′ nucleotides. If the 5′ nucleotide is located in a helix, then the method with T4 DNA ligase should be used. A basic protocol for using T4 RNA ligase to generate circular RNA is described next; this protocol works well for circularization of RNAs containing more than 70 nucleotides. A modified protocol using T4 RNA ligase and a DNA template works well for circularization of RNAs containing fewer than 30 nucleotides.[13]

Protocol 1: RNA Circularization Using T4 RNA Ligase

Note: The RNA should be purified by denaturing gel electrophoresis and stored in water prior to the application of this procedure.

1. The RNA in 10 m*M* Tris-HCl, pH 7.6, is incubated at 90° for 2 min followed by incubation at room temperature for 3 min.
2. A 10× buffer solution is added for a final condition of 50 m*M* Tris-HCl, pH 7.6, 10 m*M* $MgCl_2$, and 10 m*M* 2-mercaptoethanol. The mixture is incubated at 37° for 10 min.
3. Adenosine triphosphate (ATP) and dimethyl sulfoxide (DMSO) are added to a final concentration of 200 μM and 15%, respectively. T4 RNA ligase (e.g., from New England Biolabs, Beverly, MA) is added to a final concentration of 1 U/μl. The mixture is incubated at 16° or 37° for at least 1 hr.

   Note: The final RNA concentration should be ~1–5 μM. Higher RNA concentration promotes formation of dimers, whereas lower RNA concentration requires larger reaction volume to produce the same amount of circular RNA. The incubation temperature and the length of incubation (1 hr–overnight) should be optimized for each RNA. At least 50% of linear RNA should be converted to circular RNA under optimal conditions.
4. The ligated product is purified by polyacrylamide gel electrophoresis (PAGE) containing 7 *M* urea. The circular RNA migrates slower than the linear RNA on a denaturing gel. All products migrating slower than the linear RNA should be purified separately. Step 5 will be applied to determine which product is the circular RNA (as compared to, e.g., a dimer).
5. Partial alkaline hydrolysis (see protocol 2, step 1) is carried out with the purified RNA from step 4. The hydrolysis reaction is analyzed

[13] L. Wang and D. E. Ruffner, *Nucleic Acids Res.* **26,** 2502 (1998).

by denaturing PAGE. A circular RNA should produce a single major product of the same length as the linear RNA control.

The circular RNA should be assayed for its function prior to the CPA experiment. In some cases, circularization has no effect, as observed for the *Bacillus subtilis* RNase P RNA[14,15] and the R17 coat protein binding site.[16] In other instances, circularization has a moderate effect, as observed for the cleavage of pre-tRNA substrates by P RNA.[6,17] There is no general guideline in determining the usefulness of a circular RNA based on the magnitude of the circularization effect. If circularization strongly decreases the activity of the RNA but the normal 5′ and 3′ ends are not directly involved in folding or function, a different linker may be used to alleviate this problem.

Once the circular RNA passes the structure and function test, circular permutation analysis (Fig. 2A) can be attempted to examine the effect of circular permutation at all backbone positions in a single experiment. There are four steps in CPA: (A) The circular RNA is subjected to partial alkaline hydrolysis under denaturing conditions to produce all CP isomers at roughly equal amounts. (B) This mixture of CP isomers is renatured. (C) The active CP isomers are physically separated from inactive CP isomers on the basis of their folding or function. (D) The termini positions of the active CP isomers are determined either by reverse transcription or by comparing the length of CP RNA products in endonucleolytic cleavage reactions.

Protocol 2: Circular Permutation Analysis (CPA)

Note: The circular RNA should be purified by denaturing gel electrophoresis and stored in water prior to the application of this procedure.

Generation of CP Mixture

1. The circular RNA is mixed with a 5× alkaline hydrolysis buffer at a final condition of 1 m*M* glycine, 0.4 m*M* $MgSO_4$, pH 9.5.

   Note: The Mg^{2+} concentration is too low for stable folding at 100°, but the presence of Mg^{2+} accelerates the hydrolysis. $NaHCO_3$ is less ideal for CPA since it is generally used at >20 m*M*, making subsequent buffering more difficult.
2. The sample is boiled at 100° for 15–60 sec.

   Note: The length of hydrolysis needs to be optimized where no

[14] T. Pan and K. Zhong, *Biochemistry* **33,** 14207 (1994).
[15] T. Pan and M. Jakacka, *EMBO J.* **15,** 2249 (1996).
[16] J. M. Gott, T. Pan, K. A. LeCuyer and O. C. Uhlenbeck, *Biochemistry* **32,** 13399 (1993).
[17] J. M. Nolan, D. H. Burke, and N. R. Pace, *Science* **261,** 762 (1993).

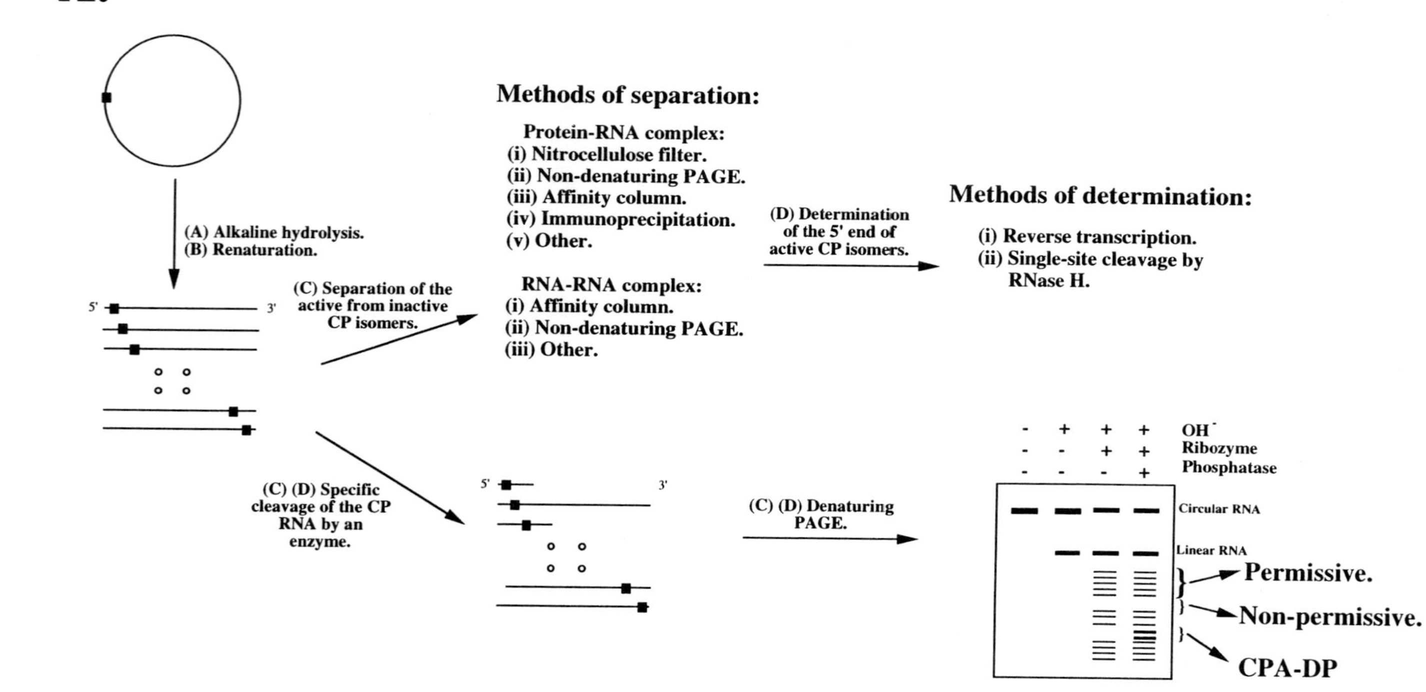

FIG. 2. (A) The general strategies of CPA. Steps C and D along the top pathway describe CP isomers that remain as the full-length molecule throughout the experiment. Steps C and D along the bottom pathway describe CP isomers that are enzymatically cleaved at a specific site. The circular RNA is shown as a circle and the linear RNAs are shown as lines. The filled square in the circular RNA represents a unique ^{32}P label. (B) The position of the ^{32}P label (*p) determines the correlation between the length of the radioactive fragment to the phosphate position in the wild-type RNA as shown for a circular RNA of 99 residues. The specific cleavage site is indicated by an arrow between nucleotides 99 and 1. The phosphate numbers refer to the phosphodiester at the 5′ side of a nucleotide (e.g., the phosphate at the cleavage site is the phosphate 1). In this example, the cleavage product from the enzymatic reaction generates a 5′-phosphate.

B.

(i) ^{32}P-label at phosphate #1.

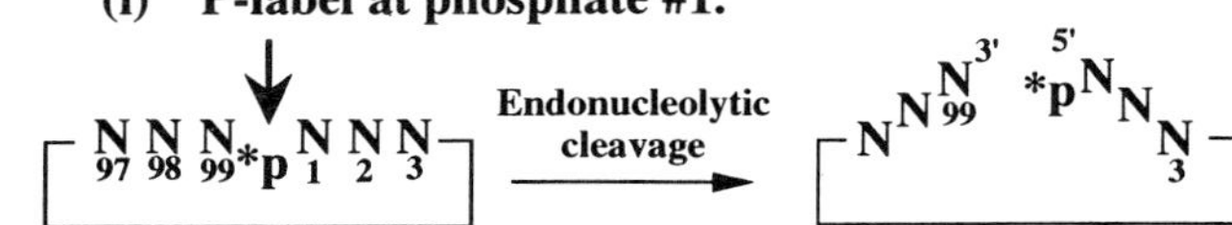

Length of RNA fragment: 1, 2, 3, ... 10, ... 50, ... 90, ... 96, 97, 98.
Phosphate # in the "wild-type" RNA: 2, 3, 4, ... 11, ... 51, ... 91, ... 97, 98, 99.

(ii) ^{32}P-label at phosphate #98.

N*p N N N N N (97, 98, 99, 1, 2, 3) — Endonucleolytic cleavage → N*pN$_{99}$ N 3' ; 5' pN N N$_3$

Length of RNA fragment: 3, 4, 5, ... 10, ... 50, ... 90, ... 96, 97, 98.
Phosphate # in the "wild-type" RNA: 97, 96, 95,.... 90, ... 50, ... 10, ... 4, 3/98, 2/99.

(iii) ^{32}P-label at phosphate #3.

N N N N N*pN (97, 98, 99, 1, 2, 3) — Endonucleolytic cleavage → N N$_{99}$ N 3' ; 5' pN N*pN$_3$

Length of RNA fragment: 2, 3, 4, ... 10, ... 50, ... 90, ... 96, 97, 98.
Phosphate # in the "wild-type" RNA: 3, 4, 5, ... 11, ... 51, ... 91, ... 97, 98, 2/99.

Fig. 2. (*continued*)

more than 25% of the circular RNA is hydrolyzed. This percentage ensures that most RNA molecules are hydrolyzed only once. Longer hydrolysis time generates RNA fragments that are shorter than the full-length linear RNA. These RNA fragments can obscure the detection of active CP isomers (protocol 3) or hybridize to CP RNAs in the mixture to interfere with the functional assay.

3. The hydrolyzed RNA is quickly cooled on ice for 4 min.

Renaturation of CP Mixture. RNA molecules are notoriously prone to form nonnative conformations that are kinetically trapped.[18] Therefore, a renaturation protocol is mandatory for optimal interpretation of CPA results. The precise details of the renaturation protocol are specific for each RNA under study. A generic protocol described later may be used to optimize the renaturation of the linear (or circular) RNA to develop a procedure suitable for the RNA of interest. The same protocol will then be used for renaturation of the CP isomers.

1. The RNA is incubated at 85–90° for 2 min followed by incubation at room temperature for 3 min.
   Note: This step should denature any alternate conformers in the absence of divalent ions.
2. A 10× $MgCl_2$ solution is added to desired Mg^{2+} concentration (recommended: 2, 10, 50 m*M*). The RNA is incubated at three different temperatures (recommended: ambient, 37° and 50°) for 10 min.
3. Other components (0.1–1 *M* monovalent ions, 0.1–1 m*M* polyamines, etc.) are added at this time. The RNA is incubated again at 37° for 5 min.
4. All samples are analyzed on a nondenaturing polyacrylamide gel.
   Note: The properly renatured RNA should migrate as a single band (suggesting conformational homogeneity) and relatively fast (suggesting compactness) on a nondenaturing gel. The stability of the tertiary RNA structures is generally Mg^{2+}-dependent.

Separation of Active CP Isomers. Any standard methods that allow physical separation of an active from an inactive RNA molecule can be applied (some examples are given in Fig. 2A). Retention by nitrocellulose filter or gel shift are useful in separating RNA–protein complexes from free RNAs. An affinity column was used to separate CP ribozymes that were capable of substrate binding.[14] The active CP RNA mixture isolated at this step is concentrated by ethanol precipitation and dissolved in water for the subsequent determination of the CP positions.

[18] O. C. Uhlenbeck, *RNA* **1,** 4 (1995).

Determination of CP Positions by Reverse Transcription. The termini positions of active CP isomers can be determined by reverse transcriptase runoff. The runoff products from the CP mixture are compared to the reverse transcriptase sequencing reaction of the circular RNA. This method can determine all termini positions in an RNA by using multiple primers at ~120–150 positions per primer. Standard protocols of reverse transcriptase sequencing and cDNA synthesis can be found in manufacturer's instructions.

Protocol 3: CPA of RNAs Cleaved at Specific Site

RNA molecules that either self-cleave or are cleaved by an endonuclease are unique cases for CPA. The cleavage reaction itself can be used to determine the termini position of active CP isomers so that a physical separation is not necessary. Steps C and D in protocol 2 are therefore combined in one experiment (Fig. 2A). The idea is to cut the circular RNA twice, first at a random site as performed by the alkaline hydrolysis and then at a specific site as performed by the catalytic cleavage. Cutting the circular RNA at two places generates fragments that are shorter than the full-length CP isomers. If the circular RNA is initially ^{32}P labeled at a single location, then the termini position of the active CP isomers can be deduced from the length of these fragments (Fig. 2B). The location of the ^{32}P label in the RNA also determines how many backbone positions can be analyzed (Fig. 2B). Since the essential aspect is to cut the RNA twice, the method of DNA oligonucleotide directed RNase H digestion to cut the RNA at a defined position[19,20] is also useful for step D in protocol 2.

1. The gel-purified, ^{32}P-labeled circular RNA is renatured as described in step B in protocol 2.
2. The ribozyme cleavage reaction is carried out under single-turnover conditions ([E] $\gg$ [S]) at one and at five half-lives ($t_{1/2} = \ln 2/k_{obs}$ at the specified [E]).

   Note: The ribozyme concentration can be adjusted to examine effects of circular permutation on different steps in the catalysis. (1) [E] $< K_b/5$ determines the effect on k_{cat}/K_m. (2) [E] $> 5\ K_b$ determines the effect on k_c. The K_c and K_b are the maximum cleavage rate at saturating [E] and the ribozyme concentration at $k_c/2$ in single-turnover reactions, respectively.
3. The reaction mixture is analyzed on a sequencing PAGE. Products

[19] J. Lapham and D. M. Crothers, *RNA* **2,** 289 (1996).

[20] Y.-T. Yu and J. A. Steitz, *RNA* **3,** 807 (1997).

shorter than the full-length CP isomers are compared to an RNA size standard to determine the termini position of the active CP isomers.

4. The perfect size standard is the circular RNA cleaved by the ribozyme prior to the alkaline hydrolysis. The gel-purified linear RNA product is then subjected to partial alkaline hydrolysis and nuclease T1 digestion. The RNA ladder produced by this method is identical to the fragments in the actual CPA experiment.

Protocol 4: CPA Coupled with Dephosphorylation

CPA-DP is a variation of CPA to attribute the effect of circular permutation to the 2′-OH group at circularly permuted positions. The resolution of standard CPA is at the nucleotide level. CPA-DP can improve this resolution to an individual functional group within the ribose phosphate backbone. In CPA-DP, half of the CP mixture from step A in protocol 2 is treated with a phosphatase (e.g., T4 polynucleotide kinase[21] to remove the 2′,3′-cyclic phosphate. The dephosphorylated CP mixture contains 2′,3′-OH at their 3′ end and is processed identically as the untreated CP mixture. The activity of a specific CP isomer is assigned in three categories (Fig. 3A). CP isomers in the first category have identical activity as the circular RNA using both mixtures. The corresponding 2′-OH groups are interpreted to make no contribution in the activity of this RNA. CP isomers in the second category have initially decreased activity as the 2′,3′-cyclic phosphate form, but have increased activity as the 2′,3′-hydroxyl form. The corresponding 2′-OH groups are interpreted to function as H-bond donors in the activity of this RNA. CP isomers in the third category have decreased activity as the cyclic phosphate form, and the activity does not increase as the 2′,3′-hydroxyl form. The corresponding 2′-OH groups are either involved in the folding or in the activity of this RNA. The quick assignment of all 2′-OH groups in these categories is extremely useful in choosing the sites to carry out the more laborious site-specific modifications.

Another advantage of CPA-DP is the detection of intermolecular interactions.[6] Some H-bond donor/acceptors in a ribozyme are changed to H-bond acceptor/donors in the transition mutants (Fig. 3B). Circular permutation of its substrate converts a 2′-OH from a potential H-bond donor to a H-bond acceptor in the form of 2′,3′-cyclic phosphate. Therefore, an altered H-bond donor–acceptor pair can be established between a specific ribozyme mutant and a specific CP isomer. On removal of the cyclic phosphate, the 2′,3′-OH terminus can function either as a donor or as an acceptor (Fig. 3A), so that increased activity will be seen for both the wild-type and the

[21] T. R. Weber, Ph.D. thesis, University of Illinois, Urbana-Champaign, (1985).

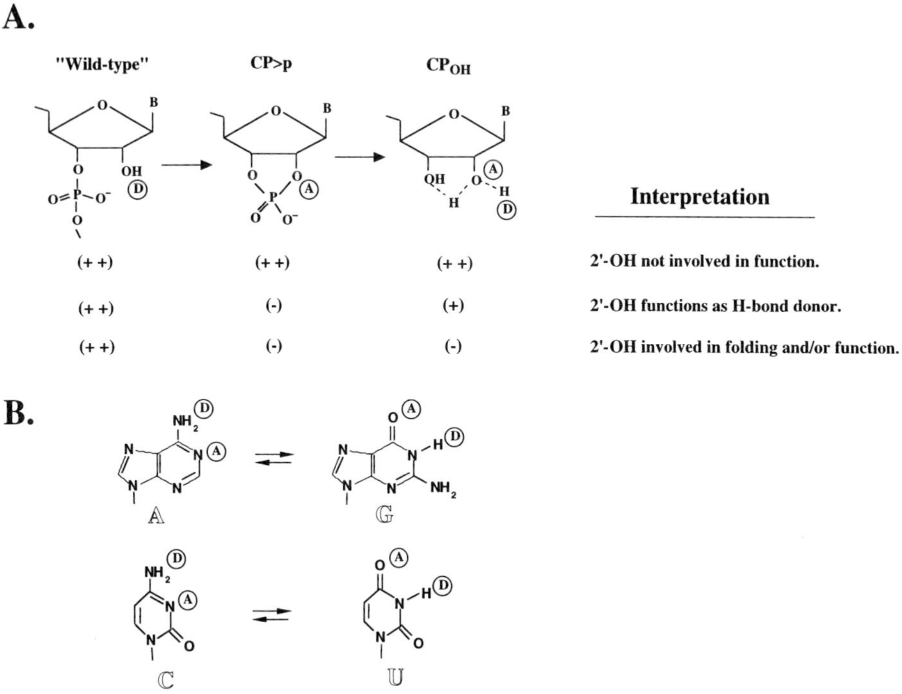

A.

"Wild-type"	CP>p	CP_{OH}	Interpretation
(+ +)	(+ +)	(+ +)	2'-OH not involved in function.
(+ +)	(-)	(+)	2'-OH functions as H-bond donor.
(+ +)	(-)	(-)	2'-OH involved in folding and/or function.

B.

A ⇌ G

C ⇌ U

FIG. 3. (A) Interpretation of CPA-DP results. The CP isomers are assigned in three categories based on the results from the 2′, 3′-cyclic phosphate form and the 2′, 3′-hydroxyl form. The 2′ position is indicated as a potential H-bond donor or acceptor in different forms. (B) Transition mutants (A ⇔ G and C ⇔ U) switch the H-bond donor–acceptor on nucleotide bases in isosteric positions. The altered H-bond donor–acceptor in the ribozyme may restore an H bond with an altered H-bond acceptor–donor in the CP substrate.

mutant ribozyme with the CP substrate as the 2′,3′-hydroxyl form. The beauty of CPA-DP is that all 2′-OH groups can be examined simultaneously for intermolecular interactions with the internal control represented by the activity of all other CP isomers in the same mixture.

1. The CP mixture from step A in protocol 2 is incubated at 37° with 1 U/μl T4 polynucleotide kinase (e.g., Amersham Pharmacia Biotech, Piscataway, NJ) in 50 m*M* Tris-HCl, pH 8.1, 10 m*M* $MgCl_2$ for 45 min to remove the 2′,3′-cyclic phosphate.
2. The reaction mixture is extracted once with phenol/$CHCl_3$ and precipitated with ethanol to remove the kinase and Mg^{2+}. The CP mixture is then dissolved in water and renatured as described under step B in protocol 2.

Structure and Function of Individual Circularly Permuted Isomers

The new 5′ and 3′ ends of a CP isomer are convenient places to incorporate unique photochemical[17,22–24] or spectroscopic probes in an RNA.[25,26] In contrast to other modified RNA, circularly permuted isomers can be easily made in large quantities for characterization by biophysical or biochemical methods. Circular permutation represents a drastic modification of the backbone in which the phosphodiester bond is broken. In addition, the phosphate is removed in a CP isomer obtained by T7 RNA polymerase transcription. Anyone interested in using circular permutation should perform CPA first and apply the CPA result in choosing the optimal CP isomer for subsequent work. Important considerations in studying a CP isomer include the choice of the CP position, the cloning strategy, the removal of the 5′-triphosphate of the CP transcript, and the stability versus the functionality.

The choice of the CP position depends on the question to be addressed. The typical application of circular permutation is to incorporate a single photo cross-linker or a fluorophore at the 5′ or the 3′ end of an RNA. Circular permutation allows easy access to an "internal" position. For this purpose, the new termini should be derived from a permissive positive as determined by CPA. If the analysis of a particular backbone position is of interest, one should choose CP positions that are nonpermissive as determined by CPA to maximize the observable effect. Because the CP isomers are synthesized as *in vitro* transcripts using T7 RNA polymerase, the 5′ nucleotide should be either a guanosine or an adenosine.[27,28] For the transcripts beginning with a pyrimidine followed by a guanosine, priming with a 5′ Py-G dimer is recommended. Do not, however, add any nucleotide to the 5′ end of the intended CP isomer to improve transcription yield, because these extra nucleotides may strongly interfere with the structure and function of the CP isomer (example 2 in later section[2]).

Cloning a CP isomer primarily depends on making a suitable DNA template from which the CP construct can be amplified by PCR. The circular RNA is often a good choice to make a full-length CP template by reverse transcription.[14] Reverse transcriptase with RNase H activity generates a

[22] A. B. Burgin and N. R. Pace, *EMBO J.* **9,** 4111 (1990).

[23] J.-F. Wang and T. R. Cech, *Science* **256,** 526 (1992).

[24] M. E. Harris, J. M. Nolan, A. Malhotra, J. W. Brown, and N. R. Pace, *EMBO J.* **13,** 3953 (1994).

[25] D. C. Lynch and P. R. Schimmel, *Biochemistry* **13,** 1841 (1974).

[26] S. A. Reines and C. R. Cantor, *Nucleic Acids Res.* **1,** 767 (1974).

[27] J. F. Milligan, D. R. Groebe, G. W. Witherell, and O. C. Uhlenbeck, *Nucleic Acids Res.* **15,** 8783 (1987).

[28] J. F. Milligan and O. C. Uhlenbeck, *Methods Enzymol.* **180,** 51 (1989).

mixture of complementary DNAs between one and two full lengths of the circular RNA. These cDNAs are excellent templates for PCR. An alternative method is to make the CP template from a tandemly repeated dimer embedded in a plasmid DNA.[17] The 3′ end of the CP isomer can be independently designed from the RNA sequence by using restriction enzymes that have nonpalindromic recognition sequence located 3′ to the site of endonucleolytic digestion and produce 5′ overhangs or blunt ends (e.g., *Fok*I[24]). Finally, a hammerhead ribozyme may be incorporated to generate the 5′ end of a CP construct that is a pyrimidine.[29,30] The hammerhead method, however, is not recommended for CP isomers whose 5′ end is located in a structured region where folding of the RNA can interfere with the cleavage by the hammerhead ribozyme.

The chemical composition of the 5′ end of a CP isomer can be an important factor in structure–function studies. When a 5′ unnatural nucleotide is needed for incorporation of a photo cross-linker or a fluorophore, priming the T7 transcription with the unnatural nucleotide produces two transcripts, one containing this unnatural nucleotide and the other a 5′-triphosphate. These transcripts are often purified together. Because the CP isomer with the unnatural nucleotide is the focus of subsequent work, no additional processing is needed. When the CP isomer is transcribed only with natural nucleotides, the 5′ end of the transcript has very high charge density due to the presence of the triphosphate. Removal of the triphosphate generates a 5′-OH group that is the same as the 5′ end of the CP isomer in the CPA experiment.

Protocol 5: Removal of 5′-Triphosphate of CP Transcripts

Note: The CP RNA containing a 5′-triphosphate should be purified by denaturing gel electrophoresis and stored in water prior to the application of this procedure. Do not attempt to dephosphorylate the RNA transcript present in the transcription mixture. Without gel purification or size-exclusion chromatography, it is difficult to remove all nucleoside triphosphates, which can severely inhibit dephosphorylation of the CP RNA.

1. Three nanomoles of the CP RNA is mixed with 50 m*M* Tris-HCl, pH 8.1, and 18 units of calf intestine alkaline phosphatase (e.g., Boehringer Mannheim, Indianapolis, IN) in 300 μl.
2. The mixture is incubated at 37° for 1 hr.
3. NaCl is added to a final concentration of 0.2 *M*. The mixture is extracted with phenol/$CHCl_3$ three times to remove the alkaline

[29] S. R. Price, N. Ito, C. Oubridge, J. M. Avis, and K. Nagai, *J. Mol. Biol.* **249,** 398 (1995).
[30] B. L. Golden, E. R. Podell, A. R. Gooding, and T. R. Cech, *J. Mol. Biol.* **270,** 711 (1997).

phosphatase. The dephosphorylated RNA is concentrated by ethanol precipitation.

4. The RNA concentration after dephosphorylation is determined by UV absorbance.

   Note: More than 90% of the RNA should be recovered in this procedure.

Because circular permutation can change the stability of the RNA relative to the wild-type form, interpretation of circular permutation effects on RNA function may not always be straightforward. The issue of RNA stability versus functionality is the same for base mutations of an RNA. A base mutation may destabilize (e.g., disruption of a Watson–Crick base pair) or stabilize (e.g., formation of a nonnative interaction) the RNA structure. Similarly, circular permutation may destabilize or stabilize the RNA structure depending on the CP position. For proper interpretation, experiments that only probe the structure of a CP isomer should be performed and compared to the results from functional assays. Chemical modification[7] and Fe(II)-EDTA hydroxyl radical footprinting[31,32] are excellent methods for the structural analysis of RNA.

Examples of Applying CPA to Study RNA Structure and Function

Folding of Tertiary RNA Structure

The proper folding of CP isomers of yeast $tRNA^{Phe}$ was analyzed by a Pb^{2+} cleavage assay.[1] Only the folded $tRNA^{Phe}$ in the presence of Mg^{2+} is accurately and quickly cleaved by Pb^{2+}.[33,34] At neutral pH and 25°, 54 of 68 analyzable CP isomers of $tRNA^{Phe}$ had comparable Pb^{2+} cleavage rates to the wild-type tRNA, suggesting that these isomers fold normally. The 14 incorrectly folded CP isomers have backbone breaks in two regions of tRNA involved in tertiary folding. These results show that CPA can be used to assess the importance of specific backbone positions for folding of a highly structured RNA.

RNA–Protein Interactions

The effect of circular permutation on binding of the coliphage R17 coat protein to its RNA binding site was studied.[16] The R17 coat protein binds

[31] J. A. Latham and T. R. Cech, *Science* **245,** 276 (1989).

[32] D. W. Celander and T. R. Cech, *Science* **251,** 401 (1991).

[33] R. Brown, J. Dewan and A. Klug, *Biochemistry* **24,** 4785 (1985).

[34] L. S. Behlen, J. R. Sampson, A. B. DiRenzo, and O. C. Uhlenbeck, *Biochemistry* **29,** 2515 (1990).

to a hairpin loop containing a single nucleotide bulge. All CP isomers outside the previously determined minimal binding site bound to the protein properly. All but one CP isomer within the binding site had strongly decreased binding affinity to the coat protein. This result shows that CPA can be used to map the minimal binding site for a protein embedded in a large RNA. For the single CP isomer with the permissive break within the binding site, only the isomer containing a 5′-OH group bound to the coat protein properly. Addition of a 5′-phosphate or several nucleotides to its 5′ end strongly decreased binding. This result shows that the precise composition of the 5′ end of a CP isomer can play an important role in its function.

CPA Coupled with Dephosphorylation (CPA-DP)

The potential 2′-OH groups in a pre-tRNA substrate involved in direct interactions with the P RNA were identified.[6] P RNA was known to interact only with the coaxially stacked T stem–loop and the acceptor stem of a tRNA. CPA of a tRNA substrate showed that all but one nonpermissive CP isomer contains backbone breaks within this region of tRNA. The single exception corresponds to a backbone position whose 2′-OH group is involved in tertiary folding of tRNA.[35] Eight 2′-OH groups were identified in the T stem–loop region as potential candidates in direct interactions with P RNA. One of the eight CP isomers had increased activity for a specific A→G mutant of P RNA, suggesting a direct ribozyme–substrate contact. These results show that CPA-DP can be used to identify potential intermolecular interactions and functional regions embedded in a large RNA. Interestingly, some of the nonpermissive CP isomers identified from the tRNA folding studies using Pb^{2+} cleavage[1] functioned properly in the P RNA cleavage assay. Therefore, different CPA results may be expected for the same molecule depending on the functional requirement of the biochemical assay.

Quantitative CPA

All CP isomers in the three examples given are classified as either permissive or nonpermissive. This simple classification relies on drawing a line through the radioactive ladder on a denaturing gel to generate valleys and peaks. The phosphorimager technology allows quantitation of individual bands within the ladder. This quantitative information can be used to classify the CP isomers into more categories to refine the interpretation of the circular permutation effect.

[35] G. Quigley and A. Rich, *Science* **194,** 796 (1976).

When the amount of CP isomers is determined by reverse transcription, a quantitative parameter referred to as *depletion factor* (DF) can be applied.[14] DF is defined as the amount of the CP isomer in the original CP mixture divided by the amount of the active CP isomer isolated from the functional assay. This ratio is then normalized to that of the CP isomer with the 5′ end at nucleotide 1 (the wild-type RNA). By this measure, the higher the depletion factor, the more detrimental the effect of circular permutation is on the function of the RNA.

The depletion factor for the CP isomers of *B. subtilis* P RNA is between 0.9 and 4.0 for substrate binding.[14] These CP isomers are classified into three categories based on their DF, taking into account the inaccuracy of the quantitation. CP isomers with DF between 0.9 and 1.5 are considered to behave like the wild-type P RNA under the assay condition. CP isomers with DF between 1.5 and 2.0 are considered to have moderately decreased affinity for the pre-tRNA substrate. CP isomers with DF > 2.0 are considered to have significantly decreased binding affinity for the substrate. The classification agrees well with the known biochemical and phylogenetic properties of this ribozyme.

When the amount of CP isomers is determined by catalytic cleavage, a parameter referred to as *cleavage factor* (CF) may be applied. CF is defined as the amount of the cleavage product originating from the CP isomer with the 5′ end at nucleotide 1 (the wild-type RNA) divided by the amount of the cleavage product at any position. If the cleavage product from the wild-type RNA cannot be quantitated, the amount of cleavage product from a CP isomer considered to have the same activity as the wild-type RNA can be used as the numerator. This definition assumes that all CP isomers are present in the original mixture in approximately equal amounts. This assumption is fulfilled for hydrolysis of most positions using the procedure described under step A in protocol 2. By this measure, the higher the cleavage factor, the more detrimental the effect of circular permutation is on the activity of the cleaved RNA.

The cleavage factors of two *in vitro* selected substrates for *Bacillus subtilis* P RNA are between 0.75 and 17 (Fig. 4). These CP isomers are classified into three categories based on their CF, taken into account the inaccuracy of the quantitation. CP isomers with CF between 0.75 and 1.25 are considered to be cleaved at the same rate as the circular substrate. This conclusion was confirmed by the time course of the cleavage reaction in the CPA experiment (T. Pan, unpublished results). CP isomers with CF between 1.25 and 4 are considered to have moderately decreased cleavage rates.

[36] T. Pan, *Biochemistry* **34,** 8458 (1995).

[37] L. Odell, V. Huang, M. Jakacka, and T. Pan, *Nucleic Acids Res.* **26,** 3717 (1998).

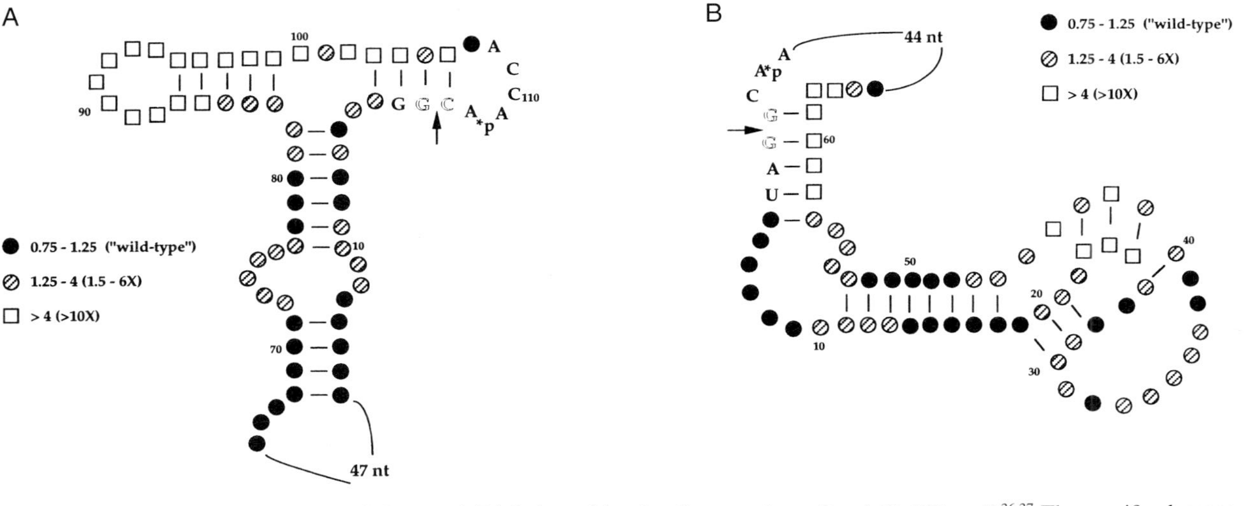

FIG. 4. CPA of selected substrates (A) 17 and (B) 8 cleaved by the ribozyme from *B. subtilis* RNase P.[36,37] The specific cleavage sites are between the highlighted nucleotides and indicated by arrows. The phosphate positions are grouped into three categories based on their cleavage factors. (●) CF, 0.75–1.25; these CP isomers are cleaved at similar rates as the circular RNA. (⊘) CF, 1.25–4; these CP isomers have a 1.5- to 6-fold decreased cleavage rate compared to the circular RNA. (□) CF >4; these CP isomers have a 10- to 400-fold decreased cleavage rate compared to the circular RNA. Nucleotides in the looped regions (47 nucleotides in 17 and 44 nucleotides in 8) can be deleted with no effect on the cleavage efficiency[15] (T. Pan, unpublished results).

The time course in CPA showed that the cleavage rates were 1.5- to 6-fold lower for the CP isomers in this category. CP isomers with CF > 4 are considered to have significantly decreased cleavage rates. The time course in CPA showed a >10-fold decrease in the cleavage rate for these CP isomers. Interestingly, all CP isomers with the termini outside the depicted regions in Fig. 4 have cleavage factors between 0.75 and 1.25 for substrate 17 and between 0.75 and 2 (with a single exception) for substrate 8. These residues can be deleted with no effect on the activity of these substrates[15] (T. Pan, unpublished results). These results indicate that CF can be used to assess quantitatively the effect of circular permutation and that large, consecutive regions of low cleavage factors may locate deletable regions within an RNA.

Acknowledgments

Research in my laboratory is supported by grants from the NIH (GM52993 and GM57880). I thank members of my laboratory for stimulating discussions on this subject.

[21] Kinetic Oligonucleotide Hybridization for Monitoring Kinetic Folding of Large RNAs

By DANIEL K. TREIBER and JAMES R. WILLIAMSON

Introduction

RNA is a versatile molecule that folds into complex three-dimensional structures to perform functions such as ligand binding and catalysis. Although the number of crystal and solution structures of RNAs is growing,[1] little is known about the mechanism of folding. The RNA folding problem, like the protein folding problem, asks how the primary sequence specifies a unique tertiary fold. An essential step toward understanding the RNA folding problem is to characterize the kinetic folding process. The relative timescales on which various folding events occur can suggest a kinetic folding pathway and may reveal the presence of potentially important kinetic intermediates. A kinetic framework of this sort aids in the design of more detailed experiments aimed explicitly at revealing the features of an RNA sequence that drive kinetic folding and, also, those features that hinder the process.

Kinetic oligonucleotide hybridization is a relatively simple *in vitro*

[1] G. L. Conn and D. E. Draper, Curr. Opin. Struct. Biol. **8,** 278 (1998).

0076-6879/00 $30.00

method for monitoring RNA folding transitions that occur with a half-time ($t_{1/2}$) of 10 sec or greater. We have used kinetic oligonucleotide hybridization to characterize the Mg^{2+}-dependent kinetic folding pathways of the *Tetrahymena* group I ribozyme and the RNA component of RNase P.[2–7] Kinetic oligonucleotide hybridization exploits the observation that many large RNAs, such as group I and II ribozymes and ribosomal RNA, require divalent metal ions, proteins, or both to acquire the native tertiary fold.[8,9] In the absence of these "folding effectors," the RNA adopts an extended, partially denatured structure consisting primarily of short-range secondary interactions.[10] The tertiary folding transitions that occur once the effector is added include compaction of the structure and formation and/or stabilization of long- and short-range interactions.[2,10,11] Structural changes in RNA, including those that accompany folding, can be visualized experimentally as a change in reactivity toward small chemical probes[12–14] and larger probes, such as complementary oligodeoxynucleotides.[15,16] Kinetic oligonucleotide hybridization takes snapshots of the kinetic folding process by monitoring time-dependent changes in accessibility to binding by short, sequence-specific oligodeoxynucleotide probes.

Overview of Kinetic Oligonucleotide Hybridization

The basic features of kinetic oligonucleotide hybridization are illustrated in Fig. 1A. First, a denatured RNA sample is prepared, and folding is initiated by the addition of a folding effector (Mg^{2+}). The denatured sample is prepared by heat-cooling radiolabeled RNA in the absence of Mg^{2+}. A low concentration of monovalent ions (2 m*M*) is present during this step

[2] P. P. Zarrinkar and J. R. Williamson, *Science* **265,** 918 (1994).
[3] P. P. Zarrinkar and J. R. Williamson, *Nature Struct. Biol.* **3,** 432 (1996).
[4] P. P. Zarrinkar and J. R. Williamson, *Nucleic Acids Res.* **24,** 918 (1996).
[5] P. P. Zarrinkar, J. Wang, and J. R. Williamson, *RNA* **2,** 564 (1996).
[6] D. K. Treiber, M. S. Rook, P. P. Zarrinkar, and J. R. Williamson, *Science* **279,** 1943 (1998).
[7] M. S. Rook, D. K. Treiber, and J. R. Williamson, *J. Mol. Biol.* **281,** 609 (1998).
[8] T. Pan, D. M. Long, and O. C. Uhlenbeck, *in* "The RNA World," (R. F. Gesteland and J. F. Atkins, eds.) p. 271. Cold Spring Harbor Laboratory Press, Plainview, New York, 1993.
[9] K. M. Weeks, *Curr. Opin. Struct. Biol.* **7,** 336 (1997).
[10] Y.-H. Wang, F. L. Murphy, T. R. Cech, and J. D. Griffith, *J. Mol. Biol.* **236,** 64 (1994).
[11] F. L. Murphy and T. R. Cech, *Biochemistry* **32,** 5291 (1993).
[12] T. Powers, G. Daubresse, and H. F. Noller, *J. Mol. Biol.* **232,** 362 (1993).
[13] A. R. Banerjee and D. H. Turner, *Biochemistry* **34,** 6504 (1995).
[14] J. A. Latham and T. R. Cech, *Science* **245,** 276 (1989).
[15] S. T. Cload, P. L. Richardson, Y.-H. Huang, and A. Schepartz, *J. Am. Chem. Soc.* **115,** 5005 (1993).
[16] K. A. LeCuyer and D. M. Crothers, *Proc. Natl. Acad. Sci. U.S.A.* **91,** 3373 (1994).

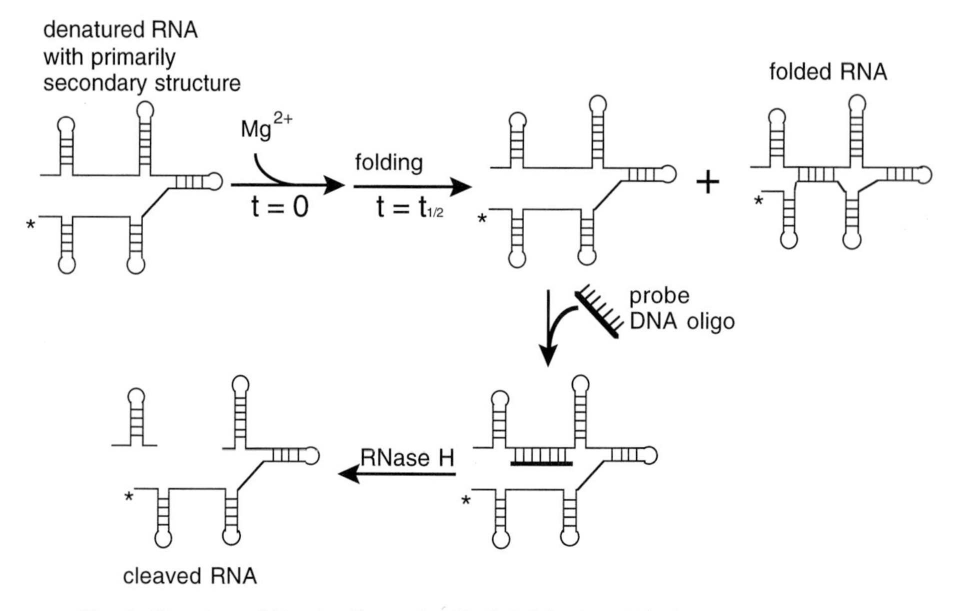

FIG. 1. Overview of kinetic oligonucleotide hybridization. (A) Conceptualized view of kinetic oligonucleotide hybridization. * indicates a radioactive label. (B) Experimental flow-chart. t_1 and t_2 indicate the folding and quench times, respectively. (C) Model kinetic folding data.

to facilitate the formation of short-range secondary structures. Once the denatured RNA has equilibrated, tertiary folding is initiated by the addition of Mg^{2+}. After addition of Mg^{2+}, both folded and unfolded molecules are present in the population. In Fig. 1A the presence of a long-range helix indicates a folded molecule.

Folded and unfolded RNAs are distinguished in a rapid quench reaction that is the cornerstone of kinetic oligonucleotide hybridization. The basic premise of the quench reaction is that folded and unfolded molecules will be differentially accessible to binding by short, sequence-specific oligodeoxynucleotide probes. For example, in Fig. 1A the folded molecules are exclusively inaccessible to probe binding due to the formation of a long-range helix at the probe binding site. RNase H, a nuclease that cleaves the RNA component of RNA : DNA hybrids,[17] is also included in the quench

[17] Z. Hostomsky, Z. Hostomaka, and D. A. Matthews, *in* "Nucleases," (S. M. Linn, R. S. Lloyd, and R. J. Roberts, eds.), p. 341. Cold Spring Harbor Laboratory Press, Plainview, New York, 1993.

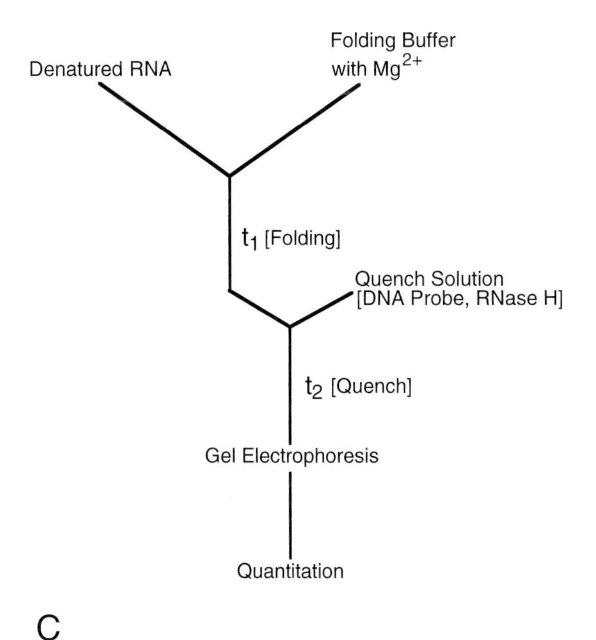

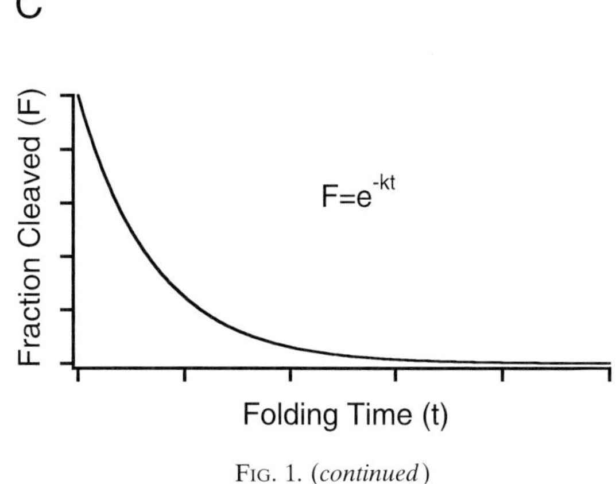

FIG. 1. (*continued*)

reaction and cleaves any accessible RNAs. For each time point, the cleaved and full-length molecules are separated by gel electrophoresis and quantitated by phosphorimager analysis. A typical experimental flowchart is shown in Fig. 1B, and Fig. 1C illustrates a model kinetic oligonucleotide hybridization data set where the fraction of cleaved RNA is plotted as a function of folding time. The rate constant for a particular folding transition is obtained by fitting the data to an exponential equation.

General Considerations

Initiation of Folding

Although we have used Mg^{2+} and other cations as the folding effectors in our studies with large ribozymes,[2,5] effectors such as proteins or other ligands are also compatible with kinetic oligonucleotide hybridization. The choice of a folding effector ultimately depends on the requirements of the particular RNA being studied. It may also be possible to initiate folding by manipulating temperature or the concentration of a denaturant.

Timescale

A primary factor that limits the range of RNAs and folding conditions that can be sampled by kinetic oligonucleotide hybridization is the folding timescale. The assay currently relies on manual pipetting to quench the folding reactions; hence, the shortest possible dead time is about 5 sec. For this reason, kinetic oligonucleotide hybridization is not well suited to measure the folding of hairpins or tRNAs, which form on the microsecond and millisecond timescales, respectively.[18,19] Despite this limitation, qualitative kinetic data for folding processes in large RNAs that occur with a half-time shorter than 5 sec can still provide valuable insight into the overall folding mechanism.[3] The use of a rapid mixing device may increase the time resolution, although other elements of the assay, such as the rate of DNA probe binding (see later section), may become limiting under these conditions.

Resolution

In principle, a large group of DNA probes could be designed to study the folding kinetics of many regions of a large RNA. In practice, however, our studies with large ribozymes indicate that the folded and unfolded states are differentially accessible to only a limited number of DNA probes.[2] This limitation of kinetic oligonucleotide hybridization necessarily restricts the resolution of the data set, but does not preclude the acquisition of useful kinetic data, especially if each domain or subdomain of an RNA can be studied with at least one probe.

[18] J. Gralla and D. M. Crothers, *J. Mol. Biol.* **73,** 497 (1973).
[19] P. E. Cole and D. M. Crothers, *Biochemistry* **11,** 4368 (1972).

The data from a kinetic oligonucleotide hybridization assay monitor a transition from a state that is accessible to both DNA probes and RNase H to a state that is inaccessible to one, or both, of these reagents. Many folding processes that lead to inaccessibility may not be as simple as the helix formation depicted in Fig. 1A. For example, a region that is single stranded in the unfolded structure may also be single stranded in the folded structure, but simply has become buried within the molecule. Likewise, accessibility does not necessarily imply no base pairing—unstable helices, which rapidly open and close, and unformed helices may be equally accessible.

Kinetic Constraints

Several kinetic conditions must be met for kinetic oligonucleotide hybridization to be a valid rapid quench assay. Here, these constraints are discussed in a general manner, and below, in the Optimization of the Quench Reaction section, more detailed experimental information is provided. The basic kinetic framework for kinetic oligonucleotide hybridization is outlined in Fig. 2. It is evident from this schematic that DNA probe (D) binding to the unfolded RNA (U) is in a kinetic competition with continued folding (U→F). The first kinetic constraint is that probe binding must be faster than the folding rate ($k_{on}[D] > k_{fold}$). This condition ensures that folding ceases once the quench is added. The rate of probe binding is optimized by adjusting the probe concentration to a point where all or most of the RNA is cleaved at the "zero" time point of a kinetic folding reaction. The zero time point is measured by adding Mg^{2+} and the quench to the RNA simultaneously.

The second kinetic constraint is that RNase H cleavage must be faster

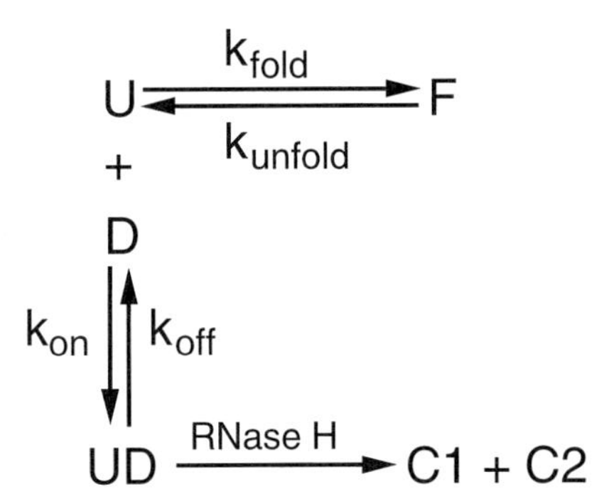

FIG. 2. Kinetic framework for kinetic oligonucleotide hybridization. U, Unfolded (accessible) RNA; F, folded (inaccessible) RNA; D, oligonucleotide DNA probe; UD, RNA–DNA complex; C1 and C2, cleavage products.

than probe dissociation. This condition ensures that unfolded molecules that are initially trapped by probe binding do not eventually fold during the course of the quench. The rate constant for probe dissociation (k_{off}) is influenced by many factors including the length and sequence of the probe, temperature, pH, and cation concentration. In our standard quench reaction [10 mM Mg^{2+}, 10 mM NaCl, 1 mM dithiothreitol (DTT), 50 mM Tris (pH 8.0), 37°] we have found that 10 nucleotide probes have a sufficiently small k_{off} to permit efficient cleavage. The rate of RNase H cleavage is influenced by a variety of factors including the kinetic constants k_{cat} and K_m and the enzyme (E) concentration. Because the quench reactions are performed under conditions of trace RNA (~1 nM) and excess enzyme, the RNA concentration is not a critical parameter. Under our quench conditions, moderate concentrations of RNase H (60 nM) permit the cleavage reaction to be complete in less than 30 sec, which is rapid relative to the rate of 10-mer probe dissociation.

The final kinetic constraint is that unfolding must be slow compared to the timescale of the quench reaction. If this condition is not met, then initially folded molecules may be cleaved during the course of the quench. Our experience with large ribozymes has been that a 30-sec quench is rapid relative to the unfolding rate that is observed under folding conditions. For new RNAs, however, this assumption must be tested. The simplest test is to subject fully folded RNA to the quench reaction for increasing periods of time. If unfolding is slow, then minimal cleavage should be observed within the first 30 sec. In our studies, only 10–20% cleavage of fully folded ribozymes is observed after a 2-min quench.[2] If this is not the case, however, then the RNase H concentration should be increased to reduce the quench time.

Excessive DNA probe concentrations can also result in the cleavage of folded molecules during the quench. In this instance it is likely that probe binding is so fast that "breathing," rather than true unfolding, is being monitored. For this reason, the probe concentrations in our quench reactions generally do not exceed 300 μM.

Methods

Oligonucleotide Screen

The first stage in designing a kinetic oligonucleotide hybridization experiment is to identify sequences that are accessible to probe binding when the RNA is unfolded, and inaccessible when the RNA is folded. The best strategy is to test several different 10-mer DNA probes that collectively target nearly the entire RNA sequence. The relative accessibility of each probe to the unfolded and folded RNA is tested in two quench reactions. In the first, accessible positions in the *unfolded* RNA are identified by adding Mg^{2+} and the quench solution to the RNA simultaneously. This reaction mimics the zero time point in a kinetic folding assay. To maximize the signal, each quench reaction contains high concentrations of the probe DNA (300 μM) and RNase H. In the second reaction, the quench is added to *folded* RNA that has been preequilibrated with Mg^{2+}. This reaction mimics the end point of a kinetic folding assay. An optimal probe gives complete cleavage of the unfolded RNA and little or no cleavage of the folded RNA. It is important to note that since the quench reaction contains Mg^{2+}, it is possible that some regions may fold more rapidly than probe binding and score as inaccessible at the zero time point.

A probe screen we performed to study the Mg^{2+}-dependent folding of the *Tetrahymena* group I ribozyme is shown in Fig. 3.[2] The secondary structure of the ribozyme is shown in Fig. 3A, and the regions targeted in the screen are numbered 1–21. Results of the diagnostic quench reactions are shown in Fig. 3B. For each probe, quench reactions were performed on RNAs that were preincubated for 0 or 10 min in the presence of Mg^{2+}. Although a majority of the probes give little or no cleavage at the 0- and 10-min time points, it is clear that the folded RNA is less accessible to several of the probes than the unfolded RNA. Importantly, the cleavage products are all of the expected length, illustrating that the cleavage reaction is highly specific. Probes 1, 15, 14, and 19, which target paired (P) regions 3 and 7 (P3, P7) have nearly ideal activity, resulting in almost complete cleavage at 0 min and little or no cleavage at 10 min. Probes 12, 13, 20, and 21 have less ideal, but still useful, behavior with ~50% of the RNA cleaved at 0 min and little cleavage at 10 min. This battery of useful probes has allowed us to study the kinetic folding of the two structural domains (P4–P6, P3–P7) that constitute the catalytic core of the ribozyme.[2–4,6,7]

Reagents

- TE buffer [10 m*M* Tris-Cl (pH 8.0), 1 m*M* Na_2EDTA]
- 10 × transcription buffer [1 ×: 50 m*M* Tris-Cl (pH 8.0), 0.01% Triton X-100, 5 m*M* DTT, 2 m*M* spermidine]
- 5 × Nucleoside triphosphates (NTPs) [1 ×: 0.2 m*M* ATP, 2 m*M* CTP, 2 m*M* GTP, 2 m*M* UTP]
- [α-^{32}P]ATP, 3000 Ci/mmol, 10 μCi/μl (New England Nuclear, Boston, MA)
- T7 RNA polymerase (prepared in house)
- *Tetrahymena* ribozyme template DNA (*Sca*I-digested pT7L-21[20])

[20] A. J. Zaug, C. A. Grosshans, and T. R. Cech, *Biochemistry* **27,** 8924 (1988).

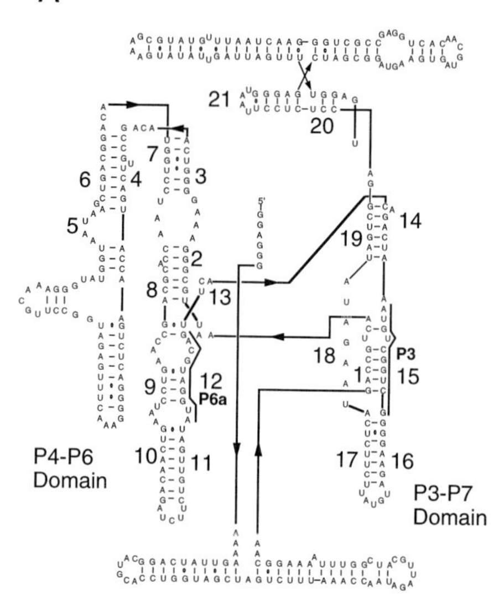

FIG. 3. DNA probe screen. (A) Secondary structure of the L-21 *ScaI Tetrahymena* group I ribozyme. The regions targeted by complementary 10-mer DNA probes are indicated numerically (1–21). Probes 2–13 target the P4–P6 domain; probes 1, 14–19 target the P3–P7 domain; probes 20 and 21 target the P9 peripheral extension. The sequences targeted by probes 12 and 15 (which are discussed throughout the text) are indicated by bold lines. Probe 12 targets paired region 6a (P6a), and probe 15 targets P3. (B) Results of the probe screen. [Adapted with permission from P. P. Zarrinkar and J. R. Williamson, *Science* **265,** 918 (1994). Copyright 1994 American Association for the Advancement of Science.] Ribozyme RNA was incubated for 0 or 10 min under folding conditions (10 m*M* Mg^{2+}, 37°) and subjected to a quench reaction containing DNA probe (300 μM) and RNase H (0.1 unit/μl). The position of full-length ribozyme is indicated in the control lane (CTRL).

RNase inhibitor (40 units/μl) (Promega, Madison, WI)
RNA elution buffer [TE buffer with 0.3 *M* sodium acetate (pH 5.2)]
100% Ethanol
1 μmol scale synthesis of each 10-mer DNA probe, deprotected and lyophilized (Operon, Alameda, CA)
10 m*M* NaCl
Sep-Pak C_{18} cartridges (Waters, Milford, MA)
HPLC-grade acetonitrile
RNase H (Amersham, formerly United States Biochemical), 5 units/μl
10 × Folding and quench buffer [1 ×: 50 m*M* Tris-Cl (pH 8.0), 10 m*M* $MgCl_2$, 10 m*M* NaCl, 1 m*M* DTT]
Stop dye [80% formamide, 100 m*M* Na_2EDTA (pH 8.0), 0.02% bromphenol blue (w/v), 0.02% xylene cyanol (w/v)]

B

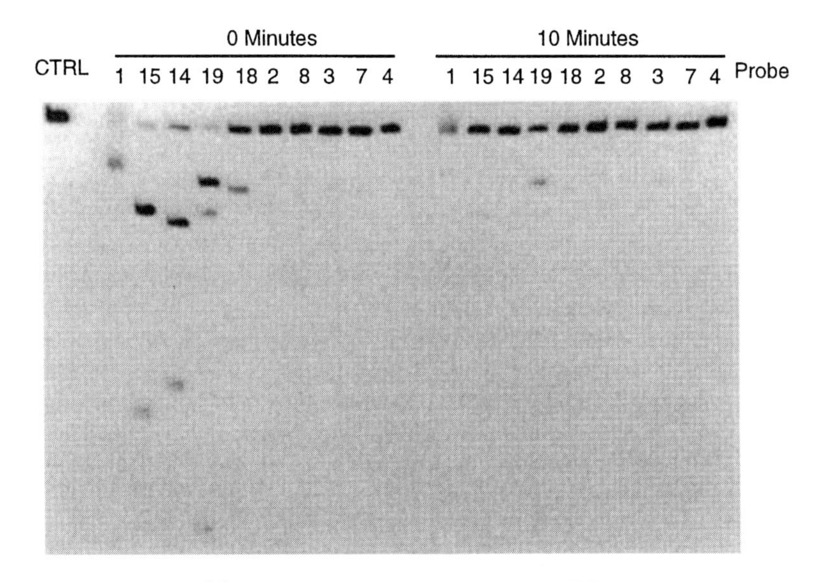

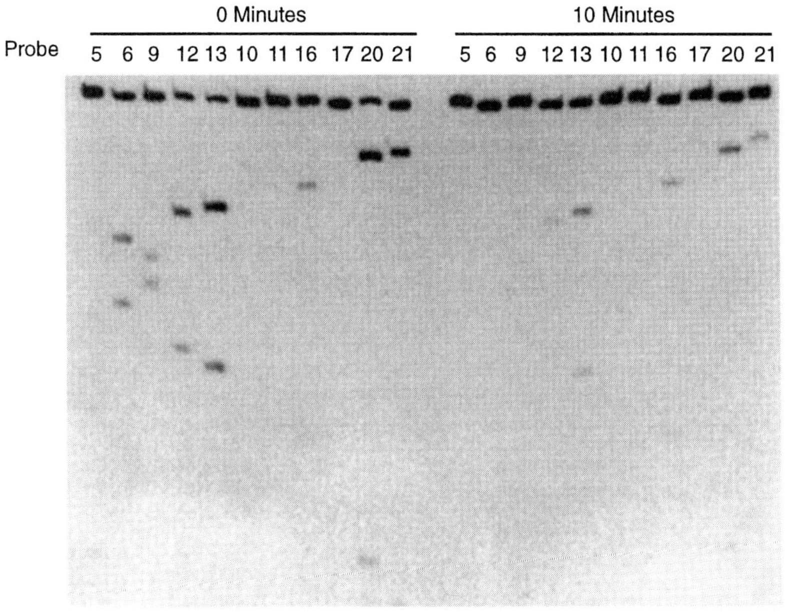

FIG. 3. (*continued*)

Denaturing polyacrylamide gel mix [6% and 20% acrylamide (w/v) (29:1 acrylamide:bisacrylamide), 8 *M* urea, 1 × TBE (90 m*M* Tris–borate, 2 m*M* Na_2EDTA, pH 8.0)] and gel running buffer [1 × TBE]
Milli-Q (Millipore Bedford, MA) deionized, distilled water for preparation of all solutions

Equipment

1.5- or 0.5-ml (depending on preference) siliconized polypropylene flip top assay tubes (Bio-Rad Hercules, CA) for all steps involving RNA
Water bath set to 37° with a secure rack to loosely hold several 1.5-ml tubes
Heat block set to 95°
Polyacrylamide gel electrophoresis (PAGE) equipment (glass plates, 0.4- and 1.5-mm combs and spacers, gel box, aluminum heat sink, and power supply)
PhosphorImager (Molecular Dynamics, Sunnyvale, CA or equivalent)
Vacuum centrifuge (Savant, Marietta, OH)

Procedure

1. Radiolabeled RNA is prepared by standard *in vitro* transcription using T7 polymerase and linearized plasmid DNA containing a T7 promoter upstream of the ribozyme coding sequence. Transcriptions (25-μl scale) contain 0.1 μg/μl template DNA, 1 × transcription buffer, 1 × NTPs, 50 μCi [α-^{32}P]ATP, 1 units/μl RNase inhibitor, and T7 RNA polymerase (amount must be optimized) and are incubated for 3.5 hr at 37°. These conditions yield body-labeled ribozymes with a specific activity of 1000 Ci/mmol. For shorter RNAs with fewer A residues, end labeling may be required to achieve a similar specific activity. Transcription is terminated by adding an equal volume of stop dye, and the full-length RNA is purified by electrophoresis through a 0.4-mm-thick 6% denaturing polyacrylamide gel. The RNA is visualized by autoradiography and excised using a new razor blade. The gel slice is placed in a disposable 15-ml polystyrene conical tube, and the RNA is eluted overnight at 4° into 1 ml of RNA elution buffer. The eluted RNA is ethanol precipitated, washed twice with 70% (v/v) ethanol, dried briefly by vacuum centrifugation, and resuspended in 100 μl of TE buffer. The RNA concentration is determined by Cerenkov counting.

2. Crude oligonucleotide probes (1 μmol) are resuspended in 100 μl of water and an equal volume of stop dye. The DNA is purified by electrophoresis through a 1.5-mm-thick 20% denaturing polyacrylamide gel. Full-

length product is visualized by UV shadowing, and the excised gel slice is placed in a disposable 50-ml polypropylene conical tube. The DNA is eluted overnight at room temperature in 10 ml of 10 m*M* NaCl. The eluted DNA is desalted and concentrated by loading on to a Sep-Pak C_{18} column that was prerinsed with acetonitrile followed by water. The loaded column is washed with water, and the DNA is eluted with 50% acetonitrile and concentrated to dryness by vacuum centrifugation. The purified DNA is resuspended in 100 μl of TE buffer and the concentration is determined spectrophotometrically.

3. 2 × Quench solutions (1 × folding and quench buffer, 600 μM DNA probe, 0.2 units/μl RNase H) are prepared in 1.5-ml tubes immediately prior to use and stored on ice. Two 10-μl aliquots are required for each probe—one for the 0-min time point and one for the 10-min time point. Five microliters of 2 × folding and quench buffer is added to the 0-min 2 × quench solutions such that the RNA added to these tubes is exposed to the final folding and quench conditions simultaneously. The quench solutions should be placed in the 37° water bath exactly 3 min prior to use.

4. Ribozyme RNA (4 n*M* in TE buffer) is annealed by heating at 95° for 1 min followed by incubation at 37° for 3 min. The Na^+ ions in TE facilitate the formation of short-range secondary structure. Fifty-microliter annealing reactions in 1.5-ml tubes are convenient for most applications and give reproducible results. The 95° step is performed in a water-filled heat block that accommodates 1.5-ml tubes.

5. For the 0-min time points, 5 μl of the annealed RNA is added to the 0-min quench solution, the quench reaction is incubated for 30 sec at 37°, and 14 μl of stop dye is added to terminate the reaction. Stop dye contains a high concentration of EDTA, which chelates Mg^{2+} and inactivates RNase H. Stop dye also serves as the denaturing gel electrophoresis sample loading buffer.

6. For the 10-min time points, 10 μl of the annealed RNA is added to 10 μl of 2× folding and quench buffer (prewarmed), and folding is allowed to proceed for 10 min at 37°. Ten microliters of the folding reaction is then added to an equal volume of the 10-min quench solution. After a 30-sec incubation at 37°, the quench reaction is terminated with 14 μl of stop dye.

7. Full-length and cleaved RNAs are separated by electrophoresis through a 0.4-mm-thick 6% denaturing gel. The gels are imaged and quantitated by using a phosphorimager. The fraction cleaved is the fraction of the total radioactivity (full-length + cleaved) that appears in the cleavage products. One cleavage product will appear for end-labeled RNA, and two for body-labeled, RNA. For body-labeled RNAs, we generally quantitate only the large cleavage product and then multiply by a factor that corrects

for the radioactivity in the small cleavage product. Since the RNA is body labeled with [^{32}P]ATP, the correction factor is determined by the relative number of A residues in the two cleavage products.

Comments

For RNAs that fold more slowly than the *Tetrahymena* ribozyme, the 10-min folding period should be extended as necessary. Our choice of 10 min was based on published folding conditions that yield fully active ribozyme.[21] It is important that the quench solution contains the folding effector (Mg^{2+} in the preceding procedure). Unfolding may occur if the folding effector is diluted on quenching. To confirm that the quench reactions are specific for the region of the RNA being targeted, the size of the cleavage products should be determined by electrophoresis with size standards. It is critical to use RNase H from Amersham (formerly United States Biochemical). RNase H from other suppliers lacks the activity and specificity required for kinetic oligonucleotide hybridization. Optimal annealing conditions and the folding effector will need to be determined for any new RNAs. A stationary rack in the water bath that loosely holds several 1.5-ml tubes is strongly recommended because it allows the large number of manipulations to be performed rapidly.

Optimization of Quench Reaction

The second stage in designing a kinetic oligonucleotide hybridization experiment is to optimize the quench reaction for any potentially useful DNA probes that were identified in the initial screen. The most important parameter that must be optimized is the DNA probe concentration. An ideal probe concentration satisfies the kinetic constraints described earlier and gives complete cleavage when Mg^{2+} and the quench are added to the RNA simultaneously. To optimize the quench, a series of probe concentrations (spanning ~2–4 orders of magnitude) is tested in quench reactions that mimic the 0-min reactions in the probe screen. The RNase H concentration, temperature, and quench time are held constant in these experiments.

Figure 4A shows the gel image from a titration experiment for probe 15. As the probe concentration is increased exponentially from 0.003 to 300 μM, the fraction of full-length RNA (F) decreases dramatically. L and S1 are the expected cleavage products, whereas S2 arises from a secondary cleavage of S1 that occurs after the primary cleavage event. The secondary cleavage is caused by partial complementarity between probe 15 and S1.[2]

[21] D. Herschlag and T. R. Cech, *Biochemistry* **29,** 10159 (1990).

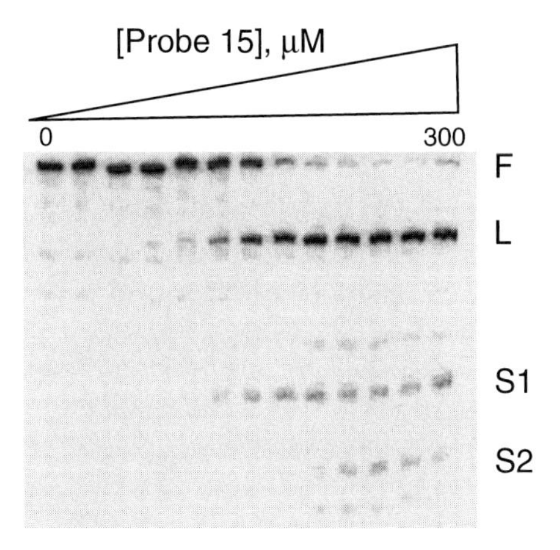

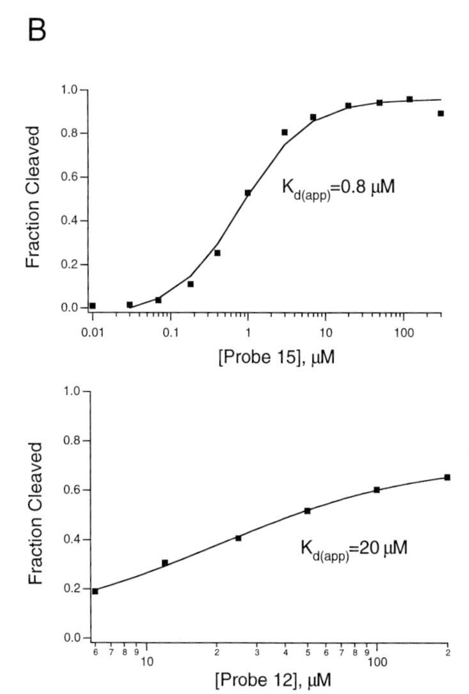

FIG. 4. Optimization of the DNA probe concentration in the quench reaction. (A) Gel image of titration data for probe 15. Ribozyme RNA was subjected to quench reactions containing 0–300 m*M* probe 15 and 0.1 units/μl RNase H. The positions of full-length RNA (F) and the long (L) and short (S1 and S2) cleavage products are indicated. (B) Titration data for probes 15 (upper) and 12 (lower) was fit to Eq. (1) to generate binding curves and $K_{d(app)}$ values.

The data are plotted in Fig. 4B (top) and show that probe 15 has nearly ideal behavior, with almost complete cleavage at concentrations greater than 20 μM. The data approximate a binding isotherm and can be fit to the binding equation [Eq. (1)]:

$$f = \frac{[\mathrm{D}]}{[\mathrm{D}] + K_{\mathrm{d(app)}}} \quad (1)$$

where the fraction cleaved (f) is related to the probe concentration (D) and the apparent dissociation constant ($K_{\mathrm{d(app)}}$). $K_{\mathrm{d(app)}}$ is not a true dissociation constant but, rather, reflects the competition between binding and folding (Fig. 2). The $K_{\mathrm{d(app)}}$ for probe 15 is 0.8 μM, and we use between 25 and 50 μM concentrations in a typical kinetic oligonucleotide hybridization experiment.

Some useful probes may not exhibit ideal behavior. For example, the titration for probe 12 is shown in Fig. 4B (lower). This probe targets a region of the RNA that folds on a much faster timescale than the region targeted by probe 15. As a result, much higher probe concentrations are required for efficient cleavage ($K_{\mathrm{d(app)}} = 20\ \mu M$), and complete cleavage is not achieved. Some additional cleavage may be rescued by increasing the probe concentration to the millimolar range; however, as pointed out earlier, such excessive probe concentrations can result in the apparent unfolding of folded molecules during the quench. We typically use 200 μM probe 12 in a kinetic oligonucleotide hybridization experiment, and the kinetic data are interpreted qualitatively (see later section). Incomplete cleavage at high probe concentrations can also be due to an inhomogeneous "unfolded" state that contains both accessible and inaccessible conformations.

Reagents

- 4 nM ^{32}P-labeled *Tetrahymena* ribozyme RNA (specific activity $\geq$ 1000 Ci/mmol) in TE buffer
- 800 μM DNA probe in TE buffer
- 10 $\times$ Folding and quench buffer
- Stop dye
- RNase H (5 units/μl), Amersham
- 6% Denaturing polyacrylamide gel mix and gel running buffer

Equipment

- 1.5- or 0.5-ml (depending on preference) siliconized polypropylene flip top assay tubes (Bio-Rad)

Water bath set to 37°
Heat block set to 95°
PAGE equipment
PhosphorImager (Molecular Dynamics, or equivalent)
Data analysis software for nonlinear least squares analysis (such as Igor by Wavemetrics)

Procedure

1. 2 × Quench solutions (1 × folding and quench buffer, 2 × DNA probe, 0.2 units/μl RNase H) covering the desired range of DNA probe concentrations are prepared in 10-μl aliquots and stored on ice in 1.5-ml tubes. Because the quench reactions for the titration mimic the zero time point, 5 μl of 2 × folding and quench buffer is added to each 10-μl aliquot. Quench solutions are preincubated at 37° for 3 min prior to initiating the quench reaction.
2. Ribozyme RNA (4 n*M* in TE buffer) is annealed by heating at 95° for 1 min followed by incubation at 37° for 3 min.
3. Annealed RNA (5 μl) is added to the quench solution and the reaction is incubated for 30 sec at 37°. Stop dye (14 μl) is added to terminate the reaction.
4. Full-length and cleaved molecules are separated and quantitated as earlier, and the data are fit to Eq. (1) using curve-fitting software.

Comments

In a well-optimized quench reaction, >85% cleavage at the zero time point and <20% cleavage after equilibration under folding conditions should be observed. Inefficient cleavage at the zero time point can result from rapidly folding RNA, inhomogeneous unfolded RNA, or insufficient RNase H/quench time. Significant cleavage at equilibrium may imply that a large fraction of the RNA is misfolding, an excessive DNA probe concentration is leading to cleavage of folded molecules, the quench reaction is too long, or insufficient Mg^{2+} is present during folding (see the Thermodynamics section later). Quench time courses on both unfolded and folded RNAs can help to diagnose these problems. The concentrations of RNase H and DNA probes and the length of the quench can then be optimized. We have always performed the quench reactions with 0.1 units/μl of RNase H for 30 sec at 37°. These values may be flexible; however, Mg^{2+} cannot be omitted from the buffer since it is required for RNase H activity.

There may be instances when a component (such as a salt) or a property (such as temperature) of the folding reaction inhibits the quench reaction. Whenever the folding conditions are changed, the efficiency of the quench should be tested. We have found the quench reaction to be quite robust, withstanding greater than 0.5 *M* urea.[6,7] One desirable feature of kinetic oligonucleotide hybridization is that the folding and quench reactions are performed in separate tubes. If some property of the folding reaction is found to inhibit the quench, then a simple solution is to add a smaller volume of the folding reaction to the quench, thereby diluting the inhibitory property. The Mg^{2+}, however, is never diluted on quenching because reduced Mg^{2+} can result in rapid unfolding and inappropriate cleavage.[2] Quenches should also be reoptimized for any mutant RNAs because the kinetic folding behavior can be significantly different from wild type.[3,6,7] The quench conditions may also have ill effects on the folding reaction. RNase H has a strict requirement for Mg^{2+} and a reducing agent.[17] If these reagents, for example, lead to unfolding of the RNA, then kinetic oligonucleotide hybridization may not be applicable.

Kinetic Folding Assays

To monitor kinetic folding, aliquots of a folding reaction are quenched at various times ($\geq$ 5 sec) after the addition of Mg^{2+}. The fraction cleaved as a function of folding time is then fit to an exponential equation to obtain the apparent first-order rate constant. Figure 5 shows Mg^{2+}-dependent kinetic folding data for both wild-type and mutant *Tetrahymena* ribozymes. Figure 5A shows the gel image for wild-type experiments with probes 15 (targeting P3–P7) and 12 (targeting P4–P6), and the data are fit to exponential equations in Fig. 5B. The probe 15 data are ideal, with a change in the fraction cleaved of ~0.9. The apparent first-order rate constant for this process (1.2 min^{-1}) was obtained by fitting the data to the rate equation [Eq. (2)]:

$$f = f_0 + f_1 e^{-k_1 t} \tag{2}$$

where f is the fraction cleaved, f_0 is f at equilibrium, f_1 is the total change in f, k_1 is the apparent first-order rate constant, and t is time in minutes. These data suggest that the P3 helix is formed and/or stabilized at 1.2 min^{-1}.

Probe 12, as discussed, does not behave ideally, and this is reflected by the small change in f (~0.4) in Fig. 5B. A rapid folding phase occurs between 0 and 5 sec that is followed by a second, much slower phase. Although the fast phase cannot be measured with sufficient time resolution to determine a rate constant, it is clearly much faster than P3–P7 formation. Hence, these *qualitative* results are useful because they place P4–P6 formation ahead of P3–P7 formation in the kinetic folding pathway.

The probe 12 data are an example of multiple phase behavior, with members of the population folding at different rates. Data of this sort are described best by a double exponential equation [Eq. (3)]:

$$f = f_0 + f_1 e^{-k_1 t} + f_1 e^{-k_2 t} \tag{3}$$

where f_0 is the value of f at equilibrium, f_1 and f_2 are the changes in f for the two phases and k_1 and k_2 are the first-order rate constants for the two phases.

In Fig. 5C the folding of a mutant ribozyme (P4–P6 mutant) with sequence changes in the P4–P6 domain[22] is compared to the wild type. For the P4–P6 mutant, the rates of P4–P6 (probe 15) and P3–P7 (probe 12) formation are nearly identical (0.5 min^{-1}), and are slower than either rate measured for the wild type. These data illustrate that kinetic oligonucleotide hybridization can distinguish between rate constants that differ by a factor of 2. Furthermore, because the mutations slow P4–P6 formation, probe 15 now behaves ideally, giving nearly complete cleavage at a relatively low concentration (20 μM).

Reagents and Equipment

Same as in Optimization of Quench Reaction section.

Procedure

1. 2 × Quench solution (1 × folding and quench buffer, 2 × DNA probe, 0.2 units/μl RNase H) for each time point is prepared in 10-μl aliquots and stored on ice in 1.5-ml tubes. Five microliters of 2 × folding and quench buffer is added to the zero time point quench. Each quench solution is preincubated at 37° for 3 min prior to initiating the quench reaction—this requires that quench solutions be transferred to the water bath throughout the folding time course.
2. Ten-microliter aliquots of 2 × folding and quench buffer for time points <30 sec are prewarmed to 37°. To avoid timing and pipetting conflicts, these short reactions are performed individually, whereas the longer time points are performed in a "one-pot" folding reaction.
3. Ribozyme RNA (50 μl of 4 nM in TE buffer) is annealed by heating at 95° for 1 min followed by incubation at 37° for 3 min.
4. Annealed RNA (5 μl) is added to the zero time point quench solution and the reaction is incubated for 30 sec at 37°. Stop dye (14 μl) is added to terminate the reaction.
5. For time points <30 sec, 10 μl of annealed RNA is mixed with the 10-μl aliquot of 2× folding and quench buffer. After the desired

[22] F. L. Murphy and T. R. Cech, *J. Mol. Biol.* **236,** 49 (1994).

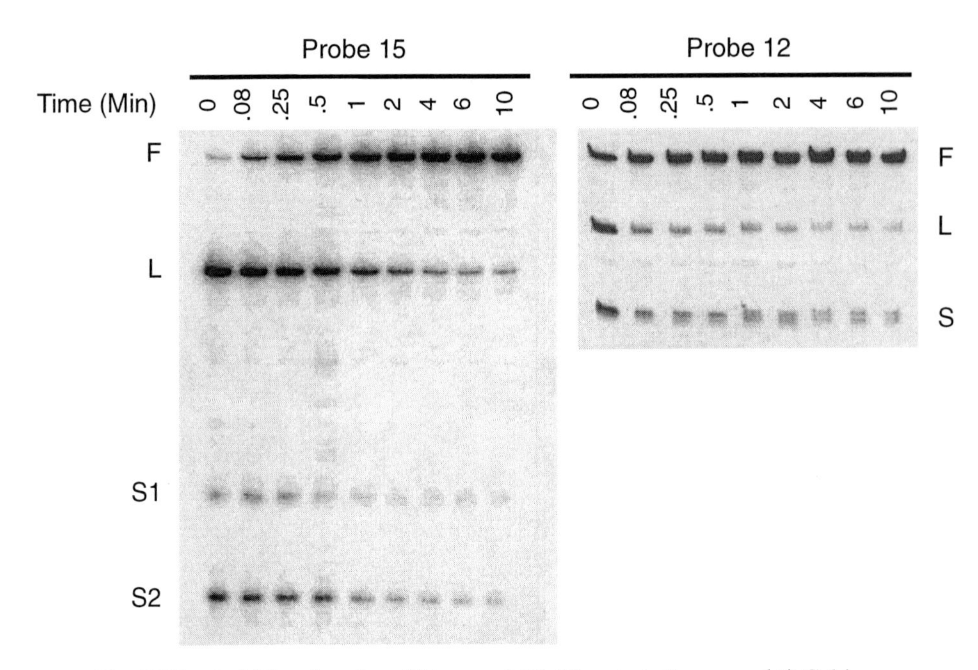

FIG. 5. Kinetic folding data for wild-type and P4–P6 mutant ribozymes. (A) Gel image of kinetic folding data for the wild-type ribozyme. Folding reactions in 10 m*M* Mg^{2+} at 37° were quenched at the times indicated with DNA probes (50 μM probe 15 or 200 μM probe 12) and 0.1 unit/μl RNase H. Positions of cleavage products are indicated as in Fig. 4. (B) The kinetic folding data from (A) are fit to exponential equations. The probe 15 data fit to a single exponential Eq. (2) with $k = 1.2\ \text{min}^{-1}$. The probe 12 data are fit to a double exponential Eq. (3). Due to timescale limitations, the rate constant for the fast phase cannot be determined accurately, but 10 min^{-1} is the lower limit. $k = 0.7\ \text{min}^{-1}$ for the slow phase, which accounts for ~20% of the total folding amplitude. (C) Kinetic folding data for the P4–P6 mutant ribozyme is compared to the wild-type data from (B). The mutant data for probes 15 and 12 fit to a single exponential, and the two rate constants are nearly identical ($k = 0.5\ \text{min}^{-1}$).

folding time, 10 μl of the folding reaction is added to the quench solution, and the quench reaction is incubated and terminated as was done earlier.

6. For time points >30 sec, 50 μl of 2 × folding and quench buffer (prewarmed to 37°) is added to 50 μl of annealed RNA. At the

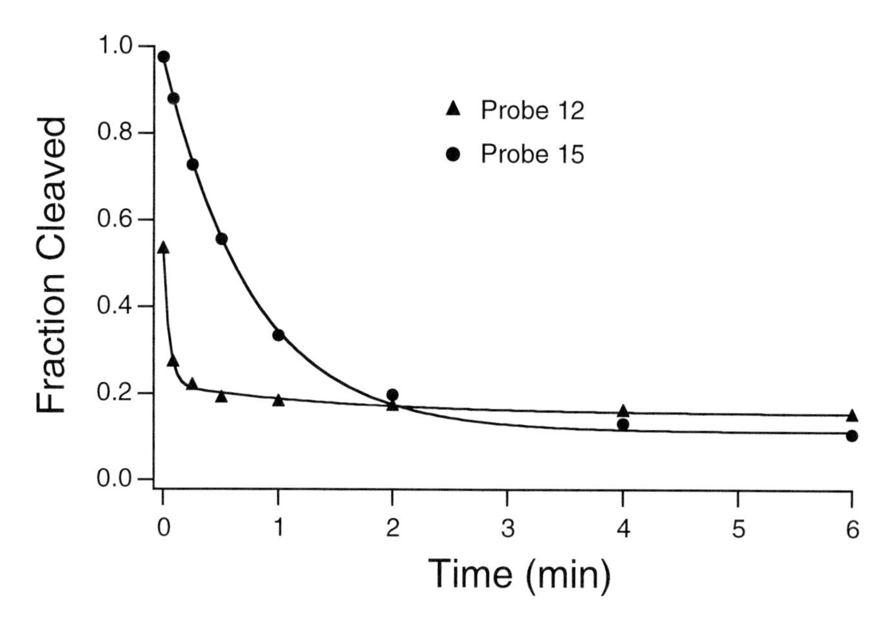

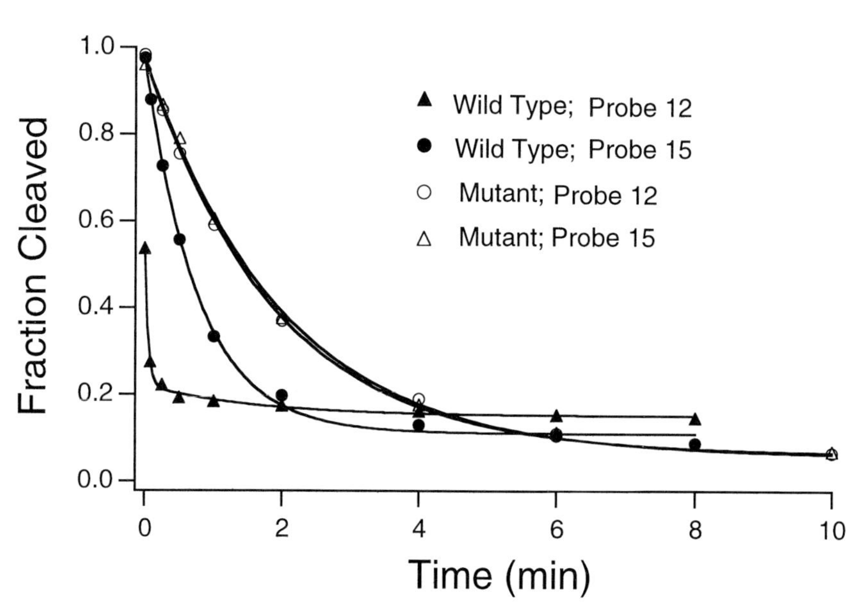

FIG. 5. (*continued*)

desired folding times, 10-μl aliquots are removed, quenched, and terminated as described earlier.

7. Full-length and cleaved molecules are separated and quantitated as earlier, and the data are fit to Eq. (2) or (3) with curve-fitting software.

Comments

A dry run is recommended to practice the rapid pipetting that is required for the kinetics assays. The use of multiple pipetting devices and a stationary rack that loosely holds 1.5-ml tubes is critical as well. If folding is performed at temperatures other than 37°, the quench should still be at 37°. The efficiency of the quench should be tested under these conditions. When folding is performed at low Mg^{2+} concentrations, 2 mM NaCl should be substituted for the chelating agent Na_2EDTA in the annealing buffer (TE).

Related Applications of Kinetic Oligonucleotide Hybridization

Thermodynamics

Thermodynamic constants can be measured by monitoring the folding–unfolding equilibrium as a function of the Mg^{2+} concentration or temperature. For large RNAs, the Mg^{2+} concentration required to achieve half-maximal folding $[Mg^{2+}]_{1/2}$ is a useful indicator of stability.[23] In a kinetic oligonucleotide hybridization experiment, the RNA is incubated with various concentrations of Mg^{2+}, and the fraction cleaved at equilibrium is determined in a quench reaction. Because the rate of approach to equilibrium may depend on the Mg^{2+} concentration, it is important to measure folding kinetics at all concentrations as well. The quench solutions should be adjusted such that the final Mg^{2+} concentration in the quench reaction is 10 mM. Equilibrium folding data for the P3 helix (probe 15) of both wild-type and P4–P6 mutant (see earlier section) *Tetrahymena* ribozymes is shown in Fig. 6. The data were fit to the binding equation [Eq. (4)]

$$f = f_0 - \frac{f_1\,[Mg^{2+}]^n}{[Mg^{2+}]^n + K_{d(app)}} \tag{4}$$

where f is the fraction cleaved at equilibrium, f_0 is the fraction cleaved in the absence of Mg^{2+}, f_1 is the total change in f, $K_{d(app)}$ is the apparent equilibrium dissociation constant for Mg^{2+} binding, and n is the Hill coefficient, a measure of cooperativity and the minimum number of Mg^{2+} ions

[23] B. Laggerbauer, F. L. Murphy, and T. R. Cech, *EMBO J.* **13,** 2669 (1994).

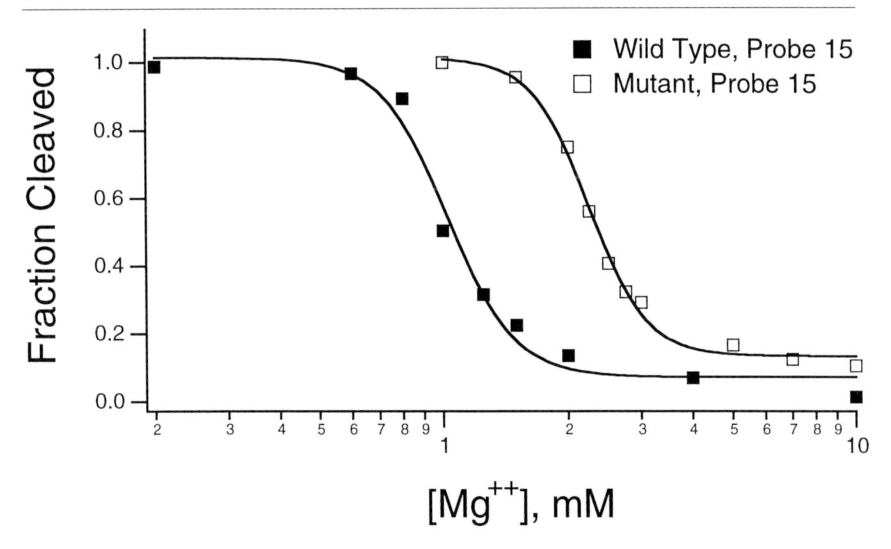

FIG. 6. Thermodynamic folding data for wild-type and P4–P6 mutant ribozymes. Ribozymes were equilibrated for 20 min with various concentrations of Mg^{2+} at 37° prior to initiation of quench reactions containing probe 15 (50 μM for wildtype, 20 μM for mutant) and 0.1 unit/μl RNase H. The data were fit to binding Eq. (4) to generate values for $[Mg^{2+}]_{1/2}$ and the Hill coefficients. Wild-type $[Mg^{2+}]_{1/2}$ = 1.0 mM; mutant $[Mg^{2+}]_{1/2}$ = 2.2 mM. Hill coefficients: wild type, n = 5.4; mutant, n = 6. 2 mM NaCl was substituted for 1 mM Na_2EDTA in the annealing buffer to ensure accurate free Mg^{2+} concentrations during folding.

involved in the folding transition. Both RNAs fold cooperatively as a function of Mg^{2+} (Hill coefficients: wild type, n = 5.4; mutant, n = 6), and the mutant RNA is destabilized, requiring additional Mg^{2+} for folding (wild type $[Mg^{2+}]_{1/2}$ = 1.0 mM; mutant $[Mg^{2+}]_{1/2}$ = 2.2 mM). The wild-type values are in agreement with data collected by other equilibrium methods.[24,25]

In Vitro Selection

In vitro selection strategies require a step where only molecules with a desired property "survive" and can be amplified. Kinetic oligonucleotide hybridization is suited for folding-based selections because unfolded molecules are cleaved, whereas folded molecules are not cleaved and survive. We have used kinetic oligonucleotide hybridization to select for mutant *Tetrahymena* ribozymes with a fast-folding P3–P7 domain.[6] Kinetic oligonu-

[24] D. W. Celander and T. R. Cech, *Science* **251,** 401 (1991).
[25] B. Sclavi, S. Woodson, M. Sullivan, M. R. Chance, and M. Brenowitz, *J. Mol. Biol.* **266,** 144 (1997).

cleotide hybridization-based thermodynamic folding selection strategies can be imagined as well.

Comparison of Kinetic Oligonucleotide Hybridization to Other Methods for Studying RNA Folding Kinetics

Several other valuable methods exist for the study of RNA folding kinetics. The list of methodologies includes synchrotron hydroxyl radical footprinting (SHRF),[25,26] UV cross-linking,[27] protein binding,[28,29] electrophoretic mobility,[30] fluorescence,[31,32] UV and circular dichroism (CD) spectroscopy,[33,34] chemical modification protection,[12,13] and gain of catalytic function.[7,34] Like kinetic oligonucleotide hybridization, each of these methods has both strengths and limitations. Fortunately, many of these methods are complementary and can be used in conjunction to perform a rigorous kinetic analysis.

SHRF, fluorescence, and UV and CD methods have time resolutions (millisecond) that are superior to kinetic olignucleotide hybridization. Whereas SHRF also provides nucleotide resolution, the transitions observed in the spectroscopic methods can be difficult to interpret structurally. Fluorescence is currently limited by the difficulty of preparing internally labeled large RNAs and has been valuable in the study of tRNA[31] (and other small RNAs) and substrate binding to large ribozymes.[32] SHRF is perhaps the most powerful of all methods but is currently limited by access and labor-intensive optimizations.

Electrophoretic mobility, UV cross-linking, and modification protection have time resolutions that are less than or similar to kinetic oligonucleotide hybridization. Electrophoretic mobility assays can allow different kinetic folding intermediates to be resolved; however, additional methods are required to characterize the intermediates structurally.[30] UV cross-linking and modification protection can provide more detailed structural information than kinetic oligonucleotide hybridization but are considerably more labor intensive.

[26] B. Sclavi, M. Sullivan, M. R. Chance, M. Brenowitz, and S. A. Woodson, *Science* **279,** 1940 (1998).
[27] W. D. Downs and T. R. Cech, *RNA* **2,** 718 (1996).
[28] K. M. Weeks and T. R. Cech, *Science* **271,** 345 (1996).
[29] R. T. Batey and J. R. Williamson, *RNA* **4,** 984 (1998).
[30] J. Pan and S. Woodson, *J. Mol. Biol.* **280,** 597 (1998).
[31] D. C. Lynch and P. R. Schimmel, *Biochemistry* **13,** 1841 (1974).
[32] P. C. Bevilacqua, R. Kierzek, K. A. Johnson, and D. H. Turner, *Science* **258,** 1355 (1992).
[33] E. J. Maglott and G. D. Glick, *Nucleic Acids Res.* **25,** 3297 (1997).
[34] T. Pan and T. R. Sosnick, *Nature Struct. Biol.* **4,** 931 (1997).

With these comparisons in mind, kinetic oligonucleotide hybridization is a reasonable first step in monitoring the folding kinetics of a large RNA, especially if several useful probes can be identified. Then, in conjunction with other methods such as CD and SHRF, more detailed questions can be addressed that require increased time and structural resolution.

Acknowledgments

This work was supported by the Rita Allen Foundation and the Skaggs Institute for Chemical Biology. The authors wish to acknowledge Patrick Zarrinkar, who developed kinetic oligonucleotide hybridization during his thesis work, for his important contributions to this work and for critical comments on the manuscript.

[22] Time-Resolved Synchrotron X-Ray Footprinting and Its Application to RNA Folding

By CORIE Y. RALSTON, BIANCA SCLAVI, MICHAEL SULLIVAN, MICHAEL L. DERAS, SARAH A. WOODSON, MARK R. CHANCE, and MICHAEL BRENOWITZ

Introduction

The ability of RNA molecules to form uniquely folded, compact tertiary structures is critical to their biological function. Determining the mechanisms of RNA folding is essential to understanding their binding and catalytic functions in diverse cellular processes such as translation and splicing. In particular, catalytic RNAs derive their enzymatic ability from the formation of discrete tertiary structures composed of multiple domains.[1–3] Individual RNA domains can fold on timescales ranging from milliseconds to minutes.[4] Characterizing RNA folding pathways and deducing their common features is an important challenge that will require the concomitant application of a variety of structural and functional techniques.

Nucleic acid footprinting assays map the solvent accessible surface of DNA or RNA by probing their protection from nucleases or modifying reagents. The development of quantitative footprinting protocols allows

[1] T. R. Cech, *in* "The RNA World," (R. F. Gesteland and J. F. Atkins, eds.), pp. 239–269. Cold Spring Harbor Laboratory Press, New York, 1993.
[2] T. R. Cech, D. Herschlag, J. A. Piccirilli, and A. M. Pyle, *J. Biol. Chem.* **267**(25), 17479 (1992).
[3] T. R. Cech and D. Herschlag, *Nucleic Acids Mol. Biol.* **10,** 1–7 (1996).
[4] For a review, see D. E. Draper *Nature Struct. Biol.* **3,** 397 (1996).

0076-6879/00 $30.00

isotherms and kinetic progress curves to be obtained for individual sites within a nucleic acid.[5–7] Protocols for quench-flow DNase I footprinting established the feasibility of conducting footprinting kinetics studies on the millisecond timescale.[6] Since the first DNase I footprinting studies of DNA–protein interactions,[8] a wide range of enzymatic and chemical cleavage agents has been successfully applied to the study of RNA and DNA structure and dynamics and their interaction with proteins. Of the plethora of nucleases available to investigators, perhaps none is more widely used than the hydroxyl radical (·OH).[9,10]

The key advantage to ·OH footprinting is that the radicals are small enough to provide sensitivity to conformational changes with single base pair resolution. In footprinting assays, ·OH can be generated through the radiolysis of water,[11,12] or by catalysts such as Fe-EDTA[9] or peroxonitrous acid.[13] Synchrotron X-ray "footprinting" is a new technique that allows time-resolved structural analysis of conformational changes of nucleic acids using ·OH generated by the passage of a bright synchrotron x-ray beam through water. The solvent-accessible surface of a nucleic acid is probed by the ·OH mediated cleavage of the phosphodiester backbone. The extremely high brightness of the synchrotron X-ray beam allows millisecond X-ray exposures to be used in footprinting experiments. Thus, the entire solvent-accessible surface of a nucleic acid can be mapped with single base resolution, on timescales as short as milliseconds.

The technique has been successfully applied to the study of the *Tetrahymena thermophila* group I intron, yielding information on folding rate

[5] M. Brenowitz, D. F. Senear, M. A. Shea, and G. K. Ackers, *Methods Enzymol.* **130,** 132 (1986).

[6] M. Hsieh and M. Brenowitz, *Methods Enzymol.* **274,** 478 (1996).

[7] D. Strahs and M. Brenowitz *J. Mol. Biol.* **244,** 494 (1994).

[8] D. J. Galas and A. Schmitz, *Nucleic Acids Res.* **5,** 3157 (1978); A. Schmitz and D. J. Galas, *Nucleic Acids Res.* **6,** 111 (1979).

[9] T. D. Tullius, B. A. Dombroski, M. E. Churchill, and L. Kam, *Methods Enzymol.* **155,** 537 (1987); T. D. Tullius and B. A. Dombraski, *Proc. Natl. Acad. Sci.* U.S.A. **83,** 5469 (1986); W. J. Dixon, J. J. Hayes, J. R. Levin, M. F. Weidner, B. A. Dombroski, and T. D. Tullius *Methods Enzymol.* **208,** 380 (1991).

[10] D. C. Celander and T. R. Cech *Science* **251,** 401 (1991); D. C. Celander and T. R. Cech, *Biochemistry* **29,** 1355 (1990).

[11] M. A. Price and T. D. Tullius *Methods Enzymol.* **212,** 194 (1992).

[12] J. Franchet-Beuzit, M. Spotheim-Maurizot, R. Sabattier, B. Blazy-Baudras, and M. Charlier *Biochemistry* **32,** 2104 (1993).

[13] P. A. King, V. E. Anderson, J. O. Edwards, G. Gustafson, R. C. Plumb, and J. W. Suggs, *J. Am. Chem. Soc.* **114,** 5430 (1992); P. A. King, E. Jamison, D. Strahs, V. E. Anderson, and M. Brenowitz, *Nucleic Acids Res.* **21,** 2473 (1993).

constants within all its major domains.[14,15] Synchrotron X-ray footprinting is currently being applied to the study of the structure and folding of other RNA molecules, the structure of DNA, DNA–protein interactions, and protein folding.[16] An article has been published in this series describing the development of synchrotron ·OH footprinting and its application to RNA folding.[17] In this article we report the advances that have been made in the application of synchrotron X-ray footprinting to the study of RNA folding at National Synchrotron Light Source (NSLS) beam line X-9A under the auspices of the Albert Einstein Center for Synchrotron Biosciences. The instrumentation and protocols that have been developed are applicable to other experimental systems and new X-ray footprinting facilities that might be established.

Photon Flux and Absorption

Calculation of Beam Flux

Beam line X-9A at the NSLS at Brookhaven National Laboratory is a bending magnet beam line producing white light over an energy range of 3–30 keV. The photon flux at the beam line can be calculated using bending magnet radiation curves and is dependent on the magnet strength and the energy and number of the electrons in the storage ring.[18] Currently, the NSLS operates at two ring energies, 2.54 and 2.8 GeV. At 2.54 GeV, the beam current decays from ~300 mA at the time of injection to ~150 mA over a 12-hr period. Similarly, the beam current decays from ~250 mA at injection to ~120 mA at the 2.8-GeV ring energy.

The flux incident on the sample is obtained by multiplying the bending magnet radiation curve by the transmission curves for each absorber be-

[14] B. Sclavi, S. Woodson, M. Sullivan, M. R. Chance, and M. Brenowitz, *J. Mol. Biol.* **266,** 144 (1997).

[15] B. Sclavi, M. Sullivan, M. R. Chance, M. Brenowitz, and S. A. Woodson, *Science* **279,** 1940 (1998).

[16] M. R. Chance, B. Sclavi, S. Woodson, and M. Brenowitz, *Structure* **5,** 865 (1997); M. R. Chance, M. Brenowitz, M. Sullivan, B. Sclavi, S. Maleknia, and C. Ralston, *J. Synchr. Rad.* **11,** 7 (1998).

[17] B. Sclavi, S. A. Woodson, M. Sullivan, M. R. Chance, and M. Brenowitz, *Methods Enzymol.* **295,** 379 (1998).

[18] Spectral curves for bend magnet radiation are available at the web page maintained by the Center for X-ray Optics at the Lawrence Berkeley Laboratory: http://www-cxro.lbl.gov/optical_constants.

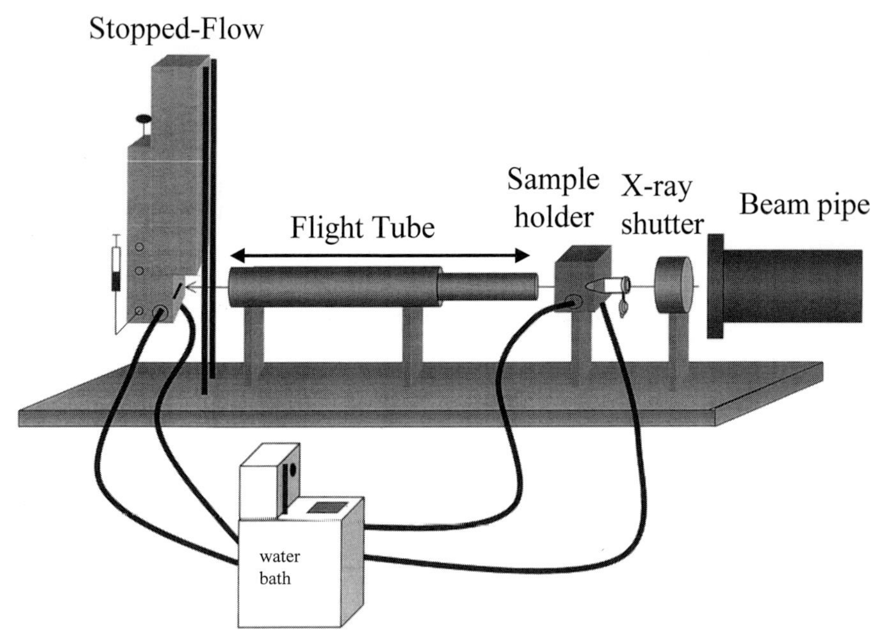

FIG. 1. Schematic representation of the experimental setup within the beam line X-9A hutch used to conduct synchrotron X-ray footprinting experiments. For studies using the rapid mixer, the X-ray shutter and sample holder are removed from the path of the beam, and the flight tube is extended to meet the beam pipe.

tween the front end of the beam line and the sample.[19] In the current configuration of X-9A, two beryllium windows, 254 and 500 μm thick, cap each end of the "beam pipe," a tube under high vacuum ($>10^{-8}$ Torr) extending from the ring wall into the experimental hutch (Fig. 1). An aluminum coating on the exit window is 25 μm thick. After exiting the beam pipe, the photons travel through an air path of 40 cm before striking the exposure cell of the stopped-flow apparatus (Fig. 1). The radiation curve calculated for X-9A at 2.8-GeV operation and 250-mA beam current is shown in Fig. 2A (solid line). Figure 2A also shows the calculated flux incident on a sample after passing through the two beryllium windows, the aluminum coating, and an air path of 40 cm (Fig. 2A, dash–dot line). It

[19] Transmission curves were also obtained from the Center for X-Ray Optics web page; these transmission curves do not take into account scattering. However, comparison with transmission curves obtained by using the mass attenuation coefficient showed that scattering by water in the energy range of interest resulted in less than a 1% difference in the calculation of flux absorbed by the sample.

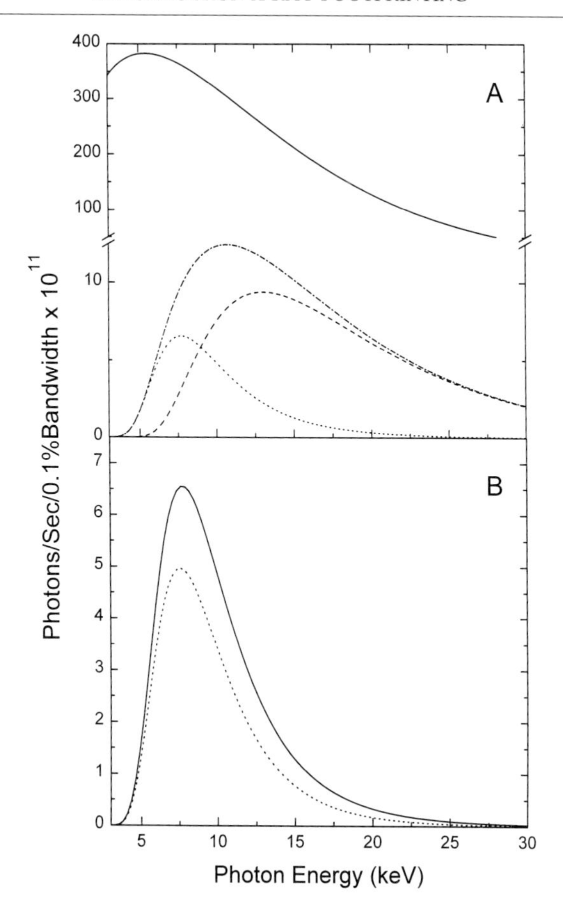

FIG. 2. (A) Solid line: the bending magnet radiation curve for X9A for 2.8-GeV operation and 250-mA beam current; dash–dot line: the calculated flux incident on the sample after passing through the two beryllium windows, the aluminum coating, and an air path of 40 cm; dashed line: the flux transmitted through the sample assuming a 1-mm-deep aqueous solution; dotted line: the flux absorbed by the sample. (B) Comparison of absorbed flux for 2.8- (solid line) and 2.54-GeV (dotted line) operation at 250 mA.

can be seen that the beam line optics absorb more than 99% of the efflux from the ring and that the peak energy is shifted from ~5 to 10 keV.

The flux transmitted *through* the sample is calculated by multiplying the flux incident on the sample by the transmission curve for a 1-mm water path length (Fig. 2A, dashed line). The difference between the incident flux and the transmitted flux is the amount absorbed by the sample (Fig. 2A, dotted line). Integration of the absorption curve gives the total number of photons per second absorbed by the sample. For a ring energy of 2.8

GeV and a beam current of 250 mA, 5.5×10^{14} photons are absorbed per second by a 10-μl sample of 7-mm^2 cross-sectional area, with the absorption maximum near 7.5 keV. This corresponds to 0.7 joules (J) absorbed per second by a 10-μl sample. Fig. 2B shows a comparison between the absorbed flux at 2.8- and 2.54-GeV operation. At comparable beam currents, 2.8-GeV operation yields about 1.4 times greater flux than 2.54-GeV operation.

Steady-State Hydroxyl Radical (·OH) Concentration

For the range of X-ray energies produced at NSLS beam line X-9A, the interaction between the X rays and water is dominated by the photoelectric effect.[20] In this interaction, the energy of an incoming photon is transferred to an electron, which is ejected from the water molecule. The ionized water molecule reacts with water to produce ·OH, yielding 287 radicals for every 10 keV of energy thermalized in solution.[20] The radicals can interact with a nucleic acid molecule or recombine to form H_2O_2 with a second-order rate constant of $5 \times 10^9\ M^{-1}s^{-1}$. As previously described,[17] the steady-state concentration of ·OH can be estimated from the photon flux. A flux of 10^{14} photon/sec, as calculated for beam line X-9A operating at 2.54 GeV and 250 mA, corresponds to a dose rate of $1.6 \times 10^{-2}\ M$/sec, which in turn yields a steady-state [·OH] of 1.2 μM.

Dose Response

A parameter key to the successful conduct of a synchrotron X-ray footprinting experiment is the amount of X-ray exposure ("dose") received by the RNA sample. Cleavage of the RNA must be controlled so that, on average, each molecule that is cleaved is cleaved only once. Exposures resulting in 10–30% RNA cleavage fulfill this requirement.[5] The correspondence between exposure to the X-ray beam and resulting cleavage of the nucleic acid is experimentally determined by a dose–response calibration. In these experiments, RNA labeled at one end with ^{32}P is exposed to the X-ray beam for a series of exposure times at constant beam energy and current.

Plotting the fraction of uncut *Tetrahymena* L-21 *Sca*I ribozyme as a function of time yields the dose–response curve.[21] The fraction of uncut RNA is measured because this is conveniently assayed by quantitating the

[20] N. V. Klassen, *in,* "Radiation Chemistry Principles and Applications," Austin, (I. Farhatazis and M. A. Rodgers, eds). VCH Publishers, Austin, Texas, 1987.

[21] Dose–response curves obtained using the sample stand and electronic shutter (Fig. 1) or in the quench-flow apparatus (Fig. 4) under similar operating conditions are identical within experimental error. The dose–response curves shown were obtained using the sample stand and electronic shutter.

decrease in density of the most slowly migrating band following denaturing polyacrylamide gel electrophoresis (PAGE) separation of the samples.[17] Figure 3A compares the dose–response relationships observed at 2.54-GeV (squares) and 2.8-GeV (circles) operation. The semilogarithmic relationship observed between RNA cleavage and X-ray exposure is predicted by Poisson statistics.[5] X-ray beam exposures of 10–25 and 3–9 msec for 2.54-

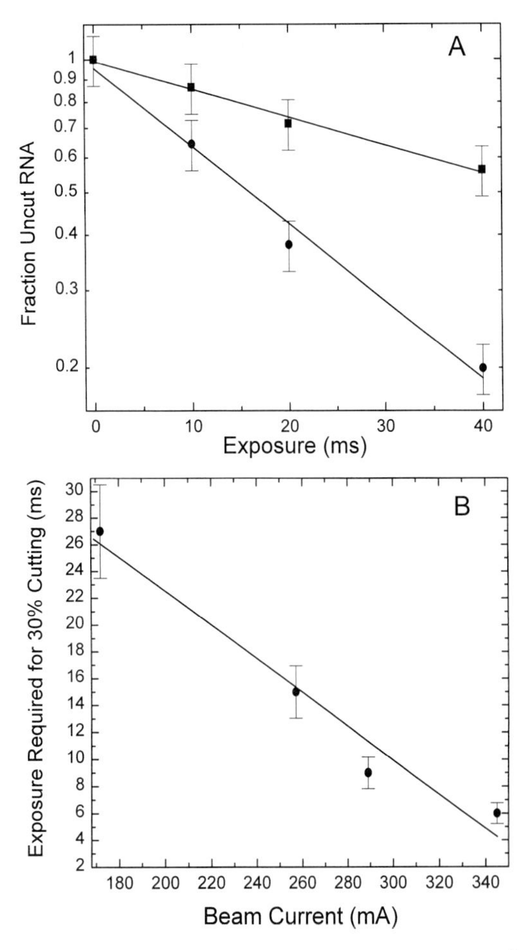

FIG. 3. (A) Dose-response curves for the L-21 *Sca*I RNA in 10 mM Mg^{2+} under operating conditions of 2.54-GeV operation and 250 mA (squares) and 2.8-GeV operation and 240 mA (circles). The error in each dose–response experiment was 13%. (B) Plot derived from several dose–response curves at 2.54-GeV operation, showing exposure necessary for 30% cleavage under different operating conditions.

and 2.8-GeV operation, respectively, are appropriate for footprinting the *Tetrahymena* ribozyme at the indicated ring current.

Because the ·OH cleavage statistics are dependent on the length of the RNA, a dose response is determined for each RNA studied by synchrotron x-ray footprinting. In addition, a calibration curve is required relating RNA cleavage to beam current at each of the two ring currents used at the NSLS since variations in the beam energy and current result in variable photon flux impacting samples. Figure 3B shows the experimentally determined X-ray exposures required to achieve 30% RNA cleavage of the *Tetrahymena* ribozyme as a function of beam current at 2.54 GeV. Because RNA cleavage and beam current are linearly related within the uncertainty of the measurements, a simple scaling factor is used to standardize beam exposure over the course of an experiment or series of experiments.

Experimental Methods

Sample Storage

At beam line X-9A all radiolabeled materials are stored within the experimental hutch. To protect the RNA from background X-ray exposure, both the RNA stock solutions and exposed samples are stored in acrylic-covered lead-lined boxes (USA/Scientific Plastics, Ocala, FL). One lead box resides in a freezer that is within the experimental hutch (which in turn is covered by 1/16-in. lead sheathing) for sample storage. A second custom-ordered lead box that is 5 in. deep, resides on the worktable to allow easy access to samples. Samples are kept at 4° using cold-pack microfuge tube holders (USA/Scientific Plastics) or maintained at other temperatures using a temperature-regulated block placed within the lead box.

Electronic Shutter for Manual Mixing Experiments

For equilibrium studies and hand-mixing kinetic studies, an electronic shutter (Vincent Associates, Rochester, NY) impervious to the white light X-ray beam has replaced the "guillotine" gravity-driven shutter used previously.[15] The electronic X-ray shutter consists of a platinum/iridium alloy plate, 6.2 mm in diameter and 1 mm thick, capable of blocking X rays up to 30 keV in energy. A T-132 controller placed outside the hutch operates the shutter. Exposures ranging from 7 msec to 167 min can be programmed. The X-ray shutter is placed in front of a sample stand adjacent to the beam pipe (Fig. 1). Lead shielding minimizes X-ray scattering from the front of the shutter. The stand and the shutter are both aligned at the start of an experiment in order to ensure that the beam uniformly impacts the sample. The sample holder and shutter are mounted on slides so that they can be moved out of the beam path for rapid-mixing experiments.

In typical equilibrium, dose–response or hand-mixing kinetics experiments, 10 μl of sample is aliquoted into a microfuge tube and held by surface tension at the bottom. The tube is placed horizontally in a sample holder machined from a block of aluminum through which water from a temperature-regulated bath is circulated. Nominal temperature control of $\pm 0.1°$ can be maintained with this apparatus. Following placement of the sample tube in the holder, the beam line enabling procedures are initiated and the safety shutter is opened, allowing the X-ray beam to enter the hutch. The electronic shutter is then triggered, followed by closure of the safety shutter and retrieval of the exposed sample.

Rapid-Mixing X-Ray Footprinting

To conduct a rapid-mixing experiment, the X-ray shutter and sample holder are removed from the path of the beam. A steel "flight tube" is extended so that it is flush against the beam pipe at one end and against the modified Kin-Tek quench-flow apparatus (KinTek Corporation, Austin, Texas) on the other end (Fig. 1). In the Kin-Tek quench-flow apparatus, an exposure chamber[17] aligned with the X-ray beam has replaced the sample loop selector (Fig. 4). Repositioning the exposure chamber to the back of the apparatus minimizes the air path of the beam by allowing the quench flow to be placed closer to the beam pipe. This arrangement maximizes the X-ray flux impacting the sample and minimizes the X-ray scattering that contributes to background degradation (see later discussion).

The exposure chamber is now being machined from Vesbel, a material capable of withstanding prolonged exposure to the white light X-ray beam. The side of the cell facing the X-ray beam is covered with a thin Kapton window. The chamber dimensions are as previously described.[17] The sample flow train of the quench-flow device is siliconized (Surfa-Sil from Pierce Chemical Company, Rockford, IL) and treated with an anti-RNase agent (RNAse-ZAP from Ambion, Austin, TX) on a regular schedule to minimize the amount of absorbed RNA and sample degradation.

The flight tube and lead shielding placed around the exposure chamber and on the face of the quench-flow apparatus minimize the scattered radiation generated as the X-ray beam passes through air (Fig. 1). The water surrounding the flow train of the apparatus additionally shields the RNA during an experiment. It is essential that a shield be placed over the syringe containing the [^{32}P]RNA solution prior to enabling the beam line. A steel

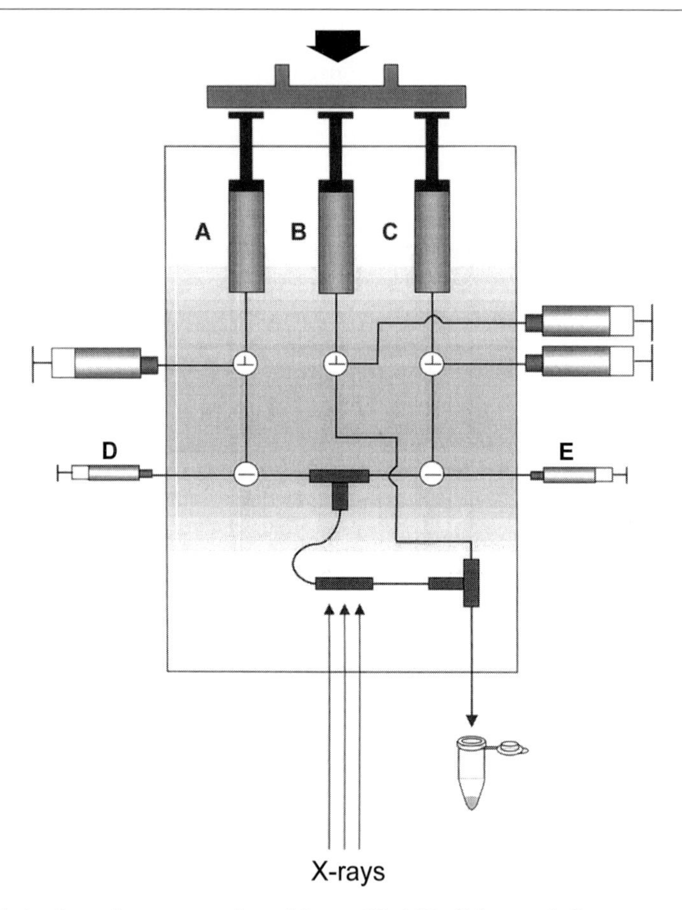

FIG. 4. A schematic representation of the modified Kin-Tek quench-flow apparatus. Note that the entire sample flow train is immersed in the circulating water bath, ensuring efficient and uniform temperature equilibration.

tube (similar in thickness to the flight tube) mounted on a slide provides sufficient shielding yet is easily moved to provide access to the syringe.

The time a sample is exposed to the X-ray beam is now determined by programming the rate at which the sample flows through the exposure chamber. This protocol more reliably exposes the entire RNA sample than the "push–pause" procedure previously described that it replaces.[17] Sample exposure to the X-ray beam is dependent on the beam energy and current (as discussed earlier), length of the RNA, and the integrity of the RNA sample. Minimally background degraded [^{32}P]RNA samples allow lower

X-ray exposures to be used because sufficient experimental signal-to-noise ratios can be achieved with less cleavage.

Solutions

TE buffer. 10 mM Tris-HCl, pH 7.5, 1 mM EDTA
TEN buffer. 10 mM Tris-HCl, pH 7.5, 1 mM EDTA, 250 mM NaCl
CE buffer. 10 mM sodium cacodylate, pH 7.5, 0.1 mM EDTA, pH 8.0
Mg^{2+} buffer. 10 mM sodium cacodylate, pH 7.5, 0.1 mM EDTA, pH 8.0, 20 mM $MgCl_2$
Precipitation buffer. 1.5 M sodium acetate, pH 5.0, 0.25 mg/ml tRNA in TE buffer
2× Gel loading buffer. 10 M urea, 0.2× TBE (0.089 M Tris–borate, 0.089 M boric acid, 2 mM EDTA, pH 8.0), 0.1% (w/v) bromphenol blue and xylene cyan.

Preparation of RNA

RNA to be used in X-ray footprinting experiments is prepared by 5′-end labeling of *in vitro* transcribed RNA with polynucleotide kinase as is generally used for biochemical analysis of RNA.[22] The RNA is prepared by T7 transcription from DNA templates followed by enzymatic removal of the 5′-triphosphate with calf alkaline intestinal phosphatase. The RNA is end labeled with [γ-^{32}P]ATP and polynucleotide kinase, followed by purification by PAGE. Gel slices containing the RNA are cut from the gel and soaked in TEN buffer at 4° overnight. The buffer is decanted to a collection tube and the gel slices washed with an additional 1.0 ml of TEN buffer. The combined TEN buffers are filtered and mixed with three volumes of absolute ethanol. The RNA is precipitated at −20° for 4–8 hr, pelleted by centrifugation, dissolved in 200 μl TE buffer and sequentially extracted with phenol followed by chloroform.[23] The precipitation is repeated a second time and the RNA dissolved in 30–50 μl TE buffer and the specific radioactivity of the RNA determined by liquid scintillation counting.

The [^{32}P]RNA is brought to the beam line in 1-μCi aliquots to a maximum of 10 μCi pursuant to the dispersible radioactivity safety protocol in use at beam line X-9A.[24] In addition, aliquotting of the RNA reduces

[22] A. J. Zaug, C. A. Grosshans, and T. R. Cech *Biochemistry* **27,** 8924 (1988).

[23] T. Maniatis, E. F. Fritsch, and J. Sambrook, "Molecular Cloning: A Laboratory Manual." Cold Spring Harbor Laboratory, Cold Spring Harbor, New York, 1982.

[24] Copies of this safety protocol can be obtained from the Center for Synchrotron Biosciences. This protocol details the procedures and materials required for the transport and handling of dispersible radioactivity at the NSLS. Shipment of radioisotopes is subject to federal and state regulations.

nonspecific degradation manifest as background in the gel autoradiograms. Typically, the 1-μCi aliquots of the RNA are diluted to 20 μl using the buffer to be used in the planned experiments since prolonged storage of highly concentrated solutions of [^{32}P]RNA often results in increased degradation and elevated background densities in the footprint autoradiograms. These aliquots of [^{32}P]RNA are subsequently used to prepare the experimental samples, as described later.

Protocol for Manual Mixing X-Ray Footprinting Experiment

The X-ray shutter is used for experiments that do not require rapid mixing of the RNA solution. Such experiments include determining dose–response relationships, equilibrium titration experiments, and slow timescale kinetics experiments. The following example protocol is a slow time-scale kinetics experiment for the Mg^{2+} initiated folding of the *Tetrahymena* ribozyme.

1. An aliquot (typically 9 μl) of [^{32}P]RNA diluted in assay buffer to an activity of ~25–100,000 dpm is placed in a microfuge tube. One microliter magnesium buffer is added to the RNA solution to initiate the folding reaction yielding a final volume of 10 μl. For example, addition of 1 μl of a 100 m*M* $MgCl_2$ solution yields a final concentration of 10 m*M* Mg^{2+}. The solutions are mixed by a quick flick of the closed tube and the samples brought to the base of the tube if necessary by a brief centrifugation. The closed microfuge tube is placed in the sample stand.
2. The experimenter exits and interlocks the hutch. The beam line safety shutter can be opened after the 20-sec interlock safety pause.
3. After a time period several seconds shy of the desired reaction time, the safety shutter is opened allowing the X-ray beam to enter the hutch. The electronic X-ray shutter is immediately activated to expose the sample, thus minimizing exposure of the RNA to scattered radiation. The electronic shutter remains open for the programmed time as determined from dose–response and calibration curves, as described earlier. The 20-sec interlock pause sets a minimum dead time for a manual experiment at ~30 sec. The exposure times typically range from 10 to 100 msec, depending on the length of the RNA.
4. The safety shutter is closed. The experimenter enters the hutch and retrieves the sample.
5. An equal volume of 2× loading buffer is added to the sample. Alternatively, precipitation buffer and three volumes of ethanol can be added to the sample. All samples are kept in a cold pack within a lead-lined box in the hutch both before and after exposure. Upon completion of the

experiment, the samples are placed in the lead-lined freezer in the hutch until they are transferred to a laboratory for processing and analysis.

Protocol for Rapid-Mixing X-Ray Footprinting Experiment

The protocol for conducting Mg^{2+}-initiated quench-flow experiments has been modified from that previously described.[17] The [^{32}P]RNA aliquots are stored in the lead-lined freezer in the hutch until ready for use. In addition to the experimental data points, control samples include RNA not exposed to the X-ray beam, RNA to which Mg^{2+} is not added, and RNA to which Mg^{2+} is added and allowed to equilibrate (i.e., a "fully folded" control). The latter two samples are exposed to the X-ray beam in the same manner as the experimental samples.

1. The circulating water bath is connected to the Kin-Tek apparatus, switched on, and allowed to equilibrate at the desired temperature of the experiment.
2. The drive syringes A, B, and C (Fig. 4) and the sample loop, reaction loop, and exposure chamber are rinsed successively with water and methanol and dried by pulling a vacuum through the sample flow train.
3. The first samples to be run are the "no-Mg^{2+}" controls. For these samples, CE buffer is loaded into drive syringes A and C. Ethanol is loaded into middle syringe B. The valves are positioned so that the syringe is open only to the tubing below. The stepping motor is advanced until the plunger touches the syringes and the tubing between the syringes and the lower valves is filled with buffer. The mixing and reaction loops are sequentially rinsed with water and methanol and vacuum dried.
4. An aliquot of RNA (to a maximum specific radioactivity of 1 μCi) is removed from the freezer, heated to 95° for 1 min, then diluted to a final volume of 200 μl in CE buffer. This volume of [^{32}P]RNA is sufficient for ~15 samples.
5. A 1-ml sterile disposable syringe is filled with CE buffer and attached to inlet D. A second syringe is filled with the [^{32}P]RNA solution and attached to inlet E (Fig. 4).
6. The solutions in D and E are pushed through the tubing up to marks that have been made on the sample loops indicating 10-μl volumes. The sample valves are opened toward the drive syringes and the lead sleeve placed over the RNA containing syringe to protect it from scattered X-rays.
7. A microfuge tube containing 15 μl precipitation buffer is attached to the exit line by a cap that has had a hole punched in it. The affixed tube is placed in a holder positioned under the quench flow.
8. The experimenter exits and interlocks the hutch. After 20 sec, the beam line safety shutter can be opened and the experiment initiated.

9. Prior to allowing the X-ray beam to enter the hutch, the mixing time and exposure protocols are entered into a program downloaded to the motor controller. A "push–pause–push–push–push" timing sequence is used, slightly different from our original protocol.[17] The first "push" mixes the solutions driven by syringes D and E (the RNA solution and the Mg^{2+} buffer) by turbulent flow. The "pause" is the reaction time. The second "push" moves the mixed sample through the X-ray exposure chamber. The rate of fluid flow determines the duration of the X-ray exposure. The third "push" moves the sample to tip of the exit line, and the final "push" expels the sample into the collection tube. The final push is conducted at a slower speed in order to minimize splashing of the radioactive sample around the collection tube. The reaction time is the only variable changed over the course of a kinetics progress curve.

10. For reaction times <4 sec, the safety shutter is opened before the timing sequence is initiated. For reaction times >4 sec, the timing sequence is initiated and the safety shutter is opened just prior to the initiation of sample exposure, using a handheld timer as a guide. The safety shutter is closed immediately after the sample enters the collection tube.

11. The experimenter enters the hutch and retrieves the sample. The effluent from the quench flow consists of 200 μl of [^{32}P]RNA and buffer and 100 μl ethanol. An additional 500 μl ethanol is added to the sample to initiate precipitation and the sample is stored in a cold pack within the lead-lined box in the hutch.

12. The mixing and reactions loops are rinsed successively with water and methanol and then vacuum dried in preparation for the next sample.

13. For the remainder of the samples in the Mg^{2+}-initiated folding experiment, the CE buffer in syringes A and D is removed and replaced with Mg^{2+} buffer.

14. Steps 6 through 12 are repeated for the remainder of the samples. For each 15 samples, a new aliquot of RNA must be prepared as described in step 4.

15. The equilibrated, "fully folded" control samples provide evidence that the folding reaction has gone to completion. For these samples, an aliquot of RNA is diluted to 45 μl in CE buffer. Then 5 μl of 100 m*M* Mg^{2+} buffer is added as a drop on the lid of the microfuge tube. The sample is heated to 95° for 1 min, then immediately spun down. This causes the Mg^{2+} to reach the solution when it is still warm, resulting in fast folding of the RNA. The fully folded control samples are loaded through inlet E and run as in steps 6 through 12.

16. On completion of the entire experiment, the samples are placed in the lead-lined freezer in the hutch until they are transferred to a laboratory for processing and analysis.

Data Reduction and Analysis

Analyses of the results of x-ray footprinting experiments are conducted as previously described in detail.[17] Briefly, the radiolysis reaction products are separated using PAGE. A storage phosphor screen and associated imager (Phosphorlmager, Molecular Dynamics, Sunnyvale, CA) are used to acquire a digital image of the electrophoretogram for densitometric analysis. Band intensity is quantitated using the Molecular Dynamics ImageQuant or equivalent image analysis software. The intensity within a protected region and a reference region is quantitated for each lane on the gel. The reference region is a set of bands that accounts for variability in sample loading in the individual lanes, but that shows the same degree of protection throughout the folding process. The densitometric results are exported to a spreadsheet (Microsoft Excel) for further processing and analysis. Each protected region is divided by the corresponding reference region in the same lane.

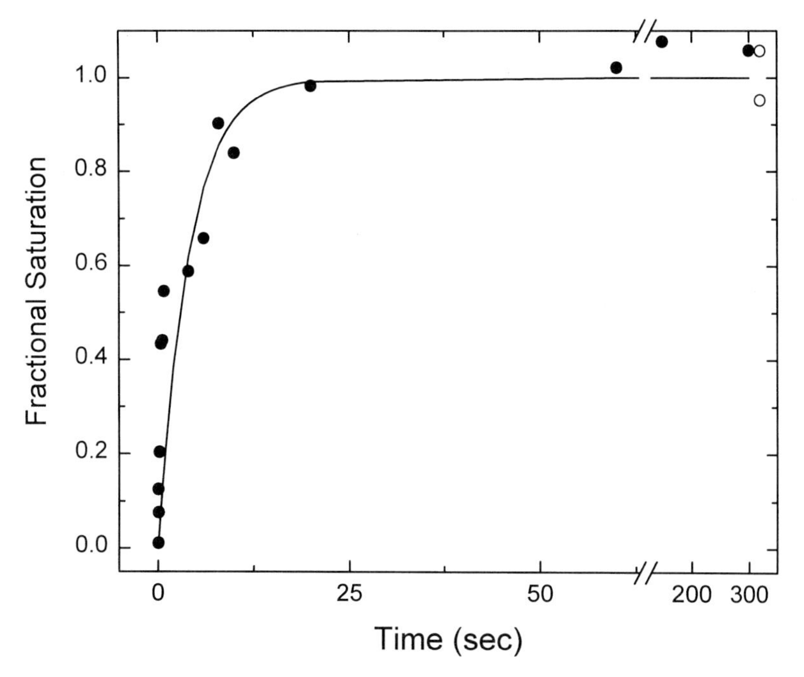

FIG. 5. A representative kinetics progress curve determined for the protection that occurs at bases 118–120 of the *Tetrahymena* group I intron RNA. Fractional protection is plotted as a function of reaction time following addition of $MgCl_2$ to a final concentration of 10 m*M*. The equilibrium folded control sample is shown by the open symbols plotted for convenience at 310 sec. The solid line depicts the best fit of the data to a single exponential with $k = 0.2 \pm 0.1\ sec^{-1}$.

The resulting data points are plotted as a function of reaction time (for a kinetics experiment) or concentration (for an equilibrium experiment). A preliminary fit of the data to the appropriate model is conducted within Excel using the embedded program Magestic. Further analysis of progress or equilibrium curves can be accomplished using any number of nonlinear least-squares analysis programs and packages.

Figure 5 shows a representative synchrotron X-ray footprinting progress curve obtained for the Mg^{2+}-initiated folding of the *Tetrahymena* group I intron. The protection quantitated to obtain this curve represents bases 118–120 within the P4–P6 domain of the ribozyme and appears at a rate 0.2 sec^{-1}. This protection is believed to represent a tertiary contact with bases within the peripheral domain P9.1 that show a comparable rate of protection.[15] In contrast, protections resulting from formation of the core of the P4–P6 domain occur at ~1 sec^{-1}, those due to folding of the P5c domain at 2 sec^{-1}, and formation of the catalytic core (domains P3–P7) at 0.03 sec^{-1}.

Conclusion

Synchrotron hydroxyl radical footprinting can be used to study folding of RNA on a millisecond timescale, and with single base resolution. Although this article focuses specifically on the application of synchrotron footprinting to RNA folding, a large number of systems are amenable to study by this technique. Studies of DNA–protein interactions, Mg^{2+}-dependent DNA isomerization reactions, and protein folding are currently under way by the Center for Synchrotron Biosciences.

[23] Analysis of Global Conformation of Branched RNA Species Using Electrophoresis and Fluorescence

By DAVID M. J. LILLEY

Introduction

Branch points are very common features of natural RNA molecules. These include asymmetrical bulges and helical junctions of various kinds, and an IUBMB nomenclature of these species is presented in Ref. 1. Natu-

[1] D. M. J. Lilley, R. M. Clegg, S. Diekmann, N. C. Seeman, E. von Kitzing, and P. Hagerman, *Eur. J. Biochem.* **230,** 1 (1995).

0076-6879/00 $30.00

rally occurring junctions include perfect three- or four-way junction (3H and 4H junctions, respectively), although these are more frequently complicated by the presence of one or more unpaired bases. The hammerhead ribozyme discussed later provides a good example of a more complicated three-way RNA junction, whereas the hairpin ribozyme is based on a perfect four-way junction in the natural viral RNA sequence. It is likely that the stereochemical characteristics of these helical junctions will have a major influence on the overall conformation of the RNA species in which they occur, and we have sought to study their geometry and conformational transitions over a number of years.

Methods for Study of Global Folding Transitions

The structure of branched nucleic acids is not easily studied by the conventional approaches of crystallography and nuclear magnetic resonance (NMR), despite the high information content of these methods. Lower resolution approaches have therefore proved valuable in gaining insight into the global structures of branched RNA and DNA species, and in the study of conformational transitions. In this laboratory we have developed a combined approach that takes information from a number of techniques, especially comparative gel electrophoresis and fluorescence resonance energy transfer (FRET), in combination with other methods where possible, including chemical probing, functional group modification, cryoélectron microscopy, and molecular modeling.

Comparative Gel Electrophoresis

The electrophoretic mobility of nucleic acids in polyacrylamide gels is very sensitive to the global shape of the migrating molecule. A classic example of this is the retarded mobility of DNA fragments on distortion of the linear structure by the introduction of phased oligoadenine sequences that cause significant curvature of the axis.[2–6] The advantage of the electrophoretic method is its simplicity and speed, allowing a number of sequence variants to be rapidly explored for example.

Theoretical Basis of Gel Electrophoresis. A difficulty in the interpretation of electrophoretic experiments arises from the absence of a complete

[2] J. C. Marini, S. D. Levene, D. M. Crothers, and P. T. Englund, *Proc. Natl. Acad. Sci. U.S.A.* **79,** 7664 (1982).
[3] H.-M. Wu and D. M. Crothers, *Nature (Lond.)* **308,** 509 (1984).
[4] S. Diekmann and J. C. Wang, *J. Mol. Biol.* **186,** 1 (1985).
[5] P. J. Hagerman, *Biochemistry* **24,** 7033 (1985).
[6] H.-S. Koo, H.-M. Wu, and D. M. Crothers, *Nature (Lond.)* **320,** 501 (1986).

theoretical understanding to provide a sound basis for interpretation. However, a number of theories provide at least qualitative agreement with experimental data.[7–9] Most of these are derived from the idea of the nucleic acid reptation,[10] in which the DNA moves through the gel under the influence of the electric field, in a "tube" created by the matrix. Lumpkin and Zimm[8] derived a relationship between the rate of migration (μ) and the end-to-end distance of the molecule:

$$\mu = \frac{Q}{\zeta}\left\langle \frac{h_x^2}{L^2} \right\rangle \tag{1}$$

where Q is the charge on the molecule, ζ is the frictional coefficient for translation along the tube, L is the contour length of the molecule, and h_x is the component of the end-to-end vector $\mathbf{h}$ in the direction of the electric field. The brackets indicate an average over an ensemble of configurations. Because branching, kinking, or bending of the nucleic acid will reduce the end-to-end distance, the theory predicts that such distortion will lead to a reduced mobility in the gel. In a later extension of this theory, Levene and Zimm[9] improved the agreement with experiment by the introduction of cross-interaction between the bendability of the DNA and the elastic properties of the gel matrix. An alternative approach to the explanation of retarded mobility by curved DNA was taken by Calladine, Drew and coworkers,[11,12] in which they calculated the probability of intersection of the cylindrical outer envelope of a superhelix intersecting with randomly located gel fibers. The radius of the cylinder increases with the curvature, thus elevating the probability of obstruction to electrophoretic motion.

Electrophoretic Retardation of Bulged RNA Species. An example of the retardation of kinked RNA species is shown in Fig. 1A. A series of RNA duplexes of the same length containing central oligoadenine bulges (a consecutive stretch of unopposed adenine bases, A_n) where n runs from 0 to 7 exhibit increasing retarded electrophoretic mobility in polyacrylamide, consistent with increasing axial distortion with the size of the bulge. Axial kinking by bulges has also been confirmed by FRET[13] (refer ahead to Fig. 4) and transient electric birefringence.[14] The 3′ untranslated region (UTR)

[7] L. S. Lerman and H. L. Frisch, *Biopolymers* **21,** 995 (1982).
[8] O. J. Lumpkin and B. H. Zimm, *Biopolymers* **21,** 2315 (1982).
[9] S. D. Levene and B. H. Zimm, *Science* **245,** 396 (1989).
[10] P. G. de Gennes, *J. Chem. Phys.* **55,** 572 (1971).
[11] C. R. Calladine, H. R. Drew, and M. J. McCall, *J. Mol. Biol.* **201,** 127 (1988).
[12] C. R. Calladine, C. M. Collis, H. R. Drew, and M. R. Mott, *J. Mol. Biol.* **221,** 981 (1991).
[13] C. Gohlke, A. I. H. Murchie, D. M. J. Lilley, and R. M. Clegg, *Proc. Natl. Acad. Sci. U.S.A.* **91,** 11660 (1994).
[14] M. Zacharias and P. J. Hagerman, *J. Mol. Biol.* **247,** 486 (1995).

of the U1A mRNA contains a double-bulge species (two singly opposed seven-base bulges, separated by four basepairs),[15] which exhibits pronounced retardation in polyacrylamide. A duplex containing one of these bulges is included in the gel shown in Fig. 1A, and has a mobility consistent with a 90° bend in the axis.[16] It is important to demonstrate that the retardation is not due to a simple point of flexibility. Figure 1B shows the electrophoretic mobility of a series of RNA duplexes of constant overall length, containing a UTR bulge separated from an A_5 bulge by a segment of duplex comprising an increasing number of base pairs. If the bulges have a fixed direction this should generate a precise dihedral angle between the outer duplex segments, and the overall shape of the molecule will vary with the length of the central duplex. This is reflected by the sinusoidal modulation of electrophoretic mobility with central spacer length (Fig. 1C); the fastest species should have a dihedral angle of around 180°, and the slowest 0°. The data can be well fitted by application of the Lumpkin–Zimm theory [Eq. (1)], where the end-to-end distance is calculated using this model for the structure.[16]

Comparative Gel Electrophoresis of Branched Nucleic Acids. The first application of the electrophoretic method to branched nucleic acids was in the study of the global conformation of four-way DNA junctions.[17–19] In this approach we generate all the possible species having two extended arms, and compare their electrophoretic mobility in polyacrylamide. Typically the longer arms might comprise 40 bp, while the shorter arms are 15 bp in length; the latter is a good balance between shortening the length while maintaining stability. The number of combinations of n taken from m ($^{m}C_n$) is given by

$$^{m}C_n = \frac{m!}{(m-n)!n!} \tag{2}$$

and thus there are three species that are derived from a three-way junction (each having two long and one short arm) and six that may be obtained from a four-way junction (each having two long and two short arms). The different mobilities of the species are analyzed on the assumption that the

[15] C. W. van Gelder, S. I. Gunderson, E. J. Jansen, W. C. Boelens, M. Polycarpou-Schwarz, I. W. Mattaj, and W. J. van Venrooij, *EMBO J.* **12,** 5191 (1993).

[16] R. J. Grainger, A. I. H. Murchie, D. G. Norman, and D. M. J. Lilley, *J. Mol. Biol.* **273,** 84 (1997).

[17] G. W. Gough and D. M. J. Lilley, *Nature (Lond.)* **313,** 154 (1985).

[18] J. P. Cooper and P. J. Hagerman, *J. Mol. Biol.* **198,** 711 (1987).

[19] D. R. Duckett, A. I. H. Murchie, S. Diekmann, E. von Kitzing, B. Kemper, and D. M. J. Lilley, *Cell* **55,** 79 (1988).

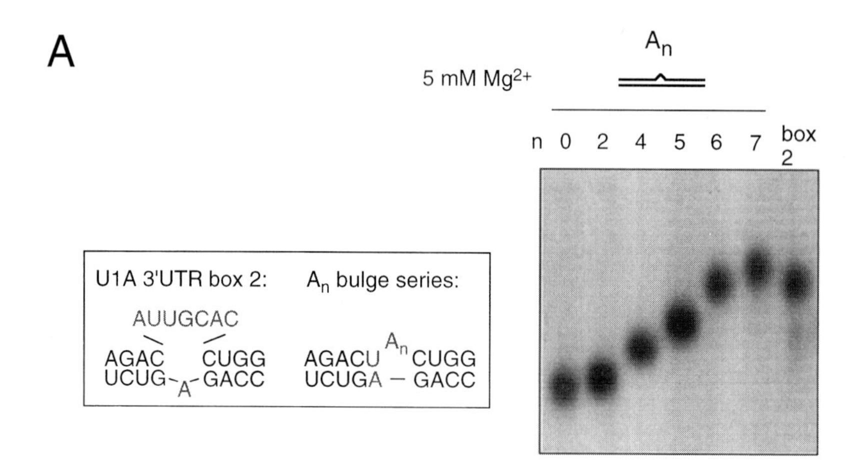

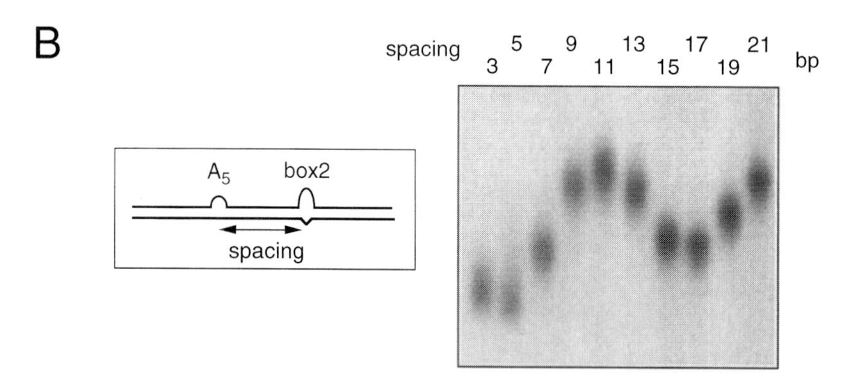

FIG. 1. Electrophoretic analysis of a series of RNA duplexes containing base bulges. (**A**) Gel electrophoresis of a series of 69-bp RNA duplexes containing an A_n bulge (where n = 0, 2, 4, 5, 6, or 7) or the U1A 3′ UTR box 2 asymmetric bulge. The radioactive RNA species were made by transcription using T7 RNA polymerase. The central sequences are shown in the insert. The RNA species were electrophoresed in a 15% polyacrylamide gel in the presence of 89 m*M* Tris–borate (pH 8.3), 5 $MgCl_2$, and the dried gel subjected to autoradiography. Note that the U1A 3′UTR box 2-containing duplex has closely similar mobility to that of the A_6-bulged duplex. (**B**) Gel electrophoresis of a series of 69-bp RNA duplexes containing a U1A box 2 sequence variably spaced (spacing between 3 and 21 bp) with respect to an A_5 bulge. The radioactive RNA was electrophoresed in 15% polyacrylamide in the presence of 89 m*M* Tris–borate (pH 8.3), 10 m*M* NaCl, 2 m*M* $MgCl_2$, and the dried gel subjected to autoradiography. Note the marked sinusoidal modulation of the mobilities, consistent with a duplex containing two defined bends with a dihedral angle determined by the spacing between them. (**C**) The measured electrophoretic mobilities in (B) were plotted as a function of spacer length; the error bars indicate the widths of the electrophoretic bands. These data were fitted by nonlinear regression analysis to the Lumpkin–Zimm theory[8] (the continuous line), using a model in which the bend angles were variables.

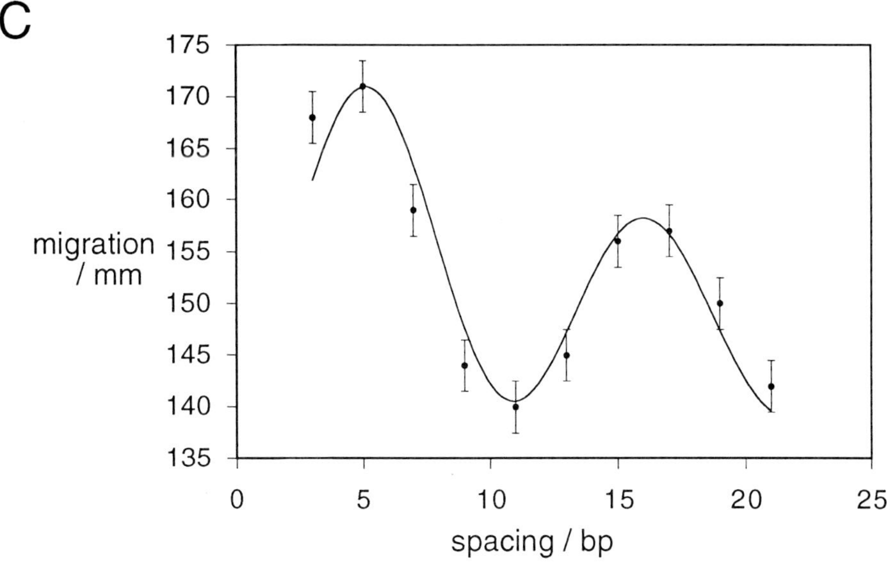

FIG. 1. (*continued*)

mobility will reflect the end-to-end distance of the two long arms, which is directly related to the angle subtended between these arms.

In the case of DNA, these experiments were originally performed either by the ligation of appropriate reporter arms to short arms,[18] or by the shortening of longer arms by restriction cleavage.[19] For RNA it is necessary to synthesize all the component strands (i.e., 16 oligonucleotides for a four-way junction) and create the required long-short arm species by hybridization of the appropriate strands. Because RNA synthesis is presently difficult for oligonucleotides of length greater than 40 nucleotides, we have generally adopted the approach of using hybrid DNA–RNA–DNA species, such that the important core of the molecule comprises RNA, while the outer reporter arm segments are DNA.[20–22] We ensure that the molecules constructed comprise RNA for all the region of interest (for example, the hammerhead core region) plus at least 5 bp into the reporter helices. Initially we were concerned that the DNA–RNA junction might influence the results, since

[20] G. Bassi, N. E. Møllegaard, A. I. H. Murchie, E. von Kitzing, and D. M. J. Lilley, *Nature Struct. Biol.* **2,** 45 (1995).

[21] D. R. Duckett, A. I. H. Murchie, and D. M. J. Lilley, *Cell* **83,** 1027 (1995).

[22] A. I. H. Murchie, J. B. Thomson, F. Walter, and D. M. J. Lilley, *Mol. Cell* **1,** 873 (1998).

any kink at this junction would add vectorially to the angle between the long arms that we are trying to deduce. However, by varying the length of the RNA helix, by introducing a hybrid DNA–RNA helix, and even by synthesizing a set of species composed entirely of RNA we concluded that this did not affect our conformational conclusions in the case of the hammerhead ribozyme.[20,23]

An example of the analysis of a four-way RNA junction is illustrated in Fig. 2A. This junction has four arms called B, H, R, and X; these are just labels for convenience although the historical origin of these names lies in the restriction sites used in the analysis of DNA junctions of the same sequence.[19] The six long–short arm species have been electrophoresed side by side in an 8% w/w polyacrylamide (29 : 1 acrylamide : bisacrylamide ratio) in a buffer containing 90 m*M* Tris–borate (pH 8.3), 1 m*M* $MgCl_2$. With metal ion-containing buffers it is important to recirculate at >1 liter/hr. In general these gels are run at room temperature, although temperature-dependent effects can be studied by used of glass plates with serpentine water flow from a circulating bath. For the four-way RNA junction we see that the six long–short arm species do not have equal mobility, clearly excluding a tetrahedral disposition of the four arms. Instead, the pattern of mobilities is described by fast, slow, slow, slow, slow, fast, which is consistent with a structure based on pairwise coaxial stacking of helical arms, with a 90° angle between the two helical axes. This generates four equivalent long–short arm species where the angle between the long arms is 90° (BR, BX, HR and HX), and two species where the angle is 180° (BH and RX). Since the fast species are BH and RX, this indicates that the junction has folded by the stacking of B on H arms, and R on X arms. There is an alternative conformer that would be based on B on X and H on R stacking, that must be of lower stability for this sequence. The comparative gel electrophoresis method can be readily used to study the effect of ionic conditions on the global conformation of the junction. If the magnesium ion concentration is elevated to 5 m*M* we see that the mobility pattern changes to fast, intermediate, slow, slow, intermediate, fast (Fig. 2B), indicating that the helical axes have rotated while preserving the choice of stacking partners. The fast species remain BH and RX (180°), but the four remaining species no longer have equal mobility. These become split into the intermediate species (BR and HX, obtuse angle) and the slow species (BX and HR, acute angle). This shows that the junction has rotated into an antiparallel conformation, similar to that adopted by DNA junctions,[19] although the pattern suggests a somewhat less symmetri-

[23] G. S. Bassi, A. I. H. Murchie, and D. M. J. Lilley, *RNA* **2,** 756 (1996).

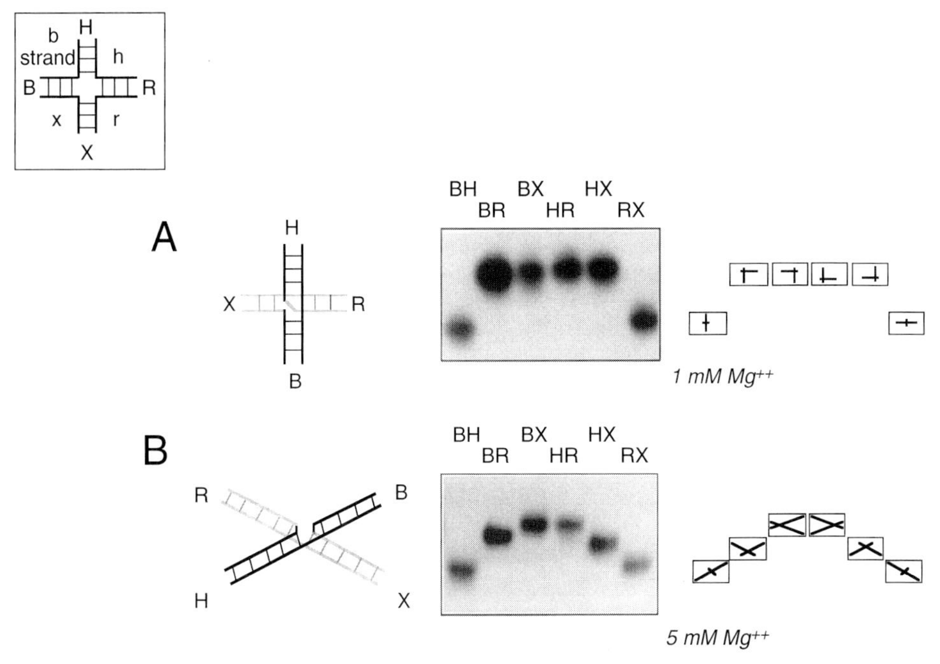

FIG. 2. Analysis of the global structure of a four-way (4H) RNA junction as a function of magnesium ion concentration, using comparative gel electrophoresis.[21] The junction has the central sequence of junction 1,[19] and it comprises four arms labeled B, H, R, and X, assembled from four component strands labeled b, h, r, and x as shown. This sequence is used to derive the six species with two long and two short arms. The electrophoretic mobilities of the six long–short arm species are compared under different conditions. (**A**) Analysis of global structure in the presence of 1 m*M* magnesium ions. The six long–short radioactively labeled junction species were electrophoresed in an 8% polyacrylamide gel in 90 m*M* Tris–borate (pH 8.3), 1 m*M* $MgCl_2$ at room temperature. The autoradiograph of the gel is shown. The six species for junction 3 migrate with a pattern indicative of a coaxially stacked structure based on B on H and X on R stacking, with a 90° angle between the axes (shown left). This leads to the formation of the six long–short species (indicated right). (**B**) Analysis of global structure in the presence of 5 m*M* magnesium ions. The six long–short radioactively labeled species were electrophoresed in an 8%polyacrylamide gel in 90 m*M* Tris–borate (pH 8.3), 5 m*M* $MgCl_2$ at room temperature. The pattern of mobilities indicates that the same coaxial stacking is preserved, but the axes have rotated in an antiparallel direction (shown left). This leads to the formation of the six long–short species (indicated right).

cal structure in the case of RNA. These conclusions have been supported by a recent FRET study.[24]

An example of the application of comparative gel electrophoresis to a slightly more complex structure is shown in Fig. 3. The hammerhead ribo-

[24] F. Walter, A. I. H. Murchie, D. R. Duckett, and D. M. J. Lilley, *RNA* **4,** 719 (1998).

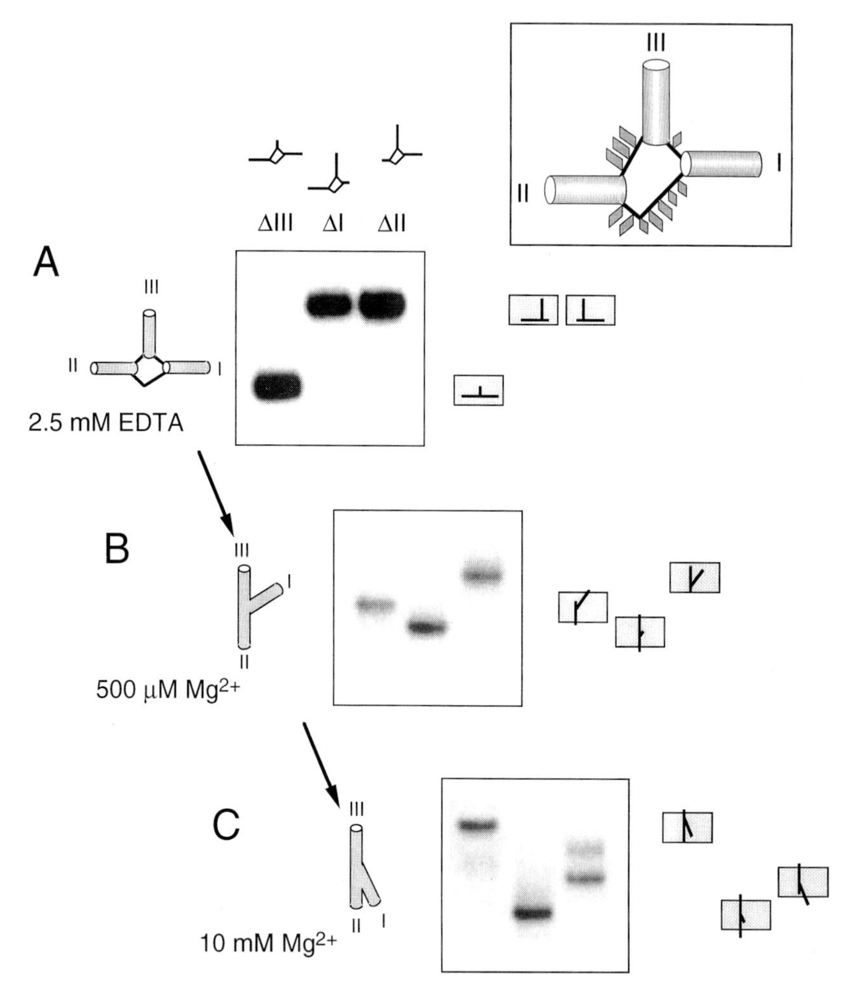

FIG. 3. Conformational transitions in the hammerhead ribozyme analyzed by comparative gel electrophoresis.[20] The hammerhead ribozyme (inset) is a kind of three-way helical junction, in which the helices are named I, II, and III as shown. The three possible species comprising two long and one short arm have been derived, and are named according to the shortened arm (e.g., species ΔI has a shortened arm I and long arms II and III). The three radioactively labeled species have been electrophoresed in polyacrylamide gels in 90 m*M* Tris–borate (pH 8.3) with either EDTA to chelate metal ions or added magnesium ions. (**A**) Global structure in the presence of 2.5 m*M* EDTA. Under these conditions the structure (indicated left) is extended, giving the fast, slow, slow pattern of mobilities interpreted on the right. (**B**) Global structure in the presence of 500 μ*M* magnesium ions. The structure has undergone a change to one in which the largest angle is now that between helices II and III (indicated left), giving the intermediate, fast, slow pattern of mobilities interpreted on the right. (**C**) Global structure in the presence of 10 m*M* magnesium ions. The structure has undergone a further change in which helix I has rotated into the same quadrant as helix II (indicated left), giving the slow, fast, intermediate pattern of mobilities interpreted on the right.

zyme is a kind of three-way RNA junction, where the helices are connected by single-stranded segments that form the conserved core of the ribozyme[25,26]; it is a $HS_1HS_7HS_3$ junction. This structure was studied by the electrophoretic technique[20] using a construct in which the core (except C_{17}, to prevent self-cleavage) plus the inner 10 bp of each helical stem were made from RNA, and the remainder of the molecule was DNA. The three species with two long and one short arm were constructed (named according to the shortened arm—thus the species ΔI is shortened in arm I), and their relative electrophoretic mobility studied as a function of conditions. In the absence of added divalent ions [90 m*M* Tris–borate (pH 8.3), 2.5 m*M* EDTA] the three species with one shortened arm clearly do not migrate with equal velocity. Species ΔI and ΔII have similar mobilities, while the ΔIII species is significantly faster. These results would be consistent with a configuration of arms equivalent to the conventional depiction of the hammerhead, where helices I and II subtend a relatively large angle, while the other two interarm angles are smaller. This suggests that the hammerhead adopts an extended structure in the absence of added ions, perhaps with a rather unstructured core region. On addition of a low concentration of magnesium ions [90 m*M* Tris–borate (pH 8.3), 500 μ*M* $MgCl_2$] the electrophoretic pattern changes significantly, indicating that the global structure of the ribozyme has changed. The slowest species is now that with a shortened helix II (ΔII), the fastest is the ΔI, and the ΔIII species displays intermediate mobility. The relative mobilities indicate that in the presence of 500 μ*M* magnesium ions, the global structure changes such that the largest angle is that between helices II and III, and the smallest is that between helices I and III. On increasing the magnesium ion concentration further [90 m*M* Tris–borate (pH 8.3), 10 m*M* $MgCl_2$], there is a second change in conformation. Although the ΔI species remains the fastest of the three, the mobilities of ΔII and ΔIII become reversed, such that the ΔIII species is now the slowest. The results suggest that the angle between helices II and III remains the largest, but that the angle between arms II and I is now the smallest. The global shape of the fully folded conformation is in good agreement with that observed in the crystal,[27,28] and confirmation of the two-stage ion-dependent folding pathway has been obtained by FRET[29] as discussed later. Using this simple and rapid technique a number of folding mutants were easily identified.[23]

[25] U. C. Uhlenbeck, *Nature* (*Lond.*) **328,** 596 (1987).

[26] D. E. Ruffner, G. D. Stormo, and O. C. Uhlenbeck, *Biochemistry* **29,** 10695 (1990).

[27] H. W. Pley, K. M. Flaherty, and D. B. McKay, *Nature* **372,** 68 (1994).

[28] W. G. Scott, J. T. Finch, and A. Klug, *Cell* **81,** 991 (1995).

[29] G. S. Bassi, A. I. H. Murchie, F. Walter, R. M. Clegg, and D. M. J. Lilley, *EMBO J.* **16,** 7481 (1997).

Fluorescence Resonance Energy Transfer

FRET provides a completely alternative approach to the study of global conformation in branched nucleic acids, free of all the assumptions that go into the interpretation of the comparative gel electrophoresis data. This spectroscopic method can give distance information in the range of 10–100 Å, which is unavailable using other solution methods, and the data are in some ways complementary to those available from other techniques, such as the shorter range distance information available from NMR. Although the method has a long history,[30–32] it has seen a resurgence of interest for nucleic acid studies in the last decade. We turned to FRET to study the structure of the four-way junction in DNA,[33–35] and because this time it has been applied to a series of nucleic acids.[13,22,24,29,36–47] In the most common approach, two different fluorophores (such as fluorescein and tetramethylrhodamine) are attached to the termini of different helical arms. Excitation energy can be transferred from the fluorophore with the higher energy of excitation (donor) to that of lower energy (acceptor) in a radiationless process. The efficiency of energy transfer (E_{FRET}) between the two dyes is strongly distance dependent, and by comparing these efficiencies for all the possible end-to-end vectors the global structure can be deduced.

[30] T. Förster, *Ann. Phys.* **2,** 55 (1948).
[31] L. Stryer and R. P. Haugland, *Proc. Natl. Acad. Sci. U.S.A.* **58,** 719 (1967).
[32] R. H. Fairclough and C. R. Cantor, *Methods Enzymol.* **48,** 347 (1978).
[33] A. I. H. Murchie, R. M. Clegg, E. von Kitzing, D. R. Duckett, S. Diekmann, and D. M. J. Lilley, *Nature* **341,** 763 (1989).
[34] R. M. Clegg, A. I. H. Murchie, A. Zechel, C. Carlberg, S. Diekmann, and D. M. J. Lilley, *Biochemistry* **31,** 4846 (1992).
[35] R. M. Clegg, A. I. H. Murchie, A. Zechel, and D. M. J. Lilley, *Biophys. J.* **66,** 99 (1994).
[36] R. A. Hochstrasser, S. M. Chen, and D. P. Millar, *Biophys. J.* **45,** 133 (1992).
[37] R. M. Clegg, A. I. H. Murchie, A. Zechel, and D. M. J. Lilley, *Proc. Natl. Acad. Sci. U.S.A.* **90,** 2994 (1993).
[38] J. L. Mergny, A. S. Boutorine, T. Garestier, F. Belloc, M. Rougee, N. V. Bulychev, A. A. Koshkin, J. Bourson, A. V. Lebedev, B. Valeur, N. T. Thuong, and C. Hélène, *Nucleic Acids Res.* **22,** 920 (1994).
[39] P. R. Selvin and J. E. Hearst, *Proc. Natl. Acad. Sci. U.S.A.* **91,** 10024 (1994).
[40] T. Tuschl, C. Gohlke, T. M. Jovin, E. Westhof, and F. Eckstein, *Science* **266,** 785 (1994).
[41] F. Stühmeier, J. B. Welch, A. I. H. Murchie, D. M. J. Lilley, and R. M. Clegg, *Biochemistry* **36,** 13530 (1997).
[42] F. Stühmeier, D. M. J. Lilley, and R. M. Clegg, *Biochemistry* **36,** 13539 (1997).
[43] P. S. Eis and D. P. Millar, *Biochemistry* **32,** 13852 (1993).
[44] E. A. Jares-Erijman and T. M. Jovin, *J. Mol. Biol.* **257,** 597 (1996).
[45] M. S. Yang and D. P. Millar, *Biochemistry* **35,** 7959 (1996).
[46] S. M. Miick, R. S. Fee, D. P. Millar, and W. J. Chazin, *Proc. Natl. Acad. Sci. U.S.A.* **94,** 9080 (1997).
[47] N. G. Walter, K. J. Hampel, K. M. Brown, and J. M. Burke, *EMBO J.* **17,** 2378 (1998).

Instrumentation. Steady-state fluorescence measurements can be made on a good fluorimeter, such as those manufactured by SLM-Aminco or ISS (Urbana). In this laboratory we use an SLM-Aminco 8100, equipped with Glan-Thompson polarizers. Measurements of FRET efficiency are performed under photon counting conditions, with the polarizers crossed at the magic angle (54.7°) in order to remove polarization artifacts. Fluctuation of lamp intensity is corrected using a concentrated rhodamine B solution as a quantum counter.

Theoretical Basis of FRET. The theory of FRET as applied to nucleic acids has been described in detail by Clegg[48] in an earlier volume in this series. On the absorption of a photon of appropriate energy, a fluorophore is excited into a higher electronic singlet state. Because the transition occurs in a time that is very much faster than nuclear motion, the molecule is transiently in a higher vibrational level of this state, which then rapidly decays into the lowest vibrational level. The excited state can then become deactivated by the emission of a photon of lower frequency, which is the well-known stokes shift of fluorescent emission. The excited state can also be depopulated by a variety of other processes, including collisional quenching and intersystems crossing to a triplet state with subsequent phosphorescent emission possible, all of which lead to a reduction in the fluorescent quantum yield. However, in the presence of another fluorophore a further mechanism of loss of energy is possible due to the coupling between the transition moments, i.e., the emission dipole of the donor fluorophore and the excitation dipole of the acceptor fluorophore. This resonance leads to a deactivation of the donor and an activation of the acceptor. The process can be observed in a number of ways, especially by the reduction in the fluorescent quantum yield of the donor, or by the increased fluorescent emission (sensitized emission) from the acceptor. These properties can be readily measured in the steady state. An example is given in Fig. 4, showing the fluorescence spectra (normalized to the donor peak at 519 nm) of a series of duplex RNA molecules (labelled with donor and acceptor at the 5′ ends) containing a central oligoadenine bulge. As the size of the bulge increases, it progressively kinks the RNA, shortening the end-to-end distance and thus increasing the efficiency of energy transfer. It can be seen that the magnitude of the acceptor peak at 560 nm increases with bulge size. Alternatively, the deactivation of the donor can be studied by the shortened lifetime of the excited state of the donor, using time-resolved measurements, and in principle information about multiple conformations and their equilibria is available from these experiments. Some examples of the application of time-resolved measurements in nucleic acids can be

[48] R. M. Clegg, *Methods Enzymol.* **211,** 353 (1992).

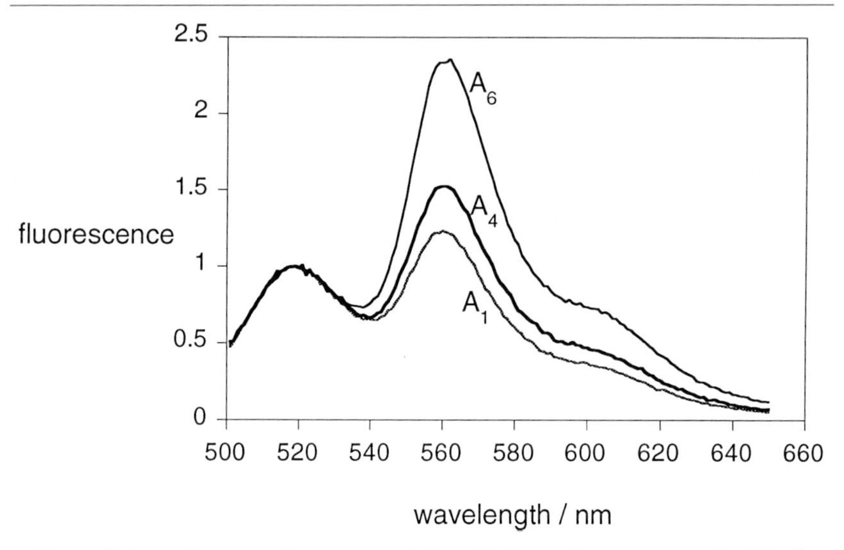

FIG. 4. Enhanced acceptor fluorescent emission with increasing energy transfer. The fluorescence emission spectra (excitation at 490 nm) for three 18-bp RNA duplexes containing oligoadenine bulges of increasing size. Each of the duplex species is terminally labeled with 5′ fluorescein and Cy-3 fluorophores. The spectra have been normalized to the fluorescein emission peak at 520 nm. The kinking due to the bulges results in a closer approach of the two fluorophores, observed as an increased fluorescent emission from the acceptor at 560 nm due to more efficient energy transfer.

found in Refs. 34, 36, 43, and 46, but the remaining part of this section concentrates on steady-state measurements.

Förster showed that the rate of energy transfer (rate constant k_{FRET} is given by

$$k_{\mathrm{FRET}} = \left[\frac{9 \ln 10 \Phi^{\mathrm{D}} \kappa^2 J(\lambda)}{128\pi^5 N \eta^4 \tau_{\mathrm{D}} R^6}\right] \tag{3}$$

This contains the sixth power of the distance R between the fluorophores, and this is the origin of the use of the method for obtaining distance information; τ_{D} is the excited state lifetime of the donor in the absence of an acceptor, Φ^{D} is the fluorescent quantum yield of the donor, η is the refractive index of the medium, N is Avogadro's number, and κ is related to the relative orientation of the two transition dipole moments. $J(\lambda)$ is the normalized spectral overlap integral, given by

$$J(\lambda) = \frac{\int_0^\infty \phi^{\mathrm{D}}(\lambda)\varepsilon^{\mathrm{A}}(\lambda)\lambda^4 d\lambda}{\int_0^\infty \phi^{\mathrm{D}}(\lambda) d\lambda} \tag{4}$$

where ϕ^{D} is the spectral shape of the fluorescent emission of the donor and ε^{A} is the molar absorbance of the acceptor at each wavelength (λ). Thus the rate of energy transfer depends on the extent of overlap between the emission spectrum of the donor and the excitation spectrum of the acceptor.

Equation (3) can be simplified to

$$k_{\mathrm{FRET}} = \tau_{\mathrm{D}}^{-1}\left(\frac{R_0}{R}\right)^6 \quad (5)$$

where R_0 is a characteristic distance for the donor–acceptor pair at which energy transfer is half maximal efficient, given by

$$R_0^6 = 8.8 \times 10^{-28}\,\Phi^{\mathrm{D}}\kappa^2\eta^{-4}J(\lambda) \quad (6)$$

The chief uncertainty in the interpretation of FRET data lies in the dependence on the relative orientation of the two transition moments, i.e., the magnitude of the κ^2 term; κ is the scalar product of the two dipole vectors, and κ^2 can take values from 4 (colinear vectors) to 0 (orthogonal vectors). However, for the situation where the dyes undergo flexible reorientation during the lifetime of the excited state, κ^2 averages to 2/3. Provided that at least one of the fluorophores is mobile this is a good assumption, and this can be estimated by measurements of anisotropy.

In general, what is measured in a FRET experiment is the efficiency of energy transfer (E_{FRET}). This corresponds to the quantum yield for energy transfer, and is the proportion of photons absorbed by the donor that leads to excitation of the acceptor by energy transfer.

$$E_{\mathrm{FRET}} = \frac{k_{\mathrm{FRET}}}{\sum k_{\mathrm{deact}} + k_{\mathrm{FRET}}} = \frac{k_{\mathrm{FRET}}}{\tau_{\mathrm{D}}^{-1} + k_{\mathrm{FRET}}} \quad (7)$$

where k_{deact} are the rate constants for all the processes leading to deexcitation of the donor other than FRET. Substitution of Eq. (5) leads to

$$E_{\mathrm{FRET}} = \frac{1}{\left[1 + \left(\frac{R}{R_0}\right)^6\right]} \quad (8)$$

Thus provided we can measure E_{FRET}, then in principle we can obtain molecular distance information in our system.

Normalized Acceptor Ratio Method. In this laboratory we usually prefer to measure efficiencies of energy transfer from the enhancement of acceptor fluorescence, using the method devised by Clegg,[48,49] because this gives

[49] R. M. Clegg, A. I. H. Murchie, A. Zechel, C. Carlberg, S. Diekmann, and D. M. J. Lilley, *Biochemistry* **31,** 4846 (1992).

greater sensitivity. The emission at a given wavelength (ν_1) of a double-labeled sample excited primarily at the donor wavelength (ν') contains emission from the donor, emission from directly excited acceptor and emission from acceptor excited by energy transfer from the donor:

$$\begin{aligned} F(\nu_1\nu') \propto \{&\varepsilon^{D}(\nu')\Phi^{A}(\nu_1)E_{FRET}da + \varepsilon^{A}(\nu')\Phi^{A}(\nu_1)a \\ &+ \varepsilon^{D}(\nu')\Phi^{D}(\nu_1)d[(1 - E_{FRET})a + (1 - a)]\} \\ = &F^{A}(\nu_1\nu') + F^{D}(\nu_1\nu') \end{aligned} \tag{9}$$

where d and a are the molar fraction of molecules labeled with donor and acceptor, respectively, superscripts D and A refer to donor and acceptor, respectively; $\varepsilon^{D}(\nu')$ and $\varepsilon^{A}(\nu')$ are the molar absorption coefficients of donor and acceptor, respectively; and $\Phi^{D}(\nu_1)$ and $\Phi^{A}(\nu_1)$ are the fluorescent quantum yields of donor and acceptor, respectively. Thus the spectrum contains the components due to donor emission [$F^{D}(\nu_1, \nu')$ i.e., the first term containing $\Phi^{D}(\nu_1)$] and those due to acceptor emission [$F^{A}(\nu_1, \nu')$ i.e., the latter two terms containing $\Phi^{A}(\nu_1)$].

The first stage of the analysis involves subtraction of the spectrum of RNA labeled only with donor, leaving just the acceptor components, i.e., $F^{A}(\nu_1, \nu')$; an example of the extraction of the acceptor spectrum is shown in Fig. 5. The pure acceptor spectrum thus derived is normalized to one

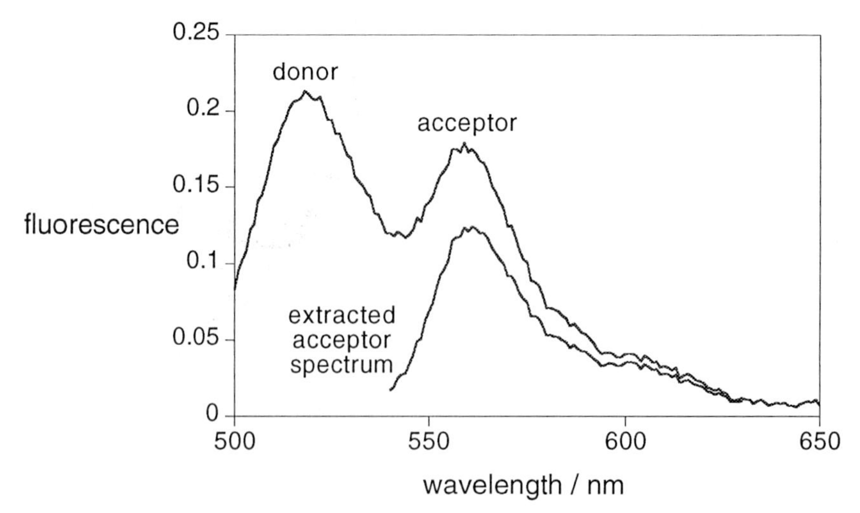

Fig. 5. Extraction of the acceptor spectrum. An example of the removal of the donor component from the spectrum of a three-way (3H) RNA junction labeled with fluorescein and Cy-3 fluorophores. The complete spectrum shows clear emission peaks due to the donor and acceptor components at 520 and 560 nm, respectively. A spectrum of the corresponding RNA labeled only with fluorescein has been normalized to the donor peak at 520 nm and subtracted, leaving a spectrum containing only the contribution from acceptor emission.

from the same sample excited at a wavelength (ν'') at which only the acceptor is excited, with emission at ν_2. We then obtain the normalized acceptor ratio:

$$(\text{ratio})_A = \frac{F^A(\nu_1\nu')}{F^A(\nu_2\nu'')} = \left\{ E_{\text{FRET}} \cdot d \cdot \frac{\varepsilon^D(\nu')}{\varepsilon^A(\nu'')} + \frac{\varepsilon^A(\nu')}{\varepsilon^A(\nu'')} \right\} \frac{\Phi^A(\nu_1)}{\Phi^A(\nu_2)} \tag{10}$$

where E_{FRET} is directly proportional to $(\text{ratio})_A$, and can be easily calculated since $\varepsilon^D(\nu')/\varepsilon^A(\nu'')$ and $\varepsilon^A(\nu')/\varepsilon^A(\nu'')$ are measured from absorption spectra, and $\Phi^A(\nu_1)/\Phi^A(\nu_2)$ is unity when $\nu_1 = \nu_2$. An analogous normalization procedure for the measurement of efficiency from donor deactivation has been presented[48] and used successfully for the analysis of the global structure of a four-way DNA junction.[49] However, it is generally less sensitive than the acceptor ratio method, as well as having certain other disadvantages.

Anisotropy Measurements. Measurement of fluorescence anisotropy is routinely performed to provide information on the flexibility and mobility of attached fluorophores. This arises because the extent of depolarization of emitted light depends on the degree of movement of the fluorophore during the lifetime of the excited state. The most useful parameter for measuring this is the anisotropy (r), given by

$$r = \frac{F_\parallel - F_\perp}{F_\parallel + 2F_\perp} \tag{11}$$

where $F_\parallel$ is the fluorescence measured with the polarizers parallel and $F_\perp$ is the equivalent measurement with the polarizers crossed. This can take values between 0 and 0.4, such that the smaller the value, the more mobile is the fluorophore. In practice, measurement of anisotropy is complicated by differences in the sensitivity of the detection system for vertically and horizontally polarized light. To overcome this, further measurements are made using horizontally polarized excitation, and the anisotropy is corrected by calculating

$$r = \frac{F_{\text{VV}} - GF_{\text{VH}}}{F_{\text{VV}} + 2GF_{\text{VH}}} \tag{12}$$

where the subscripts V and H refer to fluorescence with vertical and horizontal polarizers respectively, in the order excitation, emission, and G is the ratio $F_{\text{HV}}/F_{\text{HH}}$. The anisotropy of the donor can be used to obtain E_{FRET}, because less rotation of the emission dipole can occur during a lifetime that is shortened by energy transfer. Since emission from an acceptor excited by energy transfer is depolarized, acceptor anistropy can also be used as a further method for the observation of FRET, and this has been applied to the study of the structure of the four-way DNA junction.[35]

Construction of Doubly-Labeled Species and Choice of Donor-Acceptor Pair. Fluorophores can be coupled to nucleic acids in a variety of ways, attached to bases or phosphates. They can be introduced internally, but the majority of work has employed flurophores attached terminally. They can be conjugated postsynthetically, typically as succinyl esters that are reacted with primary amine groups attached to the end of the nucleic acid via an alkyl linker, or they can be coupled as phosphoramidites as the final step of the synthesis. The most commonly used donor flurophore in FRET experiments with nucleic acids is fluorescein (Fig. 6A). This is available as a phosphoramidite, though some care in the choice of the compound is needed, because two structural isomers of the dye are available, and isomeric purity is spectroscopically desirable. A major advantage of fluorescein is that when it is linked to the 5′ terminus via a six-carbon linker the fluorophore is very mobile, with an anisotropy generally in the range of 0.10–0.14. This removes much of the uncertainty associated with the orientation of the transition moments, and κ^2 is probably close to 2/3 using fluorescein. The choice of acceptor is more variable. Tetramethylrhodamine (Fig. 6B) was used in many earlier studies,[13,33–35,37,40–43,45,46] having an excitation at 560 nm and a good spectral overlap with fluorescein giving an R_0 of about 45 Å. However, in general this must be conjugated to the nucleic acid postsynthetically, and these reactions are very inefficient at the pH range accessible with RNA. A detailed photophysical study has shown that rhodamine experiences a number of distinct environments when attached at the 5′-phosphate of a double-stranded DNA helix via a C_6 linker.[50] Hexachlorofluorescein has been used as an acceptor in some studies,[47] but we have found that the fluorophore can be unstable during the purification of the dye-labeled nucleic acid leading to significant spectral changes (Unpublished data). More recently we[22,29] and others[44] have used indocarbocyanine-3 (Cy-3) as the acceptor of choice (Fig. 6C). This fluorophore is spectrally quite similar to tetramethylrhodamine, although the overlap with the emission spectrum of fluorescein is better, giving a value of R_0 of around 56 Å[44] (R. Grainger and D.M.J.L., unpublished data, 1999). The quantum yield is also higher than that of tetramethylrhodamine, making the application of the $(\text{ratio})_A$ method easier. It is excited at 547 nm and has an emission maximum at 560 nm; spectral deconvolution when paired with fluorescein is straightforward (see the example in Fig. 5). Cy-3 can be attached to the 5′

[50] G. Vamosi, C. Gohlke, and R. M. Clegg, *Biophys. J.* **71,** 972 (1996).

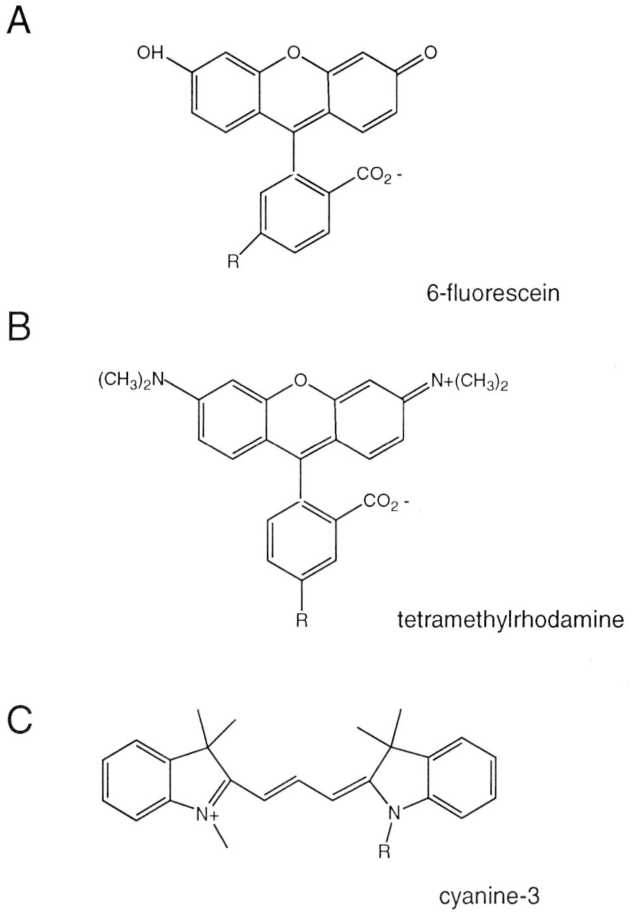

A

6-fluorescein

B

tetramethylrhodamine

C

cyanine-3

FIG. 6. Structures of some fluorophores commonly used in FRET studies of nucleic acids. (**A**) Fluorescein. This is extensively used as the donor in these experiments. The 5-fluorescein isomer is shown. (**B**) Tetramethylrhodamine. This dye has been used as an acceptor from fluorescein in many studies. (**C**) Indocarbocyanine-3, Cy-3 has a number of advantages as an acceptor from fluorescein.

terminus of RNA with good efficiency as a phosphoramidite, and there are no special purification problems. When attached via a C_3 linker the attached dye is very constrained, with anisotropy values around 0.33, we think that the fluorophore is probably fixed on the end of the helix close to the axis [D.M.J.L., R. Grainger, and D. G. Norman, unpublished data, 1999].

In this laboratory all RNA species used in FRET studies are generated by chemical synthesis of component strands followed by hybridization in

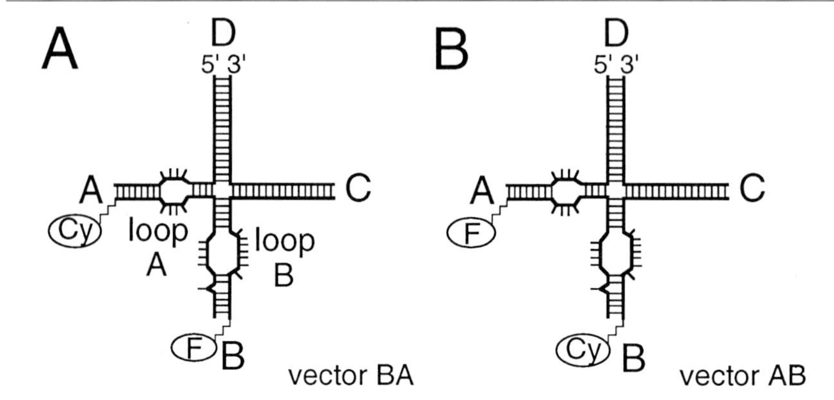

FIG. 7. Analysis of the global conformation of the hairpin ribozyme using FRET.[22] In its natural viral context, the hairpin ribozyme comprises a four-way (4H) RNA junction in which two of the arms carry loops (the A and B loops). Ribozyme cleavage occurs within the A loop. In the FRET experiments the adenine at the cleavage position is replaced by deoxyadenosine, to prevent self-cleavage from occurring during the experiment. (**A** and **B**) All the combinations of the ribozyme with fluorescein and Cy-3 attached to different 5′ termini were synthesized and assembled. There are six such vectors, each of which can be studied in both directions. The BA vector $B_{(fluorescein)}$-$A_{(Cy\text{-}3)}$ is illustrated in (A), while the reversed AB vector $A_{(fluorescein)}$-$B_{(Cy\text{-}3)}$ is shown in B. (**C**) Comparison of FRET efficiency values of (E_{FRET}) for one set of six end-to-end vectors in the presence of 10 μM magnesium ions. Under these conditions all six values are relatively low, indicating an extended structure. The slightly lower values for AD and CB are consistent with a 90° crossed structure based on A on D and B on C stacking (indicated right). (**D**) E_{FRET} values in the presence of 10 mM magnesium ions for the same set of end-to-end vectors. Under these conditions the structure had clearly changed, with BA exhibiting highest efficiency. This is consistent with the formation of a structure in which the end-to-end vector for helices A and B is the shortest such distance (indicated right). This is an antiparallel structure that is somewhat distorted by the close association between the A and B loops. (**E**) E_{FRET} values in the presence of 10 mM magnesium ions for the reversed set of vectors. Note that the pattern of FRET efficiencies is closely similar to the first set of vectors under the same conditions (D).

appropriate combinations and purification. RNA species are synthesized using ribonucleotide phosphoramidites with 2′-*tert*-butyldimethylsilyl (TBDMS) protection. Fluorophores are coupled to the 5′ terminus as phosphoramidites, and the average efficiency of fluorophore conjugation is typically 97%. Oligoribonucleotides are deprotected in 25% ethanol/ammonia solution at 55° for 6 hr (dye labeled) or 12 hr (unlabeled). TBDMS groups are removed by treatment in 0.5 ml 1 M tetrabutylammonium fluoride in tetrahydrofuran for 16 hr at 20° in the dark with agitation. The deprotected RNA is desalted by gel filtration followed by ethanol precipitation. All RNA species are purified by electrophoresis in polyacrylamide gels (usually 20%) containing 7 M urea; fluorescently labeled species are signifi-

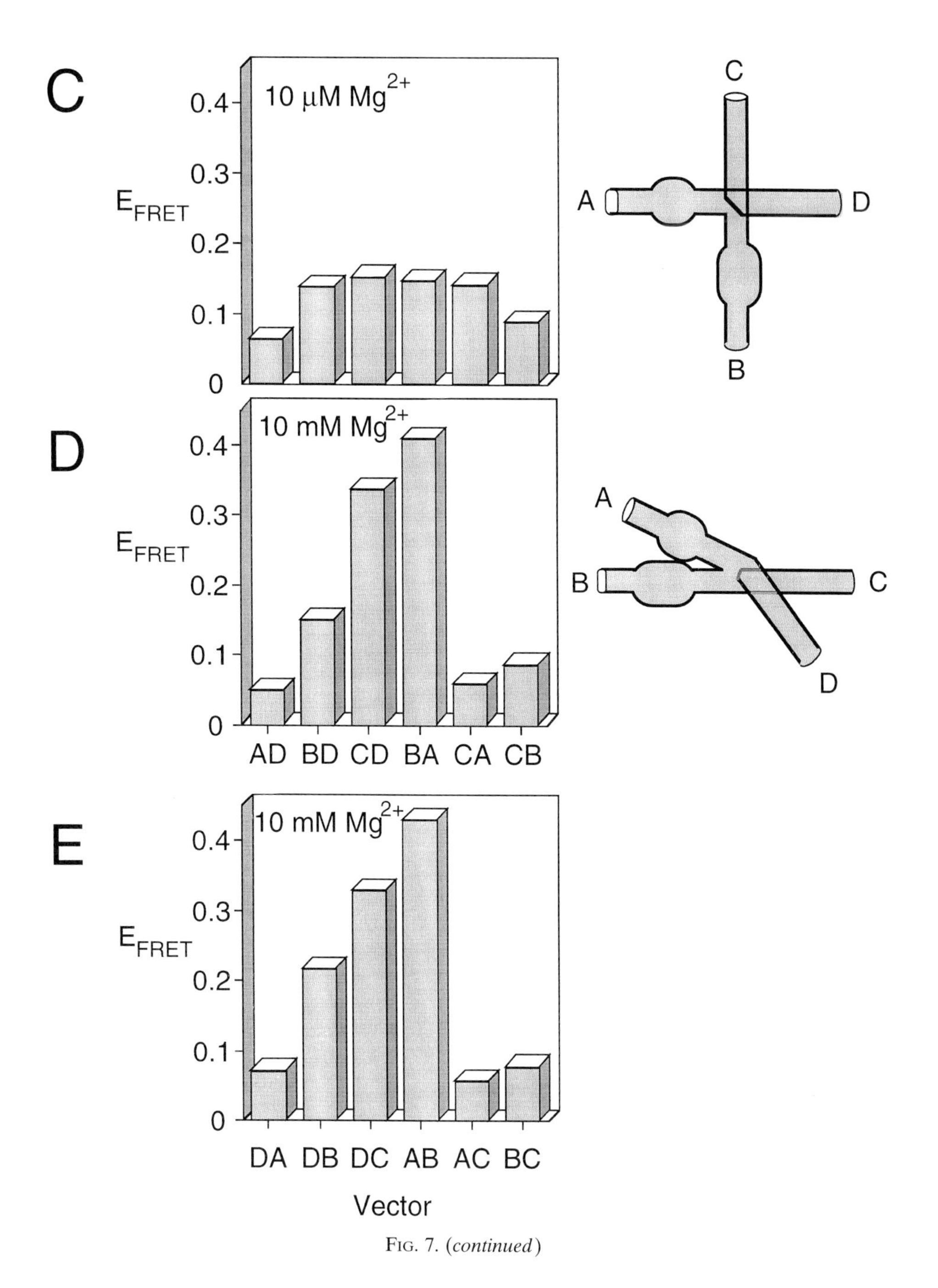

FIG. 7. (*continued*)

cantly retarded in the gel system. Bands are excised, and the oligonucleotides electroeluted into 8 *M* ammonium acetate, and recovered by ethanol precipitation. Fluorescently labeled oligonucleotides are further purified by reversed-phase HPLC (C_{18} column, eluted with a linear gradient of 100 m*M* ammonium acetate/acetonitrile, with a flow rate of 1 ml/min). The required stoichiometric combinations of fluorophore-labeled and unlabeled strands are then heated in 90 m*M* Tris-borate (pH 8.3), 25 m*M* NaCl for 10 min, at 80°, followed by slow cooling over several hours. The doubly labeled species are purified by electrophoresis in a polyacrylamide gel under nondenaturing conditions at 4° at 150 V in 90 m*M* Tris–borate (pH 8.3), 25 m*M* NaCl with recirculation at >1 liter/hr. The fluorescent species are recovered by band excision and electroelution.

Comparative Approach. In general we prefer to avoid the calculation of absolute distances from FRET measurements. When possible we have used a comparative approach, where the efficiencies of energy transfer are compared for a series of closely similar molecules. Usually these are the different pairwise end-to-end vectors derived from a branched nucleic acid with arms of similar length. Thus for three- and four-way junctions there are, respectively, three and six different end-to-end vectors that can be compared, and the variation of these distances with conditions (often metal ion concentration) can be followed. To try to ensure constant conditions for the different species, we maintain a constant terminal sequence, and we have found that when the dyes are attached to a terminal 5′ CpC sequence they are spectroscopically well behaved. A further check on the reproducibility of the measurements can be made by reversing the positions of the fluorophores for each end-to-end vector. Thus if a given vector has fluorescein attached to arm A and Cy-3 attached to arm B (vector AB in our nomenclature), a second species BA can be constructed, where the fluorescein is now attached to arm B and the Cy-3 to arm A. This requires *de novo* synthesis and construction of all the molecules, and is therefore a good indication of experimental error. Moreover, since the dyes are now attached to different arms, this should reveal any effect arising from local environment. In general, we find that there is very good agreement between the two sets of measurements. An example of the vector comparison method applied to the hairpin ribozyme is given in Fig. 7.[22,51] It can be seen that the structure is significantly dependent on the presence of magnesium ions, undergoing a structural transition as the concentration is increased. In its fully folded form the efficiencies for the different end-to-end vectors vary from 0.04 up to 0.4 for the closest vector. Reversal of each of the vectors gives a pattern of efficiencies that is closely similar, giving further confidence

[51] F. Walter, A. I. H. Murchie, J. B. Thomson, and D. M. J. Lilley, *Biochemistry* **37,** 14195 (1998).

in the measurements and the conclusions derived from them. The variation of efficiency for a given vector can also be studied as a function of conditions. Figure 8 shows an example, following the shortening of the closest vector of the hairpin ribozyme as a function of a magnesium ion concentration. These data can be fitted to various models for ion-induced folding. Frequently the folding can be followed by the simultaneous shortening of one vector and the lengthening of another, which again gives extra confidence in the FRET method. For example, the first folding transition of the hammerhead ribozyme (Fig. 9) can be followed by the shortening of the vector between the ends of the I and III stems, and by the lengthening of the vector between the ends of the II and III stems.[29] Some metal ions, such as cobalt(III) hexammine, cause significant quenching of the donor fluorescence, and therefore the spectra require correction using a Stern–Volmer plot whereby the fractional quenching is plotted as a function of quencher concentration.

Other Methods for Studying Global Structure in RNA

RNA folding is a difficult problem, and information from any possible source can be valuable. Data from chemical probing can be valuable. Tertiary structure can be examined by the accessibility of the backbone to

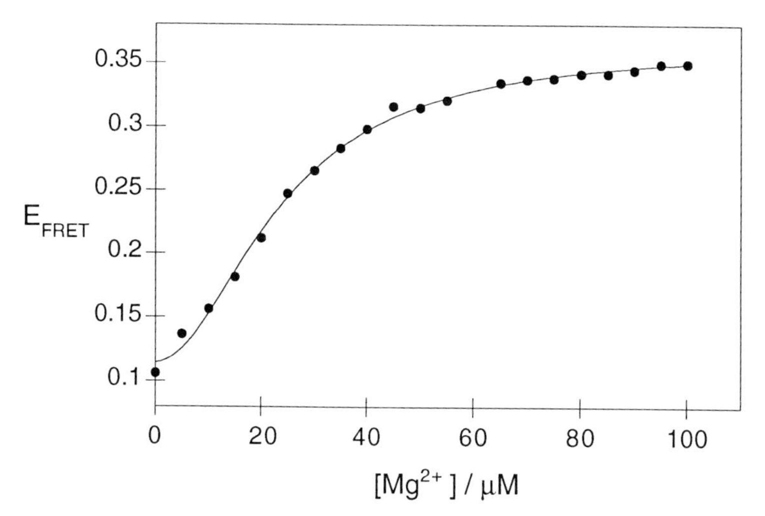

FIG. 8. Ion-induced folding of the hairpin ribozyme over the range 0–100 μM magnesium ions followed by the change of E_{FRET} for the shortest distance A-B.[22] The experimental data (●) were fitted by regression to a simple binding model where the cooperative binding of two ions to the RNA induces a global structural change from conformation 1 to conformation 2. The line shows the fits to this model.

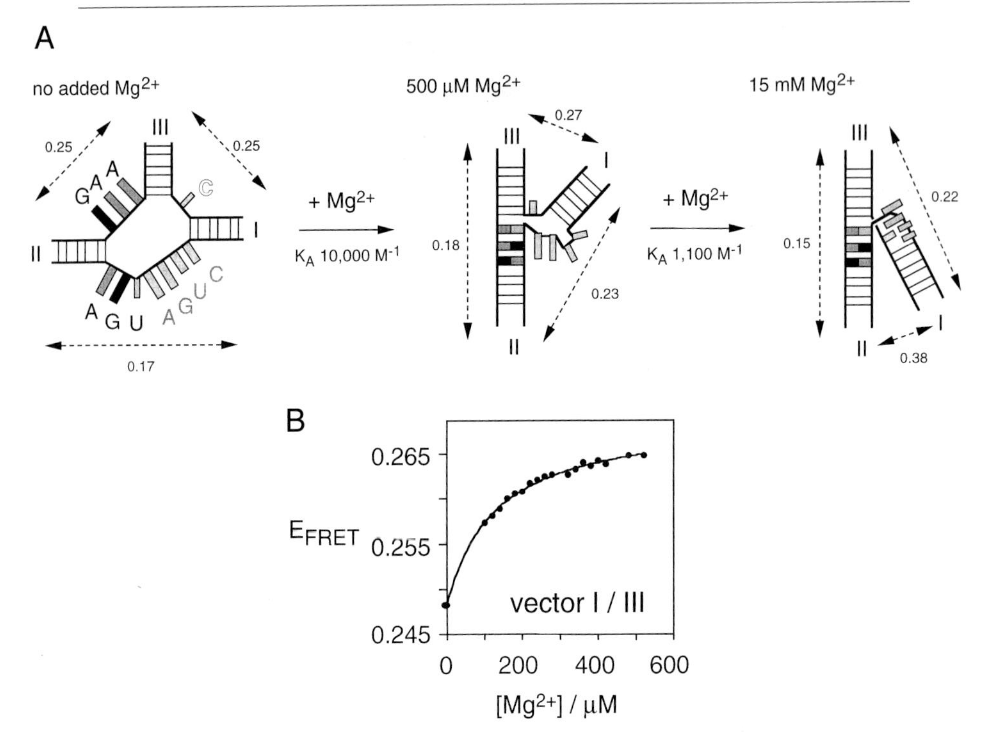

FIG. 9. The folding of the hammerhead ribozyme followed by FRET.[29] The FRET efficiencies for the three end-to-end vectors were measured as a function of magnesium ion concentration, from which the two-stage folding process first deduced from comparative gel electrophoresis (see Fig. 3) was confirmed. (**A**) The folding pathway. The scheme shows the initial, intermediate, and final stages of the ion-induced folding process, with the end-to-end E_{FRET} values indicated alongside the broken double-headed arrows. The first stage of the folding (500 μM magnesium ions) involves the lengthening of the II-III vector, and is likely to correspond to the folding of domain II by the formation of the consecutive G•A mispairs. Addition of further magnesium ions induces the folding of the uridine turn by the CUGA sequence. This brings about a rotation of helical arm I, such that it comes close to helix II in the final structure; thus the I-II distance gives the highest E_{FRET} value in the ribozyme under any condition. The magnesium ion concentration range over which the final structure is formed correlates closely with cleavage activity, and thus this conformation is likely to represent the potentially active ribozyme. The global shape of the final structure is in good agreement with that observed in the crystal.[27,28] (**B**) Titration of the energy transfer for the I-III vector over the 0–500 μM magnesium ion concentration range, corresponding to the first transition. The data can be well fitted to the binding of a single magnesium ion with an apparent association constant in the range of 10,000 M^{-1}.

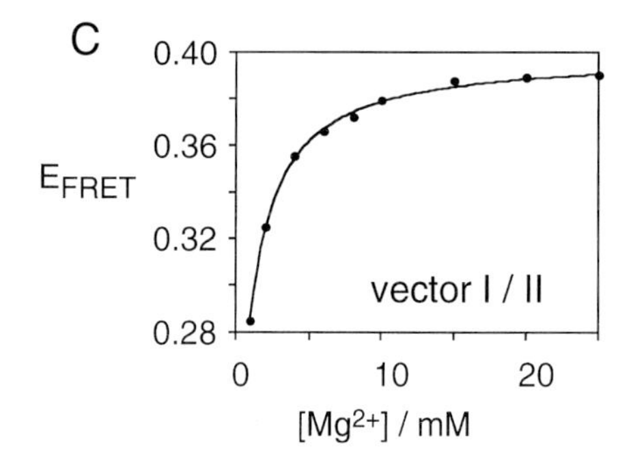

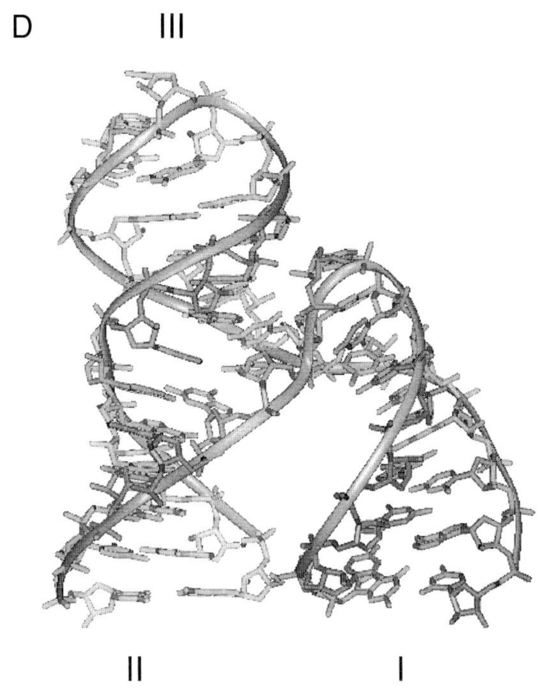

FIG. 9. (*continued*) (**C**) Titration of the energy transfer for the I-II vector over the 1–25 mM magnesium ion concentration range, corresponding to the second transition. These data can also be fitted to the binding of a single magnesium ion, with an apparent association constant of close to 1100 M^{-1}. (**D**) Crystal structure of the hammerhead ribozyme. This image was generated from the coordinates of Pley *et al.*[27] using the view equivalent to that shown in the folding scheme. It can be seen that the global shape is the same as the form of the ribozyme when fully folded in magnesium ions.

probes such as hydroxyl radicals generated by Fe(II)-EDTA; this has been successfully applied to the group I ribozyme for example.[52] In addition, metal ion complexes can give important information on metal ion binding sites, exemplified by the use of uranyl-induced photocleavage[20] and terbium fluorescence[53] in the hammerhead ribozyme, and lead cleavage[54] and hydroxyl ions generated by the Fenton reaction using coordinated ferrous ion in the group I ribozyme.[55] Functional group substitution can be extremely revealing in combination with all the methods discussed earlier in order to probe requirements for correct folding. This can include sequence changes (mutation), phosphate modification (e.g., phosphorothioate, methyl phosphonate), sugar modification (e.g., deoxyribose, arabinose), and base modification (too numerous to list examples). In addition, specific reporter groups can be introduced into the RNA in order to follow the folding process, such as the use of fluorescent 2-aminopurine substitution in the hammerhead ribozyme.[56] Chemical cross-linking is another valuable source of structural information, that has been applied to the hammerhead[57] and hairpin[58] ribozymes.

Other physical approaches are important. One additional method used prominently in the analysis of branched RNA structures has been transient electric birefringence.[14,59–61] Electron microscopy has been applied to RNA structure,[62] and cryoelectron microscopy is another promising approach.[63] Molecular modeling can also be important in testing and refining stereochemical ideas.

While information derived from individual methods such as FRET is extremely useful in the study of RNA folding, the value of combining data from many sources cannot be overemphasized. This can include the electrophoretic and spectroscopic methods, and adding in data from functional studies, chemical probing and so on. But reference to molecular modeling and crystallographic data when available can add greatly to the

[52] F. L. Murphy and T. R. Cech, *Biochemistry* **32,** 5291 (1993).

[53] A. L. Feig, W. G. Scott, and O. C. Uhlenbeck, *Science* **279,** 81 (1998).

[54] B. Streicher, U. Vonahsen, and R. Schroeder, *Nucleic Acids Res.* **21,** 311 (1993).

[55] C. Berens, B. Streicher, R. Schroeder, and W. Hillen, *Chem. Biol.* **5,** 163 (1998).

[56] M. Menger, T. Tuschl, F. Eckstein, and D. Porschke, *Biochemistry* **35,** 14710 (1996).

[57] S. T. Sigurdsson, T. Tuschl, and F. Eckstein, *RNA* **1,** 575 (1995).

[58] D. J. Earnshaw, B. Masquida, S. Müller, S. T. Sigurdsson, F. Eckstein, E. Westhof, and M. J. Gait, *J. Mol. Biol.* **274,** 197 (1997).

[59] K. M. A. Amiri and P. J. Hagerman, *Biochemistry* **33,** 13172 (1994).

[60] M. W. Friederich and P. J. Hagerman, *Biochemistry* **36,** 6090 (1997).

[61] Z. Shen and P. J. Hagerman, *J Mol. Biol.* **241,** 415 (1994).

[62] Y. H. Wang, F. L. Murphy, T. R. Cech, and J. D. Griffith, *J. Mol. Biol.* **236,** 64 (1994).

[63] J. Dubochet, M. Adrian, I. Dustin, P. Furrer, and A. Stasiak, *Methods Enzymol.* **211,** 507 (1992).

insight available. As an example, our studies of the ion-induced folding of the hammerhead ribozyme[20,23,29] have been enormously aided by the availability of crystal structures of the RNA.[27,28,64] Our feeling is that it is the combination of the data from the different techniques that is usually most revealing with a problem as complex as RNA folding.

Acknowledgments

I thank all the present and past members of this laboratory who have participated in the studies of branched RNA structure and whose data are included in this review, especially Derek Duckett, Alastair Murchie, Gurminder Bassi, Frank Walter, and Richard Grainger. I happily acknowledge a long-standing and stimulating collaboration with Dr. Bob Clegg on the application of FRET to nucleic acid structural analysis. The work in this laboratory is funded by the Cancer Research Campaign and the Biotechnology and Biological Sciences Research Council.

[64] W. G. Scott, J. B. Murray, J. R. P. Arnold, B. L. Stoddard, and A. Klug, *Science* **274,** 2065 (1996).

[24] Application of Circular Dichroism to Study RNA Folding Transitions

By TOBIN R. SOSNICK, XINGWANG FANG, and VALERIE M. SHELTON

Introduction

Circular dichroism (CD) spectroscopy has been widely used to study secondary structure formation in both proteins and nucleic acids.[1] Modern spectropolarimeters in conjunction with computers and multiple reference spectra enable the researcher to assign secondary structure content with reasonable accuracy.[2] Since the discovery of ribozymes, the study of RNA folding has extended into the characterization of the tertiary organization. Although CD cannot provide structural detail about the tertiary organization, it can be used to monitor RNA tertiary folding transitions that may not be observable by absorbance spectroscopy.[3] CD in combination with the nonionic denaturant urea can be used to characterize secondary and

[1] G. D. Fasman, "Circular Dichroism and the Conformational Analysis of Biomolecules." Plenum Press, New York, 1996.

[2] D. M. Gray, S. H. Hung, and K. H. Johnson, *Methods Enzymol.* **246,** 19 (1995).

[3] T. Pan and T. R. Sosnick, *Nature Struct. Biol.* **4,** 931 (1997).

0076-6879/00 $30.00

tertiary RNA folding transitions, providing information on both the equilibrium free energy and relative amount of surface buried in the folding transition. This article describes a basic outline for using CD and analyzing Mg^{2+} and urea-dependent folding transitions to determine thermodynamic and kinetic parameters of RNA folding and the effects of mutations.

Measurement of Circular Dichroism Spectrum

The measurement of an RNA spectrum is similar to the measurement of a protein spectrum. Useful spectral information is readily obtainable between 200 and 320 nm. Both buffers (e.g., 20 m*M* Tris-HCl) and the RNA itself absorb strongly in the far UV and measurements below 200 nm are difficult due to high absorbance. In the region above 200 nm, the maximal absorbance is due to the nucleotide bases, which have a broad maximum in the region around 260 nm. Higher sample concentrations are desirable because this will increase the size of the CD signal. However, if concentrations are too high, too much light is absorbed and the signal quality deteriorates. The optimal absorbance is in the region between 0.3 and 0.8 absorbance unit (AU).

Nearly all features in the RNA spectrum are broad and instrumental resolution can be set to 2 nm although data can be sampled every 0.5 or 1 nm. This increases the amount of light passing through the sample, hence increasing signal quality. The use of high-quality UV quartz rather than highly absorbing glass cuvettes also will help extend measurements farther into the UV. It is important to measure buffer blanks in the same cuvette orientation for proper background subtractions both for the CD and absorbance measurements. For a 20 μg/ml RNA sample with an $A_{260} \sim 0.5$ in a 1-cm path length cuvette, a 100-nm scan at 2-nm resolution with reasonable signal quality can be taken in about 5 min using an instrumental time constant of 4 sec (Fig. 1).

Modern commercially available CD spectrometers, for example those made by Jasco (Easton, MD), are capable of simultaneous CD and absorbance measurements. The amplitude of the applied photomultiplier voltage (high tension) directly relates to absorbance of the sample and is recorded along with the CD signal. This enables one to measure transitions by both probes without any additional measurements.

The CD spectrum should be put on an absolute scale normalized to the number of nucleotides in order that proper comparisons to other spectra can be made. This can be calculated using the formula for the molar circular dichroic absorption:

$$\Delta\varepsilon(\text{units of cm}^2\ \text{mmol}^{-1}) = \theta/(32{,}980CLN) \tag{1}$$

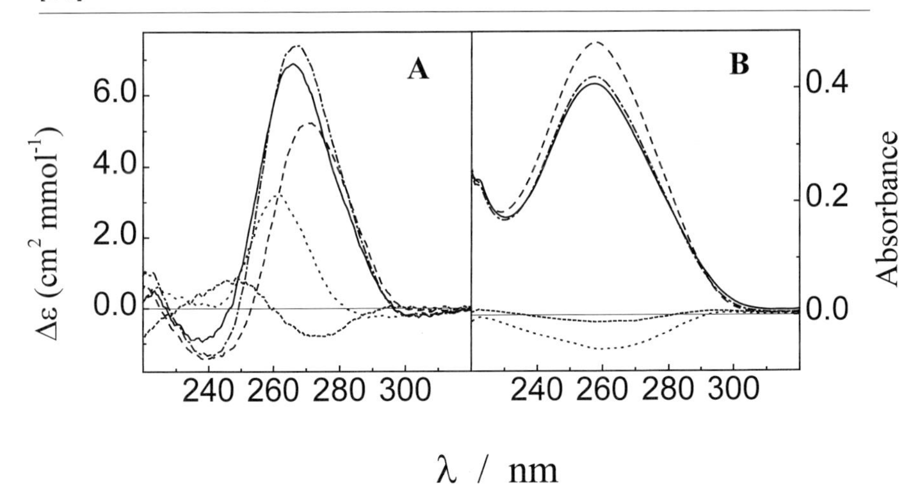

FIG. 1. Circular dichroism and absorbance spectra of the specificity domain of *Bacillus subtilis* P RNA in 3 *M* urea, 20 m*M* Tris, pH 8.1, at 37°. (A) CD and (B) absorbance spectra of the unfolded state (no $MgCl_2$, dash lines), intermediate state (0.45 m*M* $MgCl_2$, dash-dot lines), and native state (20 m*M* $MgCl_2$, solid lines) and the difference spectra of I-U (dot lines) and N-I (short dash lines) are shown. The addition of 3 *M* urea reduces the ellipticity at 260 nm of the unfolded state indicating the disruption of residual helical structure. However, 3 *M* urea does not significantly affect the spectra of either the intermediate or native states.

where θ is the measured (raw) CD amplitude in mdeg; C is the sample concentration in mol/liter; L is the cell path length in cm; and N is the number of nucleotides of the RNA.

Measurement of Mg^{2+}-Induced Folding Transitions

The tertiary folding of many RNAs is intimately coupled to the binding of divalent cations, typically Mg^{2+}. Near-UV circular dichroism can be used to follow the Mg^{2+} dependence of the folding transitions.[3] To determine the number of transitions and appropriate wavelengths to monitor each, an initial coarse Mg^{2+} dependence of the CD spectrum should be conducted. The spectrum of unfolded, Mg^{2+}-free RNA should be measured followed by a series of spectra at increasing Mg^{2+} concentrations. This can be done expediently by adding small volumes of a concentrated Mg^{2+} solution to a single sample and correcting for the resulting dilution. The initial unfolded state of the RNA can be obtained after a renaturing step involving heating to 90° for 2 min to remove any residual structure formed during purification steps.

We have found that the two-domain ribozyme P RNA from *Bacillus subtilis* folds with two distinct structural transitions, an unfolded to intermediate, U-to-I, transition and an intermediate to native transition.[3] The U-to-I transition occurs at micromolar Mg^{2+} concentrations and is readily monitored by changes in absorbance (A_{260}) and CD at 260 nm ($\Delta\varepsilon_{260}$). These signals are primarily sensitive to the formation of helical structure. The second, tertiary transition occurs in millimolar Mg^{2+} range with a Hill coefficient of about 4. The formation of the tertiary structure of RNA is, to a large extent, the result of structural changes involving the ribose–phosphate backbone and non-Watson–Crick base pairing of nucleotides.[4,5] In P RNA, this tertiary transition has very minimal absorbance changes and is only easily measurable by changes in the CD spectrum. For P RNA, the most sensitive wavelengths to the tertiary transition are in the region between 275 and 290 nm. The structural origins of these spectral changes are unknown and the wavelength to be monitored for a given RNA should be chosen empirically based on its particular spectral properties. Figure 2 shows the Mg^{2+} titration for the isolated specificity domain of P RNA.

The transition midpoint may be sensitive to buffer conditions. For example, the intermediate-to-native transition of P RNA, as measured by activity and CD, shifts from ~1 to ~2 mM when the trisHCl concentration is increased from 20 to 50 mM at 37°.[6] The Hill coefficient, however, remains unchanged. The increase in the midpoint may be due to trisHCl chelating to and reducing the free Mg^{2+} concentration.

Thermodynamic measurements require that the system be reversible and in equilibrium during the measurement if meaningful parameters are to be extracted. Tertiary RNA folding transitions can occur in seconds to hours depending on the specific RNA and the experimental conditions (e.g., temperature, Mg^{2+} and urea concentration). Hence, one must exercise extreme care to ensure that the system has equilibrated at each measurement point. Slow transitions often have a high activation enthalpy and the reaction can be greatly accelerated with increased temperature. For example, an activation enthalpy of 50 kcal/mol will result in ~18-fold increase per 10°.

After determining the Mg^{2+} ranges of each folding transition, and the wavelengths that are most sensitive to each transition, a more finely spaced titration should be conducted so that accurate thermodynamic parameters can be determined. Although one can correct for dilution of the RNA resulting from the addition of concentrated Mg^{2+} solutions, the use of the

[4] D. W. Celander and T. R. Cech, *Science* **251,** 401 (1991).
[5] J. H. Cate, A. R. Gooding, E. Podell, K. Zhou, B. L. Golden, C. E. Kundrot, T. R. Cech, and J. A. Doudna, *Science* **273,** 1678 (1996).
[6] T. Pan, X. Fang, and T. R. Sosnick, *J. Mol. Biol.* **268,** 721 (1999).

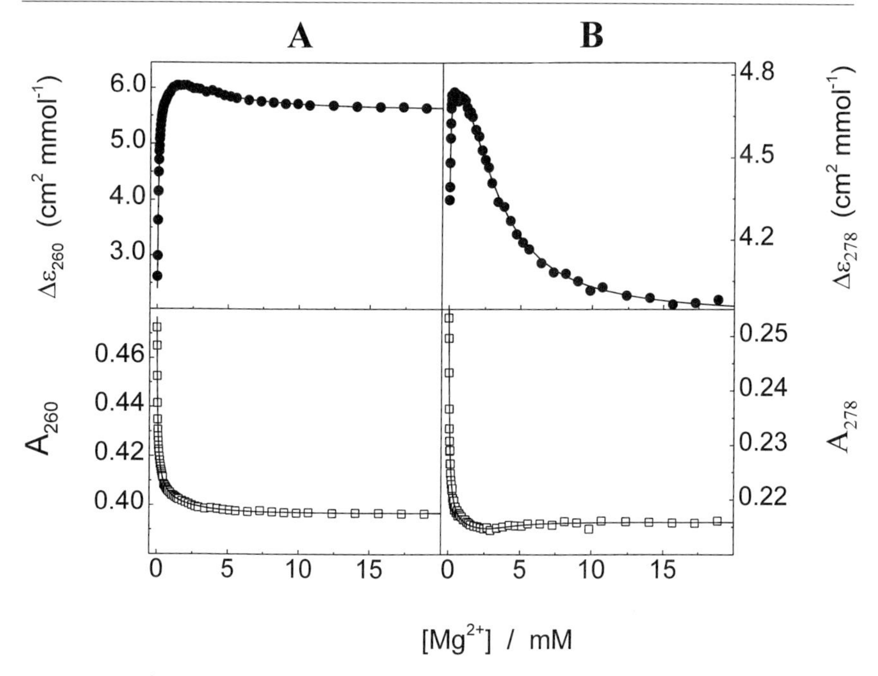

FIG. 2. Mg^{2+}-induced folding. Mg^{2+} dependence of the structural forms of the specificity domain of *B. subtilis* P RNA in 3 *M* urea at 37°, 20 m*M* Tris, pH 8.1, are monitored by CD (filled circles) and absorbance (open squares) at (A) 260 nm (helical structure) and (B) 278 nm (tertiary structure). Fitting with Eq. (6) (solid curves) results in $K_{D1} = 0.064 \pm 0.002$ m*M* ($n_1 = 1.1 \pm 0.1$) and $K_{D2} = 3.4 \pm 0.4$ m*M* ($n_2 = 2.0 \pm 0.5$).

least added volume will increase the accuracy of the measurement. Hence, for two transitions occurring at highly disparate Mg^{2+} concentrations, multiple Mg^{2+} stock solutions (e.g., 10 m*M* and 1 *M* $MgCl_2$) should be used to optimize coverage of each transition without requiring the addition of large volumes of solutions.

The sensitive titration may require that more than 50 separate measurements be conducted, each following the addition of small quantities (<10 μl) of the concentrated Mg^{2+} stock solutions. Standard manual methods of addition (i.e., open sample compartment, add solution, mix, close sample compartment, wait for N_2 flush) are time consuming and often inaccurate due to baseline shifts. Rapid additions without opening the sample compartment can be made through a small-bore HPLC-style capillary tubing connected to calibrated glass syringe. To ensure thorough mixing, a small magnetic stir bar (<9 mm long) is placed in a 1-cm path length cell. If the spectrophotometer is not so equipped, a small electronic stirrer plate that fits in the bottom of the cuvette holder can be purchased (NGS Precision Cells, Farmingdale, NY). To further increase accuracy, the glass syringe can be attached to manual push-button repeating dispenser (Hamilton Co., Reno, NV) that can reliably dispense 1/50 of the syringe volume per increment. To further automate the titration process, an electronically controlled titrator can be programmed and interfaced to the spectrophotometer.

Analyzing a Mg^{2+} Folding Transition

Two parameters, the dissociation constant, K_d, and Hill coefficient, n, can be obtained from Mg^{2+} titration experiments. The transition midpoint, $Mg^{2+}_{1/2}$ (units of M) is related to the dissociation constant by $(Mg^{2+}_{1/2})^n = K_d$. For simplicity, $Mg^{2+}_{1/2}$ will be written as K_D in this article. The Hill coefficient identifies the minimal number of Mg^{2+} ions involved in the transition and provides a measure of the cooperativity of the transition. A Hill coefficient of unity can be attributed to multiple, independent site binding of approximately the same K_d, so that the total number of Mg^{2+} ions bound is unknown. This number can be determined by other methods including equilibrium dialysis.[7-9]

Specific Mg^{2+} ion binding sites have been observed in tertiary RNA structures such as tRNA (reviewed in Ref. 10) and the *Tetrahymena* group I ribozyme.[5] This suggests that the specific binding of metal ions is required for the folding of most tertiary RNAs. However, the actual number of required Mg^{2+} may be different from the number observed in the crystal structure. Furthermore, certain modified tRNAs may form tertiary structures in the presence of high concentrations of monovalent ions.[11,12] This latter result suggests that alternative interpretations for Mg^{2+}-induced folding are possible including polyelectrolyte theory, which proposes cations localize around and neutralize the highly charged phosphate backbone.[13,14] However in P RNA, the K_D *increases* on the addition of monovalent cations (e.g., Na^+).[3] This indicates that Mg^{2+} must overcome the monovalent cations to drive a tertiary folding transition as a result of the monovalent cations preferentially interacting with the less structured forms. For these

[7] A. Stein and D. M. Crothers, *Biochemistry* **15,** 157 (1976).

[8] M. Bina-Stein, and A. Stein, *Biochemistry* **15,** 3912 (1976).

[9] J. A. Beebe, J. C. Kurz, and C. A. Fierke, *Biochemistry* **35,** 10493 (1996).

[10] T. Pan, D. M. Long and O. C. Uhlenbeck, *in* "The RNA World" R. F. Gesteland and J. F. Atkins (eds.), pp. 271–302. Cold Spring Harbor Press, Plainview, New York, 1993.

[11] P. E. Cole, S. K. Yang, and D. M. Crothers, *Biochemistry* **11,** 4358 (1972).

[12] D. M. Crothers, P. E. Cole, C. W. Hilbers, and R. G. Shulman, *J. Mol. Biol.* **87,** 63 (1974).

[13] G. S. Manning, *Q. Rev. Biophys.* **11,** 179 (1978).

[14] M. T. Record, Jr., C. F. Anderson, and T. M. Lohman, *Q. Rev. Biophys.* **11,** 103 (1978).

reasons, a single transition with no observable intermediates and a Hill coefficient of greater than 1 probably represents the specific binding of n Mg^{2+} ions.

For a single nth order transition between states A and B,

$$A + n\,Mg^{2+} \overset{K_d}{\longleftarrow} B \tag{2a}$$

the dissociation constant and Mg^{2+} midpoint can be written as

$$K_d = (K_D)^n = \frac{[Mg^{2+}]^n[A]}{[B]} \tag{2b}$$

The Mg^{2+}-dependent equilibrium free energy, $\Delta G^\circ_{A\rightarrow B}$, between states A and B is the logarithm of the ratio of their populations and can be written

$$\Delta G^\circ_{A\rightarrow B} = -RT \ln K_{eq}^{A\rightarrow B} = -RT \ln \frac{[B]}{[A]} = -RT \ln \left(\frac{[Mg^{2+}]}{K_D}\right)^n$$
$$= -nRT \ln \frac{[Mg^{2+}]}{K_D} \tag{3}$$

where R is the gas constant (1.987 cal/K/mol) and T is the absolute temperature (K). The CD signal as a function of Mg^{2+} can be fit to a binding curve

$$\theta(Mg^{2+}) = \theta_A + \frac{[Mg^{2+}]^n}{[Mg^{2+}]^n + (K_D)^n}(\theta_B - \theta_A) \tag{4}$$

where θ_A and θ_B are the signals of initial state A and final state B, respectively. For two well-separated transitions A→B and B→C, the signal, $\theta(Mg^{2+})$, can be fit as the sum of two single transitions

$$\theta(Mg^{2+}) = \theta_A + \frac{[Mg^{2+}]^{n_1}}{[Mg^{2+}]^{n_1} + (K_{D1})^{n_1}}(\theta_B - \theta_A)$$
$$+ \frac{[Mg^{2+}]^{n_2}}{[Mg^{2+}]^{n_2} + (K_{D2})^{n_2}}(\theta_C - \theta_B) \tag{5}$$

where n_1 and K_{D1}, and K_{D2} are the Hill coefficients and dissociation constants of the first and second transitions, respectively, and θ_A, θ_B, and θ_C are the CD signals of states A, B, and C (Fig. 2). For two overlapping transitions where all three species are significantly populated, the signal should be fit to reflect this linked equilibrium according to

$$\theta(Mg^{2+}) = \frac{\theta_A + \theta_B([Mg^{2+}]/K_{D1})^{n_1} + \theta_C([Mg^{2+}]/K_{D1})^{n_1}([Mg^{2+}]/K_{D2})^{n_2}}{1 + ([Mg^{2+}]/K_{D1})^{n_1} + ([Mg^{2+}]/K_{D1})^{n_1}([Mg^{2+}]/K_{D2})^{n_2}} \tag{6}$$

Equation (6) reduces to Eq. (5) when $K_{D1} \ll K_{D2}$.

Characterizing Folding Using Urea

The nonionic denaturant urea can be used to denature both secondary and tertiary RNA structures. A urea titration can determine the free energy and size of a folding transition. At the simplest level, urea denatures RNAs by preferentially stabilizing the unfolded state, however, the precise mechanism is unknown. Urea may promote unfolding either by forming hydrogen bonds with newly exposed carbonyls and nitrogens on the bases, increasing the structure of water making the exposure of hydrophobic regions (e.g., the aromatic rings) less favorable, or weakly binding to the RNA with the number of urea binding sites increasing upon unfolding.[15]

Regardless of which modes are operational, we have observed that the stability of secondary and tertiary RNA structures have a linear dependence on urea concentration at a fixed Mg^{2+} concentration just as observed for proteins

$$\Delta G^\circ(\text{urea}) = \Delta G^\circ_{H_2O} + m^\circ[\text{urea}] \tag{7}$$

For proteins, the slope m° is proportional to the amount of denaturant-sensitive surface buried on folding and is directly related to the size of the protein.[16] Recent studies in our laboratory indicate that this is true for RNAs as well, as long as all transitions are accounted for.[16a,16b]

A urea titration can be performed in a similar manner to a Mg^{2+} titration. The CD signal at the appropriate wavelength is measured for a series of samples at varying urea concentrations over an extended range, for example, 0–7 *M*. The required concentration range will depend on the size of the RNA, with larger molecules typically requiring a smaller range since the width of their unfolding transition is narrower as their m° values are larger. The measured range must also be wide enough that accurate pre- and posttransition baselines can be observed so that proper fitting and analysis can be performed.

Unlike a Mg^{2+} titration, generally one cannot add enough concentrated urea solution to cover the required urea concentration range. To overcome this problem, two alternative titration methods can be employed. In either method, two stock solutions, one at low and one at high urea concentration, are made each containing the same RNA concentration. This ensures that when the two stock solutions are mixed, the RNA concentration will remain constant and the CD signal will not need to be corrected for varying RNA concentration. A convenient method to make up these two equiconcentration solutions is to make up a stock of 10× concentrated RNA in either the low or high molarity urea solution. One part of the concentrated sample

[15] G. I. Makhatadze and P. L. Privalov, *J. Mol. Biol.* **226,** 491 (1992).
[16] J. K. Myers, C. N. Pace, and J. M. Scholtz, *Protein Sci.* **4,** 2138 (1995).
[16a] V. Shelton, T. R. Sosnick, and T. Pan, *Biochemistry* **39,** (2000).
[16b] X. Fang, T. Pan, and T. Sosnick, *Biochemistry* **39,** (2000).

is diluted with nine parts of each of the low and high molarity urea solutions to produce the same RNA concentration in the two differing urea stock solutions.

In the first titration method, individual samples at varying urea concentrations are made and measured separately. These samples are made by mixing the two stock solutions at varying ratios. To make up 21 separate 1.5-ml samples, at least 15 ml of each stock solution need to be made. In the second method, a single sample initially containing the dilute urea solution is measured after each addition of a small aliquot (e.g., 50 μl) of the concentrated urea solution. This method produces more accurate data using about 5- to 10-fold less sample. The size of each aliquot determines the urea concentration spacing of the data. As with the Mg^{2+} titration, this can be done expediently through a small-bore HPLC-style capillary tubing connected to calibrated glass syringe. A portion of the cumulative sample must be removed partway through the titration to prevent overflow of a cuvette. With viscous urea solutions, extreme care must be taken to ensure adequate mixing of the different density solutions. A simple method to test for proper mixing is to confirm that the trace for dilutions going from low to high urea concentration matches that for dilutions from high to low concentration. This procedure also confirms that the folding is reversible as required for the extraction of meaningful thermodynamic parameters.

A modification of the second method involves the removal of a small aliquot prior to the addition of an equivalent volume aliquot so that the total sample volume remains constant. This variation reduces the total amount of sample required while increasing the range of the urea titration for a given number of additions. A two-syringe programmable titrator capable of both removal and additions can greatly facilitate the implementation of this variation (Fig. 3A).

Analyzing Urea Folding Transitions

To extract the free energy and $m°$ values from a urea titration for a transition between states A and B, the data should be fit using a nonlinear least squares program according to

$$\theta(\text{urea}) = \frac{\theta_A + \theta_B e^{(-\Delta G_{H_2O} - m°[\text{urea}])/RT}}{1 + e^{(-\Delta G_{H_2O} - m°[\text{urea}])/RT}} \tag{8}$$

where θ_A and θ_B are the baseline values of states A and B, respectively (Fig. 3A). The origin of this equation is based on the relationship

$$\Delta G°(\text{urea}) = \Delta G°_{H_2O} + m°[\text{urea}] = -RT \ln K_{eq}(\text{urea}) = -RT \ln \frac{[B]}{[A]}$$

$$= -RT \ln \left[\frac{\theta_A - \theta(\text{urea})}{\theta(\text{urea}) - \theta_B} \right] \tag{9}$$

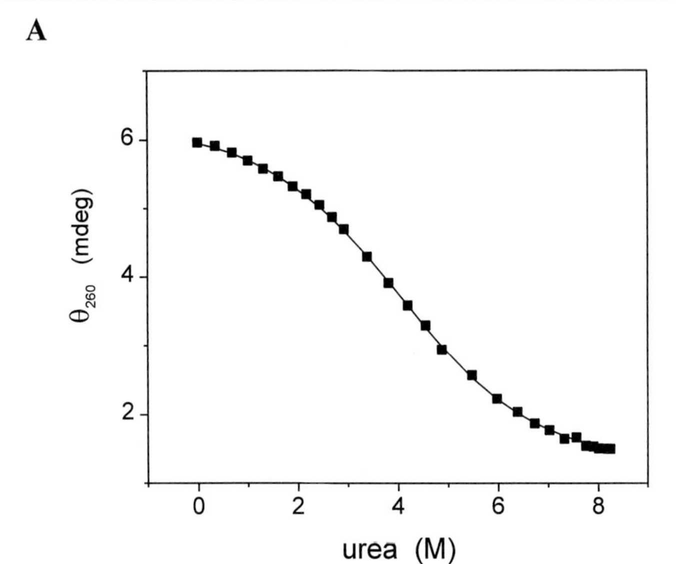

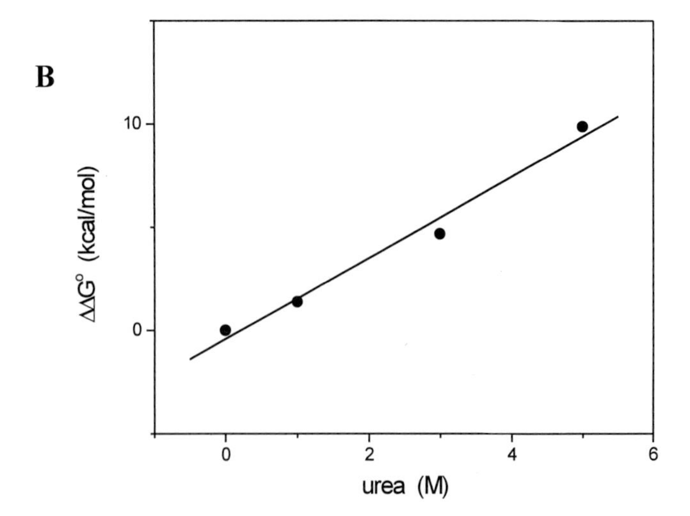

FIG. 3. Urea titrations. (A) CD signal at 260 nm (base stacking) versus urea for 2 μM self-complementary RNA duplex decamer (5′-GUAAUAUUAC-3′) at 20°, 20 mM sodium cacodylate, 0.5 M NaCl, pH 6.6. Fitting with Eq. (10) results in $\Delta G^{\circ}_{H_2O} = -11.0 \pm 0.2$ kcal/mol and $m^{\circ} = 0.74 \pm 0.04$ kcal/mol/M. Titration was performed using the "remove X μl/add X μl" protocol. (B) Free-energy changes as a function of urea concentration for P RNA from *B. subtilis*[3] at 37°, 20 mM Tris, pH 8.1, calculated from the shift in the K_D according to Eq. (12) using a Hill coefficient of 4.

The baseline values may either be a constant value or may be approximated as a line (e.g., θ_A (urea) = m_A [urea] + $\theta_A^{H_2O}$).

An analogous equation to Eq. (8) can be written for a homodimeric system A $\leftrightarrow$ 2B:

$$\theta(\text{urea}) = \theta_A + \frac{(K_{eq}^2 + 8K_{eq})^{1/2} - K_{eq}}{4}(\theta_B - \theta_A) \tag{10}$$

with the concentration dependent equilibrium constant defined as $K_{eq} = [B]^2/[A]C_T$, where C_T is the total RNA concentration. Figure 3A shows a urea titration of a self-complementary RNA duplex decamer fit to this equation.

An alternative method for fitting a urea titration is to calculate $\Delta G°$ at each of the original data points. The resulting $\Delta G°$ versus urea plot (Fig. 4B, middle) is fit to the linear relation shown in Eq. (7). Although this is intuitively reasonable, this procedure suffers from a number of deficiencies. To convert the CD signal to free energy, the pre- and posttransition baseline regions must be chosen. A reasonably good fit to the calculated $\Delta G°$ values is possible with a poorly chosen baseline, however, this will result in an incorrect $m°$ value. Furthermore, the $\Delta G°$ values calculated at the extrema of the titration will have very high uncertainties since $\Delta G°$ is calculated from the logarithm of the ratio of a large to small number (this ratio can even be negative and produce a nonsensical $\Delta G°$ value). Hence, unless the errors in data are propagated correctly, the fitting of data converted to $\Delta G°$ will statistically weight the data incorrectly. This will compromise the validity of the statistical uncertainties for $\Delta G°$ and $m°$ generated by the fitting routine. For these reasons, it is preferred that the fitting routine fit the original CD data and the baselines simultaneously.

An alternative method to determine the $m°$ value of a transition is to measure the K_D at varying concentrations of urea. Using Eq. (3), the decrease in free energy due to the presence of urea at any given Mg^{2+} concentration can be written

$$\Delta\Delta G°(\text{urea}) = \Delta G°(\text{urea}) - \Delta G°_{H_2O} = -RT\ln\left(\frac{K_D^{H_2O}}{K_D(\text{urea})}\right)^n$$
$$= -nRT\ln\frac{K_D^{H_2O}}{K_D(\text{urea})} \tag{11}$$

where n is the Hill coefficient of the transition. A plot of $\Delta\Delta G°$ (urea) versus urea concentration can be fit directly to the linear relation shown in Eq. (7). Figure 3B illustrates this method with P RNA from *B. subtilis.*[3]

For two separated or overlapping transitions from A$\rightarrow$B and B$\rightarrow$C with

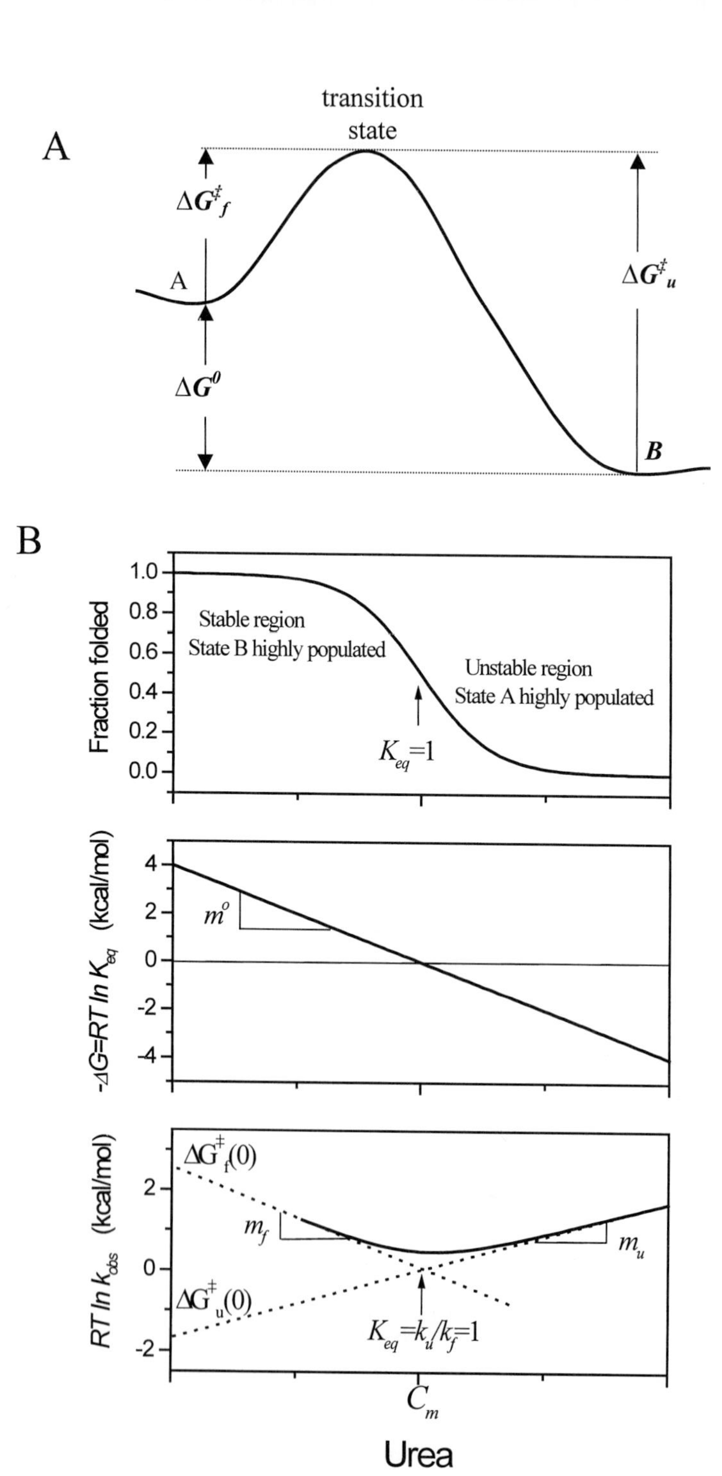

A
transition state
$\Delta G^{\ddagger}_{f}$
A
ΔG^{0}
$\Delta G^{\ddagger}_{u}$
B
B
1.0
0.8
0.6
0.4
0.2
0.0
Fraction folded
Stable region
State B highly populated
Unstable region
State A highly populated
$K_{eq}=1$
4
2
0
-2
-4
$-\Delta G=RT\ ln\ K_{eq}$ (kcal/mol)
m^{o}
$\Delta G^{\ddagger}_{f}(0)$
m_f
m_u
$\Delta G^{\ddagger}_{u}(0)$
$K_{eq}=k_u/k_f=1$
$RT\ ln\ k_{obs}$ (kcal/mol)
C_m
Urea

urea-dependent stabilities $\Delta G^\circ_{A\to B}$(urea) and $\Delta G^\circ_{B\to C}$(urea), respectively, the signal can be fit according to

$$\theta(\text{urea}) = \frac{\theta_A + e^{-\Delta G^\circ_{A\to B}(\text{urea})/RT}\left(\theta_B + \theta_C e^{-\Delta G^\circ_{B\to C}(\text{urea})/RT}\right)}{1 + e^{-\Delta G^\circ_{A\to B}(\text{urea})/RT}\left(1 + e^{-\Delta G^\circ_{B\to C}(\text{urea})/RT}\right)} \quad (12)$$

where θ_A, θ_B, and θ_C are the ellipticities of states A, B, and C. Again, these baseline values may be urea dependent and be approximated as a line.

Kinetic Measurements: Chevron Analysis

The denaturant dependence of a kinetically two-state, all-or-none folding reaction between two states, A and B,

$$A \underset{k_u}{\overset{k_f}{\rightleftharpoons}} B \quad (13)$$

produces a "chevron" or V-shaped curve with its vertex at the midpoint of the equilibrium transition[17] (Fig. 4B, bottom). The dependence of the ΔG° on urea concentration is commonly described by the linear relationship shown in Eq. (7). Equations (14a,b) describe the analogous linear dependence of the activation free energy, ΔG^+, for kinetic folding (f) and unfolding (u) reactions:

$$\Delta G^\ddagger_f(\text{urea}) = -RT \ln k_f^{H_2O} - m_f[\text{urea}] + \text{constant} \quad (14a)$$
$$\Delta G^\ddagger_u(\text{urea}) = -RT \ln k_u^{H_2O} - m_u[\text{urea}] + \text{constant} \quad (14b)$$

The denaturant dependence of the activation energies for folding and unfolding reactions are given by m_f and m_u, respectively, which represent the amount of urea-sensitive surface area buried at each stage of the reaction. The m_f parameter represents surface burial from the starting A state to

[17] C. R. Matthews, *Methods Enzymol.* **154,** 498 (1987).

FIG. 4. Urea dependence of thermodynamic and kinetic parameters. (A) Free-energy diagram of the starting state A, transition state, and ending state B along with free-energy differences between each. (B) Urea dependence of the fraction folded (top), ΔG° (middle), and observed activation energy, $\Delta G^\ddagger$ (bottom). The observed rate, k_{obs}, is the sum of the folding, k_f, and unfolding rates, k_u, at a given final urea concentration. For example, the left side of the chevron arm represents a refolding experiment under native conditions where the folding rate is much greater than the unfolding rate so that $k_{obs} \approx k_f$. At the denaturation midpoint, C_m the folding and unfolding rates are equal and $k_{obs} = 2k_f$ giving the bottom of the chevron a rounded appearance.

the transition state, whereas m_u represents burial from the transition state to the ending B state. Modifications appropriate for dimeric systems can be found in Ref. 18.

The constant term in Eq. (14a,b) represents the logarithm of an appropriate frequency of attempts (attempt frequency) to cross the kinetic barrier. Canonical transition state theory written as $k^{TST} = k_B T/h \exp(-\Delta G^{\ddagger}/k_B T)$ offers only a first approximation, predicting an upper limit for the rates of barrier crossing steps. The application of reaction rate theory to macromolecular reactions in solution has proven to be a difficult task.[19] The quantum mechanical prefactor ($k_B T/h \sim 10^{13}$ sec^{-1}) overestimates by many orders of magnitude the attempt frequencies of a polymer moving in high-friction media such as aqueous solutions.[20–22] The appropriate prefactor is currently unknown and may depend inversely on solvent viscosity according Kramer's theory.[19]

When equilibrium and kinetic folding reactions are limited by the same barrier and are effectively two state (only the starting and ending states significantly populated), it is possible to calculate the equilibrium values for the change in free energy and surface burial from kinetic measurements according to $\Delta G^{\circ}_{H_2O} = \Delta G_u^{\ddagger H_2O} - \Delta G_f^{\ddagger H_2O}$ and $m^{\circ} = m_u - m_f$, derived from Eqs. (9) and (14). When the kinetically determined values for $\Delta G^{\circ}_{H_2O}$ and m° account for all the energy and surface burial in the equilibrium transition, no stable kinetic intermediates are significantly populated that bury any urea-sensitive area in either the folding or unfolding pathways. This provides strong evidence for applicability of a two-state model for the reaction. In this circumstance, any kinetic probe will measure the (single) rate of the reaction, and the most convenient probe can be used.

Even when the folding reaction is not an ideal two-state reaction, the dependence of the rate on denaturant concentration provides a measure of the surface buried or exposed in the transition. For a reaction where the transition state buries more urea-sensitive surface area than the starting state, the dependence of the folding activation energy, the m_f parameter, will be positive. However, for a reaction where the RNA is in a kinetic trap that must undue a considerable number of nonnative interactions in order for folding to proceed, the m_f parameter, will be negative.[3]

[18] M. E. Milla and R. T. Sauer, *Biochemistry* **33,** 1125 (1994).

[19] H. A. Kramers, *Physica* **7,** 284 (1940).

[20] C. M. Jones, E. R. Henry, Y. Hu, C. K. Chan, S. D. Luck, A. Bhuyan, H. Roder, J. Hofrichter, and W. A. Eaton, *Proc. Natl. Acad. Sci. U.S.A.* **90,** 11860 (1993).

[21] D. K. Klimov and D. Thirumalai, *Phys. Rev. Lett.* **79,** 317 (1997).

[22] B. J. Berne, M. Borkovec, and J. E. Straub, *J. Phys. Chem.* **92,** 3711 (1988).

Effects of Mutations on Kinetics

To probe the folding pathway, the modern RNA biochemist has the ability to specifically modify individual nucleotides. The chevron formalism provides a powerful framework for evaluating the effects of mutations on the kinetic barriers on a two-state folding reaction (Fig. 5). If the structural feature modulated by the mutation is fully formed in the transition state, then the free energy of the transition state and the equilibrium stability will be altered equally. The unfolding rate will not change; only the folding will be affected and the left-side, folding arm of the chevron will be displaced

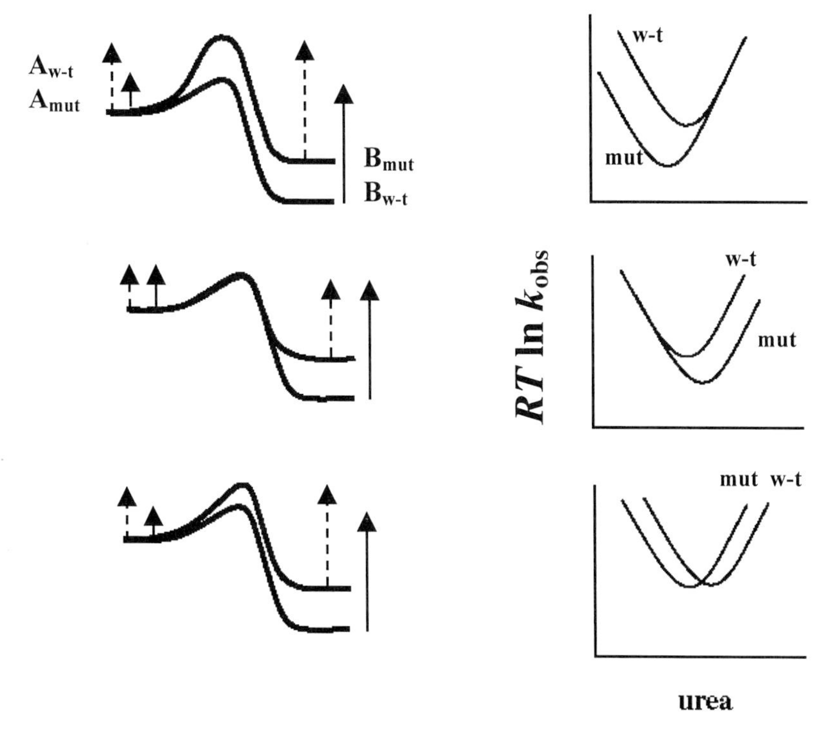

FIG. 5. Possible effects of mutations on folding and unfolding rates in a two-state reaction: Φ analysis. Left: Free-energy diagrams for wild-type (w-t) and mutant (mut) species. Right: Corresponding chevron plots. Activation energies ($\Delta G^{\ddagger}$) for folding and unfolding of the wild-type sequence (solid) and mutant (dashed) are indicated by the height of the arrows. A mutation may be expressed kinetically as a decrease in folding rate ($\Phi_f = 1$, top), an increase in unfolding rate ($\Phi_f = 0$, middle), or a combination of both depending ($0 < \Phi_f < 1$, bottom) on whether the mutated residue's interaction is present, absent, or partially formed in the transition state of the rate-limiting step.

vertically (Fig. 5, top). Conversely, if the folding rate is unchanged for a mutation, then the structure is not formed at the site of the substitution in the transition state and the right-side, unfolding arm of the chevron is displaced vertically (Fig. 5, middle).

This strategy, termed Φ analysis,[17,23] can extend to any structural/energetic feature. The extent of the specific interaction present in the transition state can be quantified by the Φ_f parameter, which ranges from 0 (completely absent) to 1 (100% native interaction). For a given mutation, Φ_f is $\Delta\Delta G_f^{\ddagger}/\Delta\Delta G^{\circ}$, the ratio of the change in folding activation energy to the change in stability, and this can be easily obtained from the chevron graph. In this manner, one can characterize the extent a perturbed structural feature is formed at the rate-limiting step.

This interpretation of Φ_f values assumes that there is a one-to-one relationship between the energy and the amount of native interaction present in the transition state. However, a nonnative, but still stabilizing interaction in the transition state (e.g., nonnative AU base pair) can also yield a nonzero Φ_f value. Generally, one cannot distinguish between a native versus nonnative interaction. A Φ analysis applied to a double mutant ("a double-mutant cycle") can partially overcome this problem as it can assay for the amount two adjacent nucleotides interact with each other in the transition state.[23] This interaction is more likely to represent a native interaction because it focuses only on the two adjacent nucleotides.

Likewise, there is ambiguity for fractional Φ_f values (Fig. 5, bottom). These may represent heterogeneous (multiple) transition states having either zero or completely formed interactions. Alternatively, a fractional Φ_f value may result from a relatively homogenous transition state population having only a partially formed interaction. Last, note that Φ analysis assumes that a single mutation does not change the pathway significantly. This generally seems to be valid in protein systems. However, certain destabilizing mutations in an α-helical coiled coil protein do significantly alter the folding pathway causing folding to begin at other regions of the molecule.[23a]

Conclusion

The use of CD in combination with Mg^{2+} or urea titrations can provide a wealth of information about both secondary and tertiary structural transitions in RNAs. Although the structural origin of the CD changes in tertiary transitions is unclear, it can provide a probe to follow these transitions.

[23] A. R. Fersht, A. Matouschek, and L. Serrano, *J. Mol. Biol.* **224,** 771 (1992).

[23a] L. B. Moran, J. P. Schneider, A. Kentsis, G. A. Reddy, and T. R. Sosnick, *Proc. Natl. Acad. Sci. U.S.A.* **96,** 10699 (1999).

With the use of computer-controlled titrators, a rapid amount of data can be taken and accurate values for stability, denaturant m° values, and Hill coefficients can be obtained over a wide variety of conditions. Hence, CD provides a nice complement to site-resolved methods such as complementary oligonucleotide hybridization,[24] hydroxyl radical footprinting,[25] or chemical modification[26] methods as described in previous folding studies of large ribozymes.

Acknowledgments

We thank Prof. Tao Pan without whose collaboration these studies would not have been conducted. This work was supported by grants from the NIH (R01GM57880 to T.R.S. and Tao Pan) and from the Cancer Research Foundation (T.R.S.).

[24] P. P. Zarrinkar and J. R. Williamson, *Science* **265,** 918 (1994).
[25] B. Sclavi, S. Woodson, M. Sullivan, M. R. Chance, and M. Brenowitz, *J. Mol. Biol.* **266,** 144 (1997).
[26] A. R. Banerjee and D. H. Turner, *Biochemistry* **34,** 6504 (1995).

[25] Fluorescence Assays to Study Structure, Dynamics, and Function of RNA and RNA–Ligand Complexes

By NILS G. WALTER and JOHN M. BURKE

Introduction

After absorbing a photon, some molecules, called fluorophores, reradiate energy with a different wavelength than the exciting light; if emission occurs with a delay in the nanosecond time range, this process is called fluorescence (if emission is longer lived, it is referred to as phosphorescence). Fluorescence-based assays have been increasingly used during the past 35 years to study structure–function relationships in biological macromolecules. The emission of fluorophores is highly sensitive to their immediate and, in some cases, distant environment, making them excellent probes to measure local as well as global structures and their changes. A low detection level (typically in the nanomolar concentration range, and under certain conditions, using instrumentation for spatial resolution, even down to a single molecule) and the ability to continuously yield a reporter signal enable sensitive real-time monitoring of dynamic processes in solution. In many cases, such processes are at the heart of understanding the function of biopolymers.

0076-6879/00 $30.00

Proteins carry intrinsic fluorophores in the form of their tryptophan and tyrosine residues. Some, such as photoreceptors or the green fluorescent protein, even contain or bind more efficient chromophores. In addition, proteins often can be labeled site specifically with synthetic extrinsic fluorophores, using side chains that allow a specific coupling chemistry, such as the primary amino group of lysine or the thiol group of cysteine. These options have led to numerous applications of fluorescence in protein analysis, surveyed in many excellent reviews.[1–8] In particular, fluorescence assays have allowed the dissection and evaluation of structural transitions in the reaction pathway of protein enzymes.[9] This type of information has proven difficult to obtain by more traditional biophysical methods, such as nuclear magnetic resonance (NMR) spectroscopy and X-ray crystallography.

Intrinsic fluorescence from the protein component has been used to kinetically and thermodynamically characterize the formation of a number of protein–RNA complexes. Among them are complexes of aminoacyl-tRNA synthetases with their cognate tRNAs,[10–12] of a translational repressor from T4 phage with the repressed mRNA,[13] of both the nucleocapsid protein[14] and the reverse transcriptase[15] from human immunodeficiency virus (HIV) with their natural ligand tRNA(3Lys), or of the Rev protein from HIV with its RNA binding element.[16]

The discovery of ribozymes in the early 1980s has drawn our attention to the nature of RNA as a structurally and functionally dynamic biopolymer that often plays an active role in interactions with its environment. As a result, efforts have increased in recent years to directly monitor RNA structural dynamics and function in solution. The most direct way to obtain

[1] A. Coxon and T. H. Bestor, *Chem. Biol.* **2,** 119 (1995).
[2] D. M. Jameson and W. H. Sawyer, *Methods Enzymol.* **246,** 283 (1995).
[3] P. Selvin, *Methods Enzymol.* **246,** 300 (1995).
[4] A. R. Holzwarth, *Methods Enzymol.* **246,** 334 (1995).
[5] L. Brand, ed., *Methods Enzymol.* **278** (1997).
[6] D. P. Millar, *Curr. Opin. Struct. Biol.* **6,** 637 (1996).
[7] Y. C. Lee, *J. Biochem.* **121,** 818 (1997).
[8] J. H. Lakey and E. M. Raggett, *Curr. Opin. Struct. Biol.* **8,** 119 (1998).
[9] K. A. Johnson, *Curr. Opin. Biotechnol.* **9,** 87 (1998).
[10] M. Baltzinger and E. Holler, *Biochemistry* **21,** 2460 (1982).
[11] M. Fournier, C. Plantard, B. Labouesse, and J. Labouesse, *Biochim. Biophys. Acta* **916,** 350 (1987).
[12] S. X. Lin, Q. Wang, and Y. L. Wang, *Biochemistry* **27,** 6348 (1988).
[13] K. R. Webster and E.K. Spicer, *J. Biol. Chem.* **265,** 19007 (1990).
[14] Y. Mely, H. de Rocquigny, M. Sorinas-Jimeno, G. Keith, B. P. Roques, R. Marquet, and D. Gerard, *J. Biol. Chem.* **270,** 1650 (1995).
[15] S. H. Trall, J. Reinstein, B. M. Wörl, and R. S. Goody, *Biochemistry* **35,** 4609 (1996).
[16] W. C. Lam, J. M. Seifert, F. Amberger, C. Graf, M. Auer, and D. P. Millar, *Biochemistry* **37,** 1800 (1998).

this information is by using fluorophores incorporated into RNA molecules. The goal of the present review is to show examples of how such molecules can be employed to yield unique information on structure, dynamics, and function of RNA and its interactions with ligands, using widely available synthesis strategies and fluorometer instrumentation. To help the reader understand its potentials and demands we first review the basic terminology of fluorescence spectroscopy.

Fluorescence Spectroscopy and Its Potentials

Fluorescence is an interaction of light with matter.[17] Absorption of a photon by a fluorophore causes its transition from a ground to an excited electronic state. The excited state has a certain lifetime. While remaining in the excited state, the fluorophore internally converts part of its original excitation energy into molecular motion (heat). When the fluorophore reverses to its ground state, this energy loss results in the so-called Stokes shift of the emission relative to the absorption wavelength. In principle, a fluorophore can be cycled indefinitely between ground and excited states, producing a continuous fluorescence signal.

Every fluorophore has its unique absorbance and emission spectrum (wavelength distribution). Its emission spectrum is a mirror image of the absorbance, Stokes shifted to longer wavelengths. This shift makes fluorescence detection very sensitive, since emitted light can optically be separated from scattered excitation light. An absorbance spectrum is normally measured as excitation spectrum, with the excitation wavelength varied and the emission wavelength fixed. Conversely, an emission spectrum is obtained with the emission wavelength varied and the excitation wavelength fixed.

In the excited state, the fluorophore is sensitive to its environment. It can lose its excitation energy in a variety of ways, most of them radiationless, resulting in a quenched fluorescence. Quenching can occur through collisional quenching, excited state reactions, static quenching, and energy transfer. In fact, many potential fluorophores, such as the natural nucleobases, are quenched so strongly in solution that their fluorescence normally cannot be observed. Typical fluorophores to be employed for fluorescent assays have a quantum yield (the ratio of the number of photons emitted to the number absorbed) of at least 10%.

Diffusional encounters between a fluorophore and a quencher result in collisional or dynamic quenching. A common dynamic quencher is molecular oxygen, which quenches nearly all known fluorophores. In addition, it

[17] J. R. Lakowicz, "Principles of Fluorescence Spectroscopy." Plenum Press, New York, 1983.

can chemically react with excited fluorophore species, such as electronic triplet states, causing irreversible chemical decay, or photobleaching. Thus, it is frequently necessary to physically or chemically remove dissolved oxygen from a fluorophore containing solution, in particular when high sensitivity is desired.

When the fluorophore and its quencher form a nonfluorescent ground state complex, static quenching results. On absorption of a photon, this complex immediately returns to the ground state without emission of fluorescence.

A special case of quenching is fluorescence resonance energy transfer (FRET; sometimes simply called fluorescence energy transfer). Here, the energy of the excited state is transferred from the donor fluorophore to an acceptor. The transfer occurs without the appearance of a photon, and is the result of direct dipole–dipole interactions between the interacting molecules. The acceptor itself often is a fluorophore, so that its unique emission can be detected in addition to a quenched donor emission. The efficiency of FRET depends on the extent of overlap of the donor emission spectrum with the acceptor absorption spectrum, the relative orientation of the donor and acceptor transition dipoles, and the distance between the two fluorophores. This latter property enables accurate measurement of distances in the range of 10–100 Å by FRET, a distance range well suited to probing RNA molecules.

All of these molecular processes limit the average amount of time, or lifetime, a fluorophore remains in the excited state. Measurements of the lifetime can, therefore, reveal the frequency of collisional encounters with quenching agents, the rate of excited state reactions, and the efficiency of energy transfer. Typical fluorophore lifetimes are in the range of 10 nsec, necessitating the use of high-speed electronic devices and detectors. An advantage of the fast time regime of fluorescence is that it is faster by several orders of magnitude than molecular motions and conformational rearrangements; a fluorescence emission event therefore, like a snapshot, contains information on the pseudostatic molecular environment of the fluorescent probe at a given point in time. As a result, time-resolved fluorescence measurements can yield statistical information on ensembles of fluorophores and their distributions between different states.

In addition to the processes described, apparent quenching can occur due to the optical properties of the sample. For example, reabsorbed fluorescence or turbidity can result in decreased fluorescence intensities. These effects are trivial and contain very little molecular information. They have to be avoided to obtain useful information. An advantage of fluorescence lifetime measurements is their insensitivity to these apparent quenching effects.

On excitation with polarized light, the emission of a fluorophore is also polarized. This polarization or anisotropy is a result of the selection of the absorbing fluorophores according to their orientation relative to the polarization plane of the excitation light (photoselection). If the excited fluorophore were completely immobile over its lifetime, the resulting emission would be completely polarized as well. Since this is normally not the case, the remaining anisotropy reveals the average angular displacement of the fluorophore that occurs between absorption and subsequent emission of a photon. This angular displacement depends on the rate and extent of rotational diffusion during the lifetime of the excited state. Rotational diffusion properties of the excited state, in turn, depend on the viscosity of the solvent, the rotational freedom of the probe itself, and the size and shape of the molecule to which it is attached. The latter property has made anisotropy measurements an indispensable tool to monitor the formation or the structural changes of biological complexes.

Instrumentation

It is important to be aware that an observed signal may not originate from the fluorophore of interest. One may detect background fluorescence from other sample components, light leaks, Raleigh or Raman scattering, and/or ordinary stray light from particles in the sample. To ensure that indeed the fluorophore of interest produces the signal, it is necessary to check some specific features such as the emission and excitation spectra of the sample. To obtain these spectra appropriate equipment has to be used, typically a spectrofluorometer.[17] We will discuss the commercially available instrumentation in some detail to reveal its potentials and to help beginners getting involved in fluorescence-based projects.

Basic Spectrofluorometer

To avoid misinterpretation of the obtained data, some attention has to be paid to the experimental details of the utilized spectrofluorometer. The basic principle of a modern spectrofluorometer is simple (Fig. 1). Light from a source, often a continuous xenon (Xe) arc lamp with a broad wavelength spectrum, is optically focused and guided through lenses, mirrors, a monochromator (typically a diffraction grating with a variable slit to select a wavelength band of defined width), a shutter, and an optional polarizer into the sample. Sample fluorescence perpendicular to the emission beam (to minimize the excitation light contribution) is collected by a lens and guided through another shutter, an optional polarizer, and a monochromator arrangement into a photomultiplier tube (PMT; converts individual photons into an electrical current) as detector.

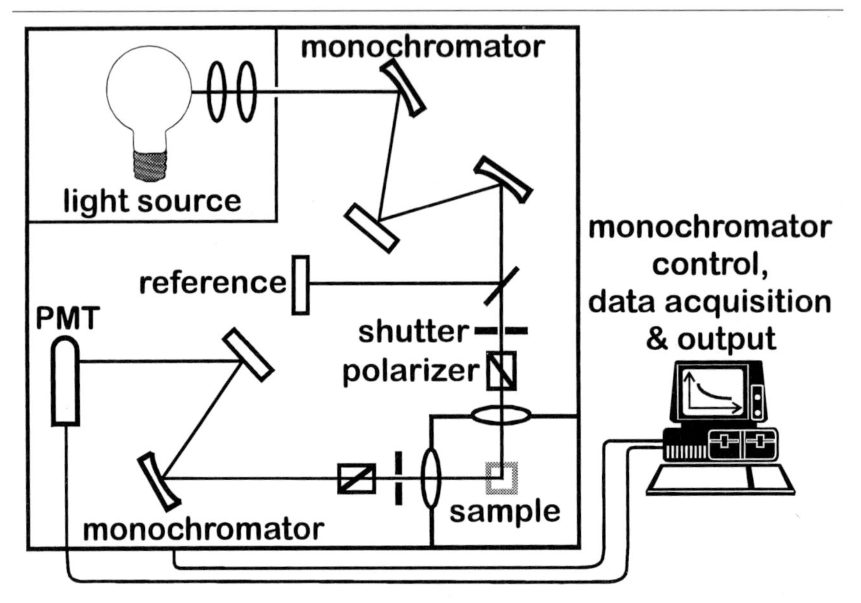

FIG. 1. The basic spectrofluorometer.

To increase the sensitivity of the signal, the wavelength and intensity of the excitation beam or the amplification of the emission signal can be varied. Because these parameters are adjusted from experiment to experiment, no absolute measurement of fluorescence intensity for a given sample is possible, and the assigned units will always be arbitrary. In addition, the observed excitation and emission spectra are dependent on the instrument on which they are recorded. Ideally, a spectrofluorometer should have an equal and constant photon output at all wavelengths, the monochromators should pass photons of all wavelengths and polarizations with equal efficiency, and the PMT should detect photons of all wavelengths with equal efficiency. However, an ideal spectrofluorometer does not exist, and variations of measured fluorescence spectra have to be expected from different instruments. To at least account for fluctuations in the excitation beam intensity, spectrofluorometers typically contain a beam splitter, passing a small fraction of the excitation light into a reference detector (Fig. 1).

Different types of commercial spectrofluorometers are available as either compact benchtop instruments or high-end modular devices that allow incorporation of alternative light sources such as pulsed lasers. In the course of projects it is often necessary to alter or extend the chosen approach, hence, versatile and easily upgradable equipment is desirable. We have performed our steady-state fluorescence experiments on an SLM/Aminco-Bowman series 2 (AB2, Spectronic Instruments, Rochester) spectrofluorometer benchtop instrument.

Optical Accessories

Depending on the desired application, accessories may be required for measuring spectroscopic qualities other than fluorescence intensity. For obtaining fluorescence lifetimes two techniques exist, the pulse method and the harmonic or phase-modulation method. In the pulse method the sample is excited by a very short light pulse, and the time-dependent decay of fluorescence intensity is recorded. Either flash lamps or pulsed lasers serve as light sources. The longer pulses (typically 2 ns) of flash lamps, on the one hand, make the correction (or deconvolution) of the observed fluorescence decay against the nonideal shape of the excitation pulse more challenging and limit the measurable lifetimes to >1 ns. Pulsed lasers with their <100-ps time pulses, on the other hand, are more demanding due to technical and cost considerations, but their versatility has led to an increasing popularity.

To obtain the entire time-resolved decay curve with sufficient accuracy, repetitive flash lamp or laser pulses have to be used. From the pulses, which have to be separated by at least five lifetimes to avoid overlap of their fluorescence responses, the complete decay curve can be constructed in two different ways. In the pulse sampling (or stroboscopic) method the gain of the PMT is briefly increased at a certain delay time after the excitation, yielding the sample fluorescence at that delay time. The fluorescence decay is obtained by sampling multiple excitation pulses, each for a different delay time. In the more widely used photon-counting method the detection system measures the time between the pulse and the arrival of the first photon. If the count rate is low enough to ensure the arrival of only a single photon per pulse, observing multiple pulses will, in total, reflect the time-resolved decay curve.

To measure lifetimes by the phase-modulation method the sample is excited with sinusoidally modulated light, and the phase shift and demodulation of the emission are used for calculating the lifetime. Light modulation is typically achieved by an ultrasonic modulator, and detection employs a cross-correlation mode. Because the lifetime is calculated indirectly from phase delay and demodulation factors, higher than first-order fluorescence decays will be averaged and both extremely short and long lifetimes can only be measured with considerable error, limiting the number of potential applications of the phase-modulation technique somewhat.

Fluorescence anisotropy measures the emission polarization from a fluorescent sample on excitation with fully polarized light, by calculating

$$A = (I_{vv} - gI_{vh})/(I_{vv} + 2gI_{vh})$$
$$g = I_{hv}/I_{hh} \quad (1)$$

where I_{vv}, I_{vh}, I_{hv}, and I_{hh} are the fluorescence intensities measured with excitation and emission polarizers subsequently in all four possible combinations of vertical (v, 0°) or horizontal (h, 90°) alignment. To measure anisotropy, a relatively simple accessory is needed, namely, polarizers in both the excitation and emission pathways. Plastic film polarizers are often included in the basic setup of a spectrofluorometer, but they might not be appropriate for a given fluorophore because they absorb UV light. Other polarizers are made of UV-transparent prisms.

In principle, two techniques are in use for measuring fluorescence anisotropies, the L- and the T-format methods. In the L format, a single emission channel is used, and the emission polarizer is shifted between parallel and vertical positions with respect to the excitation polarization. In the T format, both positions are measured simultaneously using two separate detection systems, branching off in opposite perpendicular directions from the excitation beam. In both the L and T formats, the correction factor g [Eq. (1)] has to be calculated to compensate for the polarizing properties of the optical components, especially the monochromators.

The polarization properties of the optical components can also pose a problem for measuring accurate fluorescence intensities and lifetimes; if the anisotropy of the sample is changing, the observed fluorescence signal may change solely due to the dependence of the monochromators on the light polarization. Polarizers can be used to eliminate this polarization effect. Mathematically, it can be shown that the measured intensity becomes independent of the sample polarization if polarizers in the magic angle orientation are employed (with the excitation polarizer in the vertical position and the emission polarizer oriented 54.7° from the vertical).

Accessories for Kinetic Measurements

One particular advantage of fluorescence over alternative biophysical methods is its inherent capacity to produce continuous information on a reaction, revealing its kinetics. This "time resolution" must not be confused with the observation of a time-resolved fluorescence decay that describes the loss of excitation energy by fluorescence on a nanosecond timescale.

To be able to observe its kinetics, a reaction is typically initiated by mixing two reacting species. Manual mixing takes several seconds to ensure homogeneity, which might not be appropriate to observe a fast reaction. To overcome this problem, stopped-flow accessories have been developed for many spectrofluorometers. They contain at least two pressure-resistant syringes for the reagenets. Driven by a triggered hydraulic or pneumatic pressure system they inject their reagent solutions into a mixing chamber, from where the homogenized mixture is guided into the observation cell. The reagent flow stops abruptly when the exhaust syringe hits a physical barrier, and signal acquisition begins. The dead time before a fluorescence signal can be recorded is typically in the millisecond time range.

Traditionally, rapid stopped-flow equipment allows for continuous recording at a single defined emission wavelength only. Full excitation and emission spectra can only be taken after completion of the run. A more recent advancement of the stopped-flow technique is time-resolved spectrofluorometry.[18–21] Here, rapid emission wavelength scanning is accomplished by a dispersing element in the emission pathway, such as a monochromator with a spinning multiple-slit disk, a spinning grating, or an acousto-optic tunable filter, and the dispersed emission light is analyzed either with a PMT or a diode array. Depending on the setup, successive emission spectra can be recorded with high time resolution. These multiple-wavelength data potentially provide more information on the fluorescent intermediates of a reaction than a single-wavelength measurement does. Stopped-flow equipment can also be combined with the optical accessories described earlier to measure fluorescence lifetimes or anisotropies.

A different principle to analyze fast reaction kinetics is the relaxation method.[22] Here, a mixture of reagents in equilibrium with their reaction products is subjected to a sudden jump in a thermodynamic parameter influencing the equilibrium position, e.g., temperature or pressure. From the relaxation time back to the original equilibrium position the kinetics of the underlying reaction can be derived. If the reagents differ in their fluorescence properties from the products, relaxation can be followed by fluorescence measurements to obtain kinetic information.[23,24]

Instrumentation Yielding Spatial Resolution

Like most other biophysical methods, a typical measurement in a spectrofluorometer averages signals from all detectable molecules in an analyzed bulk volume. Fluorescence can, however, also yield detailed information about the spatial resolution of fluorescence from a sample, even down to the single-molecule level. Most prominent among the available techniques are the flow cytometer, where a laser excites a flowing fluid strem[25,26]; the fluorescence microscope to spatially resolve a distribution of fluorescing

[18] P. S. Brzovic and M. F. Dunn, *Methods Biochem. Anal.* **37,** 191 (1994).
[19] M. R. Eftink and M. C. R. Shastry, *Methods Enzymol.* **278,** 258 (1997).
[20] C. D. Tran and R. J. Furlan, *Anal. Chem.* **65,** 1675 (1993).
[21] T. Hartmann and A. S. Verkman, *Anal. Biochem.* **200,** 139 (1992).
[22] M. Eigen, *Q. Rev. Biophys.* **1,** 3 (1968).
[23] S. M. Coutts, D. Riesner, R. Romer, C. R. Rabl, and G. Maass, *Biophys. Chem.* **3,** 275 (1975).
[24] D. Labuda and D. Porschke, *Biochemistry* **19,** 3799 (1980).
[25] B. H. Villas, *Cell Vis.* **5,** 56 (1998).
[26] M. C. Roslaniec, C. S. Bell-Prince, H. A. Crissman, J. J. Fawcett, P. M. Goodwin, R. Habbersett, J. H. Jett, R. A. Keller, J. C. Martin, B. L. Marrone, J. P. Nolan, M. S. Park, B. L. Sailer, L. A. Sklar, J. A. Steinkamp, and L. S. Cram, *Hum. Cell* **10,** 3 (1997).

RNA molecules in live cells[27–31a]; and fluorescence correlation spectroscopy (FCS), where temporal autocorrelation of their fluorescence bursts yields information on the translational diffusion properties of individual fluorescent molecules traversing the sharp focus of a laser coupled to a confocal microscope.[32–35] Notably, spatial resolution of fluorescence signals is also a core technology in the rapidly developing fields of high-throughput screening for drug discovery[36,37] and chip-based hybridization assays for genome and expression pattern analysis.[38,39] Again, all of these techniques may be coupled with the optical accessories described earlier to yield further information, e.g., on fluorescence lifetimes or anisotropies.

Fluorescent RNA Derivatives

The most abundant bases in RNA are guanine, adenosine, uracil, and cytosine. These natural bases, however, do not fluoresce due to strong quenching in solution and cannot be used for fluorescence assays (except at biologically irrelevant, extremely low temperatures[40]). This lack in autofluorescence enables the background-free use of site-specifically incorporated fluorophores as probes for their local environment. RNA from natural sources sometimes carries suitable intrinsic fluorophores, in particular the

[27] X. F. Wang and B. Herman, "Fluorescence Imaging Spectroscopy and Microscopy." Wiley, New York, 1996.

[28] J. S. Ploem, *in* "Fluorescent and Luminescent Probes for Biological Activity," (W. T. Mason, ed.), p. 1. Academic Press, London, 1993.

[29] F. W. D. Rost, "Fluorescence Microscopy." Cambridge University Press, Cambridge, Massachusetts, 1992.

[30] F. W. D. Rost, "Quantitative Fluorescence Microscopy." Cambridge University Press, Massachusetts, 1991.

[31] D. Lansing Taylor, and E. D. Salmon, *Methods Cell. Biol.* **29,** 207 (1989).

[31a] P. Chartrand, E. Bertrand, R. H. Singer, and R. M. Long, *Methods Enzymol.* **318,** [33] (2000).

[32] N. L. Thompson, *in* "Topics in Fluorescence Spectroscopy," (J. R. Lakowitz, ed.), Vol. 1, p. 337. Plenum Press, New York, 1991.

[33] R. Rigler, *J. Biotechnol.* **41,** 177 (1995).

[34] P. Schwille, F. Oehlenschläger, and N. G. Walter, *Biochemistry* **35,** 10182.

[35] N. G. Walter, P. Schwille, and M. Eigen, *Proc. Natl. Acad. Sci. U.S.A.* **93,** 12805 (1996).

[36] L. Silverman, R. Campbell, and J. R. Broach, *Curr. Opin. Chem. Biol.* **2,** 397 (1998).

[37] J. G. Houston and M. Banks, *Curr. Opin. Biotechnol.* **8,** 734 (1997).

[38] M. Chee, R. Yang, E. Hubbell, A. Berno, X. C. Huang, D. Stern, J. Winkler, D. J. Lockhart, M. S. Morris, and S. P. Fodor, *Science* **274,** 610 (1996).

[39] L. Wodicka, H. Dong, M. Mittmann, M. H. Ho, and D. J. Lockhart, *Nat. Biotechnol.* **15,** 1359 (1997).

[40] V. Kleinwächter, J. Drobnik, and L. Augenstein, *Photochem. Photobiol.* **7,** 485 (1968).

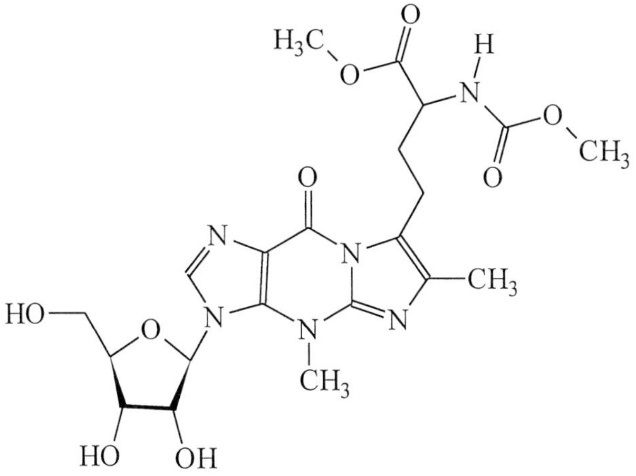

Wye nucleoside

Ex_{max} = 335 nm; Em_{max} = 443 nm

FIG. 2. Structure and fluorescence properties of the Wye base. Ex_{max}, excitation maximum; Em_{max}, emission maximum.

Wye (or Y) base (Fig. 2). This strongly modified guanine was extensively exploited in the 1980s to characterize structure, dynamics, and functional interactions of the anticodon loop of $tRNA^{Phe}$, using fluorescence intensity, anisotropy, and lifetime measurements of its Wye base.[41–46] However, the necessity to obtain the labeled RNA from natural sources and the fact that neither the nature of the fluorophore nor its attachment site can be manipulated have severely restricted applications of the Wye base.

With the invention of *in vitro* transcription using cloned RNA polymerases on defined DNA templates[47,48] and with the advancement of chemical

[41] N. Okabe and F. Cramer, *J. Biochem.* **89,** 1439 (1981).

[42] H. Paulsen, J. M. Robertson, and W. Wintermeyer, *J. Mol. Biol.* **167,** 411 (1983).

[43] W. Bujalowski, E. Graeser, L. W. McLaughlin, and D. Porschke, *Biochemistry* **25,** 6365 (1986).

[44] F. Claesens and R. Rigler, *Eur. Biophys. J.* **13,** 331 (1986).

[45] I. Gryczynski, H. Cherek, and J. R. Lakowicz, *Biophys. Chem.* **30,** 271 (1988).

[46] G. Striker, D. Labuda, and M. C. Vega-Martin, *J. Biomol. Struct. Dyn.* **7,** 235 (1989).

[47] J. F. Milligan and O. C. Uhlenbeck, *Methods Enzymol.* **180,** 51 (1989).

[48] S. van der Werf, J. Bradley, E. Wimmer, F. W. Studier, and J. J. Dunn, *Proc. Natl. Acad. Sci. U.S.A.* **83,** 2330 (1986).

solid-phase RNA synthesis,[49–52] two approaches became available in the 1980s to intently introduce fluorescent probes into a desired RNA. Subsequently, many covalent labeling strategies were developed to attach a variety of fluorophores.[53–56] We focus on fluorophores that are commercially available for site-specific attachment to chemically synthesized RNA. Although the length of chemically synthesized RNA at present is somewhat restricted due to limiting coupling efficiencies, the incorporation of a variety of additional nonnatural and modified nucleotides is possible, allowing for very flexible experimental strategies.

Labeling with Fluorophores during Chemical Synthesis

Chemical solid-phase synthesis of oligoribonucleotides has become a standard procedure in many molecular biology laboratories. Successful large-scale synthesis of RNA as long as 52 nucleotides has been described.[57] The chain is built from 3′ to 5′ end, starting from a solid-phase support, by repetitive cycles of condensation of 3′-activated and 5′-protected monomers with the growing chain. A few commercial suppliers exist, but our experience is that assembling the RNA oneself on an automated DNA/RNA synthesizer using commercially available β-cyanoethyl phosphoramidite activation chemistry is still the most reliable and versatile approach. Companies such as Glen Research (Sterling, VA), Applied Biosystems (Foster, CA), Pharmacia Biotech (Piscataway, NJ), Clontech (Palo Alto, CA), or ChemGenes (Waltham, MA) supply the required reaction chemistry together with phosphoramidites of unmodified and modified ribonucleotides and a variety of RNA modifiers including fluorescent probes. Automated DNA/RNA synthesizers are available from, e.g., Applied Biosystems, PerSeptive Biosystems (Framingham, MA), or Pharmacia Biotech. Our laboratory has had good experience with Glen Research chemistry on an Applied Biosystems 392 DNA/RNA synthesizer.

[49] K. K. Ogilvie, N. Usman, K. Nicoghosian, and R. J. Cedergren, *Proc. Natl. Acad. Sci. U.S.A.* **85,** 5764 (1988).

[50] F. Eckstein, ed., "Oligonucleotides and Analogues: A Practical Approach." Oxford University Press, United Kingdom, 1991.

[51] B. S. Sproat, *Curr. Opin. Biotechnol.* **4,** 20 (1993).

[52] R. H. Davis, *Curr. Opin. Biotechnol.* **6,** 213 (1995).

[53] A. Waggoner, *Methods Enzymol.* **246,** 362 (1995).

[54] C. Kessler, *J. Biotechnol.* **35,** 165 (1994).

[55] J. Temsamani and S. Agrawal, *Mol. Biotechnol.* **5,** 223 (1996).

[56] R. P. Haugland, "Handbook of Fluorescent Probes and Research Chemicals," (M. T. Z. Spence, ed.), 6th Ed. Molecular Probes, Eugene, Oregon, 1996.

[57] B. Sproat, F. Colonna, B. Mullah, D. Tsou, A. Andrus, A. Hampel, and R. Vinayak, *Nucleosides Nucleotides* **14,** 255 (1995).

Fluorophores that can be attached during synthesis need to be unreactive to coupling and deprotection chemistry. The currently most prominent fluorescent probes from Glen Research for attachment to RNA during synthesis are shown in Fig. 3, together with their basic fluorescence properties.

Fluorescein can be coupled to an RNA (1) 5′ terminally, in form of the chain-terminating 5′-fluorescein phosphoramidite; (2) internally, using either the fluorescein phosphoramidite with a removable 5′-DMT (dimethoxytrityl) protection group (creating an abasic site after coupling the next phosphoramidite) or fluorescein-deoxythymidine (dT) phosphoramidite (if a 2′-deoxythymidine can be tolerated); and (3) 3′ terminally, employing the fluorescein CPG solid-phase support to prime the synthesis (Fig. 3). In

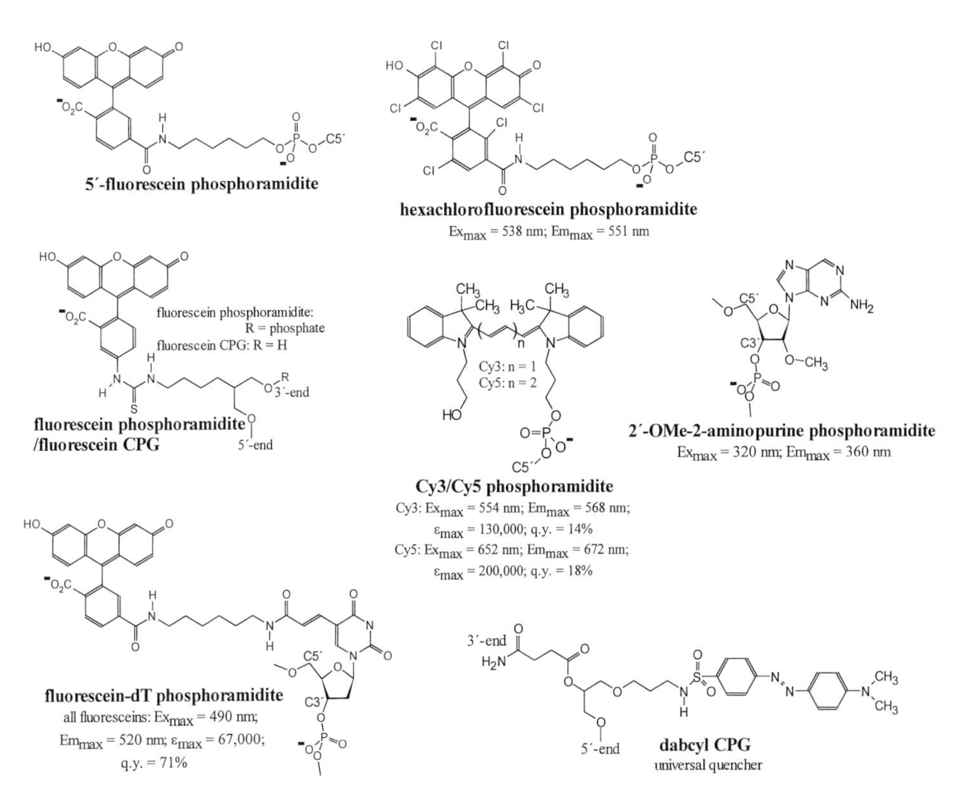

FIG. 3. Fluorophores for labeling during chemical RNA synthesis. The deprotected structures are shown, together with the names of their corresponding synthesis reagents. Typical fluorescence properties (may change after incorporation into RNA): Ex_{max}, excitation maximum; Em_{max}, emission maximum; ε_{max}, extinction coefficient at Ex_{max} [liter/(mol cm)]; q.y., quantum yield.

protein analysis, fluorescein has been the predominant green fluorophore for decades, due to its relatively high absorbance, near-optimal match to the 488-nm spectral line of the argon-ion laser, excellent fluorescence quantum yield, and good water solubility.[56] However, fluorescein is protonated below its pK_a 6.4, yielding a nonfluorescent acid form, and its relatively high photobleaching rate makes careful removal of dissolved oxygen from the reaction solution necessary to obtain a high detection sensitivity. A closely related dye, hexachlorofluorescein, is available as a phosphoramidite for 5′ end labeling of RNA (Fig. 3). Its six chlorine substituents shift its absorbance relative to fluorescein to longer wavelengths, enabling fluorescence resonance energy transfer (FRET) from the latter. We found hexachlorofluorescein to be sensitive against urea-induced hydrolysis so that oligoribonucleotides containing hexachlorofluorescein must not be purified on urea containing gels (see later discussion).

A different set of dyes to be utilized in RNA synthesis is the phosphoramidites of the cyanine derivatives Cy3 and Cy5 (Fig. 3). Even though they contain a removable MMT (4-monomethoxytrityl) protection group, Cy3 and Cy5 are not stable against repetitive synthesis cycles. They should be added at the 5′ terminus and the MMT group removed on the synthesizer. Due to its red-shifted absorbance spectrum, Cy3 is suitable as an acceptor for energy transfer from fluorescein.

2-Aminopurine (2AP; Fig. 3) is a fluorescent base analog of adenine. It base pairs with uracil in a structure isomorphous with an A-U Watson–Crick base pair. Currently, it is commercially available as the 2′-*O*-methyl-or 2′-deoxy-substituted phosphoramidite so that it can be internally incorporated only into RNA sites where modification of the 2′-hydroxyl group is tolerated. Several laboratories have worked out procedures to synthesize the ribose form of the phosphoramidite for studies in RNA.[58,59] 2-Aminopurine has the unique advantage that it directly reports on structural changes, especially in the base stacking pattern, around a single base, while most other fluorophores probe their environment in the minor or major groove of the nucleic acid double helix, depending on their attachment site.[60] In a DNA duplex, analysis of its fluorescence decay has shown that 2-aminopurine can be resolved into four differentially stacked species.[61]

One approach to analyze changes in nucleic acid structure is to observe

[58] M. Menger, T. Tuschl, F. Eckstein, and D. Porschke, *Biochemistry* **35,** 14710 (1996).

[59] B. B. Konforti, D. L. Abramovitz, C. M. Duarte, A. Karpeisky, L. Beigelman, and A. M. Pyle, *Mol. Cell* **1,** 433 (1998).

[60] D. P. Millar, *Curr. Opin. Struct. Biol.* **6,** 322 (1996).

[61] R. A. Hochstrasser, T. E. Carver, L. C. Sowers, and D. P. Millar, *Biochemistry* **33,** 11971 (1994).

an accompanying change in quenching of a site-specifically attached fluorophore. Altered fluorophore quenching may be brought about by a change in distance to a quencher and the probability of their collisional encounters. A site-specifically incorporated universal quencher that has been used for detection of DNA hybridization is dabcyl.[62] The 3′ terminus of a nucleic acid can be modified using the 3′-Dabcyl CPG solid-phase synthesis support from Glen Research (Fig. 3).

Postsynthetic Labeling with Fluorophores

Postsynthetic labeling with a site-specific fluorophore requires incorporation of a modification during synthesis that offers a unique postsynthetic coupling chemistry.[53] In use are primary alkylamino and alkylthiol modifications with different linker lengths. Site of modification can be the 5′ terminus, the 3′ terminus, and an internal position of the RNA, if the 5′ modifiers, 3′ modifiers, and a base-modified 2′-deoxythymidine are employed, respectively (Fig. 4A). The listed modifiers are stable against synthesis and deprotection chemistry. In the future, the development of additional labeling strategies can be expected, e.g., exploiting the 2′-hydroxyl groups and 3′,5′-phosphodiester linkages of RNA to attach fluorophores.

After synthesis and deprotection, primary amino groups can be specifically reacted with succinimidyl ester, isothiocyanate, or sulfonyl chloride groups on the fluorophore (Fig. 4B). Alkylthiol modifications on the RNA may be labeled using maleimide or iodoacetamide groups on the fluorophore, or in a disulfide exchange reaction (Fig. 4B). Many differnt reactive fluorophore derivatives, including appropriate labeling protocols for oligonucleotides, are available from Molecular Probes (Eugene, OR).[56] Typical labeling reactions are carried out over several hours under mild conditions at room temperature. A pH between 8.5 and 9.5 is required for coupling to alkylamino groups, since they need to be unprotonated; alkylthiols react in the range of pH 6.5–8.0. The absence of other (especially buffer) components with primary amino and thiol groups in the coupling reaction is essential to ensure specific attachment to the modified RNA. Incorporation of both an amino and a thiol group at different sites of the RNA enables two different fluorophores to be attached in separate labeling reactions, e.g., to observe FRET between them.

Two examples of reactive fluorophore derivatives to be attached postsynthetically are shown in Fig. 4C: the succinimidyl esters of pyrene and tetramethylrhodamine. Pyrene has an exceptionally long lifetime (up to

[62] S. Tyagi and F. R. Kramer, *Nature Biotechnol.* **14,** 303 (1996).

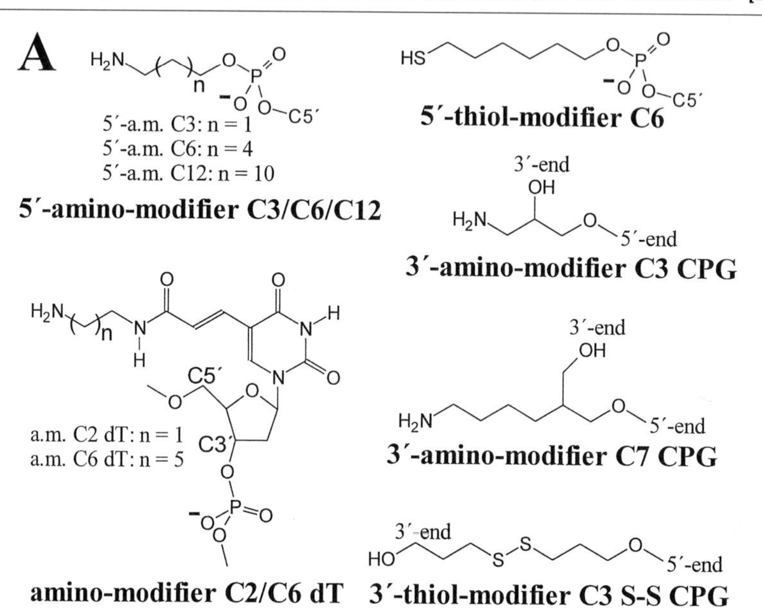

FIG. 4. Postsynthetic labeling of RNA with fluorophores. (A) Available alkylamino and alkylthiol modifiers. The deprotected structures are shown, together with the names of their corresponding synthesis reagents. (B) Labeling chemistry for alkylamino and alkylthiol modifications, respectively. (C) Exemplary fluorophores for postsynthetic labeling of synthetic RNA. Typical fluorescence properties (may change after coupling to RNA): Ex_{max}, excitation maximum; Em_{max}, emission maximum; ε_{max}, extinction coefficient at Ex_{max} [liter/(mol cm)]; q.y., quantum yield.

>100 ns) so that quenchers in its molecular environment have a particularly strong influence on the observed fluorescence. Tetramethylrhodamine is readily excited by the spectral lines of mercury-arc lamps and He–Ne lasers and is intrinsically more photostable than fluorescein.[56] It is a well-suited and widespread acceptor for FRET from fluorescein. Its conjugates with DNA have been shown to be able to populate multiple spectroscopic states, at least one of which is nonfluorescent.[63] A variety of closely related dyes with modified spectroscopic properties are available for oligonucleotide

[63] G. Vámosi, C. Gohlke, and R. M. Clegg, *Biophys. J.* **71,** 972 (1996).

B

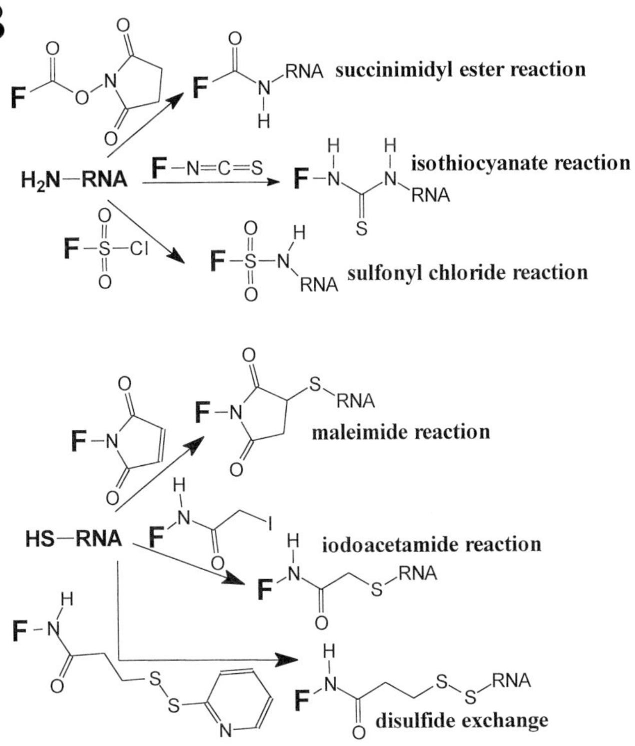
succinimidyl ester reaction
isothiocyanate reaction
sulfonyl chloride reaction
maleimide reaction
iodoacetamide reaction
disulfide exchange
H2N—RNA
HS—RNA

C

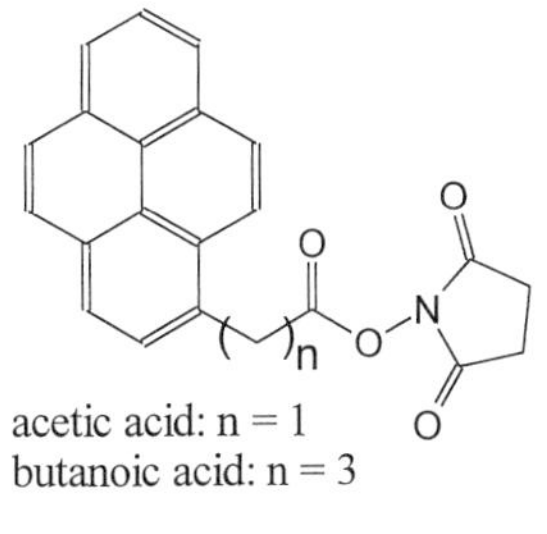
acetic acid: n = 1
butanoic acid: n = 3

1-pyreneacetic(/butanoic) acid succinimidyl ester

Ex_{max} = 340 nm; Em_{max} = 385 nm; ε_{max} = 29,000

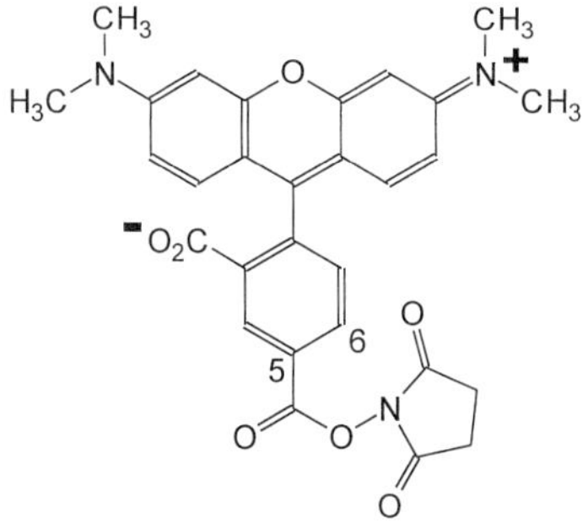

5-carboxy-tetramethylrhodamine succinimidyl ester

Ex_{max} = 554 nm; Em_{max} = 573 nm; ε_{max} = 85,000; q.y. = 28%

FIG. 4. (*continued*)

labeling from Molecular Probes (Eugene, OR), under names such as X-rhodamine, Texas Red, Rhodamine Red, and Rhodamine Green.[56]

Deprotection and Purification of Oligoribonucleotides

After synthesis, oligoribonucleotides need to be deprotected and purified. If they already contain fluorophores, these steps may require special precautions. We use a mild deprotection procedure that appears to be compatible with all fluorophores tested so far. It comprises (1) incubation for 4 hr at 65° in 1 ml of a 3:1 mixture of concentrated aqueous ammonia and ethanol (to remove the exocyclic amine protection groups), (2) drying in a lyophilizer, and (3) 20 hr of tumbling at room temperature with 800 μl triethylamine trihydrofluoride (to remove the 2′-OH-silyl protection groups).[64] Full-length RNA is recovered by precipitation with 1-butanol, drying, and purification on a denaturing (8 *M* urea) 20% polyacrylamide gel. For RNA containing hexachlorofluorescein, urea has to be omitted from the gel (see earlier discussion). Subsequent C_8 reversed-phase high-performance liquid chromatography (HPLC) in 100 m*M* triethylammonium acetate, with a linear elution gradient of 0–40% and 0–60% acetonitrile (50 min, 1 ml/min) for unlabaled and labeled strands, respectively, serves to remove material that is not fully deprotected. Under these conditions, fluorophore labeled RNA is considerably retarded relative to unlabeled RNA, due to its increased hydrophobicity. The elution gradients are chosen to elute the oligoribonucleotides between 14 and 20 min. If desired, fluorophores are coupled to alkylamino and alkylthiol groups at this stage. Subsequently, the labeled RNA is recovered by ethanol precipitation and several washes with 80% ethanol to remove residual reactive fluorophore, and is repurified by C_8-reversed-phase HPLC. One OD_{260} of the dried and resuspended HPLC peak fractions is assumed to correspond to a concentration of 37 μg/ml RNA. To obtain an accurate concentration for labeled RNA, the additional absorbances of the fluorophores at 260 nm should be taken into account with, e.g., $A_{260}/A_{492} = 0.3$ for fluorescein, $A_{260}/A_{535} = 0.3$ for hexachlorofluorescein,[65] and $A_{260}/A_{554} = 0.49$ for tetramethylrhodamine.[34] Alternatively, the RNA concentration can be calculated from the absorbance of the dye, e.g., using an extinction coefficient of 29,000 liter/(mol cm) for pyrene attached to pyrimidine bases.[66]

[64] B. Sproat, F. Colonna, B. Mullah, D. Tsou, A. Andrus, A Hampel, and R. Vinayak, *Nucleosides Nucleotides,* **14,** 255 (1995).

[65] K. P. Bjornson, M. Amaratunga, K. J. H. Moore, and T. M. Lohmann, *Biochemistry* **33,** 14306 (1994).

[66] R. Kierzek, Y. Li, D. H. Turner, and P. C. Bevilacqua, *J. Am. Chem. Soc.* **115,** 4985 (1993).

Examples of Fluorescence Assays for RNA

RNA Design

To initiate a project involving fluorescent RNA, the labeled strand has to be carefully designed. Site of attachment, linker length, and choice of fluorophore(s) have a strong influence on the observed fluorescence, enabling the rational design of probes for observing specific interactions in RNA. For example, it is well established that fluorescein and tetramethylrhodamine become specifically quenched by guanine bases in their local environment,[67,68] while pyrene is primarily quenched by pyrimidines.[69] Quenching is mediated by photo-induced electron transfer between the excited fluorophore and the base and largely depends on the frequency of their collisional encounters. Consequently, the length of the attachment linker and the hydrophobicity of the fluorophore as well as the structural flexibility of the base (that is altered on base pairing) determine the observed fluorescence signal.[70] These properties can be used to place a fluorophore at a site where it is likely to undergo a fluorescence change on a specific RNA structural change or ligand binding event of interest. Alternatively, an acceptor fluorophore or a quencher molecule can be site specifically attached to the RNA (or its ligand) to observe a similar effect. In this case, or if fluorescence anisotropy is measured to obtain information on the rotational diffusion properties of a labeled RNA, base-mediated quenching effects rather should be avoided to simplify data analysis.

General Considerations for Data Interpretation

Fluorescence measurements by necessity only yield relative values. Experiments to observe a specific RNA interaction, therefore, need to allow comparison of at least two signals, e.g., a change in fluorescence intensity, lifetime, or anisotropy upon addition or change of an essential cofactor (such as other RNA strands, ligands, or metal ions) has to be observed.

Once such a fluorescence signal change is observed, control experiments have to be performed to establish its origin, e.g., apparent quenching effects by stray or scatter light have to be ruled out. An elegant way to do so is to perform a titration with the added cofactor and record concentration dependent signal changes. The detection system parameters (including excitation and emission wavelengths) should be optimized for the highest possi-

[67] N. G. Walter and J. M. Burke, *RNA* **3,** 392 (1997).

[68] J. Widengren, J. Dapprich, and R. Rigler, *Chem. Phys.* **216,** 417 (1997).

[69] M. Manoharan, K. L. Tivel, M. Zhao, K. Nafisi, and T. L. Netzel, *J. Phys. Chem.* **99,** 17461 (1995).

[70] J. B. Randolph and A. S. Waggoner, *Nucleic Acids Res.* **25,** 2923 (1997).

ble signal quality. RNA concentrations need to be balanced between low reagent consumption and good signal quality. Ideally, several different fluorescence properties such as intensity, lifetime, and anisotropy should be measured to ensure correct interpretation of the data. Examples of successful applications of fluorescence spectroscopy in RNA biochemistry follow.

Monitoring Secondary Structure Formation by Fluorescence Quenching

Intriguing systems to study structure, dynamics, and function of RNA are ribozymes. Through their catalytic function, which involves dynamic folding transitions, they directly report on the presence of a biologically active structure. As a model system, we have studied the RNA folding pathways of the hairpin ribozyme, a reversible endonucleolytic motif from the negative strand of tobacco ringspot virus satellite and related viroid RNAs associated with plant viruses.[71–73] For biochemical studies *in vitro,* the satellite RNA has been truncated to about a 50-nucleotide ribozyme component that binds and cleaves a 14-nucleotide substrate *in trans.*[73] This construct, containing a two-way junction, has been used for targeted RNA inactivation within mammalian cells, and is the basis for experimental strategies in human gene therapy of genetic and viral diseases.[72]

To study reactions of the hairpin ribozyme, we developed a set of assays based on quenching of 3′-fluorescein-labeled substrates by a guanosine on the 5′ end of the substrate-binding strand of the ribozyme (Fig. 5A).[67] Addition of a ribozyme excess to the labeled substrate results in a decrease in steady-state fluorescence and an increase in anisotropy. We used the fluorescence quenching effect to monitor in real time the formation of the ribozyme–substrate complex, and to deduce the second-order substrate binding rate constant from a plot of the observed pseudo-first-order rate constants over a range of ribozyme concentrations.[67] Upon cleavage, the short 5′ and 3′ products rapidly dissociate so that the observed fluorescence increase reflects the cleavage rate constant. If a noncleavable substrate analog is used, its dissociation can be directly observed as a fluorescence increase after addition of a large excess of unlabeled noncleavable substrate analog as chase. From the ratio of the dissociation and binding rate constants, the equilibrium dissociation constant can be calculated to yield information on the thermodynamic stability of the ribozyme–substrate complex.[67,74]

[71] J. M. Burke, S. E. Butcher, and B. Sargueil, *Nucleic Acids Mol. Biol.* **10,** 129 (1996).
[72] D. J. Earnshaw and M. J. Gait, *Antisense Nucleic Drug Dev.* **7,** 403 (1997).
[73] N. G. Walter and J. M. Burke, *Curr. Opin. Chem. Biol.* **2,** 24 (1998).
[74] N. G. Walter, E. Albinson, and J. M. Burke, *Nucleic Acids Symp. Ser.* **36,** 175 (1997).

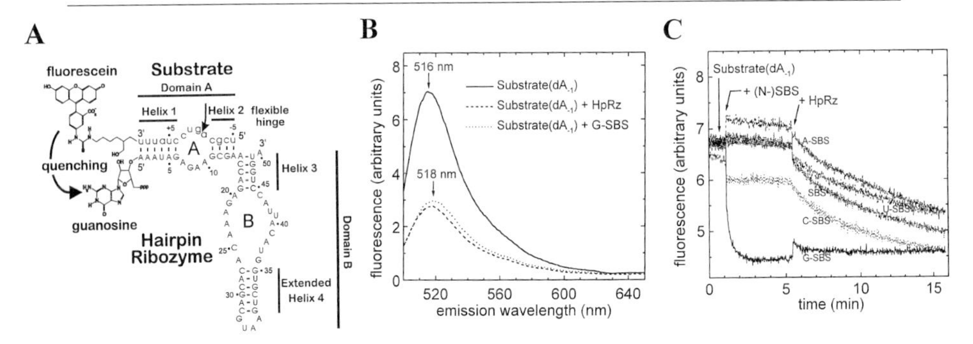

FIG. 5. Quenching assays to monitor secondary structure formation of the hairpin ribozyme–substrate complex. (A) The ribozyme–substrate complex (arrow, cleavage site in loop A). To observe fluorescence quenching on complex formation, fluorescein is coupled to the 3′ end of the substrate (small letters) so that it is located close to a dangling 5′-terminal G of the ribozyme (capital letters). (B) Steady-state fluorescence emission spectra of 10 n*M* fluorescein-labeled, noncleavable substrate analog (dA_{-1} modified) in standard reaction buffer (50 m*M* Tris-HCl, pH 7.5, 12 m*M* $MgCl_2$) at 25°, before (solid line) and after addition of a 10-fold excess of hairpin ribozyme (HpRz, dashed line) and substrate-binding strand with 5′ G (G-SBS, dotted line), respectively. Excitation was at 490 nm. After addition of HpRz or G-SBS, quenched steady-state fluorescence is observed, and the emission peak maximum is shifted slightly from 516 to 518 nm. (C) Base-specific quenching of fluorescein by guanosine. Fluorescence emission of 1 n*M* fluorescein-labeled, noncleavable substrate analog (dA_{-1} modified) under standard conditions was followed over time. After preincubation, a 10-fold excess was added of a substrate-binding strand either without (SBS) or with one of the four natural nucleosides at the 5′ end (G-SBS, U-SBS, C-SBS, A-SBS). Only G-SBS induces a quenching effect, resulting in an exponential signal decay on binding. Addition of all other substrate-binding strands leads to slightly altered, but stable fluorescence signals, due to changed scattering. After several minutes, addition of 10 n*M* hairpin ribozyme (HpRz) followed, resulting in a gradual displacement of the substrate-binding strands in the complexes by ribozyme. (From Walter and Burke.[67])

The about 55% decrease in steady-state fluorescence is also observed when an excess of isolated substrate-binding strand with a 5′-G is added to the 3′-fluorescein-labeled substrate (Fig. 5B). We were able to prove that the quenching is specifically mediated by the 5′-guanosine, because no other base on the substrate-binding strand exerts a similar effect (Fig. 5C). Figure 5C also demonstrates that the raw fluorescence signal may exhibit slight alterations that cannot be correlated with the structural change under investigation. Only careful control experiments (involving, e.g., addition or change of a cofactor) will prevent misinterpretation of the data.

Similar quenching assays have been utilized to monitor the formation and dissociation of a complex between a multiple-component ribozyme derived from a group II intron and its 5′-fluorescein-labeled substrate.[75] Turner and co-workers used quenching of 5′-pyrene-labeled substrates to study the mechanism of substrate binding to the group I intron ribozyme from *Tetrahymena*.[66,76–78] From the biphasic binding kinetics at high substrate concentration, they were able to identify an open complex as a folding intermediate on the pathway to the fully tertiary structured closed complex. Finally, in the hammerhead ribozyme, quenching of site-specifically incorporated 2-aminopurine has been employed to observe ribozyme–substrate complex formation and subtle structural changes induced by local metal ion binding.[79]

Monitoring Tertiary Structure Formation by FRET

Fluorescence resonance energy transfer (FRET) has been used as a "molecular ruler" for biopolymers to estimate, under physiological (and other) conditions, distances in the range of 10–100 Å between a donor and a matching acceptor fluorophore.[3,5,60,80–84] The efficiency E of energy transfer is defined as the fraction of donor molecules de-excited through energy transfer to be acceptor, and can be calculated from the donor fluorescence intensities in the presence (I_{DA}) and absence (I_D) of the acceptor

$$E = (1 - I_{DA}/I_D) \qquad (2)$$

Förster showed that transfer efficiency and fluorophore distance are linked by

$$E = 1/(1 + R^6/R_0^6) \qquad (3)$$

where R is the donor-acceptor distance and R_0 is the Förster distance, at which 50% of the donor energy is transferred.[85]

Using this relationship, FRET has been employed to analyze the three-

[75] P. Z. Qin and A. M. Pyle, *Biochemistry* **36,** 4718 (1997).
[76] P. C. Bevilacqua, R. Kierzek, K. A. Johnson, and D. H. Turner, *Science* **258,** 1355 (1992).
[77] D. H. Turner, Y. Li, M. Fountain, L. Profenno, and P. C. Bevilacqua, *Nucleic Acids Mol. Biol.* **10,** 19 (1996).
[78] Y. Li and D. H. Turner, *Biochemistry* **36,** 11131 (1997).
[79] M. Menger, T. Tuschl, F. Eckstein, and D. Porschke, *Biochemistry* **35,** 14710 (1996).
[80] R. M. Clegg, *Methods Enzymol.* **211,** 353 (1992).
[81] P. Wu and L. Brand, *Anal. Biochem.* **218,** 1 (1994).
[82] R. M. Clegg, *Curr. Opin. Biotechnol.* **6,** 103 (1995).
[83] C. G. dos Remedios and P. D. Moens, *J. Struct. Biol.* **115,** 175 (1995).
[84] M. Yang and D. P. Millar, *Methods Enzymol.* **278,** 417 (1997).
[85] C. R. Cantor and P. R. Schimmel, "Biophysical Chemistry," Vol. 2, p. 448. Freeman, San Francisco, 1980.

dimensional structure of the hammerhead ribozyme, by calculating the FRET efficiency E from the extracted donor and acceptor contributions to the steady-state fluorescence spectra of doubly labeled ribozyme–substrate complexes.[86] The resulting model, however, although in general agreement with the global shape as determined by X-ray crystallography, was subsequently shown to predict an incorrect relative orientation of two of its three helical arms.[87] FRET has also been used to observe the bending of RNA helices containing bulge loops of varying length,[88] and to analyze global structure changes on metal ion titration in the hammerhead[89] and hairpin ribozymes[90,91] and in RNA four-way junctions.[92]

Several problems are associated with the use of FRET for determining absolute distances in nucleic acids: (1) Quenching of the fluorophores by RNA bases complicates data analysis. Because different structures have to be compared to obtain reliable dimensions, special care has to be taken to keep the sequence close to the fluorophores and their linker lengths identical between constructs. However, because subtle tertiary structure changes also may have an influence on the local environment of the dyes, identical fluorophore quenching between constructs is difficult to ensure. (2) For calculations of FRET distances one has to assume that the probes are able to undergo free, isotropic motion.[83] The high fluorescence anisotropy values for some fluorophore–nucleic acid conjugates indicate significant interactions between the dye and either bases or the negatively charged backbone of the nucleic acid.[63,86] Such interactions can be expected to interfere with fluorophore mobility and to bias distance measurements.[83] (3) There is usually some uncertainty in the position of the FRET dyes due to the flexibility of their linker arm. These problems need to be addressed when intra- or intermolecular distances in biological macromolecules are to be measured accurately.

On the other hand, a unique advantage of fluorescence measurements over other biophysical methods is their ability to produce a continuous signal to report on dynamic processes in solution. If relative signal changes over time are observed for a single construct, kinetic rates for conformational transitions can be inferred, avoiding problems associated with abso-

[86] T. Tuschl, C. Gohlke, T. M. Jovin, E. Westhof, and F. Eckstein, *Science* **266,** 785 (1994).

[87] S. T. Sigurdsson, T. Tuschl, and F. Eckstein, *RNA* **1,** 575 (1995).

[88] C. Gohlke, A. I. H. Murchie, D. M. J. Lilley, and R. M. Clegg, *Proc. Natl. Acad. Sci. U.S.A.* **91,** 11660 (1994).

[89] G. S. Bassi, A. I. H. Murchie, F. Walter, R. M. Clegg, and D. M. J. Lilley, *EMBO J.* **16,** 7481 (1997).

[90] A. I. H. Murchie, J. B. Thomson, F. Walter, and D. M. J. Lilley, *Mol. Cell* **1,** 873 (1998).

[91] F. Walter, A. I. H. Murchie, J. B. Thomson, and D.M.J. Lilley, *Biochemistry* **37,** 14195 (1998).

[92] F. Walter, A. I. H. Murchie, D. R. Duckett, and D. M. J. Lilley, *RNA* **4,** (1998).

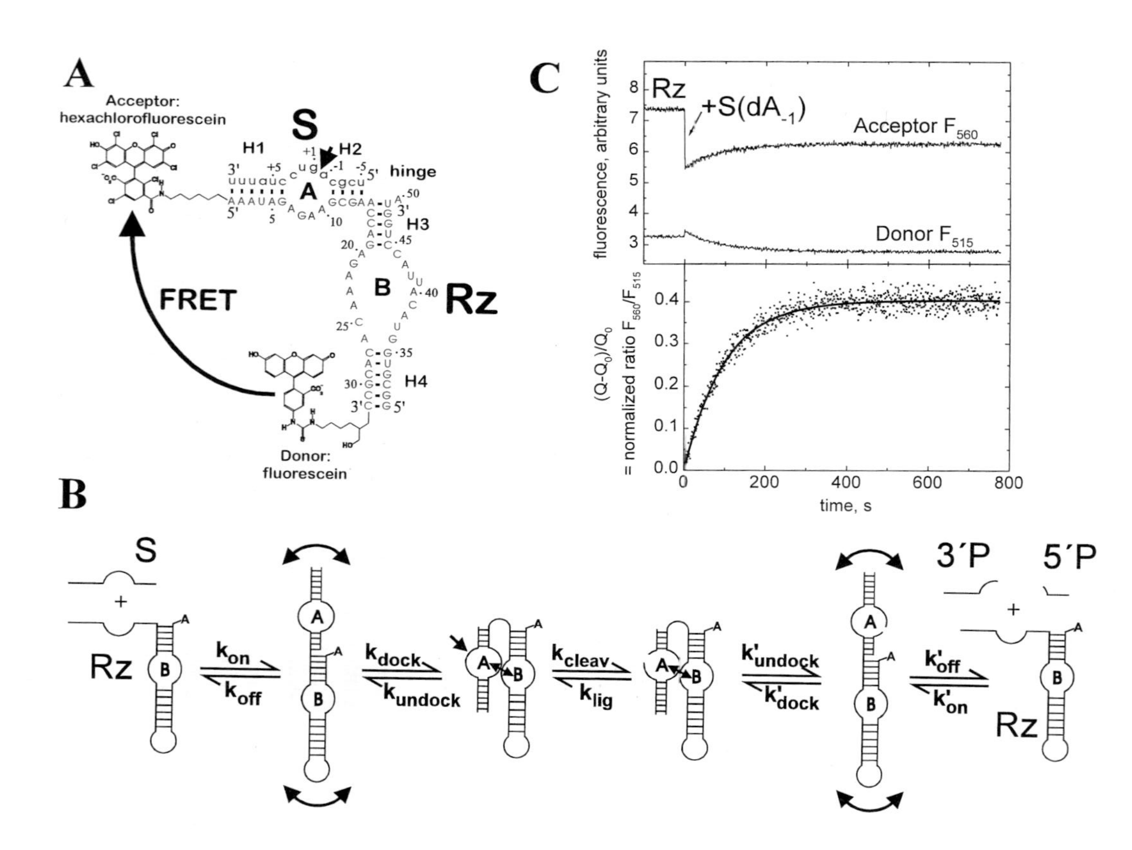

A
Acceptor:
hexachlorofluorescein
S
H1
H2
hinge
H3
A
B
Rz
H4
FRET
Donor:
fluorescein
C
Rz
+S(dA-1)
Acceptor F560
Donor F515
fluorescence, arbitrary units
(Q-Q0)/Q0 = normalized ratio F560/F515
time, s
B
S
Rz
kon
koff
kdock
kundock
kcleav
klig
k'undock
k'dock
k'off
k'on
3´P
5´P
Rz

lute distance measurements. We, therefore, developed a FRET-based assay to monitor in real time the reversible folding (docking) of the hairpin ribozyme–substrate complex from an open into a tertiary structured, closed conformation (Fig. 6).[93] In fact, we were able to measure accurately kinetic rate constants for this change in global structure under a variety of conditions, including modifications to the substrate and ribozyme, changes in metal ion composition, temperature, and pH. We found that docking precedes both cleavage and ligation reactions (Fig. 6B), but is rate limiting only for ligation. Strikingly, most modifications to the RNA or reaction conditions that inhibit cleavage do so by preventing docking, emphasizing the importance of RNA tertiary structure transitions for biological function.[93]

Of course, FRET can also be utilized to observe secondary structure formation in RNA, by placing a suitable donor–acceptor pair at sites that become close on association of complementary strands. Goodchild and coworkers have used this approach to measure the kinetics of formation and dissociation of hammerhead ribozyme–substrate complexes, and to answer

[93] N. G. Walter, K. J. Hampel, K. M. Brown, and J. M. Burke, *EMBO J.* **17,** 2378 (1998).

Fig. 6. Studying the kinetics of tertiary structure folding of the hairpin ribozyme–substrate complex by FRET. (A) The doubly labeled ribozyme–substrate complex. Fluorescein and hexachlorofluorescein are coupled as donor and acceptor pair to the 3′ and 5′ ends of the 5′ half of the two-strand ribozyme (Rz, capital letters) to enable distance-sensitive FRET (curved arrow). Short arrow, potential cleavage site in the substrate (S, small letters). (B) Minimal reaction mechanism for hairpin ribozyme catalysis as revealed by FRET. Substrate *in trans* is bound by the ribozyme into an open extended conformation, which subsequently folds into a docked bent structure, enabling loops A and B to interact. Site-specific cleavage follows (short arrow), the complex unfolds into an open complex, and the 5′ and 3′ cleavage products (5′P, 3′P) dissociate. All steps are fully reversible and can be characterized by individual rate constants. (C) Fluorescence signals over time as a result of tertiary structure folding of the ribozyme–substrate complex. The doubly labeled ribozyme, excited at 485 nm, displays a strong signal for the acceptor fluorophore at 560 nm and a weaker one for the donor at 515 nm. On manual addition of a saturating excess of noncleavable substrate analog (dA_{-1} modified), the acceptor fluorescence drops due to base-mediated quenching in the ribozyme–substrate complex. Subsequently, the acceptor signal increases, while the donor signal decreases at the same rate, indicating enhanced energy transfer efficiency between them. The normalized ratio Q of the acceptor:donor fluorescence as a measure for relative FRET efficiency was least-squares fitted with the equation $y = y_0 + A(1-e^{-t/\tau})$, yielding a first-order reaction rate constant of $1/\tau = 0.61$ min^{-1} with $A = 0.40$ and $\chi^2 = 0.00032$ (solid line). Conditions were 200 n*M* S(dA_{-1}) and 20 n*M* Rz (with a 10-fold excess of the unlabeled 3′ strand) in 50 m*M* Tris-HCl, pH 7.5, 12 m*M* $MgCl_2$, 25 m*M* dithiothreitol (DTT), at 25°. (From Walter *et al.*[93])

the question about how facilitator oligonucleotides might improve the turnover of substrate cleavage.[94]

Analyzing Conformational Isomers by Time-Resolved FRET

In principle, the strong dependence of FRET on the donor–acceptor distance [Eq. (3)] can be utilized to derive near-Ångstrom resolution of dimensions in a biopolymer. However, the intrinsic flexibility of nucleic acids and the fluorophore attachment linkers will give rise to distance distributions within the ensemble of analyzed molecules. Albaugh and Steiner[95] have demonstrated how these distance distributions can be derived experimentally from the associated multiexponential decay of the donor fluorescence. Millar and co-workers have applied this time-resolved FRET (tr-FRET) technique to the analysis of global structures of DNA three- and four-way junctions,[84,96] and were able to demonstrate that DNA four-way junctions typically can be resolved into equilibrium mixtures of two conformational isomers.[97]

We have studied the equilibrium distributions of hairpin ribozyme–substrate complexes between the active docked and inactive extended conformers by tr-FRET (Fig. 7). To avoid base-mediated quenching of the analyzed donor fluorescence (see earlier discussion), a guanosine-free sequence proximal to fluorescein in the complex was chosen (Fig. 7A). To derive distance information, two time-resolved fluorescence decays were collected, one for the sample with acceptor in place, one under identical conditions, but employing a donor-only RNA complex. Instrumentation comprised a mode-locked, 90-ps-pulse argon-ion laser for excitation at 514 nm, and perpendicular emission detection with polarizers in magic angle position, a 530-nm cutoff filter, and a single-photon counting photomultiplier.[96] The sample (150 μl) was incubated in the cuvette at measurement temperature for at least 15 min, prior to collecting >40,000 peak counts. The instrument response function to deconvolute the observed fluorescence decay was obtained using a dilute solution of nondairy coffee creamer to scatter the laser pulses. The decay of fluorescein emission in the doubly labeled complex was analyzed by a model of fluorophore distance distributions.[97]

[94] T. A. Perkins, D. E. Wolf, and J. Goodchild, *Biochemistry* **35,** 16370 (1996).

[95] S. Albaugh and R. F. Steiner, *J. Phys. Chem.* **93,** 8013 (1989).

[96] P. S. Eis and D. P. Millar, *Biochemistry* **32,** 13852 (1993).

[97] S. M. Miick, R. S. Fee, D. P. Millar, and W. J. Chazin, *Proc. Natl., Acad. Sci. U.S.A.* **94,** 9080 (1997).

$$I_{DA}(t) = \sum_k f_k \int P_k(R) \sum_i \alpha_i \exp\left[-\frac{t}{\tau_i}\left(1 + \left(\frac{R_0}{R}\right)^6\right)\right] dR \qquad (4)$$

where the first sum refers to the number of distributions, either one or two, each with fractional population f_k and distance distribution $P_k(R)$. The distribution was modeled as a weighted Gaussian,

$$P(r) = 4\pi R^2 c \exp[-a(R - b)^2] \qquad (5)$$

where a and b are parameters that describe the shape of the distribution and c is a normalization constant. Equation (4) was used to fit experimental data by nonlinear least squares regression, with a, b, and f_k for each distribution as adjustable parameters. Two distance distributions were used for analysis when a single distribution failed to give a good fit, as judged by the reduced χ^2 value and by inspection of residuals. In all such cases the inclusion of a second distribution resulted in a dramatic improvement of the fit. The intrinsic donor lifetimes τ_i and decay amplitudes α_i were determined for each set of experimental conditions by a sum-of-exponentials fit to the donor intensity decay in the donor-only complex. The Förster distance R_0 of 55 Å was calculated from the overlap of the donor emission and acceptor absorbance spectra, and by assuming free, isotropic motion of the dyes.[96] The latter assumption was supported by time-resolved fluorescence anisotropy decay experiments,[96] which revealed large-amplitude rotational motions of both fluorescein and tetramethylrhodamine, characterized by half cone angles of 28° and 42°, respectively.

Using this approach, we were able to distinguish ribozyme–substrate complexes, where domains A and B do not dock to a detectable ($\geq$2%) extent (such as those with a substrate mutation $G_{+1}A$; Fig. 7B), from those that can perform this essential tertiary structure transition, as required for catalytic activity (Fig. 7C). The equilibrium distributions between docked and extended conformers in the latter complexes were used to describe the folding energy landscape of the hairpin ribozyme, and have helped to understand the role of the interdomain junction in stabilizing docking.[97a]

Observing Ligand Binding to RNA

Several approaches exist to analyze binding of ligands to RNA by fluorescence methods. Few have made use of fluorescent RNA, while many more employ fluorescent ligands. If the ligand is a protein, its intrinsic fluorescence often can be utilized to infer binding and dissociation kinetics and the equilibrium dissociation constant (as mentioned in the Introduc-

[97a] N. G. Walter, J. N. Burke, and D. P. Millar, *Nat. Struct. Biol.* **6,** 544 (1999).

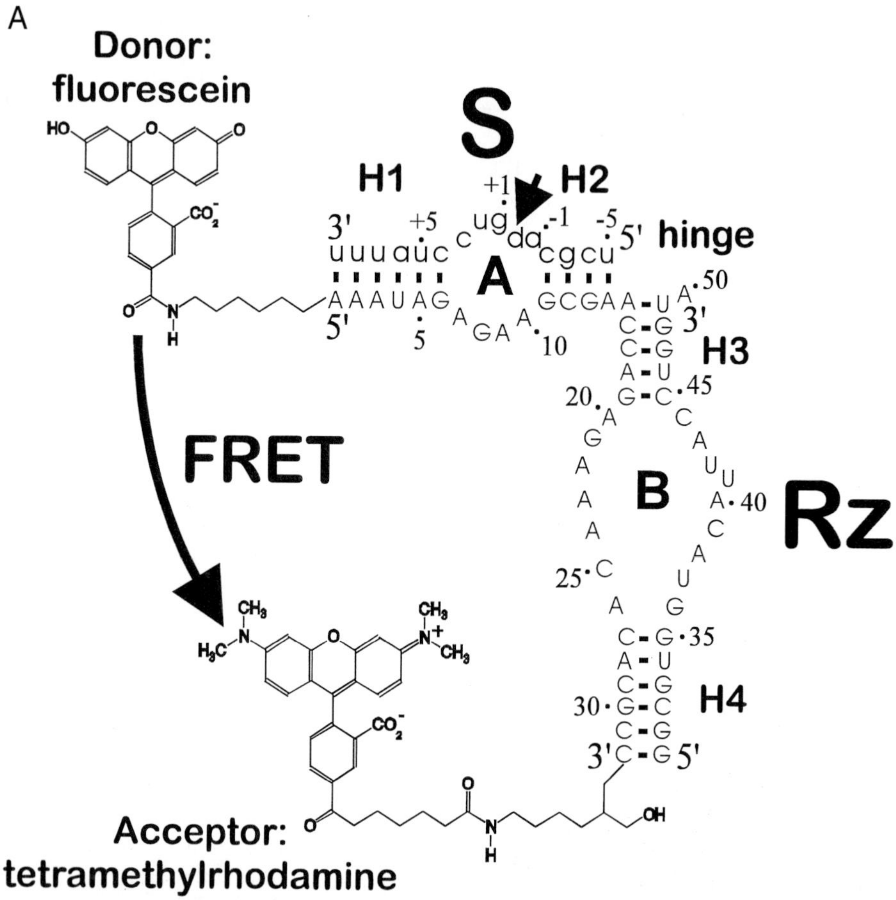

A

Donor: fluorescein

Acceptor: tetramethylrhodamine

FRET

Fig. 7. Resolution of conformer distributions of the hairpin ribozyme–substrate complex using time-resolved fluorescence resonance energy transfer (tr-FRET). (A) The basic doubly labeled ribozyme–substrate complex. Fluorescein and tetramethylrhodamine are coupled as donor and acceptor pair to opposite ends of the ribozyme (Rz, capital letters) to enable distance-dependent FRET (curved arrow). Short arrow, potential cleavage site in the substrate (S, small letters). Cleavage is blocked by a dA_{-1} modification. (B) Revealing a single conformer by tr-FRET. A $G_{+1}A$ mutant substrate impairs docking of the ribozyme–substrate complex and results in a flexible extended conformer with fluorescein (F) and tetramethylrhodamine (T) at distant ends. The time-resolved donor (fluorescein) fluorescence decay of this complex under standard conditions (50 mM Tris-HCl, pH 7.5, 12 mM $MgCl_2$, at 17.8°) is measured and the donor–acceptor distance information extracted. A single continuous, three-dimensional Gaussian distance distribution (dashed line) fits the data very closely (giving a reduced χ^2 of 1.08). (C) Revealing an equilibrium between two conformers by tr-FRET. If the complex can fold from the initial extended structure into the docked conformer (in the presence of a G_{+1} base) the donor decay data (dots) can only be fitted with the sum of two Gaussian distance distributions (χ^2 = 1.28), revealing the presence of both docked and extended conformers (deconvoluted contributions to the donor decay represented by solid and dashed lines, respectively). The relative abundance of these isomers is obtained directly from the analysis and is reflected in the relative heights of the corresponding distance distributions. tr-FRET thus defines the equilibrium constant K_{dock} and the energetic difference between docked and extended conformers in the RNA folding energy landscape, ΔG_{dock}.

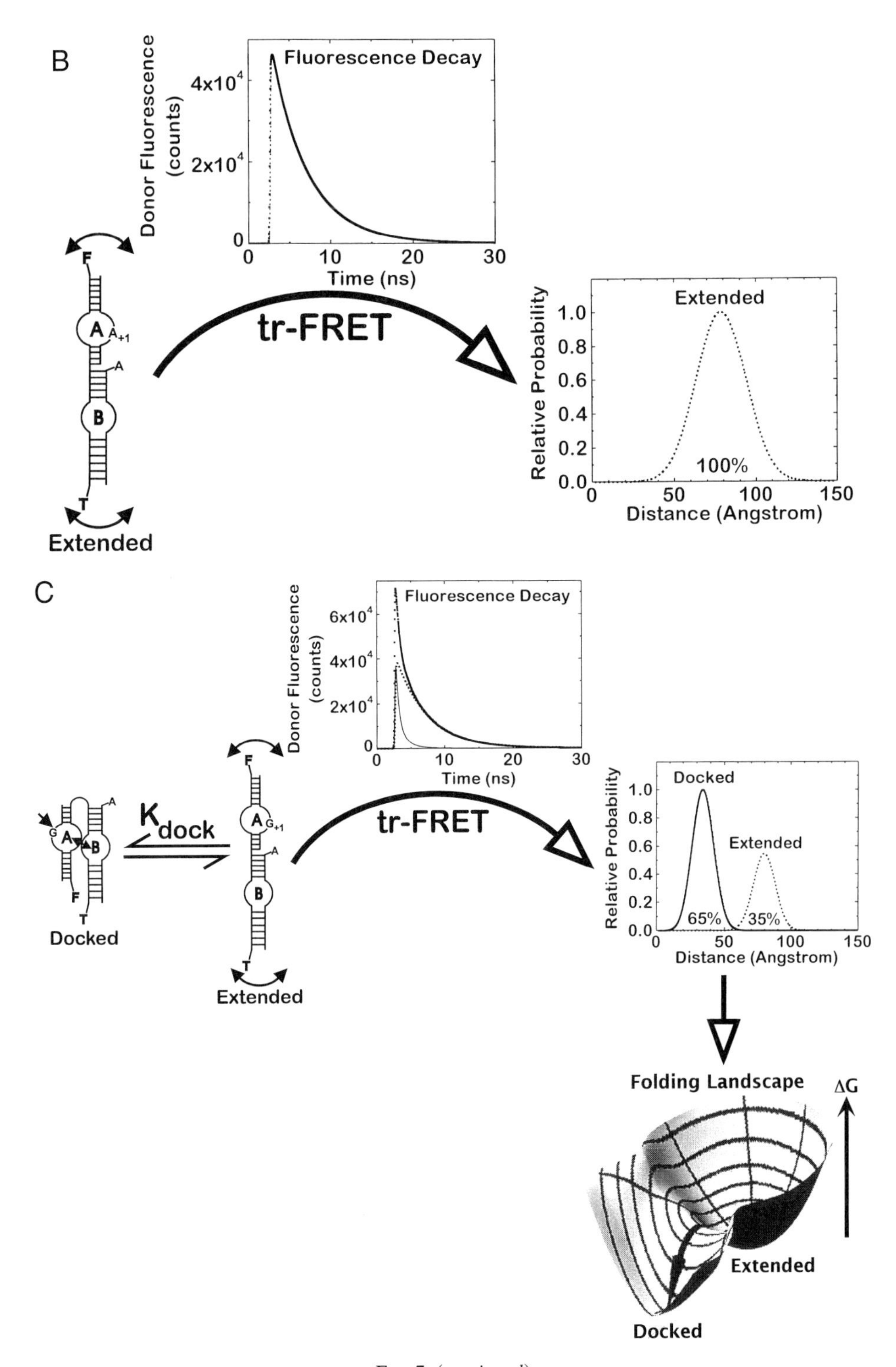

FIG. 7. (*continued*)

tion).[10–16] An example for a smaller intrinsically fluorescent ligand is riboflavin. RNA aptamers have been identified by *in vitro* selection that bind to riboflavin by virtue of a G-quartet structural motif, and the RNA–ligand interaction was characterized by fluorescence quenching in the complex.[98]

If the ligand is not fluorescent by itself, it can be labeled with a reactive fluorophore derivative, as discussed earlier for RNA. For example, aminoglycosides and small peptides labeled with tetramethylrhodamine or fluorescein were used to characterize their binding to specific RNA sequences, by measuring an increase in their fluorescence anisotropy on complex formation.[99,100]

Sensitized fluorescence from a ligand that only becomes fluorescent on binding to RNA has been employed to study binding of the lanthanide ion Tb^{3+} to the hammerhead ribozyme.[101] Horrocks and co-workers previously have shown that time-resolved fluorescence spectroscopy of lanthanide ions, especially Eu^{3+}, can yield valuable information on the binding mode of physiologic metal ions such as Mg^{2+} and Ca^{2+} to biopolymers.[102,103]

Finally, a fluorophore-labeled competitor can be used to report on the complex formation between an RNA and its ligand. Figure 8 gives an example from Preuss *et al.*, in which fluorescence from a 5′-pyrene-labeled DNA probe was employed to study interactions between Qβ replicase and various template RNAs.[104] As for all other studies on RNA–ligand complexes discussed here, the theoretical background for extracting binding affinities from the fluorescence data is well developed (for review, see for example Ref. 105).

Concluding Remarks

Researchers have only recently begun to embrace fluorescence methods as a unique tool to study structure, dynamics, and function of RNA and RNA–ligand complexes. They benefit from techniques previously developed for protein analysis and subsequently transferred to studies of DNA structure and dynamics. Combination of existing techniques and future developments in fluorescence-based technologies such as flow cytometry,

[98] C. T. Lauhon and J. W. Szostak, *J. Am. Chem. Soc.* **117,** 1246 (1995).
[99] Y. Wang, J. Killian, K. Hamasaki, and R. R. Rando, *Biochemistry* **35,** 12338 (1996).
[100] Y. Wang, K. Hamasaki, and R. R. Rando, *Biochemistry* **36,** 768 (1997).
[101] A. L. Feig, W. G. Scott, and O. C. Uhlenbeck, *Science* **279,** 81 (1998).
[102] W. D. Horrocks and D. R. Sudnick, *Science* **206,** 1194 (1979).
[103] D. Chaudhuri, W. D. Horrocks, J. C. Amburgey, and D. J. Weber *Biochemistry* **36,** 9674 (1997).
[104] R. Preuss, J. Dapprich, and N. G. Walter, *J. Mol. Biol.* **273,** 600 (1997).
[105] M. R. Eftink, *Methods Enzymol.* **278,** 221 (1997).

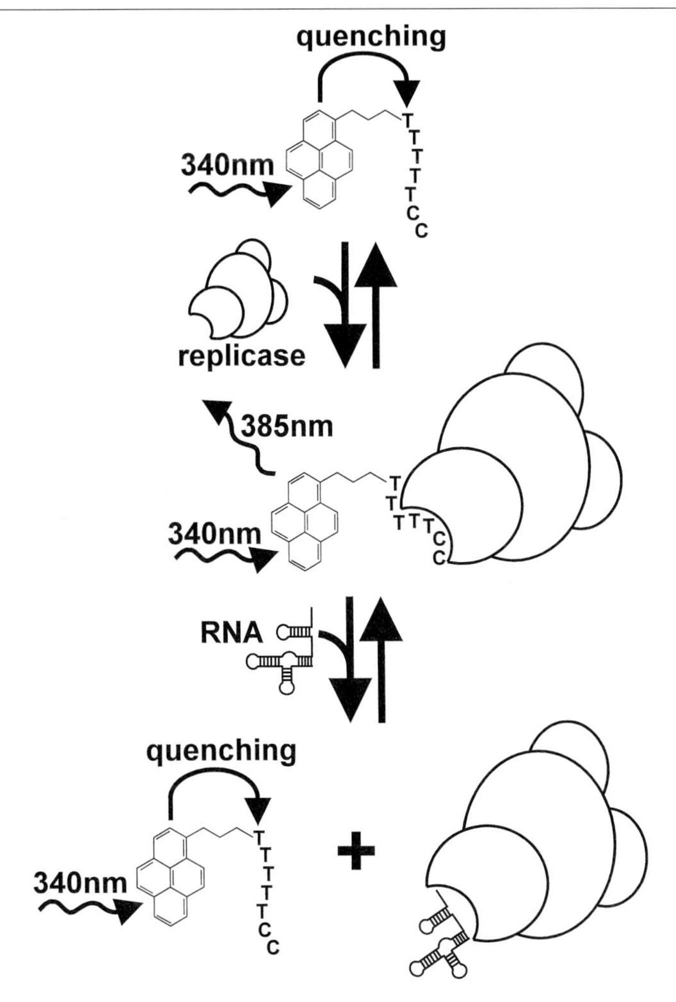

Fig. 8. Fluorometric reverse titration assay to observe binding of Qβ replicase to its template RNA. When free in solution, the 5′-pyrene-labeled DNA probe (excited at 340 nm) is strongly quenched through interactions between pyrene and its pyrimidine-rich sequence. On reversible binding of the heterotetrameric phage Qβ replicase to the probe, pyrene becomes dequenched and fluoresces with a peak around 385 nm. When template RNA is added, it partially displaces DNA probe in the complex (corresponding to the relative equilibrium binding constants of RNA and probe), resulting in a fluorescence decrease. (From Preuss *et al.*[104])

single-molecule fluorescence microscopy, high-throughput screening for drug discovery, and chip-based hybridization assays can be expected to have a strong impact on the extent to which the principles described in this review will be utilized.

Acknowledgments

This work was supported by grants from the U.S. National Institutes of Health to J.M.B., and a Feodor Lynen fellowship from the Alexander von Humboldt foundation and an Otto Hahn medal fellowship from the Max Planck Society for N.G.W.

[26] Transient Electric Birefringence for Determining Global Conformations of Nonhelix Elements and Protein-Induced Bends in RNA

By PAUL J. HAGERMAN

Introduction

Transient electric birefringence (TEB) is a sensitive method for characterizing both the conformations and the flexibilities of RNA molecules in solution. In its application to the study of RNA conformation, it is fundamentally a hydrodynamic method; RNA helices with central bends undergo more rapid rotational diffusion than do their linear counterparts because the bent molecules experience less frictional resistance. Because rotational diffusion is much more sensitive to changes in overall helix conformation than is translational diffusion, and because the relationship between the experimental diffusion constants and hydrodynamic theory is more straightforward, TEB has proven to be quite useful for quantifying both intrinsic and protein-induced bends in RNA (for a brief review, see Ref. 1). Another important distinction between rotational and translational diffusion pertains to the magnitude of the direct frictional contribution of a bound protein or element of RNA tertiary structure. For centrally placed elements, rotational diffusion is relatively insensitive to the frictional surface of the added protein or RNA structure since such elements lie close to the center of rotation. Thus, one does not generally require detailed knowledge of the shape of the bound protein or nonhelix element.

In a typical TEB experiment, an element of interest (e.g., branch, protein

[1] P. J. Hagerman, *Curr. Opin. Struct. Biol.* **6,** 643 (1996).

0076-6879/00 $30.00

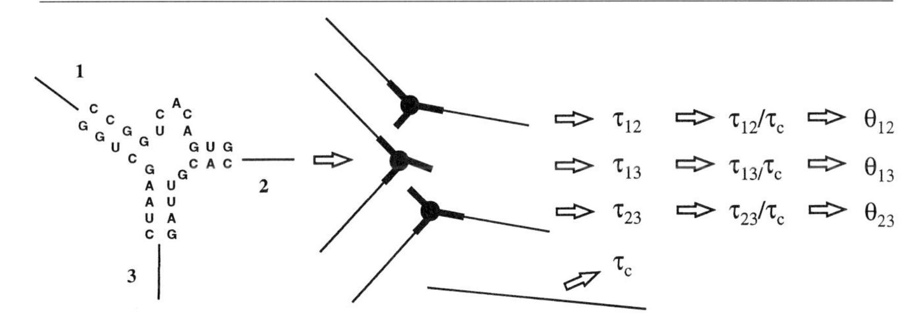

FIG. 1. Transient electric birefringence as a hydrodynamic method for defining global helix conformation. A nonhelix element (e.g., 3-helix junction) is cast in the center of an otherwise helical RNA molecule. Measurement of the rotational decay times (τ_{ij}) of the heteroduplex molecules relative to that of a control helix (τ_c) and subsequent comparison with a computational model for τ versus θ yields apparent interstem angles. Additional measurements with phased bends can yield estimates of bend flexibility.

binding site) is placed at the center of an otherwise duplex RNA molecule (Fig. 1). By comparing the birefringence decay time of the heteroduplex (or protein-bound) molecule to a linear control, a measure of the effective angle between the helix arms is obtained. As described later (τ ratio approach), the method is most effective when the lengths of the bent and control molecules are nearly identical, and the sequences are identical except for the nonhelix element itself. Such a comparison minimizes uncertainties associated with the underlying hydrodynamic models. Finally, it has been established that neither the extended arms, nor the application of the orienting field ($E \leq 10$ kV/cm), influence the geometry of the nonhelix element being investigated, even for nucleic acid species with a high degree of intrinsic flexibility.[2–4]

Principle of the Birefringence Method

Solutions containing RNA are normally optically isotropic; however, optical anisotropy can be induced in the solution by applying a brief electric

[2] K. R. Hagerman and P. J. Hagerman, *J. Mol. Biol.* **260,** 207 (1996).

[3] J. B. Mills, E. Vacano, and P. J. Hagerman, *J. Mol. Biol.* **285,** 245 (1999).

[4] M. W. Friederich, E. Vacano, and P. J. Hagerman, *Proc. Natl. Acad. Sci. U.S.A.* **95,** 3572 (1998).

field in a direction perpendicular to the optical path (Fig. 2). Following removal of the field, the RNA molecules randomize through Brownian (rotational) diffusion, and the solution again becomes optically isotropic. Anisotropy is typically detected as the difference in UV absorbance parallel and perpendicular to the electric axis ($\Delta A = A_{\parallel} - A_{\perp}$) (transient electric dichroism; TED), or in the current instance, as the difference in refractive index ($\Delta n = n_{\parallel} - n_{\perp}$). An outline of the optical principle governing birefrin-

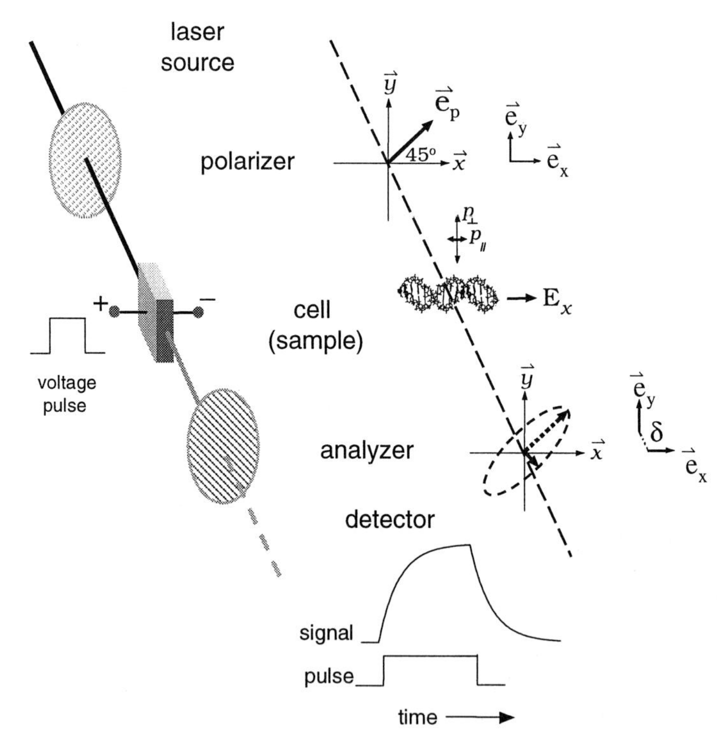

FIG. 2. Basic principle of the TEB method. In the absence of an orienting electric field, polarized light passing through the sample cell is nearly extinguished on passing through a second polarizer (analyzer) with a transmission axis that is nearly perpendicular to the direction of polarization ($\mathbf{e}_p$). The transmission axes of both polarizers are set at 45° with respect to the field axis ($\mathbf{E}_x$). On application of a brief (1-μs) orienting pulse, the RNA molecules tend to align (slightly) along the field axis. Due to the greater optical polarizability of the RNA in the plane of the bases ($p_{\perp} > p_{\parallel}$), the y component of the light is preferentially retarded. The resultant beam becomes elliptically polarized, allowing an incremental amount of light to pass through the analyzer to the detector. On removal of the field, the light passing through the analyzer returns to its steady-state level.

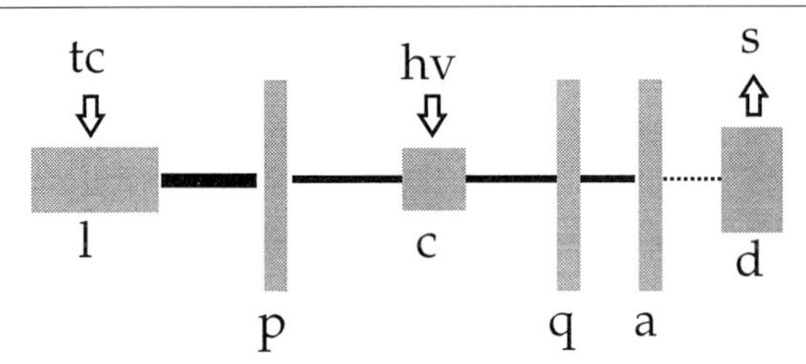

FIG. 3. Schematic of the TEB instrument currently in use in the author's laboratory. (1) Diode laser, (tc) Peltier-effect temperature clamp for the laser diode, (p) polarizer, (c) sample cell, (hv) high-voltage pulse generator, (q) quarter-wave plate, (a) analyzer, (d) detector, (s) output signal to (d) digitizing oscilloscope.

gence measurements is presented in Fig. 2. Detailed reviews of TEB theory and methods have been presented elsewhere.[1,5-7]

The decay time (τ) for loss of solution anisotropy is a strong function of the length and overall conformation of an RNA molecule. For linear RNA helices in the 100- to 200-bp range, τ varies approximately as $(\text{length})^{2.5}$, roughly an eightfold stronger length dependence in this size range than for sedimentation velocity ($s_{20,w}$) measurements. Moreover, the introduction of a 90° bend at the center of a 150-bp RNA helix results in an approximately twofold drop in τ.[8,9] One significant advantage of TEB relative to TED for hydrodynamic studies is that, by virtue of the optical setup for birefringence,[5,6] instrumental sensitivity can be made extremely high.[2]

Instrumentation

A general schematic of the birefringence instrument currently used in the author's laboratory is presented in Fig. 3. The optical arrangement is a standard quarter-wave plate configuration[5,10]; however, the instrument is otherwise substantially different than the device described earlier.[10] The basic features of the current instrument are as follows: (1) single-shot instrument signal-to-rms noise (S_{instr}/N_{rms}), ~ 800; (2) overall system re-

[5] E. Fredericq and C. Houssier, "Electric Dichroism and Electric Birefringence." Oxford University Press (Clarendon), London and New York, 1973.

[6] C. T. O'Konski, ed., "Molecular Electro-Optics. Part 1, Theory and Methods." Marcel Dekker, New York. 1976.

[7] E. Charney, *Quart. Rev. Biophys.* **21,** 1 (1988).

[8] E. Vacano and P. J. Hagerman, *Biophys. J.* **73,** 306 (1997).

[9] M. Zacharias and P. J. Hagerman, *J. Mol. Biol.* **247,** 486 (1995).

[10] P. J. Hagerman, *Methods Enzymol.* **117,** 198 (1985).

sponse time, $t_{1/2} \approx 15$ ns; (3) sample volume, 35 μl; and (4) stray light constant[5] (K_{SL}) $\approx 8 \times 10^{-5}$. Thus, a 3-min measurement (180-pulse average at 1 Hz) yields a S_{instr}/N_{rms} of $\sim 10^4$. A brief description of the instrumental components is given later; a more detailed description is to appear elsewhere (Schleif and Hagerman, in preparation).

High-Voltage Pulse Generator

The pulse generators, built in-house, employ a MOSFET-based switch, basically a modified GRX grid driver (Directed Energy, Inc.), and a standard 0- to 2000-V supply. The high-voltage pulse generator acts as a voltage follower, receiving TTL pulses from a logic device that provides a set number of pulses of specified duration and pulse frequency.

Laser Source

The source is a 5-mW TOLD9211 diode laser (Toshiba). The diode, as well as electronics for laser operation and feedback regulation, are all housed in a machined copper cylinder. The cylinder is partially enclosed in a temperature clamp consisting of a Peltier junction thermoelectric device. This temperature clamp holds the diode to within a few thousandths of a degree during operation, thus eliminating noise due to mode hopping. Mode hopping occurs as the laser cavity expands and contracts with fluctuations in ambient temperature. Domains of mode-hopping instability occur every 2–3°, and although the transitions take place in the picosecond timescale, they are accompanied by small changes in laser power that are observed as spikes or shifts in baseline output. By clamping the diode temperature at a point midway between regions of instability, this contribution to the noise is eliminated.

TEB Cell

The cell cavity is approximately 1.8 mm wide × 2 mm high × 10 mm long, and receives a constant 35 μl of sample. The sides of the cavity are platinum, and the remaining cell housing is a machined block of black Delrin. The Delrin block also encloses channels to allow circulation of coolant for temperature regulation. The windows are optical glass cylinders that have been chosen for their low residual birefringence; they are set against the faces of the cell cavity on thin beads of latex, and are held in place by foam pads. This "soft-mount" configuration has been found to eliminate much of the strain birefringence that occurs with adiabatic expansion of the buffer during the field pulse.

Detector

The heart of the detector comprises a PIN040A photodiode (UDT Instruments, Baltimore, MD) and an ultra-low-noise, high-bandwidth (1.7-GHz) operational amplifier (CLC425; Comlinear Corp., Fort Collins, CO). The detector is housed in a cast alloy case to minimize RF pickup. Moreover, the power leads for both the detector and the laser are wrapped about ferrite rings immediately prior to entering the housings.

Digitizing/Averaging Oscilloscopes

The detector output is captured and subjected to signal averaging with a LeCroy model 9310 digitizing oscilloscope (Le Croy, Chestnut Ridge, N.Y.). The averaged (binary) curves are analyzed on laboratory PCs using laboratory software (see later section).

General Considerations for the Design of Experiments

As noted, TEB measurements generally involve the placement of a nonhelix element (e.g., branch, loop) or protein binding site at the center of an RNA helix. Several strategies can be used to optimize the sensitivity of the measurements:

Central Placement of Nonhelix Element or Protein Binding Target

Because TEB decay times reflect the degree of foreshortening of the bent helix relative to a linear control molecule, the sensitivity of the method is maximal when the bend is placed at the center of the RNA helix. This behavior is analogous to the position dependence observed in gel mobility shift measurements. In fact, almost independent of the angle itself, the reduction in τ is approximately twofold greater for a centrally placed bend than for a bend placed at a point approximately one-quarter in from one end of the helix.[8] In addition, for protein binding, the reduction in sensitivity with off-center placement of the binding site is exacerbated by the physical dimensions of the protein itself (frictional contributions), which increase rapidly as the protein is moved away from the center of the RNA (see later section).

Length of RNA Helix

Four general factors influence the choice of length for the RNA helix: (1) The helix should be short enough to present a reasonably rigid framework for studying the central bend, and for ensuring decay profiles that are essentially single exponential in the absence of a bend. Thus, RNA

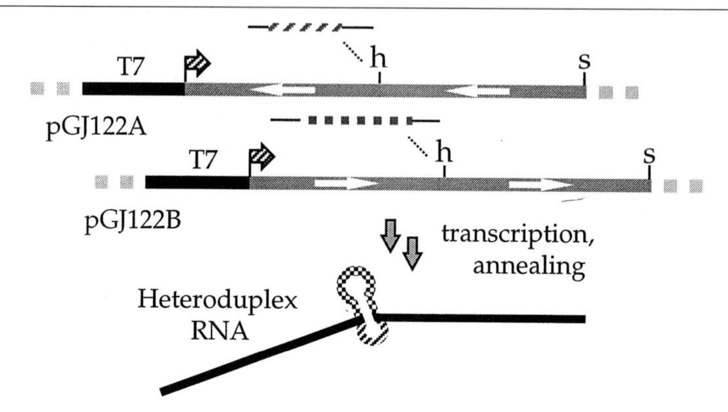

FIG. 4. Production of heteroduplex RNA molecules for TEB measurements. Plasmids pGJ122A and pGJ122B differ only in the orientation of a 136-bp template that is delimited by a downstream *Sma*I site (s). The template also has a central *Hin*dIII (h) site for insertion of additional sequence elements. Transcription from the pair of plasmids yields a pair of transcripts that, when annealed, yield heteroduplex or elongated duplex RNA molecules.

helices that are comparable to or less than one persistence length[11] (~200–250 bp) are preferred. (2) As the length of the helix is increased, so too is the propensity for intrastrand secondary structure formation that could interfere with the annealing process. (3) The total (combined) length of the helix arms should be at least four to five times the approximate linear span of the nonhelix element being studied in order to reduce any hydrodynamic contributions from short (nonextended) helices or bound proteins). (4) When the span of the nonhelix element is not known with precision, increasing the length of the helix arms, or performing measurements with several arm lengths, will reduce the uncertainty due to the span length itself.[8]

In view of these considerations, we have generally employed helix arms that are approximately 70 bp (per arm). Our standard (parent) templates (i.e., for the helix arms) yield two complementary (136-nucleotide) transcripts, that, when annealed, form a continuous helix. The templates have been placed adjacent to the T7 RNA polymerase promoter in a pair of plasmids, pGJ122A and pGJ122B,[12] which differ only in the orientation of the template (Fig. 4). The templates were originally designed to minimize the formation/stability of secondary structure in the individual transcripts.[13]

[11] P. Kebbekus, D. E. Draper, and P. J. Hagerman, *Biochemistry* **34,** 4354 (1995).
[12] K. M. A. Amiri and P. J. Hagerman, *Biochemistry* **33,** 13172 (1994).
[13] F.-U. Gast and P. J. Hagerman, *Biochemistry* **30,** 4268 (1991).

Data Analysis

In a typical transient birefringence measurement, a set of field–response curves, $I(t)$, is pointwise averaged to yield $\langle I(t)\rangle$. This quantity is converted to a fractional change in light intensity, $\langle \Delta I(t)\rangle/\langle I_\alpha\rangle$ [$= (\langle I(t)\rangle - \langle I_\alpha\rangle)/\langle I_\alpha\rangle$)], where $\langle I_\alpha\rangle$ is the pointwise average of the steady-state light intensity at a specified analyzer setting (α degrees). This (averaged) time-dependent change in light intensity is converted to the birefringence response, $\Delta n(t)$, using the following relation for a quarter-wave plate optical configuration[5]:

$$\Delta n(t) = (2 \times 10^{-8}\lambda/360l)\{\sin^{-1}[(\langle \Delta I(t)\rangle/\langle I_\alpha\rangle + 1)^{1/2} \sin \alpha] - \alpha\} \quad (1)$$

where λ is the wavelength (Å) and l is the path length (cm). For each set of measurements, the (averaged) birefringence response of the buffer, $\Delta n(t)_{\text{buffer}}$, is subtracted from $\Delta n(t)_{\text{sample}}$ to yield a corrected $\Delta n(t)$ curve. At this point, $\Delta n(t)$ represents the entire field–response curve, typically comprising 1024 data points. The portion of the curve that precedes the end of the field pulse is then eliminated; and the remainder of the curve is subjected to further analysis. Because we are principally concerned with the time dependence of $\Delta n(t)$, not its absolute amplitude, the field-free decay curve is normalized, and is analyzed as a sum of two exponentials:

$$\Delta n(t)/\Delta n(0) = \alpha_f \exp(-t/\tau_f) + \alpha_s \exp(-t/\tau_s) \quad (2)$$

In Eq. (2) α_f and α_s are the fractional amplitudes ($\alpha_f + \alpha_s = 1$) associated with the fast (f) and slow (s) components of the decay curve, and τ_f and τ_s are the corresponding decay times. In this regard, although the most general form of the birefringence response possesses five independent components,[14,15] only the fastest and slowest components are expected to have nonzero amplitudes for symmetric or approximately symmetric molecules, including those possessing significant flexibility.[15,16] Those expectations are borne out in both experimental and Brownian dynamics analyses of birefringence decay curves.[9,17]

The normalized decay curves are analyzed using the Levenberg–Marquardt (LM) method,[18] which minimizes a χ^2 statistic for the experimental and model decay curves. This is now a standard approach and is available in several commercial software packages (e.g., SigmaPlot; SPSS

[14] M. E. Rose, "Elementary Theory of Angular Momentum." Wiley, New York, 1957.

[15] W. A. Wegener, R. M. Dowben, and V. J. Koester, *J. Chem. Phys.* **70,** 622 (1979).

[16] W. A. Wegener, *J. Chem. Phys.* **84,** 5989 (1986).

[17] M. Zacharias and P. J. Hagerman, *Biophys. J.* **73,** 318 (1997).

[18] W. H. Press, W. T. Vetterling, S. A. Teukolsky, and B. P. Flannery, "Numerical Recipes in Fortran: The Art of Scientific Computing," pp. 678–683. Cambridge University Press, Cambridge, UK, 1992.

Inc., Chicago, IL). We have discussed the use of the LM method in the context of a demonstration of the relationship between instrument signal-to-noise ratio (S/N) and the standard deviations of the amplitudes and decay times for a normalized, two-component decay curve.[2] For essentially all of our studies of intrinsic or protein-induced bends in nucleic acids, $\tau_s/\tau_f \geq 4$ and $\alpha_s \geq \sim 0.2$. Within this τ_s–α_s range, the standard deviations of τ_s and α_s typically do not exceed ~4% for LM analyses of sets of averaged decay curves with individual $S_{curve}/N \geq 500$. For this situation, a set of at least five averaged decay curves (typical experimental protocol) would yield standard errors that are less than 2%. Thus, $S_{curve}/N > 500$ is a useful target for performing TEB measurements. As an example, for a single-shot signal-to-noise (S_{instr}/N) of ~800 and an S_{instr} of 1 V, an S_{curve} of ~100 mV would require an averaged S_{instr}/N of ~5000 [= 500 (S_{instr}/S_{curve})]. This S_{instr}/N can be achieved with a 40-sweep average. For a 1-Hz pulse repetition frequency, this average requires slightly less than 1 min.

Fluctuations of sample cell temperature during the course of a series of measurements can add to the variation of decay times for a set of measurements. For example, a ±0.2° variation over the period of a set of measurements can add ~0.6% to the spread in the decay times. Thus, for high-precision measurements, it is important to hold the cell temperature within narrow limits.

Analysis of Birefringence Decay Times: τ Ratio Approach

Birefringence decay times are usually interpreted in terms of a hydrodynamic representation of the molecule under study, with various parameters of the model (e.g., helix rise, hydrodynamic radius, intrinsic flexibility) being varied until the computed rotational diffusion times are in agreement with the experimental results. The interpretation is thus subject to uncertainties in both the model itself and the underlying hydrodynamic theory. On the other hand, if one compares the hydrodynamic behavior of a "test" molecule—a helix with a central bend—with a helix of identical sequence (and length) apart from the bend itself, the *ratio* of the decay times turns out to be remarkably insensitive to uncertainties in either the helix parameters or the hydrodynamic model.[19,20] In particular, for nucleic acid helices possessing a central bend, both static (equilibrium ensemble) and dynamic (Brownian dynamics) approaches yield quantitative agreement for the relationship between the τ ratio and the angle of the central bend.[8,19,20] The hydrodynamic models themselves are all based on the Stokes sphere formal-

[19] M. Zacharias and P. J. Hagerman, *Biophys. J.* **73,** 318 (1997).

[20] D. B. Roitman, *J. Chem. Phys.* **81,** 6356 (1984).

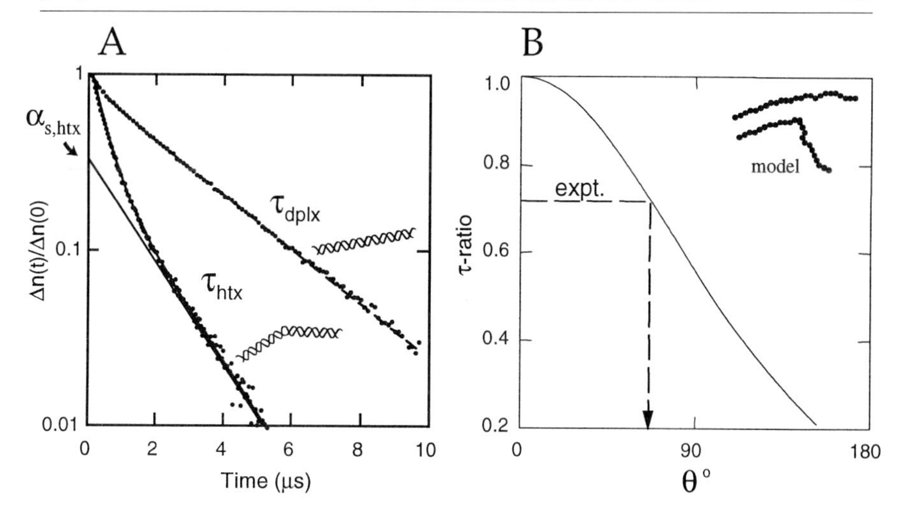

FIG. 5. Outline of the τ ratio approach. (A) Terminal decay times are determined for bend-containing (e.g., heteroduplex; τ_{bend}, τ_{htx}) and fully duplex control helices (τ_{dplx}). Each experimental profile comprises two exponential components, $\alpha_{slow(s)}$ and $\alpha_{fast(f)}$ ($\alpha_s + \alpha_f = 1$); $\alpha_{s,htx}$ is illustrated. (B) The ratio of the two decay times (τ ratio $\equiv \tau_{htx}/\tau_{dplx}$) is compared with ensemble-averaged, computed curves of τ ratio as a function of *apparent* bend angle.

ism of Kirkwood, and now include both hydrodynamic interactions and intramolecular excluded volume corrections.[8,21–23]

The standard τ ratio approach therefore involves the construction of two DNA or RNA helices of "identical" length (~ 100–200 bp), with one helix possessing a nonhelix element whose bend is to be determined, or a target sequence for protein binding. Determination of the τ ratio for this pair of molecules yields a quantitative estimate for the angle of the intrinsic (or induced) bend (Fig. 5).

The near absence of any dependence of the τ ratio on variations in helix parameters allows the τ ratio-versus-angle profile to be expressed as a simple interpolation formula derived from Monte Carlo computations (valid for both DNA and RNA helices)[8]:

$$\theta = 1.46 \times \cos^{-1}(\tau/\tau_c) + 0.005[\sin^{-1}(1 - \tau/\tau_c)]^{2.3} \quad (3)$$

where τ/τ_c is the ratio of the slow decay times for the test (τ) and control (τ_c) helices. For bends that are larger than 100–120° (interstem angles

[21] J. Garcia de la Torre and V. A. Bloomfield, *Quart. Rev. Biophys.* **14,** 81 (1981).

[22] J. Garcia de la Torre, S. Navarro, M. C. Lopez-Martinez, F. G. Diaz, and J. J. Lopez-Cascales, *Biophys. J.* **67,** 530 (1994).

[23] P. J. Hagerman and B. H. Zimm, *Biopolymers* **20,** 1481 (1981).

<60°), the τ ratio must be computed, because differences in intrinsic helix rigidity as well as the dimensions of additional, short helices or bound proteins must be taken into consideration.[8,24] Additional interpolation formulas have been published for several specific cases.[8]

In using the τ ratio analysis to examine specific nonhelix structural elements, two additional issues need to be considered; namely, uncertainty in the axial length of the nonhelix element, and the possibility that the element possesses additional flexibility. For simple elements such as bulges, the uncertainty in the axial length of the bulge is negligible relative to the length of the flanking helices. For larger elements (e.g., tRNA cores), significant uncertainty in the linear (axial) dimension can lead to a reduced level of accuracy in the apparent interhelix angle relative to the precision of the measurements, due to the greater uncertainty in the choice of the length of the reference helix.[25] A systematic analysis of the influence of errors in the estimated length on the apparent interstem angle has been presented by Vacano and Hagerman.[8]

Analysis of Flexibility of Nonhelix Elements in RNA: Phased τ Ratio Approach

Although τ ratio measurements provide a sensitive, quantitative measure of the bend angle introduced by either a nonhelix element or bound protein, the resultant angles are "apparent" in that they may arise from either "fixed" bends (helix-like flexibility) or from points of increased flexibility (Fig. 6)—both fixed and flexible bends will yield reduced τ ratios. However, if two bends are placed with varied interbend separation in the same molecule, the flexibilities of the individual bends can be ascertained (Fig. 7). For fixed (helix-like) bends, the range of τ ratios for all torsional phasings between bends is specified by the magnitudes of the single bends. For bends with increased flexibility, the amplitude of the phase variation of the τ ratios will be reduced in proportion to the added flexibility (Fig. 7). In the limit of simple isotropic flexibility (no fixed component), all phase coherence is lost.

The phased τ ratio approach assigns an average degree of flexibility for the bending element. For simple nonhelix elements (e.g., bulges, internal loops, short single-stranded regions) or for protein-induced bends, the regions of increased flexibility (as well as the bending locus itself) are usually well localized. For much larger elements (e.g., tRNA cores), assignment of the specific regions of added flexibility may require additional studies.

[24] J. W. Orr, P. J. Hagerman, and J. R. Williamson, *J. Mol. Biol.* **275,** 453 (1998).
[25] A. Frazer-Abel and P. J. Hagerman, *J. Mol. Biol.* **285,** 581 (1999).

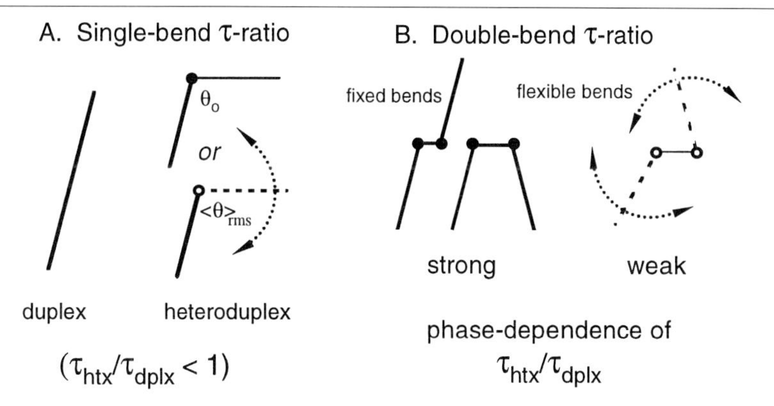

FIG. 6. Principle underlying the phased τ ratio approach. (A) Both fixed and flexible bends yield τ ratios that are less than 1. (B) RNA molecules with two bends in various torsional alignments will display a range of τ ratios that depends on the flexibility of the bend.

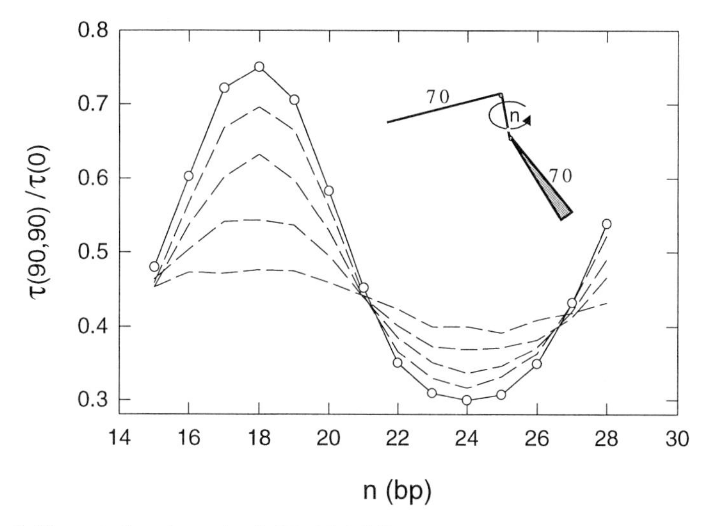

FIG. 7. Plots of phased τ ratios [$\tau(90, 90)/\tau(0)$] for a set of RNA molecules with two 90° bends separated by n bp.[8] The flanking helices are each 70 bp. $\tau(0)$ refers to a control RNA helix of length (140 + n) bp. Each bend is centered on a 10-bp region possessing various amounts of added flexibility (reduced persistence length, P): (—O—), no added flexibility (P_{bend}/P_{helix} = 1); (---), P/P_{helix} = 0.25, 0.125, 0.0625, and 0.031 in order of decreasing curve amplitude.

In any case, the method does allow one to determine whether a specific element is associated with increased conformational freedom. This approach has been used to demonstrate that in the presence of magnesium ions, the core of yeast $tRNA^{Phe}$ is essentially helix-like; that is, it possesses no more flexibility than an equivalent length of RNA helix.[4]

Protein–RNA Interactions

TEB measurements generally require micromolar concentrations of RNA molecules for adequate signal generation; therefore, the method cannot be used to study binding equilibria for proteins whose K_d values are submicromolar. On the other hand, for studies of relatively weak protein/small ligand–RNA interactions (e.g., $K_d > 2–5\ \mu M$) or for studies of stoichiometric binding leading to structural changes in RNA, the TEB approach is potentially quite useful. For example, TEB has been used to study the interaction between the HIV transactivation response element (TAR) and Tat-derived peptide.[26] In the absence of magnesium ions, the TAR bulge introduces a $50 \pm 5°$ bend in the surrounding helix; however, on binding the Tat-derived peptide (or arginine), the bend is essentially eliminated. In a separate study,[27] TEB was used to demonstrate that the ribosomal protein, S15, is capable of inducing a substantial rearrangement in the geometry of its binding target, a three-helix junction in the central domain of *Escherichia coli* 16S rRNA. Prior to S15 binding, helices H20, H21, and H22 were found to be separated by approximately equal angles (~120°). On binding of S15, helices H21 and H22 became essentially colinear, whereas helix H20 assumed a much more acute angle (~40°) with respect to helix H22.

The preceding two studies serve to illustrate the utility of the TEB approach for protein(peptide)–RNA interactions; however, several important, potentially limiting aspects of the method should be borne in mind. First, because measurements are generally performed at relatively low salt concentrations (<50–80 m*M* monovalent cation; <10–20 m*M* divalent cation), and utilize micromolar concentrations of RNA (and ligand), there exists a significant potential for protein–protein and/or protein–RNA aggregation. This phenomenon appears to be quite protein specific; many proteins do not appear to aggregate at all under the conditions employed in TEB measurements. For those proteins that do appear to aggregate, one approach that may prove useful is to perform the binding experiments at higher salt concentration, followed by simple dilution into the TEB buffer.

[26] M. Zacharias and P. J. Hagerman, *Proc. Natl. Acad. Sci. U.S.A.* **92,** 6052 (1995).
[27] J. W. Orr, P. J. Hagerman, and J. R. Williamson, *J. Mol. Biol.* **275,** 453 (1998).

Second, because binding experiments are performed in the micromolar range, nonspecific binding of protein may present a problem. This latter problem may be circumvented in some cases by performing titration curves under conditions where nonspecific binding is minimal, followed by exchange into the desired TEB buffer. In any case, one important feature of TEB is that reductions in τ ratios can only arise as a result of protein-induced bends or increases in flexibility; simple protein–protein aggregation, nonspecific binding, etc., will invariably lead to τ ratios that exceed 1.0.

Finally, as noted earlier, one important advantage of TEB relative to hydrodynamic methods that involve translational diffusion is the relative insensitivity of the method to the physical size of the protein itself. For proteins whose average radii are less than $\sim$ 10–15% of the length of the helix arm ($\leq \sim 30\ K_d$), essentially the only influence of centrally bound protein is through its influence on the angle between adjacent helices. This can be understood by noting that the bound protein is located near the center of hydrodynamic resistance, where frictional effects with solvent are minimal. For larger proteins, one can estimate the approximate contribution of the protein itself through a direct computational analysis in which the volume occupied by the protein, as well as its shape, is added to the hydrodynamic model. The same situation applies for the case of nonextended helices in multihelix junctions. For example, for a nonextended helix of ~8–10 bp, the direct hydrodynamic contribution of the stem would result in a maximum contribution to the τ ratio of 1–2% for a 60° bend.[8] For the case of bound protein, the direct hydrodynamic (frictional) contribution of the nonextended helix can be computed.

Acknowledgment

The author is indebted to Mr. Ferber Schleif for contributions to the design and fabrication of the TEB instrument. This work was supported by a grant from the National Institutes of Health (GM35305).

Section IV

Modeling Tertiary Structure

[27] Structure–Function Relationships of RNA: A Modeling Approach

By FABRICE LECLERC, BELSIS LLORENTE, and ROBERT CEDERGREN

Introduction

RNA molecules have received considerable attention in the past years because of the development and improvement of different methods applied to the study of the structure–function relationships of nucleic acids. These studies include structure determination,[1,2] chemical synthesis,[3] *in vitro* selection,[4] and computational modeling.[5] Whatever the methods used, experimental or theoretical, the approach to studying the relation between structure and binding properties or catalytic activity consists of introducing small perturbations that will affect the RNA structure (mutations, functional group modification) and/or that of its ligand (synthesis of different analogs). When studying binding properties, the experimental measurement of binding constants for different RNA ligands provides some data that can be used to compute a quantitative structure–activity relationship (QSAR) model. This mathematical model, based on the partitioning of the binding energy into different contributions, correlates the binding affinity to some specific physicochemical parameters. Therefore, three-dimensional QSAR (3D-QSAR) methods can be used to predict the affinity of new or designed small molecules that are potential ligands. The application of 3D-QSAR methods can also be extended to the study of other types of perturbations when the function concerns catalytic activity. In this case, mutations and chemical modifications at one position of the RNA itself can be considered as the interaction of a ligand, in our case a modified nucleotide, with a rigid oligonucleotide. Here, we focus on the example of a small self-cleaving RNA domain present in some plant viroids and virusoids, the hammerhead ribozyme,[6–9] which serves as a primary example of the RNA structure–function paradigm.

[1] A. Ramos, C. C. Gubser, and G. Varani, *Curr. Opin. Struct. Biol.* **7,** 317 (1997).

[2] S. R. Holbrook and S. H. Kim, *Biopolymers.* **44,** 3 (1997).

[3] N. Usnam and R. Cedergren, *Trends Biochem. Sci.* **17,** 334 (1992).

[4] S. E. Osborne and A. D. Ellington, *Chem. Rev.* **97,** 349 (1997).

[5] N. B. Leontis and J. SantaLucia, "Molecular Modeling of Nucleic Acids." ACS, Washington, DC, 1998.

[6] D. M. Long and O. C. Uhlenbeck, *FASEB J.* **7,** 25 (1993).

[7] J. Bratty, P. Chartrand, G. Ferbeyre, and R. Cedergren, *Biochim. Biophys. Acta* **1216,** 345 (1993).

0076-6879/00 $30.00

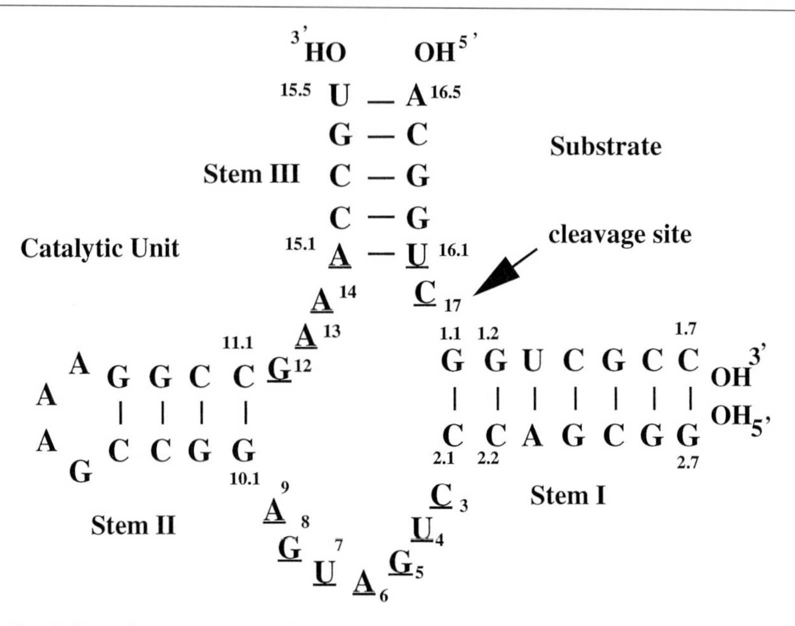

FIG. 1. Secondary structure of the hammerhead ribozyme. It consists of three base-paired stems flanking a central core of two nonhelical segments (C3–A9 and G12–A14) plus an unpaired nucleotide at the cleavage site (C17). Cleavage occurs between nucleotides 17 and 1.1. The nucleotides of the core are underlined.

Relationships between Ribozyme Structure and Catalytic Mechanism

X-Ray crystal structures[10,11] of the hammerhead ribozyme have allowed the formulation of hypotheses regarding the roles played by different functional groups in catalysis. Analysis of the activities of hammerhead analogs prepared by the substitution of modified nucleosides at key positions has been the driving force in the emergence of this model ribozyme. However, despite many mutagenesis, phylogenetic, and chemical modification studies,[12,13] the activities of some hammerhead analogs prepared by the substitution of modified nucleosides at key positions in the ribozyme core (Fig. 1) remain difficult to explain based on the crystal structures. These inconsisten-

[8] K. M. Amari and P. J. Hagerman, *J. Mol. Biol.* **261,** 125 (1996).

[9] L. Beigelman, A. Karpeisky, J. Matulic-Adamic, C. Gonzalez, and N. Usman, *Nucleosides Nucleotides* **14,** 907 (1995).

[10] H. W. Pley, K. M. Flaherty, and D. B. McKay, *Nature* **372,** 68 (1994).

[11] W. G. Scott, J. T. Finch, and A. Klug, *Cell* **81,** 991 (1995).

[12] T. Tuschl, M. M. Ng, W. Pieken, F. Benseler, and F. Eckstein, *Biochemistry* **32,** 11658 (1993).

[13] D. B. McKay, *RNA* **2,** 395 (1996).

cies have been attributed on one hand to the fact that the X-ray structures may not correspond to that of the biologically active ribozymes. On the other hand, they may not possess all characteristics of the structure of the transition state.[14] Other rationalizations include the possibility that the modified nucleosides incorporated in the hammerhead could provoke subtle changes in the conformation and that recalcitrant modifications affect low-energy hydrogen bonds whose absence does not influence activity.[15] A typical inconsistency involves position 7 of the hammerhead core (Fig. 1). Phylogenetic studies demonstrated that this position is not conserved: U is most common, but structures with A and C at this position have also been characterized.[16] Despite the existence of these variants, X-ray crystal structures propose three specific interactions at position 7: the O-2, the N-3 proton (U-specific interaction) and the O-2′ of U7 with the N-6 proton of A14, the O-6 of G8 and the N-1 of A14, respectively.[10,11] Beigelman *et al.*[9] have shown that an abasic nucleoside modification (the ribose–phosphate moiety only) at position 7 decreases the cleavage rate. Based on data from a rigorous study, Burgin *et al.*[17] reported that the catalytic activity of 12 hammerhead ribozymes analogs derived from the incorporation of modified nucleoside bases at position 7 (Table I) is not consistent with the hydrogen bond interaction that involves the O-2 atom of U7 suggested by the crystal structures. Monochromatic X-ray cryocrystallography has been used to trap a conformational intermediate of a biologically active hammerhead ribozyme that accumulates before cleavage.[18] However, the structure does not show any significant modification of the hydrogen-bonding pattern between U7 and G8 or A14 and thus does not offer any clue about the influence of U7 on the catalytic rate.

3D-QSAR Method

Basic Concepts

The preceding example illustrates the complexity of determining the precise role of chemical functional groups in the process of RNA catalysis. The complexity of structure–function relationships needs to be matched by the sophistication of the analytical platform. 3D-QSAR analyses use

[14] O. C. Uhlenbeck, *Nature Struct. Biol.* **2,** 610 (1995).
[15] P. Chartrand, F. Leclerc, and R. Cedergren, *RNA* **3,** 692 (1997).
[16] A. C. Forster and R. H. Symons, *Cell* **50,** 9 (1987).
[17] A. B. Burgin, C. Gonzalez, J. Matulic-Adamic, A. M. Karpeisky, N. Usman, J. A. McSwiggen, and L. Beigelman, *Biochemistry* **35,** 14090 (1996).
[18] W. G. Scott, J. B. Murray, J. R. P. Arnold, B. L. Stoddard, and A. Klug, *Science* **274,** 2065 (1996).

classification theory and statistical methods to quantify the importance of structural and electronic features in the biological activity of small ligands.[19] In 3D-QSAR, different combinations of atoms or pseudoatoms, when rings are involved, and their associated chemical or electronic properties are gathered into submolecular grouping called biophores. Biophores are then used to orient superimpositions of all molecules that contain them. The properties of superimposed atoms are taken as variables and an ensemble of simultaneous equations relating these variables to the activities of individual molecules is solved. The solution is a quantification of the relative contribution of each property in the activity, and permits a calculation of the theoretical activity of novel molecules.

This technique is particularly valuable in rational drug design where knowledge of preferred functional groups or, rather, the relative importance of electronic and chemical properties of atomic centers can be incorporated into the design and synthesis of new ligand effectors. Also, QSAR information can be useful in three-dimensional modeling of complexes involving the ligand, since it identifies the properties of key atomic centers as well as their three-dimensional juxtaposition in a ligand.[20,21]

Normally, QSAR analyses are performed on small ligands of biological interest because analyses of atomic attributes in macromolecules such as the hammerhead domain, where the conformation complexity is great, are too demanding for present software and hardware. Recently, however, experimentation has shown that the activity of some ribozyme analogs containing an abasic nucleoside can be rescued by the addition of the missing base to the medium.[22] The fact that the added base serves as an allosteric effector suggested a QSAR strategy to analyze structural variants of the hammerhead by assuming that the modified nucleosides used in the synthesis of the analog behave more like ligands rather than part of the macromolecule.

Catalytic Activity and Compounds

A 12-member compound set composed of the four natural nucleotides and eight different pyrimidine analogs was assembled[17] (Table I). To consider all potential contacts between the modified nucleoside and the remainder of the ribozyme, the structure of the entire nucleoside was used even

[19] A. C. Good, *in* "Molecular Similarity in Drug Design," (P. M. Dean, ed.), p. 24. Chapman and Hall, London, 1995.

[20] B. Llorente, F. Leclerc, and R. Cedergren, *Bioorg. Med. Chem.* **4,** 61 (1996).

[21] F. Leclerc and R. Cedergren, *J. Med. Chem.* **41,** 175 (1998).

[22] A. Peracchi, L. Beigelman, N. Usman, and D. Herschlag, *Proc. Natl. Acad. Sci. U.S.A.* **93,** 11522 (1996).

though only the nitrogen base or aromatic ring was varied. The cleavage rates (k_{rel}) for the ribozyme analogs expressed as a fraction of the wild-type activity were taken as a measure of the catalytic activity. Because QSAR analysis is based on the formalism of linear free-energy relationships, the activity was converted to a logarithmic form, $\log(1/k_{rel})$, that is proportional to the free-energy change that occurs during biological response.

Logical–Combinatorial Approach

QSAR analysis is performed by the APEX-3D module of the InsightII platform (Molecular Simulation Technologies, San Diego, CA), which is based on a combinatorial–logical analysis proposed by Golender *et al.*[23] In this methodology, centers refer to particular positions in the molecule that could be atoms or pseudoatoms such as ring centers, and each center is associated with a number of chemical or electronic attributes. The use of the property of a center makes the ensuing calculations independent of the properties of any particular atom. Both the centers and their associated properties are user defined. Input to APEX consists of the ligand structures, their biological activity, the centers, and their attributes. Structures are built using the InsightII Builder module and energy minimized using the molecular mechanics program Discovery (Molecular Simulation Technologies).

Biophores are local arrays of such centers which are common to a set of molecules having similar relative positions and activity values. Biophores are automatically identified by APEX (APEX-3D), and can be incorporated into a three-dimensional model of the ligand by combining the three-dimensional biophoric patterns present in the molecules under study. In addition, the final result of the analysis produces an equation of the form:

$$A = ax + by + cz + d$$

where x, y, and z are all attributes of the biophore that affect the activity of a compound; a, b, and c are the coefficients or weighting given to each attribute, A is the calculated activity of the ligand, and d, the independent term of the equation. QSAR models are achieved by a multiple linear regression algorithm. To avoid fortuitous correlations, which can be produced by this algorithm especially when the number of variables is greater than the number of compounds, the probability of chance correlations is calculated and used to select the most significant biophores.[24] Models with the best fit for properties to the biological activity have a high correlation

[23] V. E. Golender, A. B. Rozenblit, "Logical and Combinatorial Algorithms for Drug Design." Research Studies Press, 1983.

[24] J. G. Topliss and R. P. Edwards, *J. Med. Chem.* **22,** 1238 (1979).

TABLE I

CHEMICAL STRUCTURES OF 12 MODIFIED BASES SUBSTITUTED IN POSITION 7

Bases and surrogate bases	Nomenclature	k_{rel}[a]
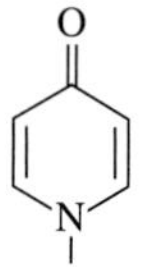	Pyridin-4-one	10.5
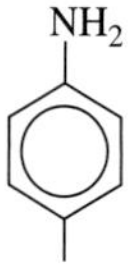	Aniline	2.5
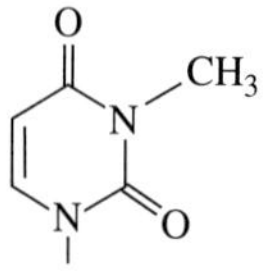	3-Methyluridine	2.1
	Phenyl	1.7
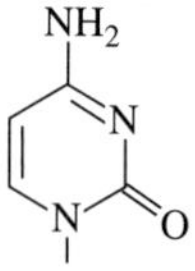	C	1.2
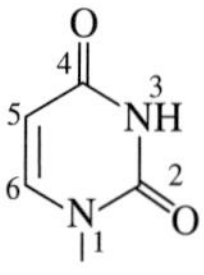	U	1
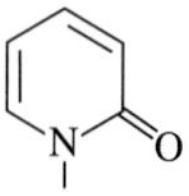	Pyridin-2-one	0.60
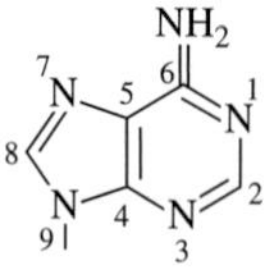	A	0.50
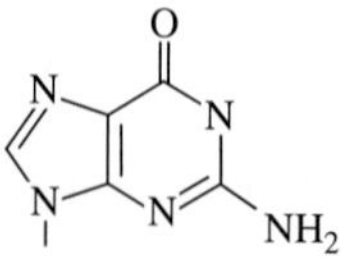	G	0.50

TABLE I (continued)

Bases and surrogate bases	Nomenclature	k_{rel}[a]
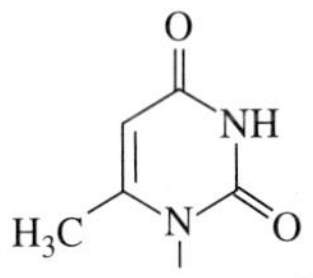	6-Methyluridine	0.23
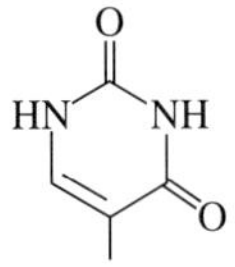	Pseudouridine	0.08
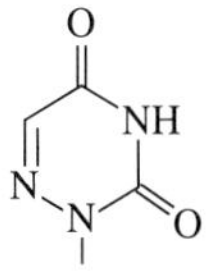	6-Azauridine	0.06

[a] k_{rel} was calculated from the activity of the analog containing each of the bases or surrogate bases divided by the activity of the analog containing U at position 7.

coefficient (r^2) and cross-validated correlation coefficient (q^2), low sums of the squared differences between the experimental activity and that calculated by the regression equation (RMSA). The cross-validation method is used to analyze 3D-QSAR model predictivity. This method consists of leaving out one compound from the data set, performing regression analysis on the remaining set, and predicting the activity of the omitted compound. The process is repeated until all compounds have been left out once. Models with highest predictive values also have low sums of squared difference between the observed and predicted activities determined by the cross-validation procedure (RMSP). The identification of good structural patterns (biophores), which classify well the molecules, requires the definition of several tasks generated by varying the parameters and properties of the atomic centers. Generally, the suggested tolerances values for the parameters in the APEX program are used; these provide that centers with properties within 20% of each other are considered equal.

3D-QSAR Models

Generation of Mathematical Models

From the general structure of a nucleotide, all nonhydrogen atoms and two pseudoatoms representing the pyrimidine and ribose rings are statistically analyzed by APEX. The properties of these atoms include electrophilicity, electronegativity, charge, hydrogen bond donor and acceptor abilities, π-population, HOMO, LUMO, hydrophobicity, refractivity, and in the case of the pseudoatoms, ring size and π electron density. In the first step, the APEX program constructs a large number of biophores composed of these atoms. We choose for subsequent analysis all biophores that were found in all 12 molecules inserted in position 7. These biophores then serve as reference points in the superimposition of the three-dimensional structures of the molecules. Evidence of the use of these reference points in the superimposition and not the overall molecular structure can be deduced from the fact that the five-membered ring of the purine is superimposed on the six-membered pyrimidine ring. The biophores selected for further study are chosen from these superimpositions based on match value of 0.84, a measure of the overall three-dimensional molecular similarity (1 being a perfect match). Selection of biophores having the highest match values is important, because it guarantees that the set of molecules possesses good three-dimensional alignments and high conformational similarities.

A quantitative structure–activity model was created using the k_{rel} assay measured for the hammerhead analogs. For the model with the highest predictive value, it was found that the variation in the activity could be rationalized by the following correlation equation:

$$\log(1/k_{rel}) = -0.25X^1_{C5} + 0.15X^2_{C2} - 0.08X^2_{C4} - 0.04X^3 + 2.63$$

where the three variables X^1, X^2, and X^3 represent the local refractivity, the electrophilicity, and the total refractivity, respectively. For each variable, the subscript gives the position of the center (atom or pseudoatom) associated with one of the three molecular properties represented.

The statistical parameters associated with the model include $n = 12$, $X = 4$, $r^2 = 0.99$, $q^2 = 0.94$, RMSA = 0.10, RMSP = 0.15, and $P = 0.04$, where n is the number of compounds, X the number of variables, r^2 the correlation coefficient, q^2 the cross-validated correlation coefficient, P the probability of chance correlation, and RMSA and RMSP the low sums of the squared difference between the observed and predicted activities determined by regression or cross-validation.

Statistical Significance and Predictivity

The statistical reliability of the model is given by the comparison of the calculated and cross–validated activity versus the observed activity for each of the 12 compounds (Fig. 2). According to these results a good predictive model is obtained. It clearly emerges that the activity is directly related to

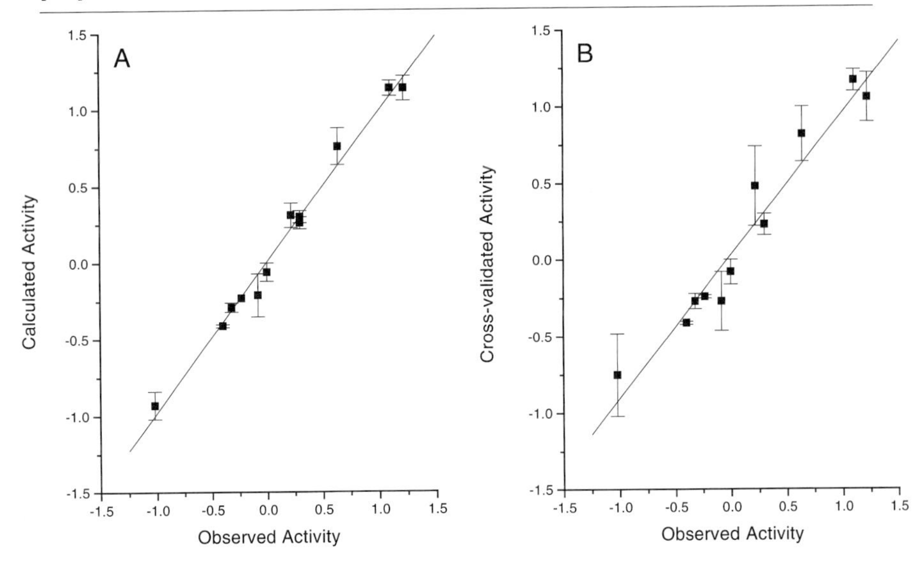

FIG. 2. Comparison of calculated and observed activities. The error bar for each point is given by the deviation of the calculated or predicted activity from the experimental activity. (A) Plot of calculated activities versus observed values. (B) Plot of cross-validated (predicted) activities versus observed values.

an increase of the molar refractivity of the substituent at position 5, the electrophilicity of the C-4 atom, the total refractivity (the total refractivity is calculated by adding all atomic refractivity contribution), and a decrease of the electron-withdrawing ability of the substituent linked to atom C-2 (decreasing the electron-acceptor ability C-2 atom). The molar refractivity of the substituent at position 5 and the electrophilicity of the C-2 atom seem to be the parameters giving the best explanations of variances in biological data, with a degree of correlation of 77 and 55%, respectively.

Table II lists the parameter values for the model equation (refractivity, C-5; electrophilicity, C-2 and C-4; total refractivity). According to this model, the most active compounds—pyridin-4-one, aniline, 3-methyluridine, phenyl, C, U, and pyridin-2-one—share a common secondary site (center of specific ligand–receptor interactions that may be present in only a subset of analyzed structures) that is associated with the molar refractivity of the substituent at the position C-5, which measures the molar volume and polarizability. This property occur two times in the equation of the model as a local property (refractivity C-5) and as a global property (total refractivity). Analysis of the electrophilicity coefficients (Table II) shows that in the active compounds, pyridin-4-one and aniline, the C-2 carbon presents no electron-acceptor ability. In contrast, the C-2 carbon of the

TABLE II
PARAMETER VALUES FOR MODEL EQUATION[a]

Bases and surrogate bases	Local refractivity (C-5)	Electrophilicity		Total refractivity
		(C-2)	(C-4)	
Pyridin-4-one	4.25	0.00	5.40	54.3
Aniline	3.45	0.00	0.00	57.4
3-Methyluridine	4.25	5.10	5.40	57.7
Phenyl	3.45	0.00	0.00	52.8
C	4.25	5.10	5.40	55.7
U	4.25	5.40	5.40	52.8
Pyridin-2-one	4.25	5.40	0.00	54.6
A	0.00	0.00	0.00	63.0
G	0.00	0.00	0.00	63.0
6-Methyluridine	0.00	5.10	5.40	58.6
Pseudouridine	0.00	6.00	5.10	52.6
6-Azauridine	0.00	6.00	6.00	50.7

[a] Different variables are calculated by the MOPAC program interfaced to APEX.

less active compounds, pseudouridine and 6-azauridine, shows the higher electrophilicity coefficient values. The fact that bases with low activity strongly contribute to the electrophilicity variable at center 2 is inconsistent with an electron-withdrawing substituent linked to the the C-2 carbon and therefore hydrogen bond formation proposed by Pley *et al.* and Scott *et al.* is not required for cleavage activity. This idea is also supported by the high catalytic activity of pyridine-4-one, aniline, and phenyl bases, which are unable to form hydrogen bonds at position 2, indicating that a 2-keto function at position 2 in pyrimidines is not essential to increase activity. The possible roles of the biophoric centers are summarized in Fig. 3 (see color insert), which shows the 12 compounds aligned according to our 3D-QSAR model and superimposed on U7 in the ground state structure of the ribozyme (for further discussion, see the subsection on the physical meaning of the 3D-QSAR model).

Design and Activity of New Modified Nucleotide Bases

Due to the large number of features of the nucleotide at position 7 that are related to the hammerhead catalytic activity, the importance of this base for activity is clear, despite the fact that it is the only phylogenetically nonconserved position within the core of the hammerhead ribozyme (Fig. 1). These features can be used to identify some structural modifications that can be considered to obtain nucleotides that could increase the activity. According to our model, the requirements for optimal activity include (1) no electron-withdrawing substituent linked to the C-2 carbon, (2) a large molar refractivity of the substituent at position 5, and (3) a strong electron-withdrawing substituent linked to the C-4 carbon. Atoms or group of atoms with greater electron polarizability substituted at position 5 should improve the activity by increasing both the molar refractivity at position 5 and the total refractivity. According to these observations and in order to optimize the activity, we combined the refractivity of the substituent at position 5 and electron-withdrawing effects at position C-4; the electrophilicity parameter at C-2 position cannot be improved, 0 is the minimum value (no electron-acceptor ability). We designed three new analogs and predicted their influence on the catalytic activity using the 3D-QSAR model obtained. The structures of these analogs and their predicted activity are summarized in Table III. The activity increases progressively from 5-nitropyridin-4-one to 5-bromopyridin-4-one to 5-sulfomethylpyridin-4-one. The three substituents at position 5 get strong attraction from dispersion interaction and maximize both the electron-acceptor ability at position C-4 and the total

TABLE III
DERIVATIVES DESIGNED AND PREDICTION OF THEIR INFLUENCE ON CATALYTIC ACTIVITY USING 3D-QSAR MODEL

Designed analogs	Predicted activity	
	$\log(k_{rel})$	k_{rel}
5-Nitropyridin-4-one 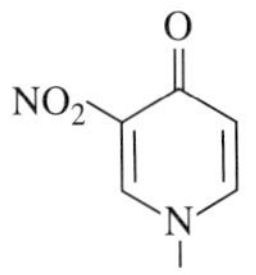	−2.16	146
5-Bromopyridin-4-one 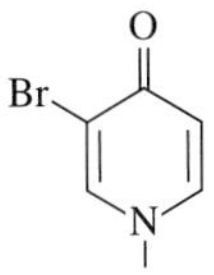	−2.31	199
5-Sulfomethylpyridin-4-one 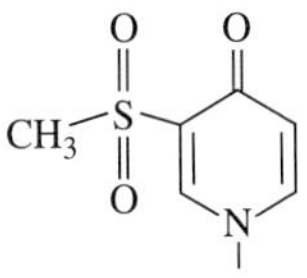	−4.09	12300

TABLE IV
GEOMETRY OF GROUND STATE AND INTERMEDIATE CONFORMATIONS AT POSITION 7

Conformation	Sugar pucker	P–O orientation (ζ/α)		P–P distance	
		G6–U7	U7–G8	G6–U7	U7–G8
Ground state	C3′-Endo, C2′-exo	(+ap, −ac)	(−ap, +sc)	5.7	6.5
First intermediate[a]	C3′-Endo, C2′-exo	(+ap, −ac)	(−ap, +sc)	6.3	6.0
Second intermediate[b]	C3′-Endo	(+ap, +ac)	(−ac, −ap)	7.1	5.8

[a] Conformational intermediate identified by Scott *et al.*[18]

[b] Conformational intermediate identified by Murray *et al.*[25] The orientation about the P–O ester bond, corresponding to the torsion angles α and ζ, is given according to the IUPAC–IUB commission (ac, anticlinal; sc, synclinal; ap, antiperiplanar). The more favorable orientation for RNA or A–DNA structures is the gauche conformation: (−sc, −sc).

refractivity; a sulfomethyl group gives the best compromise based on these parameters.

Physical Meaning of 3D-QSAR Model

Crystallographic freeze-trapping techniques have been used to isolate a conformational intermediate of a hammerhead analog that is arranged to form an in-line transition state.[25] Its structure reveals some significant conformational changes from the ground state. In the catalytic pocket and at position 7, the hydrogen bonding, base stacking (compare Figs. 3 and 4; see color insert), backbone, and sugar conformation are changed (Table IV). The three hydrogen bonds observed in the previous X-ray structures are weakened either because of an increase in the donor–acceptor distance (this is the case of the hydrogen bond involving the O-2′ of U7 with the N-6 proton of A14 which is lost) or because of the loss of directionality (where the angle between the donor, the built hydrogen, and the acceptor is less than 125 degrees). The sugar pucker and the conformation of the phosphate backbone at position 7 (in particular the orientation about the P–O ester bond between U7 and its neighbors G6 and G8) is indicative of a directed conformational change from the ground state to the first freeze-trapped intermediate identified with this last conformational intermediate. The direction of this change involves a stretching of the P–P distance between G6 and U7 and a decrease in the P–P distance between U7 and G8. It is interesting to note that an increase of the P–P distance would be facilitated by a C-2′-endo rather than C-3′-endo pucker at position 7, and would allow releasing part of the internal strain due to the unfavorable

[25] J. B. Murray, D. P. Terwey, L. Maloney, A. Karpeisky, N. Usnam, L. Beigelman, and W. G. Scott, *Cell* **92,** 665 (1998).

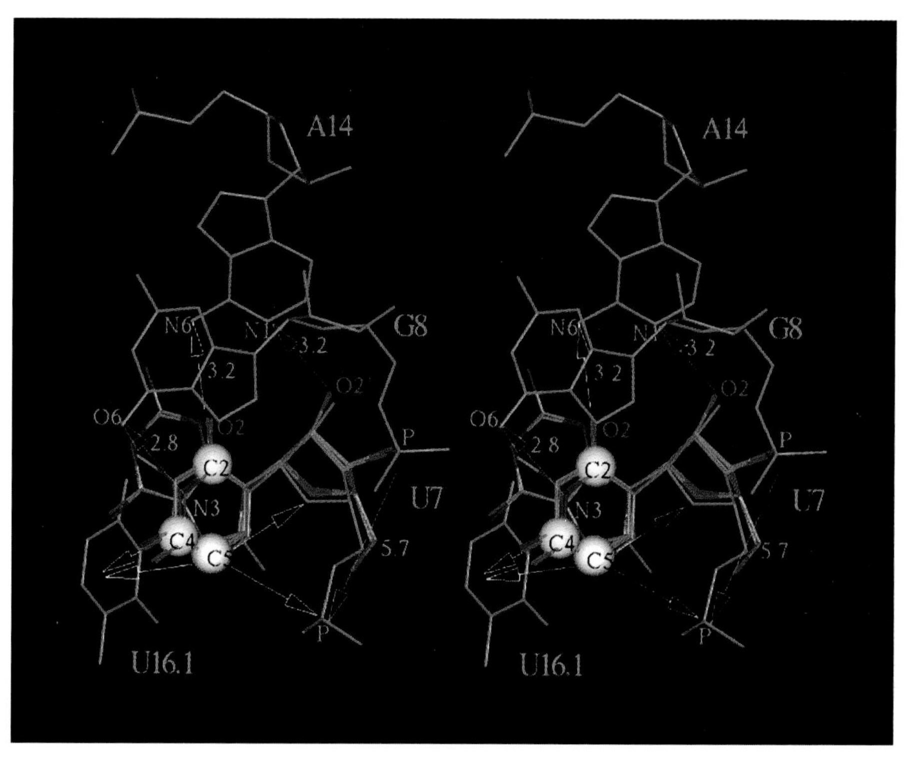

FIG. 3. Stereo view of the 3D-QSAR model inserted into the RNA region of the ground state structure of the hammerhead ribozyme[18] defined by U7 and its nearest neighbors. The 12 compounds aligned according to the 3D-QSAR model were positioned by the superimposition of the uridine from the model on U7. The biophoric centers are represented by solid spheres colored in yellow when the associated property is the electrophilicity and white when it is the refractivity. The hydrogen-bonding between U7 and A14 is indicated by arrows connecting the donor and acceptor and a numerical value for the distance. Additional arrows represent the possible role of the biophoric centers in the stabilization or destabilization of the structure through the interaction of U7 with its neighbors. The yellow arrow represents the negative effect of the electrophilicity at position C-2, the blue arrow the possible role of the electrophilicity at position C-4 on the base stacking with U16.1, and the white arrows the possible influence of the refractivity at position C-5 on the sugar and backbone conformation. The distance between the two phosphorous atoms from G8 and U7 connected by an arrow is given for information (see Table IV).

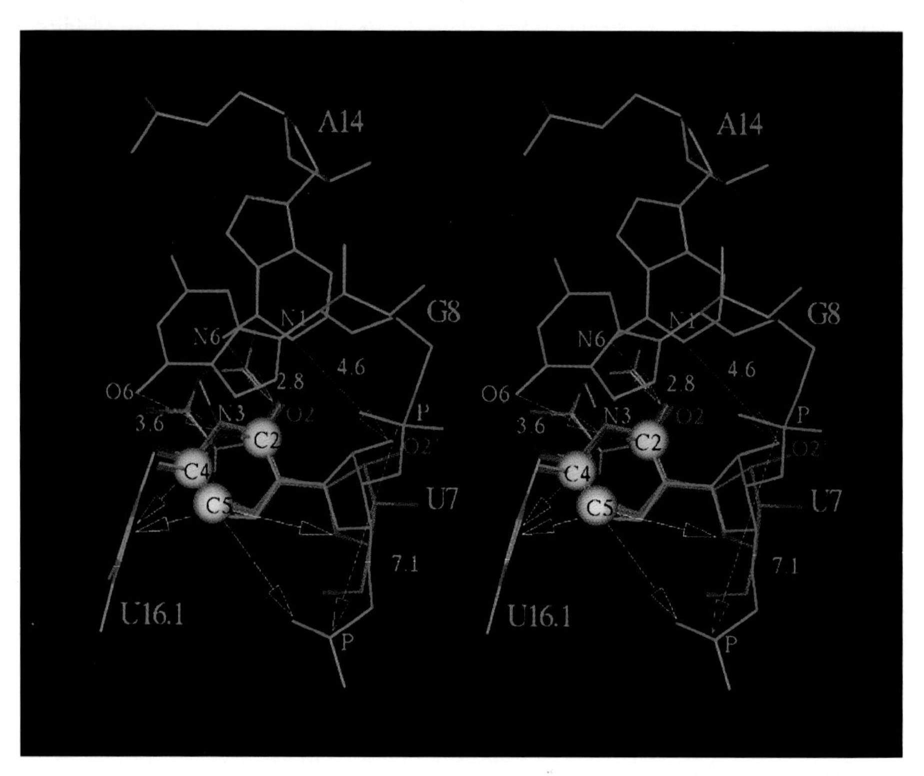

FIG. 4. Stereo view of the 3D-QSAR model inserted into the RNA region of the intermediate state structure of the hammerhead ribozyme[25] defined by U7 and its nearest neighbors. The positions of the 12 compounds from the 3D-QSAR model were obtained as described in Fig. 3. The modification of the hydrogen-bonding pattern between U7 and A14 from the ground state is indicated by green arrows and distances. The distance between the two phosphorous atoms from G8 and U7 connected by an arrow is given for comparison with Fig. 3. The arrows connected to the biophoric centers have the same meaning as in Fig. 3.

orientation about the P–O ester bond between G6 and U7 (Table IV). This hypothesis is consistent with the general trend followed by the modified nucleotides for which the improved catalytic rate is partly correlated to a preference for a C-2′-endo pucker.[17] The base stacking is also affected: in the later conformational intermediate U7 is only weakly stacked on A15.1 and completely unstacked from U16.1 (Fig. 4).

A high activity requires a low electrophilicity of the C-2 atom that is consistent with the breaking of the hydrogen bond between the C-2 oxygen of U7 and the N-6 proton of A14 to reach the transition state. The molar refractivity of the C-5 atom can interfere with the base stacking, as has been shown for 5-substituted pyrimidines.[26] It could also affect the sugar pucker or the backbone conformation, as has been shown for other modified pyrimidines.[27] Indeed, the influence of the base on the sugar pucker was demonstrated by Burgin *et al.*[17] The electrophilicity of the C-4 atom can influence the base stacking of U7 with its neighbors A15.1 and more specifically U16.1; for instance, modifications at the C-4 position are known to control the base orientation.[28] This position is also involved in the stabilization of the cleaved complex through base pairing with U4.[29] The absence of any center positioned on the N-3 atom indicates that the hydrogen bond between the N-3 proton of U7 and the O-6 oxygen of G8 is not essential as revealed by the X-ray structure. Possible roles of the different biophoric centers and their associated properties in the cleavage activity are summarized in Figs. 3 and 4. Because our 3D-QSAR model is based on a small data set, additional base modifications should be tested to determine its reliability. Nevertheless, it gives the first quantitative model that explains the role of position 7 in the structure–function relationship of the hammerhead ribozyme, and some directions for future experiments.

As noted in a previous work,[17] the base orientation, the pucker, and the backbone conformation at position 7 are interrelated factors that may all contribute at the same time to the activity. However, we do not need to understand in detail the role of these factors to make quantitative predictions. Nevertheless from our 3D-QSAR study and the analysis of the X-ray structures, we can propose a meaningful model from the physical and structural point of view based on the idea that the base modifications at this position interfere with the formation of the transition state and the stabilization of the cleaved complex. The activity can be affected in two different ways: the hydrogen-bonding pattern and/or the backbone conformation are changed in such a way that the activation barrier necessary to

[26] W. Hillen, E. Egert, H. J. Lindner, and H. G. Gassen, *FEBS Lett.* **94,** 361 (1978).
[27] W. Saenger, D. Suck, M. Knappenberg, and J. Dirkx, *Biopolymers* **18,** 2015 (1979).
[28] B. Lesyng and W. Saenger, *Naturforsch.* **C36,** 956 (1981).
[29] J. P. Simorre, P. Legault, A. B. Hangar, P. Michiels, and A. Pardi, *Biochemistry* **36,** 518 (1997).

go from the ground state to the transition state structure is increased or reduced. For example, the breaking of the hydrogen bond between the C-2′ oxygen of U7 and the N-6 proton of A14 and the increase of the P–P distance between G6 and U7 will tend to lower the activation barrier (note that the modifications should not alter the ability of the hammerhead ribozyme to fold in the correct conformation corresponding to the ground state).

Conclusion

3D-QSAR methods have been traditionally applied to the derivation of the relationship between the structural characteristics of a ligand and its biological (binding) activity and are directed toward modeling protein–ligand interactions. Because these methods do not require a knowledge of the structure of the receptor, they are well adapted to the limited structural information available for RNA and RNA ligand complexes. In this study, we have shown that these methods can also be useful in establishing the relationship between functional group modifications at a given position in a RNA structure and its biological function. They can provide quantitative models for very complex systems in which different RNA conformations may be involved. Finally, they can produce quantitative predictions for *de novo* modifications aimed at optimizing a given activity and thus they represent a very powerful tool for designing new experiments.

Acknowledgments

The authors thank Abdel Khiat for collaboration and William G. Scott for providing the coordinates of the X-ray structures. Robert Cedergren is a Fellow of the Evolutionary Biology Program of the Canadian Institute for Advanced Research. This work was supported by MRC of Canada. The article is dedicated to Robert Cedergren, his wife and his children.

[28] Computational Modeling of Structural Experimental Data

By MICHAEL A. BADA and RUSS B. ALTMAN

Introduction

The information explosion that is gripping molecular biology has challenged our traditional mechanisms for the collection, storage, and analysis

0076-6879/00 $30.00

of experimental data. In particular, it is becoming more difficult to create explanatory and predictive models that are consistent both internally and with the huge volumes of published data. The difficulty increases when a large variety of heterogeneous experimental approaches are used to gather data from multiple perspectives. A central strategy for managing this information overload is the creation of technologies that store and represent these data in novel ways. To facilitate computational processing of data, it is especially critical to develop standardized structured data formats for representing biological data.

There has been measured progress in creating methods for efficiently representing and storing biological data. There are several domains for which standardized representations of data are now routine. For example, the Protein Data Bank contains a template for reporting the results of X-ray crystallographic and nuclear magnetic resonance experiments that produce three-dimensional structures of biological molecules.[1] Similarly, GenBank contains a template for DNA sequencing experimental results and associated annotations. However, the large majority of biological experiments do not have standardized templates.[2] The results of these experiments are still predominantly disseminated in published texts accompanied by figures and tables for summary and convenience. While this format is useful for knowledge extraction by readers on a per-article basis, it does not allow for efficient integration of all data relevant to a particular topic, and it certainly is not amenable to computer-based data extraction for the purposes of further computations on these data.

The investigation into the structure of the bacterial ribosome (and other large nucleoprotein complexes) exemplifies some of the problems associated with current data dissemination. Until recently, large-scale application of crystallographic and nuclear magnetic resonance (NMR) techniques has not been successful. Thus, great efforts have been made toward the elucidation of the three-dimensional structure of the bacterial ribosome using a variety of experimental approaches, including, for example, chemical cross-linking and RNA footprinting methods.[3–5] Individual efforts typically produce a limited amount of data on a fragment of this huge macromolecular assembly. The resulting data are distributed unevenly throughout the assem-

[1] F. C. Bernstein, T. F. Koetzle, G. J. Williams, E. E. Meyer, Jr., M. D. Brice, J. R. Rodgers, O. Kennard, T. Shimanouchi, and M. Tasumi, *J. Mol. Biol.* **112**(3), 535 (1977).

[2] D. A. Benson, M. S. Boguski, D. J. Lipman, J. Ostell, and B. F. Ouellette, *Nucleic Acids Res.* **26**(1), 1 (1998).

[3] P. V. Baranov, P. V. Sergiev, O. A. Dontsova, A. A. Bogdanov, and R. Brimacombe, *Nucleic Acids Res.* **26**(1), 187 (1998).

[4] A. Huttenhofer and H. F. Noller, *EMBO J.* **13**(16), 3892 (1994).

[5] P. B. Moore, *Ann. Rev. Biophys. Biomol. Struct.* **27,** 35 (1998).

bly and have different levels of precision and accuracy. It is therefore quite difficult to create self-consistent models and to evaluate the compatibility of individual data sets with these models. As in other areas of biology, ribosomal research includes data collected using the techniques of genetics, biochemistry, biophysics, and phylogenetic analysis, and so it is increasingly difficult for researchers to keep abreast of relevant data generated using unfamiliar techniques.

To show the value of structured representations of data in dealing with these critical issues, we have built a prototype knowledge base of structural data pertaining to the 70S ribosome of *Escherichia coli,* called RIBOWEB. Diverse types of data taken principally from published journal articles are represented using a set of templates within this knowledge base, and these data are linked to each other with numerous and rich connections. Not only does this representation allow for easier and more convenient data retrieval by human users, but it facilitates automated data analysis by computer programs. We believe that formal representations of the data and models within scientific subdisciplines hold promise as a key method for delivering the next generation of scientific data resources and represent the way in which scientific data should be published in the future.

In this article, we describe the design principles behind the RIBOWEB knowledge base, illustrate how we have represented certain key types of experimental data, and describe the resulting knowledge base as it is currently publicly available. To demonstrate the utility of structured representations, we have written computer programs that use the knowledge base to create summary statistics about the density of information about the RNA and protein components of the 70S ribosome as a function of the type of experiment performed. The RIBOWEB system is also capable of interactively evaluating the consistency of particular data sets with proposed three-dimensional models of the ribosome.

Implementation

We have built our system around the concept of an *ontology.* For our purposes, the key features of an ontology are (1) a hierarchical classification of concepts (classes) from general to specific; (2) a frame, or list of attributes, for each class with a range of permitted values for each attribute; and (3) a set of relations between classes to link concepts in the ontology in more complicated ways than implied by the underlying hierarchy. The classes and relations are together referred to as the ontology. The leaves of the classification tree are termed *instances;* these represent concrete examples of the more abstract classes found in the internal part of the tree. Each attribute of an instance may have one or more corresponding values, whereas classes typically specify which attributes an instance may possess. Thus, a class may be *proteins,* but an instance would be *ribosomal protein S5 in E. coli.* The attribute "amino acid sequence" is assigned to the protein class but is given a specific value only in the context of an instance. We define a knowledge base as the combination of an ontology and an associated set of instances. The instances are most specific to the domain of interest (in this case the *E. coli* 70S ribosome), and the classes are generally relevant to a larger class of problems (such as all ribonucleoprotein complexes).

The three main classes within the RIBOWEB ontology are Reference-Information, Physical-Thing, and Data. Each journal article from which data has been entered into the knowledge base is itself entered into the ontology as an instance of Journal-Article. Each instance of a reference lists the author(s), journal name, publication year and title and contains a link to the PubMed (http://www.ncbi.nlm.nih.gov/PubMed/) Medline entry for that source. Most importantly, the data reported within this reference are represented in the knowledge base and linked to the reference with a relation called Reports. Each data instance, in turn, possesses an inverse relation called Reported-By, which specifies the reference that contains the datum. Relations such as Reports and its inverse, Reported-By, are declared in a formal way in the system so that the domain and range of these relations can be specified explicitly and enforced. This guarantees that the semantics of the knowledge base remain consistent—only Reference-Information instances can "report" a piece of Data. This link between data and reference is an important one, because it ties all high-level information about an experiment to information about its source.

In describing the findings of experiments, many of the attributes within data instances are instances within the class Physical-Thing. There are four main subclasses within the class: Organism, Molecule, Molecular-Ensemble, and Molecular-Part. Organism specifies a hierarchy for organisms for which ribosomal structural information has been collected. Even though the domain of our knowledge base is structural data for the *E. coli* 70S ribosome, other organisms are sufficiently similar (e.g., *Bacillus stearothermophilus*) such that results obtained experimentally may be applied to the *E. coli* system. Furthermore, organisms whose 16S RNA can be aligned to extract base conservation or covariation data are represented as well. Thus, this subclass provides links from each organism to all data that has been entered into the knowledge base for the organism. The concept hierarchy for Physical-Thing is shown in Fig. 1A.

The second subclass of Physical-Thing is Molecule, which is simply any chemically distinct molecule (i.e., a collection of atoms chemically bonded to form a discrete entity). Included in this subclass are the sub-subclasses of nucleic acids, proteins, and small organic and inorganic compounds that

A

- Physical-Thing
  - Molecular-Ensemble
    - 30s-Ribosomal-Subunit
  - Molecular-Part
    - 30s-Subunit-Part
      - 30s-Subunit-Head-Part
      - 30s-Subunit-Lower-Body-Part
      - 30s-Subunit-Neck-Part
      - 30s-Subunit-Platform-Part
      - 30s-Subunit-Upper-Body-Part
    - Chemical-Group
    - Nucleic-Acid-Part
      - Rna-Part
        - Mrna-Position
        - Rna-Base
          - Adenine
          - Cytosine
          - Guanosine
          - Uridine
        - Subsequence
        - Terminus
    - Protein-Part
      - Amino-Acid
    - Ribosomal-Functional-Site
  - Molecule
    - Macromolecule
      - Nucleic-Acid
        - Dna
        - Rna
          - Mrna
          - Rrna
          - Trna
      - Protein
        - Antibody
        - Nuclease
        - Ribosomal-Cofactor
        - Ribosomal-Protein
          - Large-Subunit-Ribosomal-Protein
          - Small-Subunit-Ribosomal-Protein
    - Mutant-Molecule
    - Small-Molecule

B

- Data
  - Biochemical-Data
    - Antibiotic-Resistance-Data
    - Binding-Affinity-Data
    - Cross-Linking-Data
    - Footprinting-Data
      - Chemical-Footprinting-Data
      - Enzymatic-Footprinting-Data
      - Fe-Radical-Attack-Data
    - Hydroxyl-Radical-Cleavage-Data
    - Rna-Partial-Digestion-Data
  - Phylogenetic-Data
    - Base-Conservation
    - Base-Interaction
      - Base-Pair
      - Tertiary-Interaction
  - Physical-Data
    - Electron-Microscopic-Data
    - Fluorescence-Microscopic-Data
    - Neutron-Scattering-Data
      - Neutron-Scattering-Distance-Data
      - Radius-Of-Gyration-Data
    - Nmr-Data
    - Small-Angle-X-Ray-Scattering-Data
    - X-Ray-Crystallographic-Data
  - Structural-Model-Data
    - 3-D-Model
    - Ribosomal-Protein-Position-Data

FIG. 1. (A) The complete hierarchy of physical things within the ontology. Instances of physical things serve as values for attributes of data instances, among other things. With this implementation, many convenient links between biological entities and data are created, and a standard terminology is enforced. (B) The complete hierarchy of data within the ontology. Each piece of data extracted from journal articles is entered as an instance of one of these classes.

are needed to sufficiently describe experimental data (e.g., footprinting and cross-linking agents). This ontology also can represent any mutant of a nucleic acid or a protein with deletions, insertions, and/or mutations. A collection of two or more molecules that associate largely through van der Waals interactions is entered into the ontology under a third subclass, Molecular-Ensemble. For example, the *E. coli* 30S subunit is composed of 16S rRNA and the 21 ribosomal S-proteins, so the RNA and proteins are entered as molecules, while the subunit is represented as a molecular ensemble.

The fourth subclass, Molecular-Part, is defined as a part of a Molecule or of a Molecular-Ensemble. Unlike molecules and molecular ensembles, molecular parts do not have a separate existence apart from the entity to which they belong, and they are not necessarily of a specific size. For example, a part can be a monomer, such as a base of a nucleic acid or an amino acid of a protein, or it can be a more abstractly defined part, such as a terminus of a nucleic acid or a ligand binding site of a protein. They allow a shorthand method for referring to sections of a molecule consisting of a subset of atoms. We have chosen to subdivide Molecular-Part into subclasses according to the nature of the partition. Thus, there is Nucleic-Acid-Part (e.g., specific bases, termini, and subsequences of specific nucleic acids); Protein-Part (e.g., specific amino acids of proteins); Ribosomal-Functional-Site, which is defined as an area of the ribosome that plays a significant, functional role (e.g., A-Site); and 30S-Subunit-Part, which was created to describe regions of the surface of the small subunit in order to represent results from electron microscopic experiments (e.g., parts of regions commonly referred to as the "head" and "platform"). Molecular parts are linked with the molecules and/or molecular ensembles which they constitute and, like molecules and molecular ensembles, to any relevant data which appear in the knowledge base.

All experimental data relevant to the structure of the 70S ribosome of *E. coli* are represented under Data, and each piece of data is recorded as an instance of some subclass of Data. For example, data from binding affinity and cross-linking experiments appear under the subclass Biochemical-Data, electron microscopic and fluorescence spectroscopic data can be found under Biophysical-Data, and data resulting from analysis of base covariation within multiply aligned sequences appear under Phylogenetic-Data. These data instances are the central focus in our knowledge base; the purpose of adding anything else to the knowledge base is to adequately convey the essential pieces of knowledge in a given datum. As described earlier, these include molecules, parts of molecules, organisms, and reference information. The concept hierarchy for Data is shown in Fig. 1B.

Each type of experiment that is represented in the knowledge base has been distilled to a small number of attributes (approximately 5–15) that are critical for understanding the information content of the experiment. Each attribute may be filled in with values of the appropriate type (as defined by the range of the attribute). Some attributes are mandatory, and some are optional. For example, the attributes (and their documentation) which appear in instances of the class Hydroxyl-Radical-Cleavage-Data are:

A-Priori-Target: Any entity (e.g., the whole ribosome or some subset thereof) on which an experiment is performed. Euphemistically, this is what is in the test tube during the course of the relevant experiment.

Cleavage-Site: The site of cleavage.

Cleavage-Strength: The strength with which the site of cleavage is cleaved by the cleaving agent, described in absolute terms (e.g., strong).

Radical-Generator: The chemical group (including the ligand that binds the specific radical generator to the ribosome) that is the hydroxyl radical generator. These radicals specifically attack RNA in a hydroxyl radical cleavage experiment.

Radical-Generator-Attached-To-Position-Of: The position in the *a priori* target to which the radical generator is attached. For example, the value of this attribute may be a wild type or a mutation of wild type.

Reported-By: The source (e.g., journal article) that reports the given piece of data.

By examining the attribute names, values for the attributes, and definitions for each of these, the basic conditions and results of the experiment can be understood. For example, Fig. 2 shows an instance of Hydroxyl-Radical-Cleavage-Data.[6] By looking at the values of A-Priori-Target, one can determine that the entity studied in this experiment was a complex consisting of 16S rRNA and the proteins S15, S16, S17, S20, S4, S7, S8, and a mutant of S5. We have attempted to make the names within the system self-explanatory, but some complex objects are not amenable to short, obvious names. However, the user can interrogate the knowledge base frame for any object, where they can view the documentation for the object and all information within the knowledge base for that object. For example, if the user did not understand what is meant by E-Coli-S5-Mut [S $\rightarrow$ C21], they could click on that value to go to its frame (Fig. 3), where a detailed representation records the fact that this molecule is S5 with the mutation

[6] G. M. Heilek and H. F. Noller, *Science* **272**(5268), 1659 (1996).

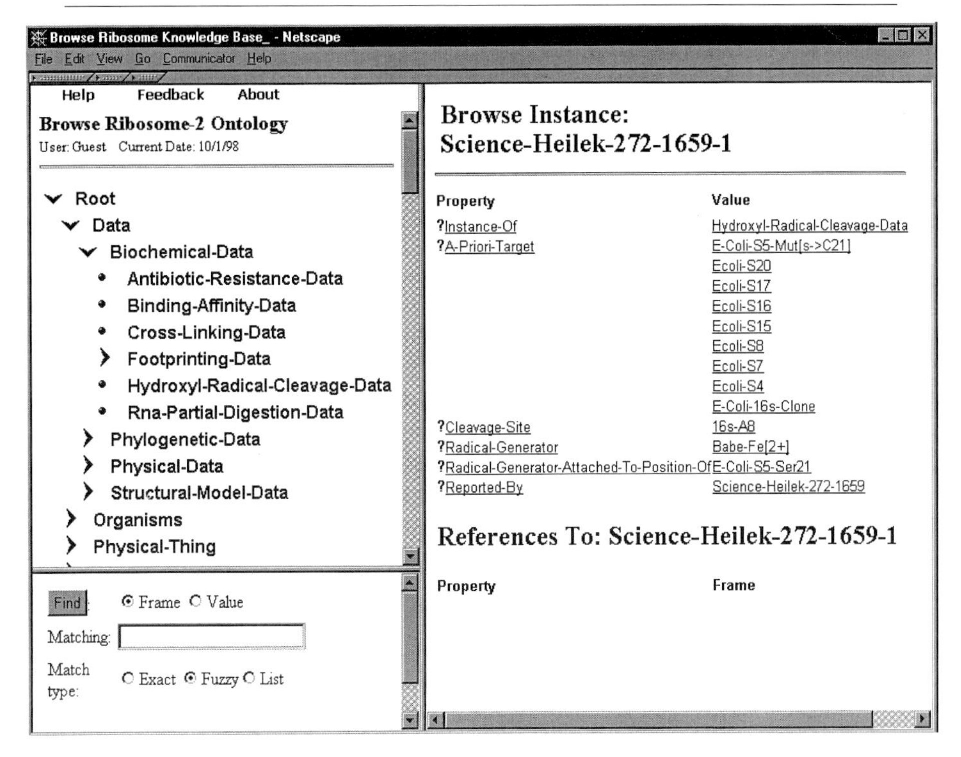

FIG. 2. A snapshot of a data instance as viewed by the SOPHIA Web browser. The upper left window shows the ontology of classes, and the large right window displays any instance(s) selected. When a class is clicked within the ontology window, a list of all instances of this class appears in the instance window. The frame for an instance is viewed by clicking on the desired instance. In the lower left window is a search engine to find specific terms within the ontology. It can search for the given term as names of instances or as values of attributes and can search for exact or approximate (fuzzy) matches.

Ser → Cys21. Also in this frame are links from E-Coli-S5-Mut [S → C21] to other instances within the knowledge base, including the wild-type molecule (Is-Mutant-Of), the mutated part (Mutated-Part), and data instances in which this molecule appears as the *a priori* target (Serves-As-Target-In).

For the instance of Hydroxyl-Radical-Cleavage-Data shown in Fig. 2, the site of cleavage is 16s-A8 (whose frame indicates that this is base A8 of 16S rRNA). The radical generator is a chemical group that was named Babe-Fe [2+] [which is Fe[2+] attached to the linker 1-(*p*-bromoacetamidobenzyl)-EDTA, or BABE], and this radical generator is attached to the position of E-Coli-S5-Ser21, or Ser21 of *E. coli* S5. However, it is actually

Browse Instance: E-Coli-S5-Mut[s->C21]

Property	Value
?Instance-Of	Mutant-Molecule
?Documentation	_E._ _coli_ S5 with the mutation Ser->Cys21
?Has-Part-Labeled	E-Coli-S5-Lys166
	E-Coli-S5-Ser129
	E-Coli-S5-Ser99
?Is-Mutant-Of	Ecoli-S5
?Mutated-Part	E-Coli-S5-Ser21

References To: E-Coli-S5-Mut[s->C21]

Property	Frame
A-Priori-Target = E-Coli-S5-Mut[s->C21]	Science-Heilek-272-1659-1
	Science-Heilek-272-1659-10
	Science-Heilek-272-1659-11
	Science-Heilek-272-1659-12
	Science-Heilek-272-1659-13
	Science-Heilek-272-1659-14
	Science-Heilek-272-1659-15
	Science-Heilek-272-1659-16
	Science-Heilek-272-1659-17
	Science-Heilek-272-1659-18
	Science-Heilek-272-1659-19
	Science-Heilek-272-1659-2
	Science-Heilek-272-1659-20

FIG. 3. The frame for an instance of Physical-Thing within the ontology. Specifically, this represents *E. coli* S5 with a Ser → Cys21 mutation, as the documentation indicates. All information that has been entered into the knowledge base pertaining to an entity can be viewed from its frame. Here, we see information about the mutant's wild-type molecule, the amino acid that was mutated, its amino acids which have been entered, and the data instances in which it appears.

attached to the cysteine to which this serine residue was mutated, as indicated by the mutant listed in A-Priori-Target. Finally, this piece of data was taken from a data source (a journal article) called Science-Heilek-272-1659 (shown in Fig. 4A). Our unique naming convention for journal articles combines the journal name, first author, volume, and first page number. The instance associated with the attribute Reports is shown in Fig. 4B; this demonstrates the degree of semantic control exerted by the system on the values of attributes.

Figure 5 shows a snapshot of an example of an instance of biophysical data.[7] More specifically, the value of the attribute Instance-Of indicates that this is an instance of Electron-Microscopic-Data. The *a priori* target, as defined above, is the *E. coli* 30S subunit. The entity being localized is the 3′ terminus of *E. coli* 16S rRNA, and the electron microscopic probe (anti-dinitrophenyl antibody) directly binds to a dinitrophenyl group. This data instance further indicates that the precision of probe binding (defined as a quantitative precision, in terms of percent, with which the electron microscopic probe binds to the site identified as being correct in the paper) is 88, and this piece of data was extracted from a journal article entered into the knowledge base as Pnas-Olson-76-3769. The values for Binding-Position-Of-Probe are sections of the three-dimensional outer surface of the small ribosomal subunit. To represent electron microscopic data in our ontology, we created an ontology for the standard sections of the 30S subunit. We first divided the subunit into five main parts corresponding to commonly used terms (head, neck, platform, upper body, and lower body), and then we divided each of these parts into octants. The unique identity of each of these octants can be specified by three coordinates: a solvent/interface coordinate (indicating if the octant is situated on the solvent or interface half of the subunit), a head/body coordinate (indicating if the octant is directed toward the head or the body of the subunit), and a platform/nonplatform coordinate (indicating if the octant is nearer the half containing the platform or on the half away from the platform). Thus, [Solvent-Head-Platform]-Platform-Part is a platform part, and of the octants that constitute the platform, it is oriented toward the solvent, toward the head, and toward the side containing the platform. This symbolic coordinate system, while admittedly of low resolution, is sufficient to represent a large fraction of the immunoelectron microscopy data collected on the ribosome. As higher resolution surface envelopes emerge, it will be necessary to create a more detailed ontology and vocabulary for describing positions on the surface of the 30S subunit.

A third example of the experimental templates in our knowledge base is the class Antibiotic-Resistance-Data, with an instance of this class shown in Fig. 6.[8] This class is used to represent experiments in which the resistance of an organism to antibiotics is measured in response to mutations intro-

[7] H. M. Olson and D. G. Glitz, *Proc. Natl. Acad. Sci. U.S.A.* **76**(8), 3769 (1979).
[8] T. Powers and H. F. Noller, *EMBO J.* **10**(8), 2203 (1991).

A

Browse Instance: Science-Heilek-272-1659

Property	Value
?Instance-Of	Journal-Article
?Authors	Heilek-G-M Noller-H-F
?Documentation	
?Journal-Name	Science
?Publication-Year	1996
?Pubmed-Link	http://www.ncbi.nlm.nih.gov/htbin-post/Entrez/query?uid=8658142&form=6&db=m&Dopt=b
?Title	Site-Directed Hydroxyl Radical Probing of the rRNA Neighborhood of Ribosomal Protein S5

References To: Science-Heilek-272-1659

Property	Frame
Reported-By = Science-Heilek-272-1659	Science-Heilek-272-1659-1
	Science-Heilek-272-1659-10
	Science-Heilek-272-1659-100
	Science-Heilek-272-1659-101
	Science-Heilek-272-1659-102
	Science-Heilek-272-1659-103
	Science-Heilek-272-1659-104
	Science-Heilek-272-1659-105
	Science-Heilek-272-1659-106
	Science-Heilek-272-1659-107
	Science-Heilek-272-1659-108

FIG. 4. (A) The frame for an instance of Reference-Information. Values for the source's authors, journal, publication year, and title can be seen here. The article's PubMed entry can be viewed simply by clicking on the value for Pubmed-Link. Furthermore, a list of the data instances from the reference that have been entered into the knowledge base can be seen in the bottom half of the screen. (B) The frame for an instance of a relation, the Reports relation. An explanation of the relation is held in the attribute Documentation. The value of Domain is the class in which the relation appears, and the value of this relation must be an instance of the value of Range. Furthermore, most links between terms within the ontology are bidirectional; the inverse of this relation is Reported-By.

duced in the 30S subunit. The instance in Fig. 6 describes an experiment with a mutant named E-Coli-16s-Mut [C → U525]. As before, a structured description of *E. coli* 16S rRNA with a C525U mutation can be found by clicking on its name. Spectinomycin and ampicillin are present in the growth medium, and the colony is grown at a temperature of 37° (the unit being specified in the documentation for the attribute Temperature). This trial measures the interaction between bases G508 and C525 of 16S rRNA, and the colony size in these growth conditions is qualified as "medium,"

B **Browse Instance: Reports**

Property	Value
?Arity	2
?Documentation	All data instances which are reported by the source (e.g., journal article) in question.
?Domain	Publication
?Instance-Of	Binary-Relation Relation
?Inverse	Reported-By
?Range	Data

References To: Reports

Frame	Property
Reported-By	Inverse
Publication	Domain-Of

FIG. 4. (*continued*)

Browse Instance: Olson-Em[e-Coli-16s-3-Prime-Terminus]

Property	Value
?Instance-Of	Electron-Microscopic-Data
?A-Priori-Target	E-Coli-30s-Subunit
?Binding-Position-Of-Probe	[solvent-Head-Platform]-Platform-Part [solvent-Head-Nonplatform]-Platform-Part [50s-Head-Platform]-Platform-Part [50s-Head-Nonplatform]-Platform-Part
?Documentation	
?Electron-Microscopy-Probe	Anti-Dinitrophenyl-Antibody
?Electron-Microscopy-Probe-Directly-Binds-To	Dinitrophenyl-Group
?Measures-Physical-Property-Of	E-Coli-16s-3-Prime-Terminus
?Precision-Of-Probe-Binding	88
?Reported-By	Pnas-Olson-76-3769

FIG. 5. The frame for an instance of Electron-Microscopic-Data, a subclass of Physical-Data. As in all frames, documentation explaining a given attribute will appear when the mouse pointer is placed over the question mark directly to the left of the attribute. This datum reports that four different segments of the 30S subunit platform bind to a probe connected to the 3′ terminus of 16S rRNA.

Browse Instance: Powers-Growth-Analysis-Ec16s[c->U525]-Amp/Spc-37

Property	Value
?Instance-Of	Antibiotic-Resistance-Data
?Antibiotics-In-Medium	Spectinomycin Ampicillin
?Colony-Size	medium
?Colony-Size-Of-Wild-Type	large
?Documentation	
?Measures-Base-Pair-Interaction-Between	16s-G506 16s-C525
?Mutations	E-Coli-16s-Mut[c->U525]
?Reported-By	Emboj-Powers-10-2203
?Temperature	37

FIG. 6. The frame for an instance of Antibiotic-Resistance-Data, a subclass of Biochemical-Data. Values that are not meaningful by themselves (e.g., numbers and qualitative terms such as "medium") are typically not entered into the ontology as part of the specific vocabulary. However, the vast majority of values of attributes are entered as such; these are visible as underlined terms here. This datum shows that spectinomycin and ampicillin moderately reduce the rate of growth in a bacteria with a mutant of C525 in 16S rRNA that replaces a GC base pair with a GU base pair.

compared to a "large" colony size for the organism with the wild-type 16S rRNA in identical conditions. These data were extracted from the journal article represented as Emboj-Powers-10-2203. Again, any of the underlined values is also an instance within the knowledge base, and the specific instances of data, physical things, and references are interlinked. For example, this instance of Antibiotic-Resistance-Data can be found starting from the frames for Antibiotic-Resistance-Data, Spectinomycin, Ampicillin, 16s-G506, 16s-C525, E-Coli-16s-Mut[C $\rightarrow$ U525], or Emboj-Powers-10-2203.

The technology for designing ontologies and delivering knowledge bases both over the Web and through interfaces that gives computer programs access to the knowledge bases has emerged during the last 5 years. Numerous software tools are available for designing ontologies, including the Web-based Ontolingua project,[9] the Windows-based Protégé program,[10]

[9] A. Farquhar, R. Fikes, and J. Rice, "The ontolingua server: a tool for collaborative ontology construction." Stanford Knowledge Systems Laboratory Technical Report No. KSL-96-26, Sept. 1996.

[10] M. Musen, S. Tu, H. Eriksson, J. Gennari, and A. Puerta, *in* "Proceedings of International Joint Conference on Artificial Intelligence" AAAI Press, Menlo Park, California, 1993.

and others.[11–13] The RIBOWEB ontology was initially developed in Ontolingua. The knowledge base is now stored in a knowledge base delivery environment developed in our laboratory called SOPHIA.[14] SOPHIA is based on the ACCESS97 database program delivered by Microsoft. It uses a relational database to store the entire ontology and knowledge base. A Web interface to SOPHIA allows browsing of its contents, and programmatic access to the data is afforded by a library of Perl functions that can be used by an application programmer to access the data with functions such as *get-all-instances-of-concept* or *get-attribute-value-of-instance.* SOPHIA also contains a simple ontology and knowledge base editor and is available on request to the authors.

Results

The knowledge base is available at http://www-smi.stanford.edu/projects/helix/riboweb/kb-pub.html and currently contains 106 classes, 165 relations, and more than 17,000 total instances. Included among the instances are 8000 data instances from approximately 170 journal articles, and these represent more than 20 different types of biological experiments, including cross-linking, RNA partial digestion, cell growth studies, binding affinity studies, footprinting, electron microscopic localization, fluorescence spectroscopy, neutron scattering, NMR, crystallographic, small-angle scattering and neutron diffraction measurements. At least half of these 20 templates have been used over a wide enough range of different published articles that we are confident that they will not have to be significantly altered. The other templates have data relatively fewer articles and thus may require refinement in terms of mandatory and optional features or overall conceptualization. Each attribute is extensively documented in terms of its semantic significance as well as the legal values that can be assigned (as in the Reports attribute, shown in Fig. 4B).

The knowledge base is useful for browsing, but its real power comes from the fact that computer programs can be written to automatically

[11] P. Karp, "The design space of frame knowledge representation systems." SRI International, Artificial Intelligence Center Technical Report No. 520, (1992).

[12] R. MacGregor, *in,* "Principles of Semantic Networks," (J. Sowa, ed.), pp. 385–400. Morgan Kaufmann Publishers, Los Altos, California, 1991.

[13] T. Mitchell, J. Allen, P. Chalasani, J. Cheng, E. Etzioni, M. Ringuette, and J. Schlimmer, *in,* "Architectures for Intelligence," (K. VanLehn, ed.). Erlbaum, Hillsdale, New Jersey, 1989.

[14] N. Abernethy and R. Altman, (19xx) "SOPHIA: providing basic knowledge services with a common DBMS." Stanford Medical Informatics Technical Report No. SMI-98-0710, May 1988.

extract and analyze data. As a demonstration of this capability, we conducted a series of computations in which the density of information along the 16S rRNA sequence was evaluated. A Perl program was written [using the application programmer's interface (API) to the knowledge base] to count the number of data instances (either in total or of a particular type of experiment) involving each RNA base in *E. coli* 16S rRNA. The program first finds all relevant data instances (for example, all cross-linking experimental results) and then examines the values of the attributes within these instances to count the occurrence of RNA bases. We display the results in two manners. First, we can create a simple linear plot of sequence versus number of hits. Second, we can use a color coding of the standard 16S rRNA secondary structure graphic to show which bases have the most and least hits. Figure 7 (see color insert) shows the overall information density for the 16S sequence, plotting the number of individual data instances mentioning each base. We used these programs to evaluate the information density for the 16S sequence from (1) all biochemical experiments (shown in Fig. 8A), (2) the subset of cross-linking experiments in the knowledge base (Fig. 8B), and (3) the subset of footprinting experiments in the knowledge base (Fig. 8C). We also computed a similar metric over the 21 proteins of the 30S subunit as a function of all available data (Fig. 8D).

Discussion

The explosion of biological data has led to increased efforts to logically represent, store, and display scientific knowledge. Several domains have successfully created standardized templates for data and their usefulness is apparent. RIBOWEB improves on these online resources in a number of ways. First, it contains templates for many types of biological data that previously had been limited to textual representation for all but the most prevalent types of data. These templates were expressly created in order to support modeling of portions of the *E. coli* 70S ribosome, but they should be useful in representing data collected on a variety of protein–RNA complexes. We have emphasized generality and semantic clarity so that the templates can be applied to other domains within the field of structural biology. We are interested in representing, for example, the structural information available about the ribosomal 50S subunit or the spliceosome.

Although our standard data representation is more limited than natural language, it contains the most salient features (particularly the results) of a given experiment. For example, the fluorescence donor and acceptor groups, their respective positions within the ribosome, the calculated distance between these groups, the efficiency of energy transfer, the *a priori* target, and the data source can all be easily discerned in a fluorescence

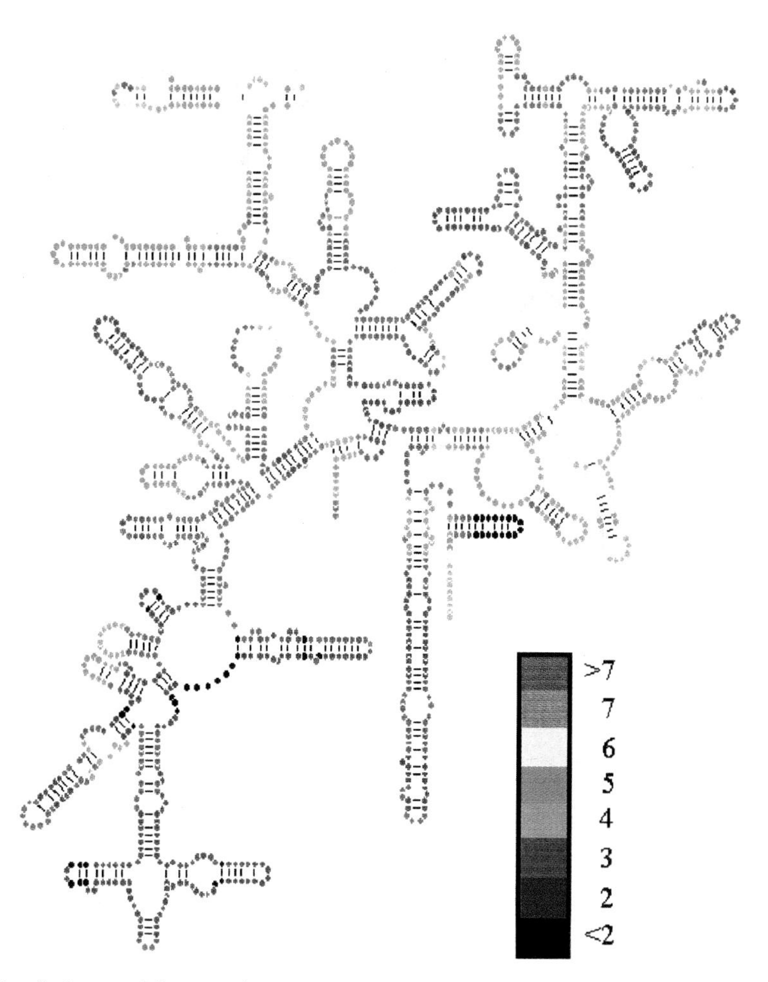

FIG. 7. A map of the secondary structure of *E. coli* 16S rRNA. Eash base has been color coded to indicate the regional information density of 16S rRNA within the knowledge base. The number of occurrences of each base in the data instances were obtained by querying the knowledge base for the individual bases or for short (≤20-mer) subsequences of which the bases were members. A moving average with a window of 9 bases was used to identify regions of the sequence with large numbers of measurements recorded in the knowledge base. Certain regions of the central domain (e.g., regions centered at approximately 525 and 710) have been intensely studied, as have been certain regions near the 3′ terminus. Conversely, there are few occurrences for large sections of the first third of the sequence. Because of the large number of data instances entered (>6000), this may be considered an approximate map of the amount of structural knowledge known by region.

spectroscopic data instance. Thus, by examining a set of attributes and their respective values, one can understand a given piece of data at its most basic level. Conversely, some aspects of experiments are not adequately represented in our knowledge base. For example, our system would not be a good place to look if a user wanted to find out the details of how a given experiment was carried out (i.e., the involved set of instructions usually listed in a Methods section); we currently do not have the capability of representing biological procedures within the framework of our ontology, although other groups have worked on this capability.[15] We have not yet added a number of relevant experimental templates to our ontology, including assays of protein translation fidelity, speed of translation, reconstitution of translation with sequential addition of cofactors, and studies of the folding and association of the two ribosomal subunits.

Because the large majority of the values for many attributes are not simply text strings but have also been entered into the ontology as instances of other concepts (most often, instances of Physical-Thing), data and the physical objects they measure are extensively cross-referenced. Using Crosslinking as an example, all physical things (e.g., RNA bases, proteins) inversely have an attribute Is-Crosslinked-In that lists all data instances in which the given entity is cross-linked with some other entity. Similarly, all small molecules have an attribute Is-Crosslinking-Agent-In whose values are data instances in which the given molecule acts as a cross-linking agent. These extensive linkages permit a very useful organization of heterogeneous data. For example, a given base of 16S rRNA may have links to data instances in which the base is a protected site in a footprinting experiment, data instances in which a conservation study has been done on the base, data instances in which the base is the site of the fluorescence acceptor in an energy transfer experiment, and mutant molecules in which the base has been deleted or mutated. Thus, within a given RNA base frame, everything that has been entered into the knowledge base about that base can be easily accessed (both in browse mode and by computer programs). In fact, a cross-linking data instance can be found starting from the attribute Serves-As-Target-In in an instance of the class Physical-Thing, from Is-Crosslinked-In in Physical-Thing, from Is-Crosslinking-Agent-In in Small-Molecule, and from Reports in Journal-Article.

The capabilities of a structured representation are illustrated by the information density computations we performed. Specifying clear criteria and filtering data instances based on these criteria are easy tasks because a unique set of attribute values determines whether a given instance fulfills the criteria (e.g., to determine which journal articles were published before

[15] C. D. Hafner and N. Fridman, *ISMB* **4,** 78 (1996).

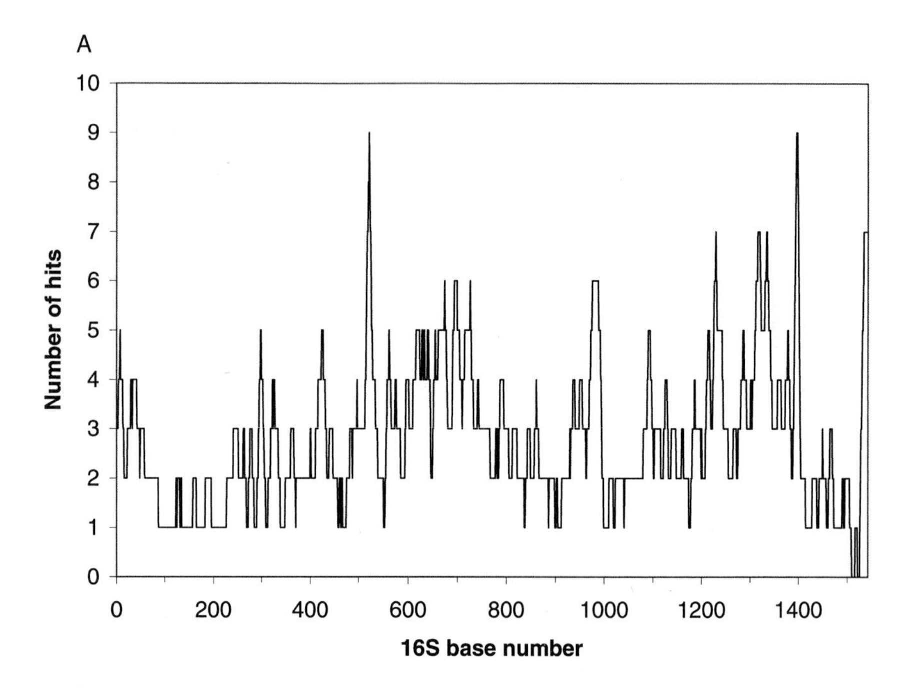

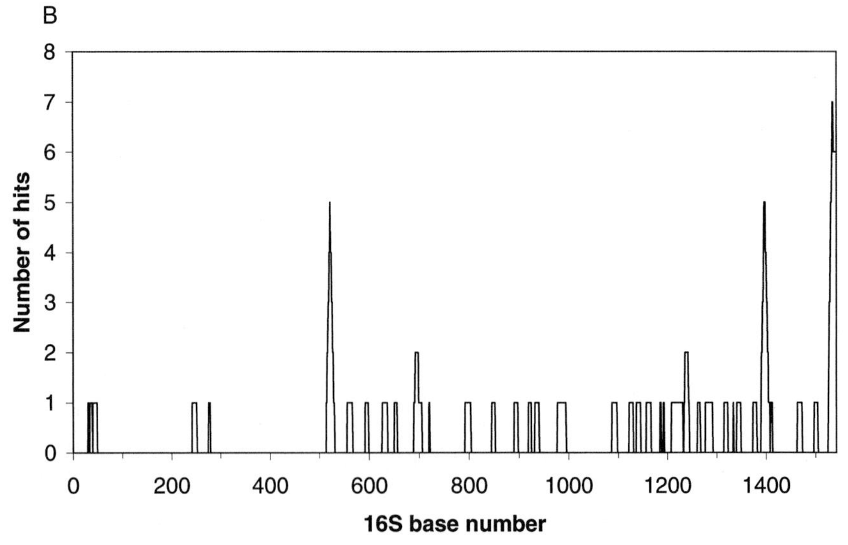

FIG. 8. Plots of the density of data as a function of sequence position for (A) biochemical data, (B) cross-linking data, and (C) footprinting data as a function of 16S sequence position. These plots show that the central domain and part of the 3′ domain (nucleotides ~1200–1400, including the Shine–Dalgarno sequence around 1400) have been most extensively studied by biochemical techniques, while less is structurally known in the intervening regions. (D) A plot of the density of data for the 21S proteins within the knowledge base. The range of hits is striking; structural knowledge of the S proteins in the context of the *E. coli* 30S subunit is clearly not homogeneous. The data show that we know about one order of magnitude more about the interactions of S7 than about the interactions of S20.

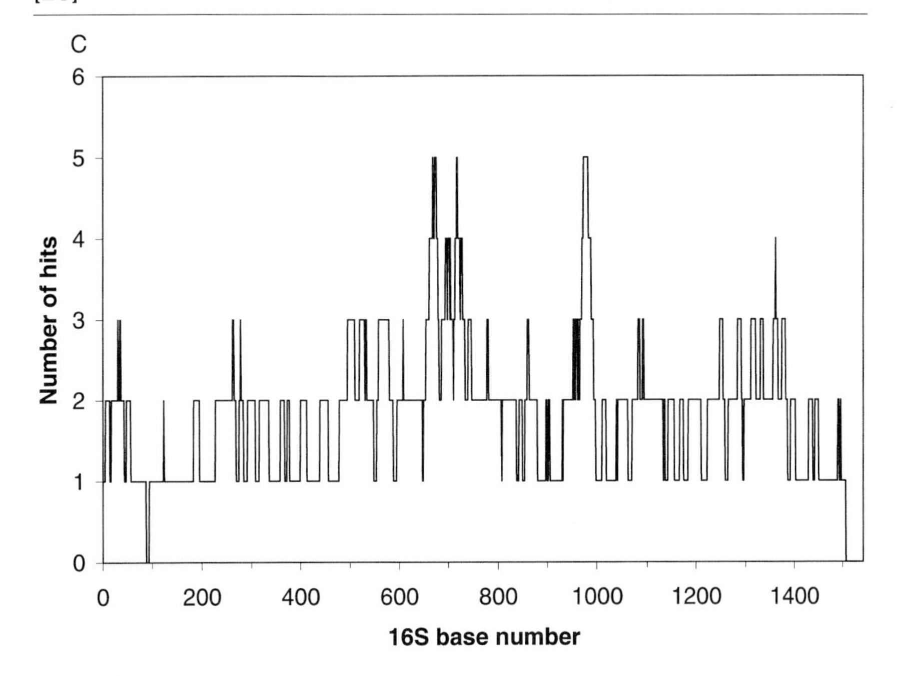

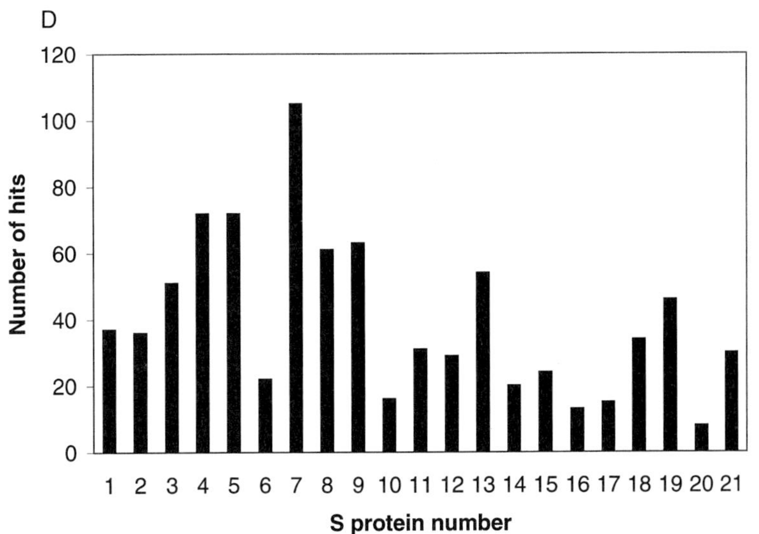

FIG. 8 (*continued*)

1990, the attribute Publication-Year is used). These information density computations can be performed in minutes, and the criteria for filtering can be altered to examine not only what is known about the 30S subunit structure but also what the principle source of data in different regions of the molecule has been.

In addition to organizing the data so that a user can interactively find and evaluate the data, structured data representations allow for automated computational analysis. A computer program can easily extract the value(s) from a given attribute of any instance and apply it to further calculations and/or evaluations. For example, the structured data representations used in our knowledge base are being put to use within the RIBOWEB system. Using the knowledge base as the basic data storage facility, RIBOWEB has a series of computational modules that can extract data from the knowledge base and interpret it for the purposes of creating input for structure-computation algorithms (such as distance geometry[16] or our probabilistic least squares algorithms,[17]) and for evaluating the compatibility of previously published structures against data sets. The availability of a large corpus of experimental data allows these programs to rapidly create reports of the compatibilities and incompatibilities between 3-D models and published data. Incompatibilities can be resolved by either discarding data (presumably with some good reason!) or by reinterpreting it. For example, we have found that the interpretation of footprinting experiments into ranges of distances between the footprinting agents and the protected sequence segments can be quite difficult, and some models are only compatible with relatively liberal distance ranges. The RIBOWEB system is accessed over the Web and allows users to explore alternative interpretations of data and their implications for structure. It supports secure login and public and private data privileges, and maintains an audit trail of all computations performed by a user.

A key feature of our system's usability is the fact that it is Web based. To use the knowledge base browser, a user does not have to obtain and install any additional software. Indeed, the Perl API can be accessed over the web using a URL mechanism that allows a remote program to access data directly from the knowledge base (as long as they have appropriate security privileges). The widespread familiarity of the World Wide Web ensures that users find navigation through our knowledge base somewhat familiar. The links are intuitive; if the user is perplexed by the meaning of any of the colored, underlined terms, they can simply click on it to go to

[16] D. C. Spellmeyer, A. K. Wong, M. J. Bower, and J. M. Blaney, *J. Mol. Graph Model* **15**(1), 18 (1997).

[17] R. Altman, *Intl. J. Human-Computer Studies* **42,** 593 (1995).

the frame for that instance, where all of the information present in the knowledge base concerning the instance is displayed. In addition, as new user interface methods become available on the web, we can take advantage of them. For example, we have written an RNA secondary structure browser using the Java language, which presents the secondary structure of an RNA molecule in a window such that individual bases or a segment of bases can be highlighted, and then the corresponding knowledge base object can be queried with the push of a button.[18] The 16S rRNA secondary structure browser and front-end to the knowledge base are shown in Fig. 9.

Our RIBOWEB knowledge base uses the scientific literature as the organizing principle for data storage. Each data instance must have an attribute Reported-By whose value is the data source (usually a journal article, but possibly some other reference type, such as Unpublished-data) in which the data were originally reported. Conversely, each publication instance has an attribute Reports that lists all data instances entered into the knowledge base from that publication. Each publication instance also provides full reference information, including a PubMed link, so as to fully credit the author(s). The Unpublished-Data type can be used by individuals who wish to submit their data to the knowledge base for further computational analysis, perhaps before publication.

Because the vast majority of data is taken directly from journal articles, the issues of data validity and consistency inherent in published literature naturally arises. Journal articles certainly may contain data which are either erroneous on their own and/or inconsistent with other published data. Data extracted from journal articles and entered into the knowledge base are represented from the perspective of the author(s) at the time of initial publication. In entering the data, we make no effort to "correct" the information in light of subsequent knowledge. Instead, we adopt the standards of traditional paper publishing—the paper remains in the public domain, although authors may subsequently report new experiments that invalidate their original findings. The discovery of inconsistencies can be partially automated, however, because the structured representation is amenable to automatic computational analysis. We have taken measures to ensure that data are entered into the knowledge base accurately, using a system of entry by one individual and review by another, and so we have confidence that the fidelity of the data in our knowledge base is fairly high. Random third-party audits are required to further assess the fidelity of our transcription processes. In the long term, we would like to create data input software that can be used by authors to directly enter and validate their data at the time of manuscript submission. Of course, computational modules that find

[18] R. M. Felciano, R. O. Chen, and R. B. Altman, *Gene* **190**(2), GC59, (1997).

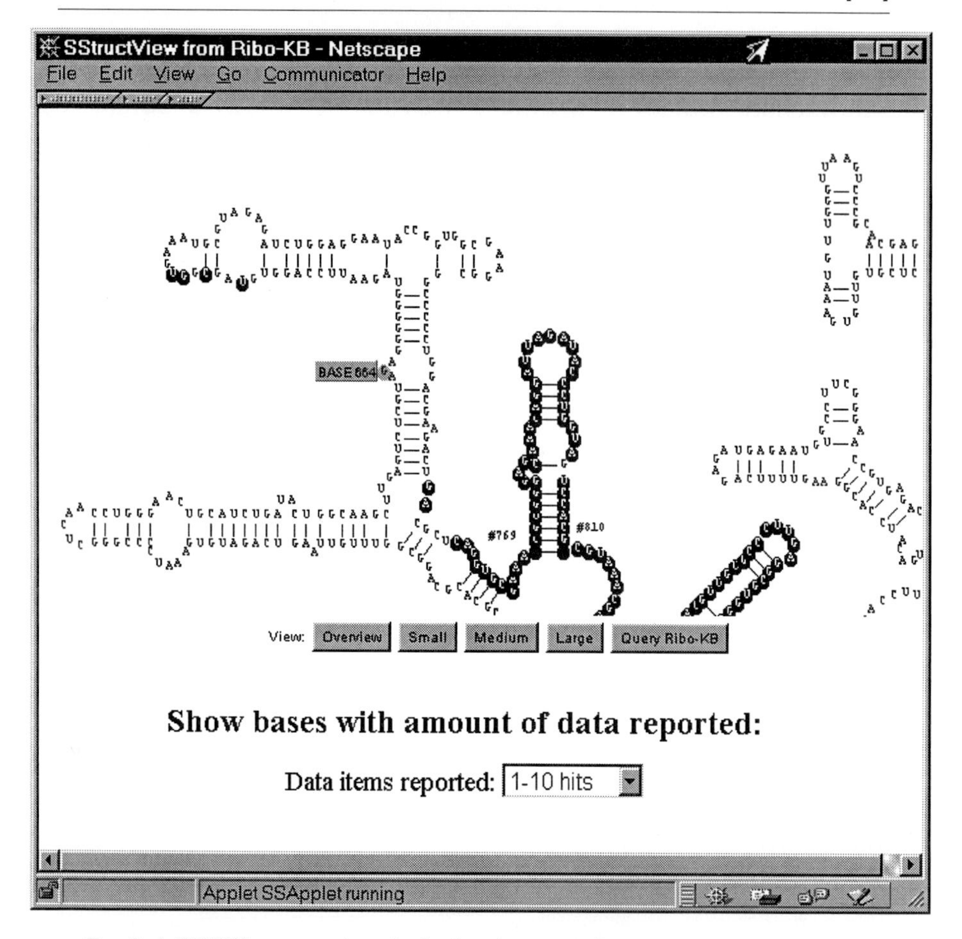

FIG. 9. A WWW browser written in the Java language allows any secondary structure to be displayed. In this screen, a section of the *E. coli* 16S rRNA secondary structure is shown in "medium" enlargement. A base has been selected for examination in the knowledge base, and those bases with a high data density are shown in white on black video.

certain pieces of data to be incompatible with others might also lead to checks of data integrity.

In summary, the purpose of the RIBOWEB knowledge base is to accurately represent and organize heterogeneous structural data *as represented in the literature.* We anticipate that the success of knowledge base data storage technologies may lead to new conventions for the publication of data in structured formats. Such a development would catalyze the creation of highly specific computational methods for literature/information retrieval and notification, automated consistency analyses between published data, and create an excellent environment for supporting individual investigators in their attempts to harness new sources of biological information.

Acknowledgments

R.B.A. is supported by NIH LM-05652, LM-06422, NSF DBI-9600637 and equipment grants from IBM and SUN. M.A.B. is supported by NIH HG-11223. We thank Harry Noller, Neil Abernethy, Richard Chen, and Ramon Felciano for useful discussions. Neil Abernethy also provided assistance in the creation of some figures.

[29] Modeling RNA Tertiary Structure from Patterns of Sequence Variation

By FRANÇOIS MICHEL, MARIA COSTA, CHRISTIAN MASSIRE, and ERIC WESTHOF

Introduction

As illustrated by the cloverleaf model of tRNA secondary structure,[1-3] attempts to extract structural information by analyzing patterns of nucleotide change and conservation in phylogenetically or functionally related sequences date back to the early days of nucleic acid sequencing. This inference process, which has become known as comparative sequence analysis (CSA) (for reviews, see Refs. 4–8), remains the most reliable way to obtain biologically pertinent secondary structures of RNA molecules in the

[1] J. T. Madison, G. A. Everett, and H. K. Kung, *Cold Spring Harbor Symp. Quant. Biol.* **31,** 409 (1966).

[2] U. L. RajBhandary, A. Stuart, R. D. Faulkner, S. H. Chang, and H. G. Khorana, *Cold Spring Harbor Symp. Quant. Biol.* **31,** 425 (1966).

[3] H. G. Zachau, D. Dutting, H. Feldmann, F. Melchers, and W. Karau, *Cold Spring Harbor Symp. Quant. Biol.* **31,** 417 (1966).

[4] R. R. Gutell, *Curr. Biol.* **3,** 313 (1993).

[5] C. R. Woese and N. R. Pace, *in,* "Probing RNA Structure, Function, and History by Comparative Analysis" (R. F. Gesteland and J. F. Atkins, eds.), pp. 91–117. Cold Spring Harbor Laboratory Press, Plainview, New York, 1993.

[6] E. Westhof and F. Michel, *in,* "Prediction and Experimental Investigation of RNA Secondary and Tertiary Foldings" (K. Nagai and I. W. Mattaj, eds.), pp. 25–51. Oxford University Press, UK, 1994.

[7] S. Baskerville, D. Frank, and A. D. Ellington, *in,* "Directed Evolutionary Descriptions of Natural RNA Structures" (R. Simons and M. Grunberg-Manago, eds.), pp. 203–251. Cold Spring Harbor Laboratory Press, Plainview, New York, 1998.

0076-6879/00 $30.00

absence of crystallographic data. Aside from the tRNA cloverleaf, the secondary structure models of ribosomal RNAs,[9–12] self-splicing introns,[13,14] the RNase P RNA[15] and many other natural or *in vitro* selected RNA molecules were all established by CSA. Phylogenetic analyses can be used to infer tertiary interactions as well. This was demonstrated as early as 1969 by Levitt who, 5 years before the crystallographic structure of tRNA became available, was able to predict one of the noncanonical base pairs and one base triple (the other proposed triple interaction turned out not to be there). Tertiary interactions identified by CSA have been incorporated into current three-dimensional models of 16S ribosomal RNA[16–19] and RNase P RNA[20,21] and even formed the main basis of a model of the three-dimensional architecture of group I self-splicing introns,[22] which has been widely used, and is now known[23] to have been fairly accurate as far as architecture is concerned.

Even though computer programs can provide invaluable help at all stages, CSA and the subsequent modeling of RNA tertiary structure still rely largely on human judgment and may conveniently, if somewhat artifically, be divided into the following steps. Within a set of related RNA sequences, segments of similar sequence are tentatively assumed to be

---

[8] F. Michel and M. Costa, *in,* "Inferring RNA Structure by Phylogenetic and Genetic Analyses" (R. Simons and M. Grunberg-Manago, eds.), pp. 175–202. Cold Spring Harbor Laboratory Press, Plainview, New York, 1998.

[9] C. R. Woese, L. J. Magrum, R. Gupta, R. B. Siegel, D. A. Stahl, J. Kop, N. Crawford, J. Brosius, R. Gutell, J. J. Hogan, and H. F. Noller, *Nucleic Acids Res.* **8,** 2275 (1980).

[10] H. F. Noller, J. Kop, V. Wheaton, J. Brosius, R. R. Gutell, A. M. Kopylov, F. Dohme, W. Herr, D. A. Stahl, R. Gupta, and C. R. Waese, *Nucleic Acids Res.* **9,** 6167 (1981).

[11] C. Branlant, A. Krol, M. A. Machatt, J. Pouyet, J. P. Ebel, K. Edwards, and H. Kossel, *Nucleic Acids Res.* **9,** 4303 (1981).

[12] C. Glotz, C. Zwieb, R. Brimacombe, K. Edwards, and H. Kossel, *Nucleic Acids Res.* **9,** 3287 (1981).

[13] F. Michel, A. Jacquier, and B. Dujon, *Biochimie* **64,** 867 (1982).

[14] R. W. Davies, R. B. Waring, J. A. Ray, T. A. Brown, and C. Scazzocchio, *Nature* **300,** 719 (1982).

[15] B. D. James, G. J. Olsen, J. S. Liu, and N. R. Pace, *Cell* **52,** 19 (1988).

[16] F. Mueller and R. Brimacombe, *J. Mol. Biol.* **271,** 524 (1997a).

[17] F. Mueller and R. Brimacombe, *J. Mol. Biol.* **271,** 545 (1997b)

[18] F. Mueller, H. Stark, M. van Heel, J. Rinke-Appel and R. Brimacombe, *J. Mol. Biol.* **271,** 566 (1997).

[19] H. F. Noller, *in,* "Ribosomal RNA" (R. Simons and M. Grunberg-Manago, eds.), pp. 253–278. Cold Spring Harbor Laboratory Press, Plainview, New York, 1998.

[20] J. L. Chen, J. M. Nolan, M. E. Harris, and N. R. Pace, *EMBO J.* **17,** 1515 (1998).

[21] C. Massire, L. Jaeger, and E. Westhof, *J. Mol. Biol.* **279,** 773 (1998).

[22] F. Michel and E. Westhof, *J. Mol. Biol.* **216,** 585 (1990).

[23] B. L. Golden, A. R. Gooding, E. R. Podell, and T. R. Cech, *Science* **282,** 259 (1998).

homologous and aligned; variable sites within or next to these conserved blocks are searched for evidence of concerted base changes ("covariation"); the statistical constraints thus uncovered are then interpreted in terms of potential secondary and tertiary base pairings; finally, efforts are made to incorporate these interactions into a coherent three-dimensional model of the RNA molecule of interest. Nevertheless, this description may be misleading, inasmuch as the entire process is a largely iterative one. Initial alignments need to be questioned, and may in fact be greatly altered, by taking potential secondary structure, as well as inferred tertiary interactions, into account; and surely enough, improved alignments lead in turn to better predictions. Another source of circularity is the tremendous reduction of the search space brought about by modeling. Critical reexamination of the sequence data at that stage can uncover biases that had been overlooked or discarded because of their marginal statistical significance in the context of a general, rather than local search. At the same time, circularity and human intervention carry of course an increased risk of getting trapped into "solutions" that are not unique ones or may even be wrong. There have been continuing efforts to automate all or part of CSA and modeling. These endeavors, which are described below, are aimed not merely at saving time, but at setting the entire inference process on a more objective footing, so as to avoid the pitfalls of human intervention.

While the sequences to be fed to CSA can come either from natural or artificially selected molecules, these two types of data sets raise different problems. In the case of natural molecules, sampling must be organized so that sequences with appropriate levels of divergence are recovered. For instance, very few sequences may be necessary to obtain a provisional secondary structure model of the molecule of interest when potential helices, rather than individual base pairs, are taken as statistical units and regarded as proven as soon as, say, two instances of substitution of any one of their base pairs by another one have been detected.[9] However, getting evidence in favor of (or against) short putative helices composed of bases that are highly conserved by evolution (because they interact with additional/alternative partners) is likely to require much larger numbers of highly divergent sequences and the same is true of most types of tertiary interactions. In the case of *in vitro* selection (SELEX) experiments, on the other hand, exploration of sequence space is according to experimental design, and whether dense or (more likely) sparse, should be approximately uniform, rather than constrained by phylogenetic history. The drawback is that the average level of divergence between functionally related molecules may be so high as to make adequate sampling unpractical, especially when several distinct families of selected molecules are present. The usual solution is to submit one or several individually selected molecules to reselection after partial randomization of their sequence.[24] However, while levels of randomization compatible with the retention of selected characteristics by a significant fraction of the population of molecules are appropriate for recovery of secondary structure and some noncanonical base–base interactions,[25] they are insufficient to ensure frequent replacement of a base triple by another acceptable one (see Refs. 7 and 26; even with a probability as low as 0.5 per site of retaining the original base, the representation of any particular alternate triple combination would remain 27 times smaller than that of the original combination in the pool to be reselected). Therefore, detection of other than binary interactions is unlikely, unless additional mutations are continuously generated,[27] elevated numbers of replication rounds are carried out,[28,29] and conditions of selection are changed from round to round in the hope of providing underrepresented base combinations with a competitive edge. These conditions differ little from those of natural selection and evolution, and *in vitro* generated phylogenies present in fact the same drawbacks as natural ones: due to the nonuniform exploration of sequence space, statistical biases in base distribution at any two sites are more likely to reflect the genealogy of the molecules that underwent selection, rather than be a consequence of intrinsic structural constraints. As explained in the next section, it is essential that this potential pitfall be taken into account when trying to interpret apparent covariations.

Importance of Sequence Alignment and Ordering

Before discussing in detail individual steps in comparative sequence analysis and the computer programs that have been written to facilitate analyses, it is important to realize where major difficulties lie. Any attempt to search for covariations rests on the assumption that all nucleotides in a particular column of an alignment are homologous, i.e., they are located either at the same place, or closer to one another than any other residue in the common three-dimensional structure of the family of molecules being analyzed. Homology is not a major issue of course when dealing with closely related molecules, since long, highly similar, uninterrupted stretches of sequence can safely be assumed to generate the same higher order structure. On the other hand, meaningful alignment of highly divergent RNA sequences can be extremely difficult. Even within an identical structural

[24] A. D. Ellington and J. W. Szostak, *Nature* **346,** 818 (1990).
[25] D. P. Bartel, M. L. Zapp, M. R. Green, and J. W. Szostak, *Cell* **67,** 529 (1991).
[26] E. H. Ekland and D. P. Bartel, *Nucleic Acids Res.* **23,** 3231 (1995).
[27] R. C. Cadwell and G. F. Joyce, *PCR Methods Appl.* **2,** 28 (1992).
[28] J. Tsang and G. F. Joyce, *Biochemistry* **33,** 5966 (1994).
[29] M. C. Wright and G. F. Joyce, *Science,* **276,** 614 (1997).

frame, sequence divergence can become considerable, one reason being that exactly or nearly isosteric combinations exist not only for the canonical base pairs, but for many other base–base interactions.[30,31,31a] Moreover, as evolutionary distance increases and insertions and deletions accumulate, not only sequence, but structure also evolves, to the point that stretches of sequence that would seem to occupy homologous locations may in fact correspond to very different structures, with distinct orientations in space.

These problems are well illustrated by self-splicing group I introns, which, despite their (highly probable) common origin and common three-dimensional ribozyme active center, share no more than eight universally conserved bases (exceptions exist in fact for at least three of these conserved sites[22,32]). As for those segments and structural components that lie at the periphery of the conserved core or between the core and splice sequences, they are clearly nonhomologous between different subgroups of group I introns.[33] In fact, the six stretches of sequence that constitute the common group I structural core can be separated from one another and the splice junctions by anything from less than 10 to more than 1000 residues.

How then could meaningful alignments of group I intron sequences.[13,14,22,33,34] ever be obtained? Like for many other RNAs with a conserved structure, the solution consisted of starting with sets of closely related molecules whose sequences could readily be aligned on the basis of similarity alone. In a second stage, sections of these separate alignments were merged by taking both patterns of base conservation and potential secondary structure into account. Finally, alignments of both the common core and subgroup-specific peripheral domains were further refined by searching for potential tertiary interactions and recurrent sequence/structure motifs as well as attempting to model higher order structure. The entire process typically requires large data sets, consisting of many sequences with all possibles levels of divergence, and reaching a stable, global alignment may take considerable time (as well illustrated in the case of the RNA component of RNase P; see Refs. 21 and 35).

As is made apparent from the preceding discussion, phylogenetic ordering is an integral part of reaching a satisfactory alignment. A more subtle reason (see Fig. 1) for sequence ordering arises from the need to know the number of base substitution events in order to estimate the possible

[30] W. Saenger, "Principles of Nucleic Acid Structure." Springer-Verlag, Berlin Heidelberg, 1984.

[31] D. Gautheret, S. H. Damberger, and R. R. Gutell, *J. Mol. Biol.* **248,** 27 (1995).

[31a] N. B. Leontis and E. Westhof, *Quant. Rev. Biophysics* **31,** 399 (1998).

[32] S. H. Damberger and R. R. Gutell, *Nucleic Acids Res.* **22,** 3508 (1994).

[33] V. Lehnert, L. Jaeger, F. Michel, and E. Westhof, *Chem. Biol.* **3,** 993 (1996).

[34] T. R. Cech, *Gene* **73,** 259 (1988).

[35] J. L. Chen and N. R. Pace, *RNA* **3,** 557 (1997).

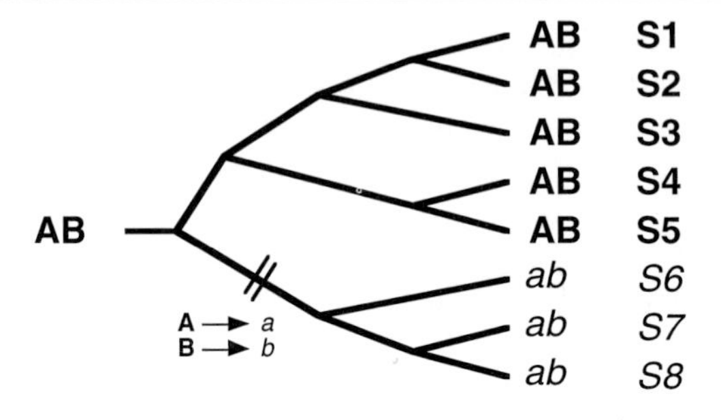

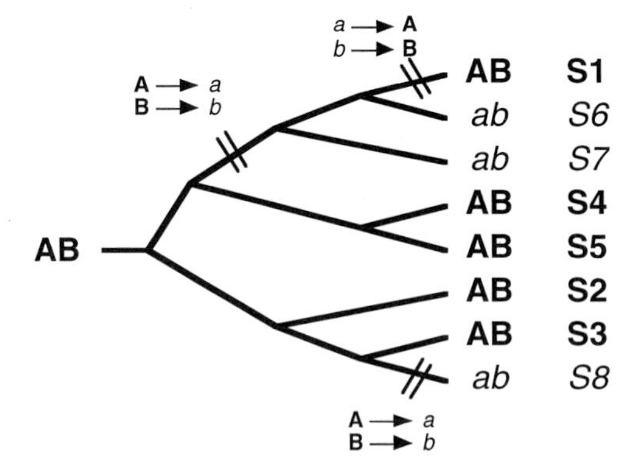

FIG. 1. Two possible interpretations of a covariation between two sites 1 and 2. In one phenotype, the sites are occupied by monomers A and B and, in the other one, by a and b. (a) Phylogenetic tree requires only two events to generate sequences S1 to S8. (b) At least six changes are required to generate the tree. Although in case (b) one can infer that the contents of sites 1 and 2 are constrained by one another, no conclusion can be reached for case (a).

significance of an apparent covariation in terms of structural constraints (e.g., discussions in Refs. 8, 22, 31, and 36). The base contents of two sites may be perfectly correlated because of a direct base–base interaction that is essential to function, so that any mutation at one site needs to be compensated by another mutation at the other site. Alternatively, an apparent correlation could arise from the chance occurrence of independent base mutation events at each of two sites in one of the deep, long branches of the phylogenetic tree, followed by stasis. The way to distinguish between

[36] S. Winker, R. Overbeek, C. R. Woese, G. J. Olsen, and N. Pfluger, *Comput. Appl. Biosci.* **6,** 365 (1990).

these possibilities is to obtain an estimate of the number of all types of events at each site, and this requires in turn at least some form of phylogenetic ordering.

The phylogeny of molecules that are unlikely to have undergone horizontal transfer may be assumed to be congruent with that of their host genomes, although the possibility remains of paralogous evolution. However, when dealing with molecules such as group I and group II introns, which are not only transposable elements, but are known to be infective on a grand scale,[37] there is no other way to retrieve their phylogenetic history than from their sequence itself (because much of the current phylogenetic tree of life was inferred from ribosomal RNA sequences, the same is true in fact to a large extent of the rRNA data sets). It may seem somewhat circular to use the same sequence set to obtain a phylogeny and search for covariation. However, phylogenetic inference and covariation analysis tend to take advantage of different subsets of sites, with slowly evolving nucleotides being the most useful for building phylogenetic trees, whereas "homoplasious" (incongruent) residues that underwent large numbers of events are the ones most likely to provide structural information.

Automation of RNA Alignment and Covariation Analysis

Automation of Sequence Alignment

A well-investigated problem is that of adding a new sequence to an already existing alignment by taking both sequence and potential secondary structure into account. The RNAlign program of Corpet and Michot[38] proceeds much as humans do, by identifying unambiguously homologous segments ("anchors") based on sequence similarity alone and then taking secondary structure into account to extend the alignment between the anchors. When the initial database of aligned sequences is large enough to include a representative sample of the family of molecules of interest, it becomes possible to build a descriptor that specifies shared features of sequence and structure on a probabilistic basis, from which a scoring system can be derived to quantify similarity. Such custom-made models can be used to write programs that will both identify new family members in nucleotide sequence databases and incorporate them into available alignments.[39,40] Even group I introns can be successfully searched in primary

[37] Y. Cho, Y. L. Qiu, P. Kuhlman, and J. D. Palmer, *Proc. Natl. Acad. Sci. U.S.A.* **95,** 14244 (1998).

[38] F. Corpet and B. Michot, *Comput. Appl. Biosci.* **10,** 389 (1994).

[39] G. A. Fichant and C. Burks, *J. Mol. Biol.* **220,** 659 (1991).

[40] N. el-Mabrouk and F. Lisacek, *J. Mol. Biol.* **264,** 46 (1996).

sequence data by this approach,[41] despite their already emphasized variability. Furthermore, as independently shown by Eddy and Durbin[42] and Haussler and colleagues,[43] methods based on stochastic context-free grammars make it possible to build optimal descriptors automatically from provisional alignments (the source code for the COVE program is available at http://www.genetics.wustl.edu/eddy/software/#cove). Unfortunately, as currently implemented, these methods fail to incorporate tertiary interactions, including the so-called "pseudoknots," which consist of contiguous canonical base pairs, but cannot be incorporated into planar, tree-like representations of secondary structure inasmuch as at least some of their constituent nucleotides belong to preexisting secondary structure loops.

All that precedes assumes that a tentative, yet reasonable alignment is available. When sequences are not too divergent and have undergone only a limited number of insertions and deletions, useful provisional alignments that can be used to search for a common secondary structure may be obtained with a program intended for multiple sequence alignment (such as CLUSTALW[44]). The problem of simultaneously aligning and folding multiple, highly divergent RNA sequences was first tackled by Sankoff,[45] but as such, the algorithm is impractical, since computation time grows exponentially with the number of sequences, and it does not find pseudoknots. Eddy and Durbin[42] were able to show that good provisional alignments can be obtained by feeding values of mutual information rather than nearest-neighbor stacking energies to the Nussinov/Zuker algorithm for prediction of secondary structure. However, the method assumes that the sequences can be globally, rather than only locally, aligned, and cannot deal with pseudoknots (however, see Ref. 46). Inversely, Gorodkin *et al.'s*[47] simplification of the Sankoff algorithm may be used to find significant local motifs of conserved sequence and secondary structure, as long as they are devoid of branches or pseudoknots.

Automation of Covariation Analysis

From a sequence alignment, it is straightforward to generate contingency tables containing frequencies of base combinations at any two sites. Either

[41] F. Lisacek, Y. Diaz, and F. Michel, *J. Mol. Biol.* **235,** 1206 (1994).
[42] S. R. Eddy and R. Durbin, *Nucleic Acids Res.* **22,** 2079 (1994).
[43] Y. Sakakibara, M. Brown, R. Hughey, I. S. Mian, K. Sjolander, R. C. Underwood, and D. Haussler, *Nucleic Acids Res.* **22,** 5112 (1994).
[44] J. D. Thompson, D. G. Higgins, and T. J. Gibson, *Nucleic Acids Res.* **22,** 4673 (1994).
[45] D. Sankoff, *SIAM J. Appl. Math.* **45,** 810 (1985).
[46] E. Rivas and S. R. Eddy, *J. Mol. Biol.* **285,** 2053 (1999).
[47] J. Gorodkin, L. J. Heyer, and G. D. Stormo, *ISMB* **5,** 120 (1997).

χ^2 analysis or mutual information[48,49] can then be used to estimate extents of divergence from the null expectation of random assortment. Alignments of up to several thousand nucleotides can readily be analyzed by the excellent BioEdit program (written by Tom Hall, http://jwbrown.mbio.ncsu.edu/RNaseP/home.html), which allows users to view the table containing mutual information for all pairs of sites either line by line (column by column) or in graphical matrix form: at appropriate levels of divergence, extended base pairings are readily picked up from the background as diagonally oriented rows of dots with a dark color (corresponding to particularly high mutual information values). Some tertiary interactions, especially the ones involving only two bases, can generate statistical biases that are strong enough to be identified in this way. However, a majority of higher order contacts such as base triples evolve rather slowly, accept partly overlapping base combinations and/or may involve nonhomologous, although adjacent, sites in different phylogenetic subgroups.[8,22,31] These contacts are likely to go unnoticed because the signals they yield are weak and tend to get lost in the pervasive background of typically high mutual information values that natural sequences generate.

What are the reasons for such a high background and is it possible to decrease it? One source of potentially misleading signals that can rather easily be dealt with is divergence in base composition. Unless this is taken into account, comparing (G + C)-rich sequences with (A + U)-rich sequences is bound to generate an apparent excess of pairs of sites with G:C, G:G and C:C combinations on the one hand and A:U, A:A and U:U on the other: especially at sites constrained to accept only purines or pyrimidines, this may lead to the illusion of covariation (not only G:C and A:U, but G:G and A:A, on the one hand, and C:C and U:U, on the other, can form isosteric base pairs[30]). However, the major source of noise in χ^2 or mutual information analyses stems from the fact that sets of natural sequences are not random samples. Rather, all members of the set are related by descent and as already pointed out, the same may be true to some extent of *in vitro* selection products whenever evolution has been allowed to occur. Hence, all but the fastest evolving sites will show some degree of statistical correlation and the problem is worsened of course by the use of "unbalanced" phylogenetic trees (those in which some subtrees are overrepresented). This is the reason why it is essential to obtain estimates of the number of substitution events that took place so as to eliminate all pairs of sites that evolve too slowly for any conclusion to be reached

[48] D. K. Chiu and T. Kolodziejczak, *Comput. Appl. Biosci.* **7,** 347 (1991).

[49] R. R. Gutell, A. Power, G. Z. Hertz, E. J. Putz, and G. D. Stormo, *Nucleic Acids Res.* **20,** 5785 (1992).

and sift out the ones for which multiple instances of concerted substitution events, especially between closely related sequences, exist. One example of this strategy can be found in Michel and Westhof,[22] who subdivided group I introns into phylogenetic subgroups so as to identify recurrent instances of conversion of one base combination into another one and obtain estimates of numbers of coordinated events (see also Refs. 8 and 31). Alternatively, one can take advantage of a known phylogenetic tree to search first for pairs of sites showing an excess of covariation (in the literal sense of coincident changes in the same phylogenetic branches, irrespective of the nature of these changes) and only afterward examine which base combinations are present in order to assess the extent of statistical bias and interpretability of observed base substitutions in a structural framework: this is the strategy that was successfully implemented by Winker et al.[36] to look for higher order structure in ribosomal RNAs. Unfortunately, their computer program was not made available and manual sifting through the output lists was necessary anyway to identify the most promising covariations by combining various criteria. In fact, no one has yet been able to implement a fully automated procedure that would successfully integrate phylogenetic and statistical analyses. However, recently developed methods that rely on the maximum weighting matching algorithm of Gamow and allow optimal combinations of base–base interactions (including pseudoknots) to be generated from matrices of base pairing likelihood scores[50,51] might be able to serve that very purpose once an appropriate scoring system has been developed.

Inferring Specific Base-Pairing Geometries from Patterns of Base Variation

Once correlated nucleotide changes that appear to reflect structural constraints rather than historical coincidence have been identified, the next step toward the building of a three-dimensional model requires that observed patterns of variation be interpreted in terms of a specific base pairing geometry. As already pointed out, most canonical base pairs readily betray themselves, if only because they can exist in an exceptionally high number of isosteric combinations (four or six, depending on whether the somewhat divergent wobble pairs are included), all of which should eventually be observed in large data sets (at least for those pairs not involved in alternate or higher order interactions). Also, sites that form a Watson–Crick pair most often possess at least one pair of immediate neighbors showing the

[50] R. B. Cary and G. D. Stormo, *ISMB* **3,** 75 (1995).

[51] J. E. Tabaska, R. B. Cary, H. N. Gabow, and G. D. Stormo, *Bioinformatics* **14,** 691 (1998).

same array of covariations, and being part of an extended helix makes for some redundancy: the average rate of change for secondary structure pairings is higher than for isolated tertiary interactions, which makes it more likely that the full set of possible variations will be observed within even a rather limited data set. These factors concur to make prediction of secondary structure much more easy and reliable than that of tertiary interactions. Nevertheless, an impressive variety of base–base interactions other than the Watson–Crick pair have been inferred by comparative analysis and many of these have been verified by base substitution experiments or, in a few cases, eventually vindicated by crystallography or NMR (reviewed by Michel and Costa[8]).

To infer the geometry of a base pairing from patterns of covariation, it is necessary to identify, among the many possible base pairing arrangements, those sets of (approximately) isosteric base pairs that best fit the available data. Drawings of isosteric or quasi-isosteric sets can be found in Saenger[30] and additional ones are discussed in more recent publications.[22,31,52–55] Still, the most revealing analyses are not predictive ones, but rather those in which known structures are taken advantage of to try and explain observed patterns of nucleotide change in large sequence datasets, as was done by Gautheret *et al.*[31] for tRNA and Leontis and Westhof[56] for 5S RNA. As emphasized by the latter authors, each type of pairing leads to a distinctive phylogenetic signature of both observed and nonobserved variations that is largely explainable based on the degree of isostery of each combination with the dominant geometry. For instance, a sheared A:G pair (A-N7 and AN6 with GN2 and GN3, respectively) may be replaced by A:A (at the expense of one hydrogen bond), but not by G:A or G:G. Moreover, a few occurrences of A:C, A:U, C:A, C:C and C:U are to be expected, but no other combination. Such full signatures that include not only frequent, but rare and missing variations are especially important to discriminate between alternate geometric interpretations of a given pattern of covariation. For example, although there are at least three possible, strictly isosteric arrangements of G:U and A:C pairs with two H-bonds each (trans Watson–Crick, bifurcated[57] and wobble, with protonation of A-N1) their minor variants (e.g., G:G and A:A in the case of the bifurcated pairings,[56]) are expected to differ. Of course, the drawback

[52] F. Michel, A. D. Ellington, S. Couture, and J. W. Szostak, *Nature* **347,** 578 (1990).

[53] L. Giver, D. Bartel, M. Zapp, A. Pawul, M. Green, and A. D. Ellington, *Nucleic Acids Res.* **21,** 5509 (1993).

[54] D. Gautheret, D. Konings, and R. R. Gutell, *J. Mol. Biol.* **242,** 1 (1994).

[55] D. Gautheret and R. R. Gutell, *Nucleic Acids Res.* **25,** 1559 (1997).

[56] N. B. Leontis and E. Westhof, *RNA* **4,** 1134 (1998).

[57] C. C. Correll, B. Freeborn, P. B. Moore, and T. A. Steitz, *Cell* **91,** 705 (1997).

of full signatures is that they require even larger data sets than are necessary to ascertain that a covariation reflects a genuine structural constraint.

Help in generating the various isosteric sets of base pairs that could account for a particular covariation can be found in the ISOPAIR computer program of Gautheret and Gutell[55] (available by ftp at igs-server.cnrs-mrs.fr, in pub/ISOPAIR). Not all of the 351 types of pairs with one or more H-bond that the program can take into account are realistic, but many irrelevant isosteric sets can be eliminated through the use of appropriate restrictive options which allow us to discard sets in which the orientation of the glycosidic bond changes from one solution to the next, or require that at least one pair per set has more than one H-bond, or else ensure uniqueness (i.e., discard sets that include base pairs not in the list of observed combinations supplied to the program; see earlier discussion of signatures). There is also the possibility for users to include base-pairing conformations other than the ones in the current version (which does not include bifurcated pairs, for instance). Finally, the program also deals with triple interactions: prediction of triples[22,31,52,58] obeys essentially the same rules as described above for simpler pairings, but tends to be more difficult because triple interactions undergo frequent rearrangements and shifts during evolution (as seen in tRNA[31,59]) and also since interactions between adjacent triple sets introduce additional statistical biases in base distribution.[31,52,60] Moreover, it must be recalled that bases that pair with two different partners in alternate states of the same molecule may display patterns of covariation essentially indistinguishable from those resulting from a genuine triple interaction (see later discussion on multiple states).

Inferring Higher Order Features: Helix Stacking and Motifs

Structural features other than base–base interactions can sometimes be identified by comparative analysis of RNA sequences. One example is the coaxial stacking of double-stranded helices, which is well known to make a major contribution to RNA architecture. As first pointed out by Woese *et al.*,[61] evolutionary conservation of the combined length of two immediately adjacent helices which themselves vary in size is very strong evidence of coaxial stacking. Examples exist in ribosomal RNA,[4] RNase P RNA,[21] and

[58] M. Levitt, *Nature* **224,** 759 (1969).

[59] T. Brennan and M. Sundaralingam, *Nucl. Acids Res.* **3,** 3235 (1976).

[60] S. Couture, A. D. Ellington, A. S. Gerber, J. M. Cherry, J. A. Doudna, R. Green, M. Hanna, U. Pace, J. Rajagopal, and J. W. Szostak, *J. Mol. Biol.* **215,** 345 (1990).

[61] C. R. Woese, R. Gutell, R. Gupta, and H. F. Noller, *Microbiol. Rev.* **47,** 621 (1983).

in those group I introns in which the P2 helix is capped by a GNRA loop: the inferred stacking of P2 on the P1 substrate helix (the one that contains the 5′ splice site) provided a major constraint for modeling of the group I ribozyme core and eventually led to the discovery that GNRA loops have RNA receptors.[22]

Some of the currently known recurrent motifs of RNA structure are sufficiently large and well defined in terms of sequence that they can be recognized with a rather high degree of confidence in RNA secondary structure models (searching for motifs is only justified of course if the combination of bases being searched is likely to be largely automous in terms of structure, i.e., will most often generate essentially the same higher order structure irrespective of its molecular context). Two well-known examples (see Fig. 2) of such motifs are the preferred receptor for GAAA loops, an 11-nucleotide consensus motif flanked by helices on both sides, and the so-called loop E motif, a 7-nucleotide compact, helix-like motif which includes three noncanonical pairs, the first one of which abuts a

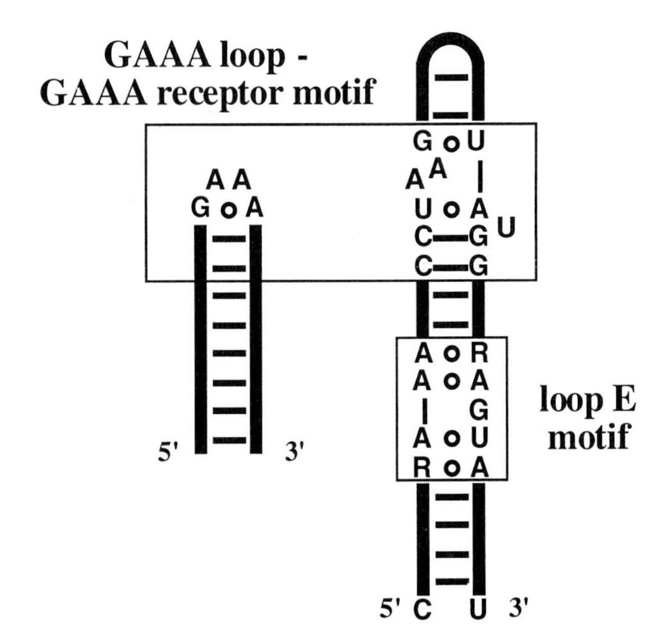

FIG. 2. Two recurrent RNA structural motifs: the GAAA tetraloop with its 11-nucleotide receptor[62] and the loop E motif from eukaryotic 5S rRNA.[57] While the former motif is known to be extensively used in RNA self-assembly,[63] the loop E motif, even though found throughout the three major kingdoms of organisms,[64] has no known receptor (either RNA or protein).

regular helix.[65,66] The atomic structure of the 11-nucleotide receptor motif in interaction with a GAAA loop was obtained by Cate *et al.*[62] shortly after the identification of the motif by comparative analysis coupled with *in vitro* selection experiments.[63,67] The 11-nucleotide motif exists and functions as a GAAA-loop receptor at several distinct locations in group I and group II self-splicing introns[63,68] and is probably present as well in the RNase P RNA of gram-positive bacteria.[69] The loop E motif gives rise to a specific UV cross-link which made possible its identification first in 5S RNA and the potato spindle tuber viroid,[70] then in the so-called sarcin stem–loop of 23S-like rRNA molecules[71] and in the hairpin ribozyme.[72] More recently, Leontis and Westhof[64] were able to identify several additional putative copies of the loop E motif in ribosomal RNAs: rather than searching individual sequences with the consensus sequence or a descriptor of the motif, they looked for its diagnostic phylogenetic signature of allowed and nonallowed variations in sets of aligned sequences.

Two programs that search sequence data for previously identified motifs are available. The RNAMOT program[73] (available by ftp at igs-server.cnrs-mrs.fr, in pub/RNAMOT) uses user-generated descriptors of a motif's consensus sequence and secondary structure to search either primary sequence databases or sets of related sequences for occurrences of that motif. RNABOB, written by S. Eddy (http://www.genetics.wustl.edu/eddy/software) does essentially the same with a different algorithm. Unfortunately, only primary sequence data can be searched by these programs, so it is left to the user to check that a possible occurrence of a given motif is compatible with the secondary structure of the molecule whose sequence is being analyzed, when this structure is known. Programs that can make use of user-supplied secondary structure models are all the more desirable in fact

[62] J. H. Cate, A. R. Gooding, E. Podell, K. Zhou, B. L. Golden, C. E. Kundrot, T. R. Cech, and J. A. Doudna, *Science* **273,** 1678 (1996).
[63] M. Costa and F. Michel, *EMBO J.* **14,** 1276 (1995).
[64] N. B. Leontis and E. Westhof, *J. Mol. Biol.* **283,** 571 (1998).
[65] B. Wimberly, G. Varani, and I. Tinoco, Jr., *Biochemistry* **32,** 1078 (1993).
[66] A. A. Szewczak and P. B. Moore, *J. Mol. Biol.* **247,** 81 (1995).
[67] M. Costa and F. Michel, *EMBO J.* **16,** 3289 (1997).
[68] M. Costa, E. Deme, A. Jacquier, and F. Michel, *J. Mol. Biol.* **267,** 520 (1997).
[69] M. A. Tanner and T. R. Cech, *RNA* **1,** 349 (1995).
[70] A. D. Branch, B. J. Benenfeld, and H. D. Robertson, *Proc. Natl. Acad. Sci. U.S.A.* **82,** 6590 (1985).
[71] Y. Endo, A. Gluck, and I. G. Wool, *Nucleic Acids Symp. Ser.* 165 (1993).
[72] S. E. Butcher and J. M. Burke, *Biochemistry* **33,** 992 (1994).
[73] D. Gautheret, F. Major, and R. Cedergren, *Comput. Appl. Biosci.* **6,** 325 (1990).

that they should make it possible to discover new recurrent RNA motifs from scratch.[74]

Among currently known RNA motifs, the complexes formed by the various GNRA loops and their preferred receptors have proved particularly useful for comparative analyses of large natural RNAs because they are sufficiently similar to have frequently replaced one another during evolution (for the same reason, it has also been possible to test experimentally a number of proposed contacts involving GNRA loops by proceeding just in the same way as for base–base interactions, i.e., combinations known to be matched or mismatched are replaced by one another and their performance is compared[63,68,75,76]). In this respect, it is important to realize that covariations involving entire motifs rather than individual base-pairing interactions may be difficult to identify and interpret by methods and programs conceived for site-by-site analyses. The need to treat motifs as evolutionary units is even more evident in cases where one motif is replaced by another one that is structurally unrelated, but nevertheless geometrically compatible (at least reasonably so). Occasional replacements of a GNRA loop–receptor interaction by conventional base pairing have been noted both in group I introns[77] and RNase P RNAs[78] and a particular GNRA loop and its receptor are known to have exchanged locations during the evolution of group II introns.[68]

Using Elements of Structure Inferred by Comparative Analysis to Model RNA Architecture

Before providing guidelines to the assembly of inferred structural components into larger models of RNA structure, critical reviewing of several well-publicized success stories of RNA structural prediction should provide a useful note of caution. The first example of an RNA tertiary pair successfully identified by comparative analysis (in this case, of a remarkably small set of sequences) was Levitt's pair between nucleotides 15 and 48 of tRNA.[58] That pair, which is most often G:C or A:U (rarely Y:R, see Table I), was correctly inferred not to be of the usual Watson–Crick type, but it was

[74] P. Gendron, D. Gautheret, and F. Major, *in* "High Performance Computing Systems and Applications," (J. Scjaeffer, ed.), pp. 323–331. Kluwer Academic Press, New York, 1998.
[75] L. Jaeger, F. Michel and E. Westhof, *J. Mol. Biol.* **236,** 1271 (1994).
[76] G. Chanfreau and A. Jacquier, *EMBO J.* **15,** 3466 (1996).
[77] L. Jaeger, E. Westhof, and F. Michel, *Biochimie* **78,** 466 (1996).
[78] C. Massire, L. Jaeger, and E. Westhof, *RNA* **3,** 553 (1997).
[79] M. Sprinzl, C. Horn, M. Brown, A. Loudovitch, and S. Steinberg, *Nucleic Acids Res.* **26,** 148 (1998).

TABLE I
COVARIATIONS BETWEEN POSITIONS 15 AND 48 IN 930 ELONGATOR tRNAs PRESENT IN tRNA DATABASE[a]

15\48	A	**C**	G	**U**
A	—	11	—	**162**
C	—	2	—	1
G	9	**721**	4	11
U	2	3	1	1

[a] From Ref. 79.

mistakenly interpreted as having the purine in the *syn* conformation, rather than being in the *trans* conformation (or of the so-called "reversed" type), with riboses arranged in opposite orientations with respect to the H-bonds, as later shown by crystallographic analyses (both arrangements lead to the same locally parallel orientation of the two interacting strands[80]): Even though the various possible *trans* Watson–Crick pairs are not strictly isosteric, they may nevertheless replace one another during evolution (e.g., discussion in Ref. 81). Another example of the difficulty of correctly inferring backbone orientation from patterns of variation is provided by attempts to guess at the structure of the Rev-binding site of human immunodeficiency virus (HIV) RNA from *in vitro* selection experiments.[25] Among selected clones, exchanges between G:G and A:A at two facing sites within an internal loop pointed to base pairing, and the pair was correctly inferred to be a symmetrical one, with the Watson–Crick sides interacting and glycosyl bonds *in trans.* But how could such a base pair exist right next to a Watson–Crick pairing? One obvious possibility was that one of the purines was in the *syn* orientation, thus allowing the interacting strands to maintain an antiparallel orientation: when additional selection experiments[53] revealed a few A:C combinations, it was concluded that the site at which pyrimidines had not yet been observed was the one with a base in *syn.* However, two independent NMR studies of the Rev-binding site complexed to its protein ligand[82,83] have shown that the purine–purine pair is accommodated by an unforeseen (except by Leclerc *et al.*[84]) local reversal of the direction of the backbone following the bulging out of a residue. (There is, nevertheless, a further twist in this story, for NMR studies of the unbound

[80] E. Westhof, *Nature* **358,** 459 (1992).
[81] L. Jaeger, E. Westhof, and F. Michel, *J. Mol. Biol.* **234,** 331 (1993).
[82] J. L. Battiste, H. Mao, N. S. Rao, R. Tan, D. R. Muhandiram, L. E. Kay, A. D. Frankel, and J. R. Williamson, *Science* **273,** 1547 (1996).
[83] X. Ye, A. Gorin, A. D. Ellington, and D. J. Patel, *Nature Struct. Biol.* **3,** 1026 (1996).
[84] F. Leclerc, R. Cedergren, and A. D. Ellington, *Nature Struct. Biol.* **1,** 293 (1994).

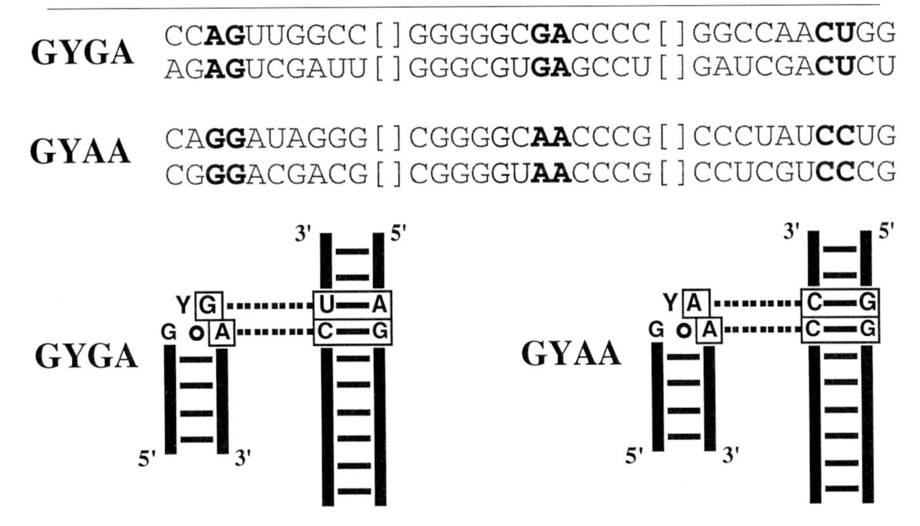

FIG. 3. Covariation rules between the tetraloops of the GYRA family and RNA helices.[22,75,86]

form of the Rev-binding site[85] point to a different conformation in which one of the purines appears to have the *syn* orientation!) A third example of a broadly correct, but wrongly detailed structural prediction is that of the interaction between GYRA loops and their receptors (see Fig. 3). RNA terminal loops with a GYAA and GYGA sequence were predicted to have specific RNA receptors consisting of consecutive canonical pairs (CC : GG and CU : AG, respectively) based on comparative analyses of group I introns sequences followed by base substitution experiments and from these data, detailed models of the interactions were published.[22,75] When these models were compared with a crystal structure of the same interaction,[86] it became apparent that the overall position and orientation of the loop within the shallow groove of its receptor helix had been correctly inferred, yet the modeled and actual interaction of the loop with its receptor had but a single hydrogen bond in common: as might have been expected, this was the one responsible for the observed covariation between the third loop nucleotide and the pair it contacts. What these three examples illustrate of course is that while comparative analysis is extremely powerful in predicting specific base–base contacts, it cannot, in principle, be expected to provide by itself information about the geometry of backbones.

Does this imply that contacts inferred by comparative analyses are

[85] R. D. Peterson and J. Feigon *J. Mol. Biol.* **264,** 863 (1996).

[86] H. W. Pley, K. M. Flaherty, and D. B. McKay, *Nature* **372,** 111 (1994).

hardly more than constraints to be fed to some sort of distance geometry or constraint satisfaction program? This might appear as the ultimate goal in objectivity and automatism. Two programs are available, YAMMP[87] and MC-SYM.[88] YAMMP exploits a pseudoatom approach with either one pseudoatom per helix or one pseudoatom per nucleotide. Appropriate potential functions have been developed. The use of spherical pseudoatoms leads to a loss in the asymmetry of the RNA fragments and, most importantly, all fine interactions that control RNA folding are not modeled. The second approach, MC-SYM, based on a constraint satisfaction algorithm, searches conformational space such that, for a given set of input constraints (secondary pairings, tertiary pairs, distances), all possible models are produced. With this methodology, Major *et al.*[88] managed, for a tRNA sequence, to generate 26 solutions which displayed the broad features of canonical tRNA structure.

Our own approach (see Fig. 4) involves the extensive use of known three-dimensional structures which are considered as building blocks for the assembly of larger RNA fragments or units. Basically, the approach can be implemented using any computerized construction program. We have developed our own set of programs on Silicon Graphics, the MANIP package.[89] The frameworks of those structures used as building blocks are held in a database that is used by the program FRAGMENT[89a] for inserting the appropriate sequence. The fragments thus produced are then assembled manually on a graphics screen using the modeling software. The resulting structure is eventually refined by restrained least-squares minimization programs (NUCLIN/NUCLSQ[90]). Molecular mechanics or molecular dynamics, although rather cumbersome and sometimes difficult to control in a meaningful way, could also be employed at this stage. The manipulations on the screen imply of course some human judgment, and inevitably reflect the stereochemical background and personal biases of the modeler. On the positive side, however, the human mind is very efficient at sorting out sets of solutions and ranking them against available data.

What is important to realize is that sequence comparisons yield consensus 3-D structure, i.e., the core of architectural constraints that all molecules of the same sequence family has to follow in order to fold in space and be functional. The method can provide the relative positions in space of motifs with respect to each other, but only with difficulty and uncertainty in the

[87] A. Malhotra, R. K. Tan, and S. C. Harvey, *Proc. Natl. Acad. Sci. U.S.A.* **87,** 1950 (1990).
[88] F. Major, D. Gautheret, and R. Cedergren, *Proc. Natl. Acad. Sci. U.S.A.* **90,** 9408 (1993).
[89] C. Massire and E. Westhof, *J. Mol. Graph. Mod.* **16,** 197 (1998).
[89a] E. Westhof, *J. Mol. Struct.* **286,** 203 (1993).
[90] E. Westhof, P. Dumas, and D. Moras, *J. Mol. Biol.* **184,** 119 (1985).

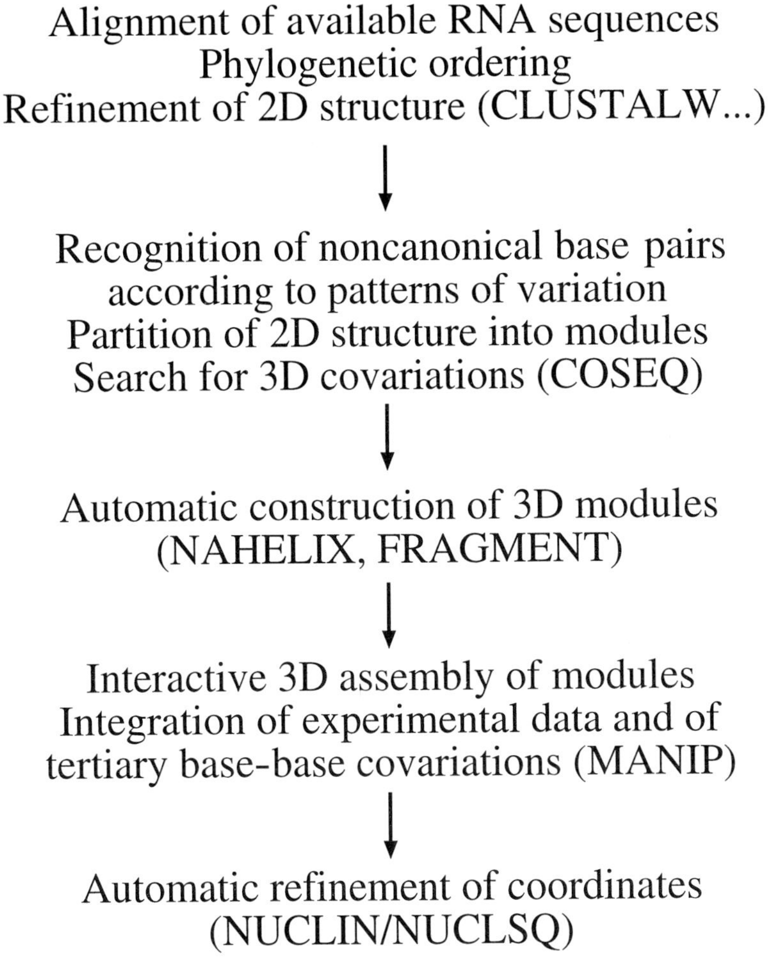

Alignment of available RNA sequences
Phylogenetic ordering
Refinement of 2D structure (CLUSTALW...)

↓

Recognition of noncanonical base pairs
according to patterns of variation
Partition of 2D structure into modules
Search for 3D covariations (COSEQ)

↓

Automatic construction of 3D modules
(NAHELIX, FRAGMENT)

↓

Interactive 3D assembly of modules
Integration of experimental data and of
tertiary base-base covariations (MANIP)

↓

Automatic refinement of coordinates
(NUCLIN/NUCLSQ)

FIG. 4. Main steps in the modeling process as implemented in the MANIP program.[89]

precise underlying atomic contacts without additional chemical probing information.

Finally, a major problem in integrating structural components inferred by comparative analysis into larger three-dimensional molecular models is the lack of information about their possible temporal coexistence. However successful statistical analyses may be in uncovering base–base interactions, they have nothing to say *a priori* about the state of the molecule on which selection for a particular base pairing arrangement exerts itself *in vivo*: this could be either the ground state, the active state, both of them, or more generally any number of alternate states in molecules like the small nuclear RNAs that undergo a succession of rearrangements.[91] To take one example, at the time the modeling of the ribozyme core of group I introns was attempted, it was still largely an educated guess that the many individual interactions identified by comparative analysis of group I intron sequences could legitimately be assembled into a single model (in fact, two closely related models pertaining to the two steps of the splicing reaction). However, evidence that a variety of group I introns actually possess well-defined structural cores that unfold in a markedly cooperative manner was soon to follow,[81,92] thus putting the entire modeling enterprise on firmer footing. Examples abound of extended canonical pairings being involved in RNA conformational switches and, more recently, a dynamic GNRA loop–receptor interaction has been identified in group II introns: the importance of experimentally assessing interactions identified by comparative analyses before attempting to combine them into a single model cannot be overemphasized.

[91] H. D. Madhani and C. Guthrie, *Ann. Rev. Genet.* **28,** 1 (1994).
[92] D. W. Celander and T. R. Cech, *Science* **251,** 401 (1991).

Author Index

Numbers in parentheses are footnote reference numbers and indicate that an author's work is referred to although the name is not cited in the text.

A

B

C

D

E

F

H

I

J

K

L

M

N

O

Q

R

S

T

U

V

W

X

Y

Z

Subject Index

A

B

D

E

F

G

H

I

K

L

L

M

N

Q

R

S

U

W

X

ISBN 0-12-182218-4